R

J

STARS
AND
RELATIVITY

IA. (Ya.) B. Zel'dovich
I. D. Novikov

Translated by Eli Arlock
Edited by Kip S. Thorne and W. David Arnett

DOVER PUBLICATIONS, INC.
Mineola, New York

Bibliographical Note

This Dover edition, first published in 1996, is an unabridged republication of the work originally published as Volume 1 of *Relativistic Astrophysics* by The University of Chicago Press, Chicago, in 1971. *Relativistic Astrophysics*, first published in two volumes, was a revised edition based on the translation of the original Russian *Relyativistskaya astrofizika* (Moscow: Izdatel'stvo "Nauka," 1967); see Editors' Foreword for details.

Library of Congress Cataloging-in-Publication Data

Zel'dovich, ÎA. B. (ÎAkov Borisovich)
    [Reliativistskaiâ astrofizika.   Volume 1. English]
    Stars and relativity / Ya. B. Zeldovich and I.D. Novikov ; translated by Eli Arlock ; edited by Kip S. Thorne and W. David Arnett.
        p.   cm.
    "An unabridged republication of the work originally published [in English] as volume 1 of Relativistic astrophysics by The University of Chicago Press, Chicago, in 1971."—T.p. verso.
    Includes bibliographical references and index.
    ISBN 0-486-69424-0 (pbk.)
    1. Relativistic astrophysics.   2. Stars.   I. Novikov, I. D. (Igor' Dmitrievich)   II. Thorne, Kip S.   III. Arnett, W. David (William David), 1940–   .   IV. Title.
QB462.65.Z4513   1996
523.01—dc20

96-35378
CIP

Manufactured in the United States of America
Dover Publications, Inc., 31 East 2nd Street, Mineola, N.Y. 11501

CONTENTS

Contents

Contents

Contents

EDITORS' FOREWORD

In our scientific and technological age Ya. B. Zel'dovich is one of those rare and valuable properties that every university and research establishment yearns for: a great physicist who is equally at home in laboratory discussions of experimental techniques, in the rarefied atmosphere of nuclear theory and elementary-particle theory, or in technology-oriented studies of shock waves and other explosive phenomena. To all these disciplines and others Zel'dovich has made fundamental contributions. For example, his most recent books (before these) are two volumes, with Yu. P. Rayzer, on the *Physics of Shock Waves and High Temperature Hydrodynamic Phenomena* (Academic Press, 1966 and 1967).

In the early 1960's Professor Zel'dovich began turning his attention to astrophysics and cosmology. Expecting that relativity would come to play an increasingly important role in astrophysics in the future, he chose as his principal research collaborator the most brilliant young relativity theorist he could find in the Soviet Union: Igor D. Novikov, who had just completed his studies at Moscow University under A. L. Zel'manov, and as a student had independently discovered Kruskal's new and powerful way of looking at the Schwarzschild solution.

Zel'dovich enjoys reporting of himself that, when he first entered astronomy from physics, his research methods were "like those of an elephant in an elegant china shop." Only later, he says, did he learn to dance without breaking the chinaware. Novikov then adds, in his colleague's defense, that the input of elephant-physicists is today a major propellant in astronomy's forward thrust.

By now, after less than ten years of astrophysical research, Zel'dovich and Novikov have come to rank among the great astrophysicists of the twentieth century. Their research contributions to the field of relativistic astrophysics make up a significant fraction of today's body of theoretical knowledge and technique. Their research is particularly remarkable for its deep physical insight; it always pierces to the heart of the physical phenomenon being studied, scattering irrelevant and minor complications off to the sides. In an era when many of us tend to get wrapped up in mathematical detail, Zel'dovich and Novikov show us how to understand our own work, as well as theirs, in exceedingly clear and elementary terms.

This spirit of the Zel'dovich–Novikov research also pervades these two volumes on Relativistic Astrophysics. As a result, in our opinion, these volumes should be a "must" on the reading lists of both practicing astrophysicists and aspiring students.

There exist a number of other books and articles reviewing special topics in relativistic astrophysics (e.g., quasistellar radio sources, gravitational collapse, relativistic stellar theory). However, this two-volume work is the only comprehensive review in existence today of the entire field of relativistic astrophysics. We think that students entering the field, and astronomers and physicists with peripheral interests in it, should begin by reading these volumes. Having thereby gained a broad overview of the field and deep insight into the relevant physical phenomena, they can then turn their attention to other, more specialized reviews and research papers.

These volumes are based in large measure on the book *Relyativistskaya astrofizika*, which Zel'dovich and Novikov wrote in 1965, 1966, and 1967. Mr. Eli Arlock made an English translation of that book for us in 1968; and then in 1969 the authors and we editors began the long collaborative process of revision, updating, and extension to new topics. The end product is two volumes that differ extensively from the Russian original. This first volume is up to date as of late 1969.

We wish to thank Zel'dovich and Novikov for the stimulating and close interaction which they have given us throughout the preparation of these two volumes. We have learned much astrophysics in the process! We also thank Eli Arlock for his valuable translation of the original Russian edition, which formed the backbone for this English edition. A number of our colleagues provided valuable assistance with smoothing out and correcting errors in the manuscript: John Dykla, Edward Fackerell, Richard Greene, James Ipser, Jonathon Melvin, Wei-Tou Ni, William Press, Richard Price, Bernard Schutz, Raphael Sorkin, Ray Talbot, and Clifford Will. We thank them; and we also thank Yvonne B. Dawson for her beautiful and speedy typing of most of the two volumes, from our illegible original manuscripts and our unintelligible magnetic tapes.

KIP S. THORNE
W. DAVID ARNETT

PREFACE TO THE ENGLISH EDITION

Astrophysics is the most rapidly evolving science of our time. The discovery of pulsars and possibly also of gravitational waves, and the formulation of new ideas on irregularities in the early stages of the Universe—these extremely important new developments have arisen since the completion (1967) of the Russian edition of our book, originally called *Relativistic Astrophysics*.

We feel strongly that the spirit of the time is against thick books written by one or two authors.

In the original Russian edition we tried to give a complete review of all important topics. Now this seems to be impossible, even with an inflated volume. In preparing this edition we have tried hard to incorporate most of the new developments; we have revised a great part of the original text; and we have added new chapters and sections. The work is now divided into two volumes. The first is devoted to general relativity, properties of matter under astrophysical conditions, and stars and stellar systems. The second volume will contain cosmology, physical processes in the early stages of evolution of the Universe, the formation of galaxies, and the building up of the structure of the Universe.

The two volumes can be read independently; references in the second volume to material given in the first are not much more frequent than references to other work. It is the aim and the spirit of the account which unify the two books.

We try to present physical pictures of the theories and processes so as to show the plausibility of the theoretical formulae and results, instead of concentrating only on exact mathematical treatments and rigorous proofs. Wherever possible, classical analogues are found for the more complicated relativistic phenomena.

Our goals are to teach the astronomer to think in terms of relativistic concepts, and to familiarize him with new ideas.

A theoretical physicist will find in these books a number of problems—solved, partially solved, and unsolved—which perhaps will interest him and induce him to work in this field. We realize that people already working actively on some branch of astrophysics may not be happy with the corresponding parts of our book. But if (or hopefully, when) they write all that is needed to give a complete picture, the total volume of such contributions will exceed by several times our already thick two volumes.

xiii

Preface to the English Edition

The English edition has been influenced strongly by two other people—the editors, Professor Kip S. Thorne and Professor W. David Arnett. A great part of Professor Thorne's stay in Moscow in the fall of 1969 was devoted to work on the books. We thank him particularly. Included are sections written fully by Thorne on topics in which he is the best expert. But not less important were many critical remarks and demands for clarification. The responsibility for the errors and misjudgments that remain lies fully on the two authors; credit for the embellishment of the English edition as compared with the first Russian edition belongs to a great extent to Thorne and Arnett. We feel the new edition to be the result of an international cooperation—with the editors, with many colleagues met at scientific symposiums, and with the authors of articles quoted. It is now up to the reader to pass judgment on whether the books achieve their goals.

We wish to thank all our colleagues who have made remarks about the first Russian edition of the book. Also, we wish to thank our research collaborators for help in preparing this English edition. In particular, we wish to express our appreciation to V. Schwartzman for his participation in the revision of chapter 13 (a number of the results presented there are due to him), to V. Imshennik and N. Nadezhin for writing §11.4, and to A. Shwarz who helped very much with the translation of the revisions and additions.

YA. B. ZEL'DOVICH
I. D. NOVIKOV

PREFACE TO THE FIRST RUSSIAN EDITION

"Relativistic astrophysics" is a term which was coined very recently: in 1963. This book on relativistic astrophysics deals with two types of systems: the Universe as a whole (cosmology); and isolated bodies, i.e., stars, galaxies, and quasars (astrophysics proper).

The need for going beyond the framework of Newtonian gravitational theory was first recognized in cosmology, where the gravitational paradox of "infinity filled with matter" was recognized even before Einstein devised his theory of relativity. After the creation of the general theory of relativity, the necessary foundations were available for the development of modern cosmology. In 1922–1924 A. A. Friedmann constructed nonstationary, relativistic cosmological models. Friedmann's remarkable work has long since become classic. In recent years our picture of the motions and forces which dominate the Universe has been revolutionized: The discovery, in 1964, of electromagnetic radiation with a 3° K blackbody spectrum provided substantive proof of Gamow's 1948 "big bang" model of the Universe. This discovery has lent greater credibility to recent calculations of the elementary-particle transformations and nuclear reactions which probably occurred at an early stage of the cosmological expansion. Also revolutionized have been theories of the births of galaxies, stars, and quasars from hot plasma in the course of the expansion of the Universe.

As opposed to cosmology, astrophysics in the narrow sense of the word remained nonrelativistic until recently. The study of conventional stars with masses less than 100 solar masses does not require the theory of relativity, at least not while the stars are in states normally observed by astronomers. The Newtonian theory of gravitation, along with thermodynamics, plasma physics, and nuclear physics, provides a complete, quantitative description of the luminosity, dimensions, and spectra of such stars; and the theory of stellar structure is in good agreement with the observational data. As it consumes its nuclear fuel, a star evolves. But this evolution, according to recent theory, will eventually lead to a state in which rapid processes and the effects of general relativity become important. Interest in these late stages of stellar evolution has increased considerably since the discovery of quasars, or "superstars" as they were originally called. Since quasars, with a total luminosity hundreds of times larger than the luminosity of giant galaxies, are extremely unusual in their properties, they have called to life a number of exotic hypotheses, ranging from matter-antimatter annihilation to the presence of negative mass. Fantastic hypotheses usually follow in the wake of any new discovery, but as a rule they are not considered very seriously. However, even attempts to explain quasars within the framework of serious, contemporary physics—e.g., as supermassive stars (stars more massive than 10^5 Suns)—lead to situations in which the effects of general relativity play a significant role. Hence, quasars are truly a subject of study for relativistic astrophysics.

These are the basic problems with which this book deals. In the exposition of each problem, the authors have attempted, so far as possible, to present a complete picture of the phenomenon and of all physical factors without limiting the discussion to purely relativistic effects, and without taking it for granted that the reader knows everything else. Perhaps the main shortcoming of this book is the insufficient consideration of the effects of magnetic fields in various phenomena. Fortunately, there exist a number of excellent monographs which treat the magnetohydrodynamics of astronomical phenomena; to mention just a few: *Cosmical Electrodynamics* by Alfvén and Fälthammar (1963), *The Fundamentals of Cosmic Electrodynamics* by Pikel'ner (1961), and *Interstellar Medium* by Kaplan and Pikel'ner (1963). We have omitted another group of important problems, namely, cosmic rays and the details of the radiation from relativistic particles and plasmas in magnetic fields. For these topics we refer the reader to the following monographs: *The Origin of Cosmic Rays* by Ginzburg and Syrovatsky (1964c), *Cosmic Radio Waves* by Shklovsky (1956), *Propagation of Electromagnetic Waves in Plasma* by Ginsburg (1962); we also refer the reader to the review of the state of the art by Ginzburg and Syrovatsky (1965).

This book does not assume that the reader has a high degree of technical competence. A very qualitative understanding of general relativity should be sufficient; it is not even necessary to have a comprehensive knowledge of the second part of the *Classical Theory of Fields* by Landau and Lifshitz (1962). The most important aspects of relativity, which have been known for a long time and which appear in most textbooks—e.g., the Schwarzschild solution for the field of an isolated mass, or Friedmann's cosmological solution—are also presented in detail here. In addition, a detailed physical interpretation of well-known formulae is given; they are somewhat "prechewed" for better digestion. We want to demonstrate to the reader all the richness of the physics included in these formulae, the methods of dealing with them, and the correspondence of the relativistic with the classical theories. In addition, the book touches upon some of the new findings of the general theory of relativity, and naturally it includes everything new that pertains to relativistic astrophysics.

The work on this book was essentially finished in the summer of 1966. Some additions were made in the course of additional work in order to include some of the findings published in late 1966 and in 1967. The rapid development of the particular branch of science which makes up the principal subject of our book created immense difficulties in its preparation. More than once were we about to bring to conclusion the entire project when new discoveries forced us to supplement and substantially expand the exposition; more than once was the submission deadline of the manuscript delayed. Even now, when this Preface is being written, the frightening thought occurs: "What will happen six months from now, when the galleys are ready; what will happen a year from now, when the book is finally printed, and when every written word becomes irreversibly solidified, while science continues to advance . . . ?" In view of all this, we have attempted to separate clearly the problems that are already solved from those which are still at the stage of being solved, and from those which are just being brought up and put on the agenda. Naturally, all our fears are concentrated upon the interpretation of astrophysical phenomena which as yet have not been adequately explained, and particularly upon the applicability (or inapplicability) of existing theoretical models for these phenomena. Insofar as the numerous models are viewed as solutions to *theoretical* problems (problems posed in connection with attempts to understand the observed phenomena, but treated thereafter as problems of theory), the models will remain correct in the future, provided that they are correct today. The reviews published by the authors, jointly and separately over the past few years, were important in the preparation of this book. However, the rapid development of science prevented these reviews from being used as extensively as was initially intended.

The authors fully realize that, along with completely reliable and long-lasting material, this book also contains assumptions and hypotheses, the majority of which will become obsolete in the next few years and may well turn out to be untenable. Faced with such a situation, we have attempted, to the best of our understanding and ability, to warn the reader as to the degree of reliability of each theory and hypothesis. Finally, some justification for our analysis of results not yet fully confirmed by observations may be that a concise exposition of questionable hypotheses stimulates their rebuttal.

The following collaborators in our research group were of great help in the writing of this book: A. G. Doroshkevich, B. V. Komberg, G. S. Bisnovati-Kogan, R. A. Syunyaev, D. K. Nadezhin, V. S. Imshenmik, V. M. Dashevsky, and V. M. Chechetkin. An awesome amount of work was done by A. A. Gus'kova in the preparation of the manuscript. Frequent discussions and exchanges of opinions with V. L. Ginzburg, I. S. Shklovsky, S. B. Pikel'ner, L. M. Ozernoy, N. S. Kardashev, E. M. Lifshitz, I. M. Khalatnikov, and many other colleagues were also of great importance. V. L. Ginzburg and L. M. Ozernoy read the book manuscript and contributed many valuable observations.

Extremely useful was the participation of one of the authors (I. Novikov) in the International Conference on Gravitation in England in the summer of 1965, and in the Third Texas Symposium on Relativistic Astrophysics in New York in January of 1967, which were attended by, among others, S. Chandrasekhar, J. A. Wheeler, W. A. Fowler, K. S. Thorne, and a number of other scientists who have made great contributions to the creation of relativistic astrophysics. Correspondence with R. H. Dicke and D. Layzer has been both important and interesting.

The credit for organizing the symposium in Byurakan in May 1966 goes to Academician V. A. Ambartsumyan. This symposium made possible meetings with J. H. Oort, M. Schmidt, E. M. Burbidge and G. Burbidge, G. Münch, C. R. Lynds, and a number of other prominent astronomers.

We would like to take this opportunity to express our most sincere gratitude to all concerned.

YA. B. ZEL'DOVICH
I. D. NOVIKOV

I THE THEORY OF GRAVITATION

1 EINSTEIN'S GRAVITATIONAL EQUATIONS

1.1 THE EQUALITY OF INERTIAL AND GRAVITATIONAL MASS

The contemporary theory of gravitation, formulated by Albert Einstein in 1916, was an extension of the special theory of relativity (STR); thus it is frequently referred to as the general theory of relativity (GTR).

The essence of STR is contained in the Lorentz transformation for spacetime coordinates, and in the corresponding transformation laws for such physical quantities as energy and momentum. All the postulates and conclusions of STR, such as the constancy of the velocity of light, the dependence of mass on velocity, mass defect and its relationship to the energy of a system, time dilation during fast motion as exemplified by the decay of particles—all of these are confirmed experimentally. Indeed, STR has by now found its place in practical engineering calculations. Thus, there are no doubts whatsoever as to the correctness of STR.

The general theory of relativity (GTR) is in an entirely different situation: Experiments which specifically confirm GTR are few at present. They include the precession of Mercury's perihelion, the deflection of light in the Sun's gravitational field, the gravitational redshift, and (as of 1968) the relativistic delay in radar signals passing near the limb of the Sun. Actually, the principal argument in favor of GTR is none of these. Rather, it is a basic fact which is known to every high school student, and which inspired Einstein's work: the proportionality of weight and mass, i.e., the equality of the acceleration of different bodies in a gravitational field.

Newton's law of gravitation, $F = -Gm_1m_2/r^2$, is very similar to Coulomb's law of electrostatics, $F = e_1e_2/r^2$. Naturally, the question arises as to why the theory of the electromagnetic field, which is formulated in Euclidean space, and GTR with its concept of spacetime curvature, are so different. Couldn't gravity also be defined as a field in Euclidean space?

It will be demonstrated in the forthcoming discussion that STR and quantum mechanics make the spacetime curvature, which is characteristic of GTR, logically unavoidable. GTR is not only the most elegant and comprehensive theory from the viewpoint of mathematics, it is also the physically necessary theory of gravitation.

As we have already emphasized, the most characteristic property of a gravitational field is the fact that it acts with complete indifference upon all

3

bodies, imparting to them equal accelerations. This is a fact that was first established by Galileo. Thus, the gravitational field, in its effects, is different from any other field known to physics.

One of the most important tasks of experimental physics, in view of current interest in the theory of gravitation, has been to measure the accuracy with which gravity imparts equal accelerations to different bodies. The principle of equal accelerations can be formulated equivalently as the principle of strict proportionality between inertial and (passive) gravitational mass: In the equations of motion for a body in a gravitational field,

$$m_i(d^2r/dt^2) = m_0\nabla U \,,$$

the inertial mass appears on the left-hand side and the gravitational mass on the right. If all bodies satisfy $m_i = am_0$, where a is a universal constant, then Galileo's law will follow, since the masses will not enter into the equation of motion at all. The constant a obviously depends only on the normalization of U; and with an appropriate choice it can be set to unity.

In 1890 Eötvös devised an exceedingly precise method to check the proportionality of the gravitational and inertial masses. In essence, his experiment consisted of the following: Any object at rest on the surface of the Earth experiences gravitational attractions not only from the Earth but also from the Sun, the Moon, and other celestial bodies. The object is also acted upon by centrifugal forces resulting from the diurnal rotation of the Earth about its axis, the annual rotation of the Earth about the Sun, and the monthly rotation of the center of the Earth about the center of gravity of the Earth-Moon system.

The attraction by planets and by other celestial bodies can be neglected. Similarly, we can neglect the centrifugal acceleration due to the motion of the Sun in the Galaxy, etc. The acceleration of the Earth's gravity is about 980 cm sec^{-2}; the centrifugal acceleration due to diurnal rotation at the latitude of Moscow is about 1.5 cm sec^{-2}. The acceleration of the Sun's gravitational field on the orbit of the Earth is about 0.5 cm sec^{-2}; the centrifugal acceleration due to the Earth's orbital motion is obviously also 0.5 cm sec^{-2}. The effect of the Moon is characterized by an acceleration of 4×10^{-3} cm sec^{-2}.

Visualize two bodies, A and B, of equal mass, balanced on the ends of a thin rod (balance arm) which is suspended at its middle by a thin fiber. The gravitational forces of the Earth, Sun, and Moon are proportional to the gravitational mass, whereas the centrifugal forces are proportional to the inertial mass.

If the two bodies have identically the same ratio of inertial to gravitational mass, the total force acting upon them will be identical. In this case the balance arm will remain in equilibrium with respect to the axis of the Earth, and also with respect to the Sun, regardless of the direction in which it is oriented. (Zero torque acts.) However, if the mass ratios are different,

4

and if the balance arm is set perpendicular to the direction of the centrifugal forces, then the forces on the two bodies will not be equal; they will produce a torque, causing the balance arm to twist beneath its supporting fiber. The centrifugal force of diurnal rotation is larger than the centrifugal force of annual rotations. However, a change in orientation of the balance arm with respect to the Sun takes place naturally during the rotation of the Earth, without a corresponding change in orientation of the balance arm relative to objects in the laboratory that surround it, or relative to the terrain of the Earth. Hence, in practice, it is most convenient to observe whether or not the balance arm is twisted by unbalanced gravitational and centrifugal forces due to the Sun and the annual rotation of the Earth.

In view of the absence of such twisting, Eötvös concluded that the ratio of gravitational to inertial mass for different bodies differs by no more than one part in 10^8. The Eötvös experiment was recently repeated by Dicke (1961a) and by Roll, Krotkov, and Dicke (1964). By placing the balance arm in a high vacuum and determining the torque on it, using a photocell and automatic feedback to prevent balance-arm twisting, Dicke was able to improve the accuracy of the Eötvös experiment. His findings agree with Eötvös's findings: the ratio between the gravitational and inertial masses of copper and lead are in agreement. But, with Dicke, the accuracy of this agreement is better than one part in 10^{10}!

Consider the significance of this finding. The inertial mass depends on energy—this is a consequence of STR. In particular, we know from STR that when two deuterium atoms are combined into one atom of helium, the inertial mass decreases by an amount equal to about 0.006 of the initial mass, in accordance with the mass defects of helium and deuterium. Precise determinations of inertial atomic masses by using a mass spectrograph on one hand, and direct measurements of the energy of nuclear reactions on the other hand, have confirmed that energy contributes to inertial mass.

What determines the (passive) gravitational mass of a body, and consequently the force which it experiences in a gravitational field? Does the gravitational force depend on the number of baryons in the body, i.e., the baryon charge, in a manner similar to the dependence of electrostatic attraction on electrical charge? Or does this force depend on the total energy of the body? For conventional matter (neither mesons nor antimatter), the baryon number and the inertial mass are approximately proportional to each other with deviations of the order of 10^{-3}. Hence with small accuracy, an Eötvös-type experiment could not have resolved the problem. However, Dicke's experimental accuracy of better than 10^{-10} leads inescapably to the conclusion that the force of gravity, like the inertial mass, is proportional to the energy of the body. Such extreme accuracy strengthens the foundations of GTR.

The Eötvös experiment shows that the attraction is not determined by the baryon charge of the body; universal gravitation thus cannot be construed as due to an analogue of the electrostatic attraction of electrical charges of opposite sign. (It is not accidental that two electric charges must have opposite signs in order to attract each other; see chapter 2.) Thus, the notion that some particles—for example, the so-called antiparticles (positrons, antiprotons, antineutrons)—can experience "antigravitation" is absolutely erroneous. From experiments on accelerators it is well known that to create antiparticles it is necessary to expend energy. This energy is the source of the mass of the antiparticle. Consequently, the antiparticle has a positive gravitational mass, exactly the same as the corresponding particle. The Eötvös and Dicke experiments provide an indirect proof of this.

Lee and Yang (1955) (see also Dicke 1962) posed the question of whether or not, along with universal gravitation, there might exist a repulsive analogue to the Coulomb force which is proportional to the number of nucleons. The Eötvös and Dicke experiments demonstrate that such a force does not exist; or, more precisely, that if it does exist, it is at least 10^7 times weaker than the gravitational force, and at least 10^{43} times weaker than the Coulomb force between two protons.[1]

It should be emphasized that, if repulsive forces due to baryon charge were to exist, GTR could be nonetheless valid. However, in such a case it would be considerably more difficult to disentangle the different forces by experiment (just as it is difficult to perform the Cavendish experiment with bodies that are electrically charged).

1.2 THE FUNDAMENTAL CONCEPT OF THE GENERAL THEORY OF RELATIVITY

To Newton it seemed obvious that physical space is Euclidean, i.e., that there exist parallel lines, that the sum of the angles in a triangle is π, that the circumference of a circle is $2\pi r$, etc., and that time flows always at the same rate everywhere.

The idea that the properties of space could be otherwise (for example, that the sum of the angles in a triangle could depend on its area) occurred considerably later. Mathematically, such spaces were first discovered and investigated by Lobachevski.

According to STR, in an inertial frame of reference the square of the four-dimensional distance (in space and time) between two infinitely close events (interval) takes the form

$$ds^2 = (cdt)^2 - (dx)^2 - (dy)^2 - (dz)^2 , \qquad (1.2.1)$$

where c is the speed of light, t is time, and x, y, z are Cartesian space coordinates. Such a system of coordinates is said to be a Galilean system.

---

1. The Lee-Yang force is incompatible with a homogeneous, isotropic cosmological model (see Vol. II).

1.2 Fundamental Concepts

Expression (1.2.1) is analogous in form to the expression for the square of the distance in Euclidean three-dimensional space as written in Cartesian coordinates (only the number of squared differentials and their signs differ). Hence, expression (1.2.1) describes for us a flat Euclidean spacetime, or, more precisely, a pseudo-Euclidean spacetime (emphasizing the special nature of time; note in expression [1.2.1] the plus sign in front of $(cdt)^2$ as contrasted with the minus signs in front of the space differentials). STR is the theory of physical processes in this flat spacetime, which we shall call Minkowski spacetime.

In this spacetime a free particle moves along a straight line (Fig. 1), called the world line of the particle. We will not pursue these matters any further because we assume that the reader is familiar with the elementary principles of STR.

Einstein's theory of gravitation was inspired by and based on the "principle of equivalence," which states that when gravity is present, as when it

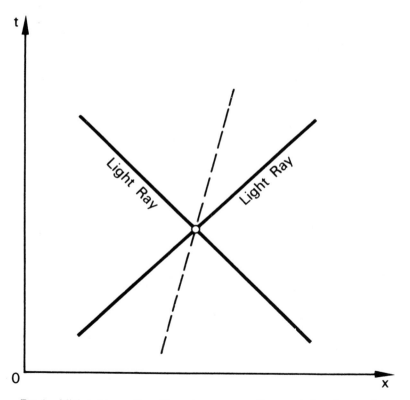

Fig. 1.—Minkowski spacetime. Here x is a space coordinate and t is a time coordinate. The dashed line is the world line of a freely moving test particle.

is absent, free particles move along extremal (geodesic) lines of spacetime—spacetime now being curved, not flat.

According to Einstein, the gravitational field is nothing more than a deviation of the properties of real spacetime from the properties of a flat manifold. It is masses that create the gravitational field—which curves spacetime. A body in this curved spacetime moves along a geodesic line, which is independent of its mass and composition. This geodesic motion in curved spacetime is perceived by us as curved motion with variable velocity. Einstein's theory postulates from the very beginning that the curvature of the trajectory and the variation of speed are spacetime properties, properties of the geodesics; and, hence, that the accelerations of all bodies must be equal. Thus the ratio of gravitational mass to inertial mass (on which the gravitational acceleration of the body depends, according to Newton) is equal for all bodies.

However, historically Einstein proceeded from more obvious assumptions based on a simple physical model of the gravitational field, which had equal accelerations for every body. This was the well-known elevator thought-experiment: In an isolated, accelerated volume, in the absence of a true gravitational field (i.e., in an elevator in space) all phenomena proceed in precisely the same manner as in a real gravitational field in which the volume either is at rest or moves uniformly.[2] Relative to the elevator, freely moving bodies undergo accelerated motion, and the acceleration is precisely the same for every body.

In the accelerated elevator (without real gravity) one can see all the effects of gravitation—for instance, a changing frequency for propagating light. In particular, compare the frequency of emitted light with the frequency which the light possesses after a time interval has elapsed. Because of the acceleration relative to an inertial frame, the velocities of the elevator and of objects at rest in it will change with time. Hence, the velocity of the light receiver at the moment the light is received will differ from the velocity of the emitter at the moment of emission. As a result of the Doppler effect this difference in velocities will produce a similar difference in frequency between the emitted and the received light. This difference will depend on the relative directions of the light ray and the acceleration.

Such an interpretation is given by an external inertial observer who knows that the elevator is being accelerated. An internal observer attributes the frequency shift to a "gravitational field" present inside the elevator. If the acceleration is directed from the emitter toward the receiver, then the light will suffer a blueshift; in the opposite case there will be a redshift. It is easy to convince oneself that a light ray should exhibit a curvature relative to the three-dimensional coordinate system which is tied to the elevator. Thus,

---

2. In order for the elevator to be at rest in a real gravitational field, an external force must be applied to it to counterbalance gravity.

phenomena in an accelerated reference frame are identical v "gravitational field."

However, only a homogeneous gravitational field with a co.. and direction throughout all space can be described in such a way. Gravi.. tional fields created by separated bodies are quite different. To cancel the Earth's gravitational field, we need elevators with different directions of acceleration at different points. Observers in the different elevators, communicating with each other, will find out about their relative accelerations, and will thereby discover the presence of a gravitational field which cannot be removed by going to any single accelerated moving coordinate system.

A true gravitational field cannot be eliminated by any coordinate transformations; Fock has particularly emphasized this in his well-known book (1961). All the same, the elevator model describes the most important properties of gravity (equality of all accelerations; the gravitational influence on light) in such a natural way that it is unreasonable to give up this model. To conserve this local use of such a model at every point in spacetime, one can introduce a transformation of the coordinate system in every region, but this transformation cannot be reduced to any particular motion in flat spacetime. This impossibility expresses the "curvature" of spacetime (see below).

The following sections will deal briefly with the mathematical methods of analyzing spacetime curvature which are required for understanding the material to follow. Readers interested in a more detailed exposition of spacetime curvature and GTR are referred to the books by Landau and Lifshitz (1962), Synge (1960), and Robertson and Noonan (1968), and also to the contribution by Zel'manov (1959*b*).

1.3   PROPERTIES OF NONINERTIAL SYSTEMS

To delineate more precisely the concept of spacetime curvature, let us first recall the peculiarities of the spatial geometry and of the flow of time in noninertial frames of reference which accelerate in flat, Minkowski spacetime. This will permit us to introduce the concepts needed for computations in curved spacetime.[3]

In Minkowski spacetime (i.e., far from all gravitating masses) as measured in an inertial frame of reference, geometry is Euclidean and time always flows uniformly. Consider, however (following Einstein 1965;[4] see also Landau and Lifshitz 1962), a rigidly rotating disk. An inertial observer, A, who

---

3. From a mathematical viewpoint, this corresponds to introducing curvilinear coordinates on a plane. The resultant formalism can then be used to compute on curved surfaces, where it is absolutely necessary to use curvilinear coordinates.

4. We refer to Einstein's collected works, the first volume of which was published in Russian in 1965. *Editors' note.*—Unfortunately, although the Russian edition of Einstein's collected works is now complete, at present there is no English edition.

is not participating in the rotation, can measure the circumference of the disk, *l*, and its diameter, *D* (for example, by measuring the circumference and diameter of a nonrotating circle directly under the rotating disk). Obviously, he finds that $l/D = \pi$. Another observer, B, located on the rotating disk, can also measure the circumference and diameter by directly applying a tape measure first to the edge of the disk, and then to its diameter.

Observer A notes that when B applies the moving tape measure to the edge of his disk, the tape measure experiences a Lorentz contraction of its length. Consequently, B packs more length units of his tape measure around the circumference of the disk, and he finds the circumference to be larger than when measured in A's inertial frame; he finds

$$l' = l/[1 - (v/c)^2]^{1/2},$$

where *v* is the velocity of the edge of the disk. When the tape measure is applied to the diameter of the rotating disk, it does not contract in length as seen by the inertial observer A, because it moves in a transverse direction. Consequently, the measurement of the diameter yields the same result as in the inertial frame: $D' = D$. Consequently, as measured by the observer on the rotating disk, the ratio of circumference to diameter is greater than π,

$$\frac{l'}{D'} = \frac{l}{D[1 - (v/c)^2]^{1/2}} = \frac{\pi}{[1 - (v/c)^2]^{1/2}} > \pi$$

—a result which violates Euclidean geometry.

Consider next the flow of time as measured by clocks on the disk's surface. The farther a clock is from the center of the disk, the greater its linear velocity of rotation and the slower it ticks as seen by the inertial observer, A. According to a familiar formula of STR, $t' = t[1 - (v/c)^2]^{1/2}$. Thus, time flows at different rates at different points on the disk. Moreover, if the angular velocity changes, the rate of time flow will also change.

But this is not all. Consider a set of clocks all riding on the disk at the same radius. The clocks move with the same speed *v;* hence, they all tick at the same rate. It is known from STR that if clocks at two different points, i and ii, on a moving body are synchronized by using light beams, then for a laboratory observer clock i will read a given time—e.g., 2 o'clock— somewhat before clock ii does (see Fig. 2).

Thus, if we were to attempt to synchronize clocks located on the circumference of a rotating disk, we would obtain the following (cf. Fig. 2).

To the laboratory observer, clock ii will appear to be running behind clock i; clock iii will run behind clock ii and even more behind clock i; and so forth. Passing around the entire circumference and returning to clock i,

---

5. Note that from the standpoint of SRT the roles of observers A and B cannot be reversed; i.e., we cannot regard B as at rest and A as rotating, because the disk experiences centrifugal forces and Coriolis forces (caused by rotation), which are absent in A's inertial frame. Frames A and B are not equivalent.

we conclude that a clock resting at the same point as clock I and synchronized with it must be set behind clock I—an obviously absurd result!

This analysis indicates that it is impossible to synchronize time on a rotating body. Not only does time flow at different rates at different points, but the very concept of synchronization does not exist.[6]

When we considered measurements of the circumference by laying out a tape measure, we were not careful about laying down all parts of the tape measure at the same "moment of time." Now we understand that the concept of "the same moment of time" is not applicable over finite regions of

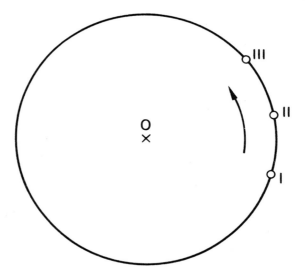

Fig. 2.—Synchronization of clocks on a rotating disk (see text)

the disk. That is why it is meaningless to speak about the properties of the disk as a whole at a given "moment" if the disk rotates with a variable angular velocity. However, for small parts of the disk one can introduce the concept of simultaneity with sufficient precision for defining the geometrical properties of those small parts of the disk, and for speaking about the deviation of their geometry from Euclidean (e.g., the deviation of the sum of the angles of a triangle from π, etc.).

Up to now we have considered the disk to be rotating with a constant angular velocity. If the angular velocity of the disk is variable, then the geometric properties of its various parts are also variable.

6. *Editors' note.*—Synchronization is used here in the sense of SRT: If a light signal proceeds from A to B and back to A, then the time of reflection at B must read the average of the times of emission and reflection at A.

Let us sum up: Even in conventional flat spacetime it is impossible to establish on an accelerated body a rigid coordinate system with a flat (Euclidean) space metric and with synchronized time. Such a coordinate system can be tied only to an inertially moving body. With the exception of special cases (for example, a rigidly rotating disk), any given frame of reference will become deformed in the course of time; its geometrical properties will change, and the rates of clocks tied to it will change.

In Newtonian physics, a rigid Cartesian frame of reference is determined by specifying at each moment of time the location of its origin and the orientation of its axes. In special relativity, to establish an accelerated reference frame one must give not only the motion of one of its points (the origin) but also the motion of all other points. We will mean by reference frame a collection of test particles (each carrying a clock) which fill all of spacetime, and which move along arbitrary paths.[7]

A similar situation occurs in GTR. The key difference is that in STR, where gravitational fields are absent, one can always pass from any noninertial frame of reference to an inertial frame of reference, and use that inertial frame throughout all of spacetime. In GTR, this cannot be done because of the curvature of spacetime.

We turn now to the mathematical formulation of these ideas. All formulae that we shall obtain will be valid not only in flat spacetime but also in the curved spacetime of general relativity, because they are local formulae and by virtue of the equivalence principle a gravitational field is locally indistinguishable from an accelerated frame.

1.4 THE MEASUREMENT OF SPACETIME INTERVALS

In an inertial frame of reference in STR, it is not absolutely necessary to use Cartesian space coordinates. Any curvilinear space coordinates (for example, spherical coordinates) can be used. The transformation from one system of space coordinates to another has the form

$$x^a = x^a(x^{1\prime}, x^{2\prime}, x^{3\prime}) . \tag{1.4.1}$$

Here the index a takes on the values 1, 2, 3; and x^1, x^2, x^3 denote the three space coordinates.

The expression for ds^2 now takes on a form different from equation (1.2.1). Instead of the three squares of the differentials of the Cartesian coordinates, we have the expression for the square of the space interval in curvilinear co-

---

7. *Editors' note.*—More precisely the "reference frame" is identified with the "collection of test particles" moving in a specified manner. A reference frame is not a coordinate system, which includes in addition a parameterization of the particles (by coordinates x^1, x^2, x^3) and a time variable x^0 (not necessarily physical time!) for each particle.

This terminology is used widely in the Soviet Union but is not common in the West.

ordinates (taken with a minus sign). In spherical coordinates this is $-dl^2 = -(dr^2 + r^2 d\theta^2 + r^2 \sin^2 \theta \, d\phi^2)$, and the squared spacetime interval is written

$$ds^2 = (dct)^2 - dr^2 - r^2 d\theta^2 - r^2 \sin^2 \theta \, d\phi^2 \, . \qquad (1.4.2)$$

In cylindrical coordinates

$$ds^2 = (dct)^2 - dr^2 - r^2 d\phi^2 - dz^2 \, . \qquad (1.4.3)$$

The transformation (1.4.1) corresponds to a passage from one three-dimensional space grid of coordinates to another. However, the frame of reference (cf. § 1.3) remains unchanged. In passing to a different frame of reference, which moves arbitrarily relative to the inertial frame, one must use a time-dependent coordinate transformation

$$x^a = x^a(x^{1'}, x^{2'}, x^{3'}, t) \, . \qquad (1.4.4)$$

The spatial part ($a = 1, 2, 3$) of this transformation defines completely the motion of the new reference system. The time part ($a = 0$) of the transformation (which involves coordinate time, not physical time; see below) can be arbitrary.

The transformation to a uniformly rotating reference frame has the form

$$x^1 = x^{1'} \cos \Omega t - x^{2'} \sin \Omega t \, ,$$

$$x^2 = x^{1'} \sin \Omega t + x^{2'} \cos \Omega t \, ,$$

$$x^3 = x^{3'} \, ,$$

where Ω is the angular velocity; or, in cylindrical coordinates,

$$r = r' \, , \quad z = z' \, , \quad \phi = \phi' + \Omega t \, . \qquad (1.4.5)$$

(The symbol t will be reserved for the time coordinate [not physical time!] in the new reference frame.) In this case, substituting equation (1.4.4) or (1.4.5) into equation (1.4.3), we find that the entire expression for ds^2 changes:

$$ds^2 = (c^2 - \Omega^2 r'^2)dt^2 - 2\Omega r'^2 \, d\phi' dt - dz'^2 - r'^2 d\phi'^2 - dr'^2 \qquad (1.4.6)$$

for the rotating disk; and

$$ds^2 = g_{ik} dx^i dx^k \qquad (1.4.7)$$

in general. Here and throughout this book any index which appears twice must be summed; latin indices take on the values 0, 1, 2, 3; x^1, x^2, x^3 denote space coordinates (generally curvilinear); $x^0 = ct$ is the time coordinate; and g_{ik} are functions of the spacetime coordinates. Each term with $i \neq k$ occurs twice in the sum (1.4.7); for example, $g_{01}dx^0 dx^1$ and $g_{10}dx^1 dx^0$. By definition, g_{ik} always equals g_{ki}.

The quantities g_{ik} are called the components of the *metric* tensor; they define the so-called spacetime metric. The specific forms of g_{ik}, as functions of the coordinates, are determined not only by the properties of spacetime but also by the particular coordinate system used, that is, by the choice of reference frame, by the choice of space coordinates in that frame, and by the choice of the method of measuring time (i.e., the time coordinate).

The expression (1.4.7) contains complete information about the geometric properties of the reference frame and about the geometric properties of space-time itself. How is this expression to be used? First, to examine the passage of time at a given point it is obviously necessary to regard x^1, x^2, x^3 as constant: $dx^1 = dx^2 = dx^3 = 0$. Then ds will be the time separation between two nearby events; i.e., it will be the interval of time multiplied by the speed of light, $ds = cd\tau$. Consequently,

$$d\tau = [(g_{00})^{1/2}/c]dx^0 . \tag{1.4.8}$$

In any reference frame formed by real bodies, it is always true that $g_{00} > 0$.

Let us now calculate the spatial distance dl between two particles at rest in our reference frame. This cannot be done by simply setting $dx^0 = 0$, x^0 = constant in expression (1.4.7). The problem with such a procedure is that the same moment of proper time does not always correspond to identical clock readings at different points in space. Therefore, before performing the computation we must determine which value of x^0 at a neighboring point is "simultaneous" with a given value of x^0 at a given point. Clock synchronization between two nearby points is performed by means of light signals.[8] We will not go into the computation here; instead, we refer anyone interested to the textbook of Landau and Lifshitz (1962). The final formula for the square of the spatial distance dl^2 is

$$dl^2 = \left(-g_{\alpha\beta} + \frac{g_{0\alpha}g_{0\beta}}{g_{00}}\right) dx^\alpha dx^\beta, \quad \alpha, \beta = 1, 2, 3 . \tag{1.4.9}$$

The terms in parentheses will be denoted $\gamma_{\alpha\beta}$. They determine the metric of the three-dimensional space in our frame of reference.[9] The volume element of this three-dimensional space is given by the expression $dV = \sqrt{\gamma}\, dx^1 dx^2 dx^3$, where $\gamma = |\gamma_{\alpha\beta}|$ is the determinant of the matrix composed of elements $\gamma_{\alpha\beta}$.

8. See n. 6.

9. *Editors' note.*—This three-dimensional space is defined only locally, in the following sense: At each event in spacetime it is an infinitesimal three-dimensional surface orthogonal to the world line (x^1, x^2, x^3) = *constant*. These infinitesimal "three-surfaces" usually cannot be meshed together to form a family of three-surfaces extending throughout spacetime. According to a theorem of differential geometry, the infinitesimal three-surfaces mesh to form finite three-surfaces if and only if the world lines (x^1, x^2, x^3) = *constant* have zero rotation—which they usually do not—or, equivalently, if and only if the angular-velocity tensor $A_{\alpha\beta}$ of equation (1.6.3) vanishes. Consequently, while the space metric $\gamma_{\alpha\beta}$ is defined at each event in spacetime, it is not usually the metric of a large-scale three-surface embedded in spacetime. Nevertheless, it *is* the metric of a (fictitious) three-surface whose geometric properties (Riemannian and Gaussian curvatures—cf. end of this section and end of § 1.8—deformations [eq. (1.6.4)], etc.) describe the geometric and kinematic properties of the reference frame.

Consider again the rotating disk as an example. Using formula (1.4.8), we obtain from equation (1.4.6) for the time interval (in the future we omit the prime on the coordinates)

$$d\tau = (1 - \Omega^2 r^2/c^2)^{1/2} dt .$$

On the axis of rotation, $r = 0$ and $d\tau = dt$. Hence the preceding formula can be rewritten in the form

$$d\tau = (1 - \Omega^2 r^2/c^2)^{1/2} d\tau_{\text{axis}} .$$

Time flows more slowly, the further removed one is from the axis of rotation.

For the element of spatial distance, we obtain from equations (1.4.6) and (1.4.9).

$$dl = \left[dr^2 + \frac{r^2 d\phi^2}{1 - \Omega^2 r^2/c^2} + dz^2 \right]^{1/2} . \tag{1.4.10}$$

Using equation (1.4.10), we find that, when r and z are constant, then $dl = (1 - \Omega^2 r^2/c^2)^{-1/2} r d\phi$, and the ratio of the circumference of the circle to its diameter is

$$\frac{l_{\text{circle}}}{D} = \frac{\pi}{(1 - \Omega^2 r^2/c^2)^{1/2}} ,$$

in accordance with the analysis of the preceding section.

For comparison, we present an example of a nonstatic reference frame (i.e., a reference frame deforming with time) in which, as for our rotating disk, the ratio of a circle's circumference to its diameter is less than π.

In Minkowski spacetime consider a collection of particles ejected in all directions with all possible speeds, from a given point in space at some moment of time. The world lines of these particles fill the interior of the light cone. The reference frame tied to the particles is called the Milne frame (after Milne, who introduced such a cosmological model).

The transformation from the spherical coordinates (r', θ', ϕ') of a non-deforming reference frame and from laboratory time, t', to Milne's space coordinates (ψ, θ, ϕ) and the proper time of the particles, t, is given by

$$t^2 = t'^2 - r'^2 , \quad \tanh \psi = r'/t' ,$$

$$\theta = \theta' , \quad \phi = \phi' .$$

Substituting these transformations into $ds^2 = c^2 dt^2 - dr'^2 - r'^2(d\theta'^2 + \sin^2 \theta' d\phi'^2)$, we obtain the expression for ds^2 in the Milne frame

$$ds^2 = d(ct)^2 - (ct)^2 [d\psi^2 + \sinh^2 \psi(d\theta^2 + \sin^2 \theta \, d\phi^2)] .$$

At a fixed moment in time, $t = const.$, we examine the curvature of an equatorial "plane," $\theta = \frac{1}{2}\pi$. The ratio of the circumference of a circle in this equatorial plane, $l = 2\pi ct \sinh \psi$, to its diameter, $D = 2ct\psi$, is greater than π: $l/D = \pi(\sinh \psi)/\psi > \pi$.

If at a given point of a two-dimensional surface the ratio of the circum-

15

ference of a small circle to its diameter, l/D, is smaller than π, then the curvature of the surface is said to be positive, and its geometry is similar to the geometry on a sphere. If $l/D > \pi$, then the curvature is negative, and the geometry is similar to the geometry on a pseudosphere (saddle-shaped surface).

Numerically, the curvature is defined by the square of the radius of curvature, a^2, which is determined in the following manner. A small triangle is drawn on the surface; the sides of the triangle are the shortest lines (geodesics) between its vertices. Let Σ denote the sum of the angles of the triangle. It can be demonstrated that the difference $\Sigma - \pi$ is proportional to the area of the triangle, S:

$$\Sigma - \pi = CS .$$

The coefficient of proportionality C is called the curvature, and the quantity $a = 1/C^{1/2}$ is the radius of curvature. If $\Sigma > \pi$, then $C = 1/a^2 > 0$, and the curvature is positive. If $\Sigma < \pi$, then $C = 1/a^2 < 0$, the curvature is negative, and a is imaginary. The smaller $|a^2|$, the larger the curvature and the more the geometry differs from Euclidean geometry.

The curvature at a given point of a three-dimensional space is determined in the following manner. A geodesic surface (analogue of a plane in Euclidean space) is drawn through the point, and its curvature is calculated. This curvature is called the Riemann curvature of space in the two-dimensional direction of the chosen surface. The curvature may be different in different directions. The curvature averaged over all directions is called the Gaussian curvature of the three-space. We will not give here formulae for computing the curvature in the general case. For important specific examples, the formulae are listed at the end of § 1.8.

Thus, a surface of constant time in the expanding Milne reference frame (see above) has negative curvature.

It is not difficult to understand the mathematical source of the non-Euclidean three-geometry of a noninertial reference frame in flat four-dimensional spacetime. The three-dimensional space of an *inertial* frame is a "flat" three-dimensional hypersurface or "slice" passing through the four-dimensional spacetime. The space of a three-dimensional *noninertial* frame results from a curved slice through four-dimensional spacetime. It is hardly surprising that the geometry of this curved slice is non-Euclidean. The situation is completely analogous to the geometry of a curved two-dimensional surface in a conventional (flat) three-dimensional space. In spite of the fact that the space is flat, the geometry on the curved surface is non-Euclidean.

1.5 Some Formulae for Curvilinear Coordinates

In STR one introduces the concept of a four-dimensional vector (four-vector) B_i as an aggregate of four quantities (functions of coordinates and time) which, under Lorentz transformations, transform as the coordinates x^i do. Examples of four-vectors are the four-velocity, $u^i = dx^i/ds$, the four-momentum, $p^i = mu^i$, and the four-potential of the electromagnetic field, A_i. A four-tensor of second rank, B_{ik}, is defined as an aggregate of quantities which transform like the product of the coordinates $x^i x^k$. An example is the electromagnetic-field tensor

$$F_{ik} = \partial A_k/\partial x^i - \partial A_i/\partial x^k .$$

Tensors of third rank and higher are defined in an analogous manner.

In inertial frames of reference in STR one uses Galilean coordinates, in which the interval takes the form (1.2.1). When passing to curvilinear coordinates in four-dimensional spacetime one must describe vectors and tensors in terms of this new type of coordinates. Most important of all is the introduction of the *covariant* and *contravariant* components of a vector.

A contravariant four-vector is any aggregate of quantities B^i (with index up) which, under a change of coordinates, $x^i = x^i(x^{0\prime}, x^{1\prime}, x^{2\prime}, x^{3\prime})$, transforms according to the rule

$$B^i = (\partial x^i/\partial x^{k\prime})B^{k\prime} . \tag{1.5.1}$$

An example of a contravariant vector is the set of coordinate differentials dx^i, since $dx^i = (\partial x^i/\partial x^{k\prime})dx^{k\prime}$.

The covariant components of the same vector B_i (with index down) are defined by

$$B_i = g_{ik}B^k . \tag{1.5.2}$$

From the determination of g_{ik} as the coefficients in equation (1.4.7), their law of transformation, $g_{ik} = (\partial x^{l\prime}/\partial x^i)(\partial x^{m\prime}/\partial x^k)g_{l\prime m\prime}$, follows. Using this law, and referring to equation (1.5.2), we obtain the law of transformation for the covariant components of a vector

$$\begin{aligned} B_i = g_{ik}B^k &= (\partial x^{l\prime}/\partial x^i)(\partial x^{m\prime}/\partial x^k)g_{l\prime m\prime}(\partial x^k/\partial x^{n\prime})B^{n\prime} \\ &= (\partial x^{l\prime}/\partial x^i)B_{l\prime} . \end{aligned} \tag{1.5.3}$$

The concept of a tensor is generalized in the same way: The transformation law for a contravariant tensor, B^{ik}, is

$$B^{ik} = (\partial x^i/\partial x^{l\prime})(\partial x^k/\partial x^{m\prime})B^{l\prime m\prime} , \tag{1.5.4}$$

and for its covariant components the transformation law is

$$B_{ik} = g_{li}g_{mk}B^{lm} = (\partial x^{l\prime}/\partial x^i)(\partial x^{m\prime}/\partial x^k)B_{l\prime m\prime} . \tag{1.5.5}$$

One also uses mixed components, which transform as

$$B^i{}_k = B^{il}g_{lk} = (\partial x^{l\prime}/\partial x^k)(\partial x^i/\partial x^{m\prime})B^{m\prime}{}_{l\prime} . \tag{1.5.6}$$

The concept of a tensor of a higher rank is generalized similarly.

The components g_{ik}, as indicated by their law of transformation, comprise a tensor. This tensor, which plays a fundamental role in the theory, is called the *fundamental metric tensor.*

The determinant

$$g = |g_{ik}| \tag{1.5.7}$$

is called the *fundamental determinant.*

The quantities g^{ik}, defined by

$$g_{ik}g^{im} = \delta^m{}_k , \tag{1.5.8}$$

where $\delta^m{}_k$ is the Kronecker delta, are called the *contravariant components* of the metric tensor. From equation (1.5.8), it follows that

$$g^{ik} = (A^{ik}/g) , \tag{1.5.9}$$

where A^{ik} is the minor of the matrix g_{ik}.

From equation (1.5.8), by using equation (1.5.5), we find that

$$B^{ik} = g^{il}g^{mk}B_{lm} . \tag{1.5.10}$$

Thus, just as the lowering of indices is performed with the covariant components g_{ik}, so the raising of indices is performed with the contravariant components g^{ik}. The mixed tensor $g^i{}_k$ is equal to the Kronecker symbol, $g^i{}_k = \delta^i{}_k$.

Consider the quantity A^iB_i. It is the scalar product of the vectors A^i and B^j, and it is unaffected by a transformation of coordinates. Specifically, the square of the length of a vector is

$$A^2 = A^iA_i . \tag{1.5.11}$$

Similarly, we can construct a scalar from two tensors

$$A^{ik}B_{ik} = A_i{}^kB^i{}_k = A_{ik}B^{ik} .$$

All three notations are equivalent. If the second tensor is the fundamental tensor, then the scalar $A^{ik}g_{ik} = A^i{}_i$ is called the *trace* of the tensor.

In a similar manner, lower-rank tensors can be formed from higher-rank tensors. For example,

$$A_{iklm}g^{mi} = A^i{}_{kli} = A_{kl} .$$

This operation is called *tensor contraction.*

In curvilinear coordinates, the concept of differentiation of vectors and tensors must also be generalized. The covariant derivatives (denoted by a semicolon) of a contravariant vector and of a covariant vector are the quantities

$$B^i{}_{;k} = (\partial B^i/\partial x^k) + \Gamma^i{}_{lk}B^l , \tag{1.5.12}$$

and

$$B_{i;k} = (\partial B_i/\partial x^k) - \Gamma^l{}_{ik}B_l , \tag{1.5.13}$$

respectively. Here $\Gamma^l{}_{mn}$ are Christoffel symbols (not tensors!) which are defined by the expressions

$$\Gamma^l{}_{mn} = \tfrac{1}{2}g^{lk} [(\partial g_{km}/\partial x^n) + (\partial g_{kn}/\partial x^m) - (\partial g_{mn}/\partial x^k)] . \tag{1.5.14}$$

Obviously, in Cartesian coordinates all the Γ^l_{mn} vanish, and covariant differentiation reduces to conventional differentiation.

Tensors of higher rank are differentiated in a similar manner:

$$B^{ik}_{;l} = (\partial B^{ik}/\partial x^l) + \Gamma^i_{ml}B^{mk} + \Gamma^k_{ml}B^{im} , \qquad (1.5.15)$$

$$B^i_{k;l} = (\partial B^i_k/\partial x^l) - \Gamma^m_{kl}B^i_m + \Gamma^i_{ml}B^m_k , \qquad (1.5.16)$$

$$B_{ik;l} = (\partial B_{ik}/\partial x^l) - \Gamma^m_{il}B_{mk} - \Gamma^m_{kl}B_{im} . \qquad (1.5.17)$$

It is important to note that from equations (1.5.12) and (1.5.14) and the expression for ds^2 one can obtain the following expression for the covariant divergence of a vector

$$B^i_{;i} = [1/(-g)^{1/2}]\partial[(-g)^{1/2}B^i]/\partial x^i . \qquad (1.5.18)$$

Finally, we give the equation in curvilinear coordinates which defines the geodesic line connecting two points in four-dimensional space (in flat space, in which we are working in this section, it is a straight line):

$$(d^2x^i/ds^2) + \Gamma^i_{kl} (dx^k/ds)(dx^l/ds) = 0 . \qquad (1.5.19)$$

In Minkowski space, as we know from STR, a body of finite rest mass moves along a straight (timelike) line; i.e., along a geodesic. Hence, equations (1.5.19) are the equations of motion of a body of finite rest mass, written in the curvilinear coordinates of a noninertial reference frame. The differential equation for a geodesic in curved spacetime has precisely the same form as equation (1.5.19) for a straight line in flat spacetime.

1.6 DYNAMIC AND KINEMATIC QUANTITIES

The quantities g_{ik} of equation (1.4.7) are constructed from derivatives of the transformation (1.4.4), which determines the motion of the reference frame used with respect to the original inertial frame. In particular, g_{ik} includes $\partial x^{i'}/\partial x^0$, the relative velocity of the two frames. Hence, it is natural that g_{ik} contains information not only about the flow of time and the geometry of the reference frame but also about its accelerations and deformation. We shall cite here only the final formulae for the computation of dynamic and kinematic quantities; for more detail see the contributions of Zel'manov (1944, 1959b). The three-dimensional vector F^α, describing the acceleration relative to the reference frame of a freely moving body momentarily at rest in it, is determined by the expression

$$F^\alpha = -c^2\Gamma^\alpha_{00}/g_{00} \qquad (\alpha = 1, 2, 3) , \qquad (1.6.1)$$

as we shall demonstrate below (eq. [1.6.1a]). The quantities Γ^α_{00} were defined in the preceding section (eq. [1.5.14]). The vector F^α forms a field of inertial forces in the reference frame. F^α is a three-dimensional vector; in working with it one must use the tensor $\gamma_{\alpha\beta}$ (see § 4 of this chapter). Remember that in order to compute the magnitude of the vector F^a (in this case a three-dimensional vector), i.e., the magnitude of the acceleration, it is necessary to form a scalar (see eq. [1.5.11]):

$$F = (F^a F_a)^{1/2} = (F^a F^\beta \gamma_{a\beta})^{1/2} \ .$$

For example, for a rotating disk, we find from equation (1.4.6)

$$F^1 = \frac{\Omega^2 r}{1 - \Omega^2 r^2/c^2} \ , \qquad F^2 = F^3 = 0 \ , \qquad F = \frac{\Omega^2 r}{1 - \Omega^2 r^2/c^2} \ .$$

The rotation of the reference frame, i.e., the field of Coriolis forces, is determined by the three-dimensional angular-velocity tensor, $A^{a\beta}$. From the angular-velocity tensor one can calculate the three-dimensional angular-velocity vector,[10] Ω^*_a:

$$\Omega^*_a = \tfrac{1}{2} \epsilon_{a\beta\gamma} A^{\beta\lambda} \ . \tag{1.6.2}$$

Here $\epsilon_{a\beta\gamma}$ is defined in the following manner: $\epsilon_{123} = (-g/g_{00})^{1/2}$; any permutation of the indices of $\epsilon_{a\beta\gamma}$ changes the sign of the component; and if at least two indices coincide, then $\epsilon_{a\beta\gamma} = 0$. The angular-velocity tensor is defined by the expression

$$A^{a\beta} = (-c/\sqrt{g_{00}})[\tfrac{1}{2}(\partial g^{a\beta}/\partial x^0) + g^{ai}\Gamma^\beta_{i0}] \ . \tag{1.6.3}$$

The scalar $\Omega^* = (\Omega^*_a \Omega^*_\beta \gamma^{a\beta})^{1/2}$ is the angle through which the reference frame rotates, at a given point, in unit proper time, $d\tau = (g_{00}dt)^{1/2}$. For a rotating disk in a cylindrical coordinate system, we have

$$A^{12} = -A^{21} = -\frac{\Omega/r}{(1 - \Omega^2 r^2/c^2)^{1/2}} \ .$$

The remaining components of $A^{a\beta}$ vanish. The angular-velocity vector has the components

$$\Omega^*_3 = \frac{\Omega}{1 - \Omega^2 r^2/c^2} \ , \qquad \Omega^*_2 = \Omega^*_1 = 0 \ .$$

In a system of reference where all $g_{0a} = 0$, all $A_{a\beta} = 0$.

If in some region $A_{a\beta} \neq 0$, i.e., the reference frame is rotating, then clocks in this region cannot be synchronized. This was demonstrated above for a rotating disk. Conversely, if $A_{a\beta} = 0$, i.e., if the reference frame is not rotating, then by a change of time coordinate, $x^{0'} = x^{0'}(x^0, x^1, x^2, x^3)$, all g_{0i} can be made to vanish, and the clocks can be synchronized.

Let us derive equation (1.6.1). If a particle is momentarily at rest in the given reference frame, then $ds^2 = g_{00}(dx^0)^2$, and the particle's four-velocity has the components

$$u^0 = dx^0/ds = g_{00}^{-1/2}; \qquad u^a = 0 \ .$$

---

10. The vector is here denoted by Ω^*_a, so that it will not be confused with the vector Ω_a of the nonrelativistic theory.

1.6 Dynamic and Kinematic Quantities

We find from formula (1.5.19) for the space components a that the inertial forces associated with the motion of the reference frame produce the acceleration

$$d^2x^a/ds^2 = -\Gamma^a{}_{00}/g_{00} .$$

Substituting the expression $ds^2 = c^2d\tau^2$ and using the symbol F^a for the inertial force, we obtain

$$d^2x^a/d\tau^2 = -c^2\Gamma^a{}_{00}/g_{00} = F^a , \qquad (1.6.1a)$$

which is equation (1.6.1).

In the specific case of a stationary metric, that is, when $\partial g_{ik}/\partial x^0 = 0$, the expression (1.6.1) can be rewritten in the form[11]

$$F_a = -(c^2/2g_{00})(\partial g_{00}/\partial x^a) . \qquad (1.6.1b)$$

Finally, the deformation of the coordinate system is described by a three-dimensional tensor $D_{a\beta}$:

$$D_{a\beta} = (c/2\sqrt{g_{00}})(\partial\gamma_{a\beta}/\partial x^0) . \qquad (1.6.4)$$

The scalar $D = D_a{}^a = D_{a\beta}\gamma^{a\beta}$ is the relative rate of volume expansion of a volume element, dV, of the reference frame, $D = (dV)^{-1}(d/d\tau)(dV)$. If the coordinate system does not deform as time passes (example, the uniformly rotating disk), then $D_{a\beta} = 0$.

Notice that F^a, $A^{a\beta}$, $\Omega^{\cdot a}$, $D^{a\beta}$ do not depend on the choice of the time coordinate. If we were to change time coordinates (in other words, select a different unit of time [different time scale] and a different origin of time at each point of the reference frame),

$$x^0 = x^0(x^{0'}, x^{1'}, x^{2'}, x^{3'}) , \qquad (1.6.5)$$

then the above quantities would not change at all. This must be true for all quantities which describe the state of the motion of the reference frame, because the transformation (1.6.5) does not affect that motion. Such quantities have been called "chronometric invariants" by Zel'manov (1956).[12] Moreover, if only the space coordinates are changed, i.e., the coordinate grid of the reference frame is drawn differently,

$$x^a = x^a(x^{1'}, x^{2'}, x^{3'}) , \qquad (1.6.6)$$

then the components of the vector F^a transform by the three-dimensional analogue of equation (1.5.1). In other words, F^a behaves as a vector in a

11. More precisely, for the validity of equation (1.6.1b) one only needs $(\partial/\partial t)(g_{0a}/g_{00}^{1/2}) = 0$.

12. *Editors' note.*—A differential geometer or Western relativity theorist would use different language. He would focus attention on the "congruence" of time lines $(x^1, x^2, x^3) =$ const., i.e., on the world lines of the test particles of the reference frame; and he would identify $A^{a\beta}$, $D^{a\beta} - \frac{1}{3}D\gamma^{a\beta}$, and D as the "rotation," "shear," and "expansion" of that congruence. Thus, the term "chronometric invariant" is equivalent to "geometric property of the congruence of time lines."

(curved) three-dimensional space: its components change, but it itself does not, and its length—the scalar F—remains the same. The scalars Ω^* and D are similarly unaffected.

Only in passing to a different reference frame, i.e., to a different state of motion of the test particles,

$$x^a = x^a(x^{0'}, x^{1'}, x^{2'}, x^{3'}) , \qquad \partial x^a / \partial x^{0'} \neq 0 ,$$

do the chronometric invariants and their scalars, F, Ω^*, D, change.

1.7 CURVATURE OF SPACETIME

In the preceding sections were described the geometric and physical properties of noninertial frames in a flat, Minkowski spacetime.

According to GTR, in the vicinity of massive bodies spacetime is curved; it is a four-dimensional Riemannian space (to be more precise, pseudo-Riemannian).[13] In a finite (not small) region of this four-dimensional space it is not possible to construct a Galilean coordinate system in which the interval has the form (1.2.1)

$$ds^2 = (dct)^2 - dx^2 - dy^2 - dz^2 .$$

However, this can be done in an infinitesimal region by introducing at a given point a freely moving (freely falling in the gravitational field) reference frame. Such a coordinate system is said to be *locally Galilean*.[14] In a locally Galilean coordinate system, no gravitational field is manifest; there exists a state of weightlessness. The mathematical possibility of selecting such a system is obviously related to the fact that a small segment of curved space is identical with its flat tangent space.

With respect to a locally Galilean coordinate system other systems, in which gravitational effects are manifest, move with acceleration; the transitions from the Galilean system at a given point to these systems are simply transitions in a small domain of spacetime from an inertial frame to noninertial frames. The forces of inertia and the forces of gravitation are locally indistinguishable. Consequently, as we have noted, all formulae for geometrical, dynamic, and kinematic quantities presented in preceding sections which have a local character (i.e., which define the properties of a coordinate system in a small domain of space at a given moment of time) will be valid not only in SRT but also for the general case of curved spacetime. The computations of lengths, time intervals, gravitational-inertial forces, rotation, etc., are

---

13. The curvature of spacetime in GTR is not necessarily related to the presence of matter or of (nongravitational) fields. As will be demonstrated later, GTR predicts the existence of gravitational waves which carry energy and cause curvature of space. Moreover, for empty, curved spacetime there exist certain nonstationary solutions which describe an anisotropic deformation of space, and which do not contain matter anywhere.

14. The number of such systems at each point is ∞^6.

performed in GTR by using the formulae of the preceding sections. Let us emphasize that now equations (1.5.19) determine, in arbitrary coordinates, not a straight line, but an extremal geodesic of curved spacetime. (In curved spacetime straight lines obviously cannot exist; geodesic curves are their analogues.)

We turn now to the mathematical methods of describing the curvature of four-dimensional spacetime. The curvature is described by a fourth-rank tensor,

$$R^i{}_{klm} = (\partial \Gamma^i{}_{km}/\partial x^l) - (\partial \Gamma^i{}_{kl}/\partial x^m) + \Gamma^i{}_{nl}\Gamma^n{}_{km} - \Gamma^i{}_{nm}\Gamma^n{}_{kl}, \quad (1.7.1)$$

called the Riemann curvature tensor. The geometrical meaning of this tensor is as follows. Let a vector from some point move along a small closed contour consisting of geodesic lines, in such a manner that the angle of the vector with each geodesic does not change (the components of the vector along the geodesic are constant). In flat spacetime the vector, upon returning to its initial point, coincides with the original vector; in curved spacetime the vector's orientation changes (but its length does not!). The change in the components of the vector A_k, as a result of its motion around a contour which encloses a small two-dimensional surface Δf^{lm}, is given by the formula

$$\Delta A_k = \tfrac{1}{2} R^i{}_{klm} A_i \Delta f^{lm}.$$

We will not pause here for a discussion of the algebraic and differential properties of the curvature tensor. We only remark that it has twenty independent components.

From the Riemann tensor we can construct a tensor of second rank by the process of contraction (see § 1.5):

$$R_{km} = R^i{}_{klm} g^l{}_i = R^i{}_{kim}. \quad (1.7.2)$$

This tensor is symmetric:

$$R_{km} = R_{mk}; \quad (1.7.3)$$

it is called the Ricci tensor. Finally, a contraction of R_{km} yields the scalar curvature of spacetime

$$R = R_{km} g^{km} = R_k{}^k. \quad (1.7.4)$$

The tensor $R^i{}_{klm}$ characterizes completely the curvature of four-dimensional spacetime. In particular, the vanishing of this tensor, $R^i{}_{klm} = 0$, in some region of spacetime is a necessary and sufficient condition for spacetime in this region to be noncurved (flat).

The vanishing of the scalar curvature, $R = 0$, or even of the Ricci tensor, $R_{ik} = 0$, is by no means sufficient for spacetime to be flat. In fact, outside matter (in vacuum) the gravitational field is determined just by the equation $R_{ik} = 0$.

1.8 THE EINSTEIN FIELD EQUATIONS AND THE EQUATIONS OF MOTION

The Einstein field equations determine a relationship between the curvature of spacetime, and the distribution and motion of matter and fields (excluding the gravitational field). They are written in the form[15]

$$R_{ik} - \tfrac{1}{2}g_{ik}R = (\kappa/c^2)T_{ik} . \qquad (1.8.1)$$

Here $\kappa = 8\pi G/c^2$ is Einstein's gravitational constant, and T_{ik} is the energy-momentum tensor which depends on the distribution and motion of matter and of electromagnetic fields (and, in principle, of other fields as well).

For a gas the energy-momentum tensor in curvilinear coordinates has the form

$$T^{ik} = (\epsilon + P)u^i u^k - Pg^{ik} . \qquad (1.8.2)$$

Here $\epsilon = \rho c^2$ is the energy density of the matter (including the rest mass of all particles) as measured in the rest frame of the matter; P is the pressure; and u^i is the four-velocity of the gas. We assume the viscosity of the gas to be small, and we also neglect all energy flux relative to the matter as being small compared with ρc^3.

The energy-momentum tensor of an electromagnetic field has the form

$$T^{ik} = -(1/4\pi)\, g_{lm}F^{il}F^{km} + (1/16\pi)\, g^{ik}F_{lm}F^{lm} , \qquad (1.8.3)$$

where F_{lm} is the electromagnetic field tensor.

For a gas, if we write out the form (1.8.2) of the stress-energy tensor in a locally Galilean coordinate system where the gas is at rest, we find

$$T_{ik} = \begin{vmatrix} \epsilon & 0 & 0 & 0 \\ 0 & P & 0 & 0 \\ 0 & 0 & P & 0 \\ 0 & 0 & 0 & P \end{vmatrix} .$$

In this frame $T_{0k} = T_{k0} = 0$ because there is no energy flow and the momentum of the gas vanishes. The spatial part of the tensor is diagonal, $T_{\alpha\beta} = P\delta_{\alpha\beta}$. The pressure is the same in every direction; this fact is called Pascal's law, and fluids which obey it are called Pascalian.

For a magnetic field directed along the x-axis (so that $H_y = H_z = 0$; $E = 0$), the stress-energy tensor has the form

$$T_{ik} = \begin{vmatrix} \epsilon & 0 & 0 & 0 \\ 0 & -\epsilon & 0 & 0 \\ 0 & 0 & \epsilon & 0 \\ 0 & 0 & 0 & \epsilon \end{vmatrix} .$$

15. For the so-called Λ term of Einstein's equations, which is omitted here, see § 1.9.

Here $\epsilon = H^2/8\pi$ is the energy density. A negative pressure (tension) equal to $T_{11} = -\epsilon$ acts along the x-axis; along the y- and z-axes a positive pressure, ϵ, acts. If the field is directed not along a coordinate axis, but in an arbitrary direction, then off-diagonal components appear in $T_{\alpha\beta}$. However, the trace $T_\alpha{}^\alpha$ (equal to $-T_{11} - T_{22} - T_{33}$ in Cartesian coordinates) is an invariant.

Let the magnetic field be chaotic, and average the stress-energy tensor over a scale many times larger than the field inhomogeneities. One will obtain

$$T_{ik} = \begin{vmatrix} \epsilon & 0 & 0 & 0 \\ 0 & \tfrac{1}{3}\epsilon & 0 & 0 \\ 0 & 0 & \tfrac{1}{3}\epsilon & 0 \\ 0 & 0 & 0 & \tfrac{1}{3}\epsilon \end{vmatrix}.$$

Thus, the pressure is Pascalian on the average; a chaotic magnetic field is similar to a gas with the special "equation of state," $P = \tfrac{1}{3}\epsilon$.

Relativistic particles moving with the speed of light in the positive direction along the x-axis produce the stress-energy tensor

$$T_{ik} = \begin{vmatrix} \epsilon & \epsilon & 0 & 0 \\ \epsilon & \epsilon & 0 & 0 \\ 0 & 0 & 0 & 0 \\ 0 & 0 & 0 & 0 \end{vmatrix}.$$

The same particles moving in the negative direction give

$$T_{ik} = \begin{vmatrix} \epsilon & -\epsilon & 0 & 0 \\ -\epsilon & \epsilon & 0 & 0 \\ 0 & 0 & 0 & 0 \\ 0 & 0 & 0 & 0 \end{vmatrix}.$$

Combining a chaotic mixture of such streams of particles, with all directions equally probable, produces again the stress-energy tensor for a relativistic gas (see above) with $P = \tfrac{1}{3}\epsilon$.

For a general stress-energy tensor let us write out the equations of conservation of energy and momentum. In STR, in the Cartesian coordinates of an inertial frame of reference, the energy-momentum tensor satisfies the relation

$$\partial T_i{}^k/\partial x^k = 0, \tag{1.8.4}$$

which, as is well known, expresses the laws of conservation of energy and momentum. The generalization of equation (1.8.4) to curvilinear coordinates is the vanishing of the covariant divergence:

$$T^k{}_{i;k} = (\partial T^k{}_i/\partial x^k) + \Gamma^k{}_{lk}T^l{}_i - \Gamma^l{}_{ik}T^k{}_l = 0 .\tag{1.8.5}$$

It is very important that the law (1.8.5) follows from the Einstein field equations (1.8.1). In fact, as is demonstrated in standard texts, the left side of the field equations identically satisfies the relation

$$(R^k{}_i - \tfrac{1}{2}g^k{}_iR)_{;k} = 0 ;\tag{1.8.6}$$

consequently, the right side must also satisfy this relation—equation (1.8.5). However, equation (1.8.5) does not express directly the law of conservation of any quantity (i.e., the constancy of a quantity in time); to obtain a conservation law, one must have an equation of the form (1.8.4), not (1.8.5).[16]

Expressions (1.8.5) are most appropriately called "equations of motion" since they express directly the laws of motion of the matter, with effects of gravity taken into account.[17] To demonstrate this for the stress-energy tensor of a gas, we select a reference frame which moves along with the matter (a "comoving reference frame"); i.e., we use Lagrangian space coordinates and the proper time of each element of the matter. If we denote by E the energy in the volume V of an element of matter, $E = \epsilon V$, then by using equation (1.8.2) we can reduce expression (1.8.5) for $i = 0$ to the form

$$dE + PdV = 0 ;\tag{1.8.7}$$

and for spatial values of i ($i = 1, 2, 3$), we can reduce it to

$$\partial P/\partial x^a - (g_{0a}/g_{00})(\partial P/\partial x^0) = (\epsilon + P)(F_a/c^2) .\tag{1.8.8}$$

Equation (1.8.7) describes the change in energy due to the work done by pressure forces during the deformation of the gas; equations (1.8.8) describe the motion of the matter in Lagrangian coordinates. Obviously, when one passes to the nonrelativistic limit where $g_{0a} \to 0$ and $\epsilon \gg P$, equations (1.8.8) become the familiar equations of Euler written in Lagrangian coordinates.

Equations (1.8.8) for the case of matter at rest are the equations of hydrostatic equilibrium

$$\partial P/\partial x^a = (\epsilon + P)(F_a/c^2) .\tag{1.8.8a}$$

Notice that the multiplier on the right side is $(\epsilon/c^2 + P/c^2) = (\rho + P/c^2)$, rather than ρ as in nonrelativistic theory.

16. Only in the case of equation (1.8.4) can we apply Gauss's theorem to the sum of the space derivatives, and then pass to a surface integral. In equation (1.8.5) we are prevented from doing so by the terms involving Christoffel symbols, which do not have the form of an ordinary divergence.

17. Expressions (1.8.4) are, analogously, the equations of motion in the absence of gravity.

Let us write in curvilinear coordinates the law of conservation of particle number. (We concentrate attention solely on types of particles which are only created or destroyed with their antiparticles, e.g., baryons.) Let n_0 be the number density of particles at a given event as measured in the comoving reference frame there. From n_0 construct the four-vector

$$j^k = n_0(dx^k/ds) . \qquad (1.8.9)$$

The vanishing of its covariant derivative (cf. eq. [1.5.18])

$$j^k{}_{;k} = \frac{1}{(-g)^{1/2}} \frac{\partial}{\partial x^k} \left[(-g)^{1/2} n_0 \frac{dx^k}{ds} \right] = 0 \qquad (1.8.10)$$

guarantees the conservation, as time passes, of the integral over three-dimensional space

$$N = \iiint n_0 \frac{dx^0}{ds} (-g)^{1/2} dx^1 dx^2 dx^3 = \iiint n_0 \frac{dx^0}{ds} (g_{00})^{1/2} dV . \qquad (1.8.11)$$

The quantity $n = n_0(g_{00})^{1/2}(dx^0/ds)$ is the number density of particles in the reference frame of the calculation (see below). Consequently, the total number of particles, N, when equation (1.8.10) is satisfied, remains constant in time.

Let us rewrite the number density of particles $n = n_0(g_{00})^{1/2}(dx^0/ds)$, using three-dimensional notation. As the first step, we construct a three-dimensional velocity vector, v^α. The interval of proper time between two neighboring events, as measured by an observer at rest in the reference frame of the coordinate system, is given by the expression

$$cd\tau = (g_{00})^{-1/2} g_{0i} dx^i .$$

The components and square of the gas velocity relative to the reference frame are given by

$$v^\alpha = (dx^\alpha/d\tau) , \qquad v^2 = \gamma_{\alpha\beta} v^\alpha v^\beta = (dl^2/d\tau^2) .$$

Using the above equations, we can rewrite the integral of equation (1.8.11) in the form

$$N = \int_V \int \int \frac{n_0 dV}{(1 - v^2/c^2)^{1/2}(1 + g_{0\alpha} dx^\alpha/g_{00} dx^0)} . \qquad (1.8.12)$$

Equation (1.8.12) differs from the conventional formula of special relativity only by the factor $(1 + g_{0\alpha} dx^\alpha/g_{00} dx^0)$ in the denominator.[18] This factor is due to the following. The integral (1.8.12) is carried out over a three-dimensional surface of constant time, x^0. However, because of the term $g_{0\alpha}$ in the metric, neighboring events on that surface are not simultaneous as measured in the reference frame of the coordinate system; rather, they are separated by the proper-time interval $cd\tau = (g_{00})^{-1/2} g_{0\alpha} dx^\alpha$.

If we were to use curvilinear four-dimensional coordinates with $g_{0\alpha} \neq 0$

---

18. The factor $(1 - v^2/c^2)^{-1/2}$ describes the Lorentz contraction of the volume element.

in STR, then, as in GTR, the formula for the number of particles would take the form (1.8.12). Thus, in GTR the law of conservation of particles is written in exactly the same form as when gravity is absent. This is hardly surprising, since gravitation obviously does not change the number of particles.

Finally, let us write an expression for the Gaussian curvature (cf. § 1.4) of the three-dimensional space of the comoving reference frame in the static case (that is, when all time derivatives vanish):

$$C_G = \tfrac{8}{3}\pi(G/c^2)\rho \,. \tag{1.8.13}$$

In the case of an isotropic deformation of matter and vanishing rotation, the Gaussian curvature is

$$C_G = \tfrac{8}{3}\pi(G/c^2)\rho - \tfrac{1}{9}D^2 \,. \tag{1.8.14}$$

These formulae are derived, for example, by Zel'manov (1959*b*).

1.9 THE COSMOLOGICAL CONSTANT

The general requirements usually placed on the equations of the theory of gravitation permit one to write a variational principle with the action in the form (where V is the four-dimensional volume)

$$S = -mc \int ds - \frac{c^3}{16\pi G}\left(\int R\,dV + \int 2\Lambda\,dV\right) \,. \tag{1.9.1}$$

The corresponding field equations have the form

$$R_{ik} - \tfrac{1}{2}g_{ik}R - \Lambda g_{ik} = (\kappa/c^2)T_{ik} \,. \tag{1.9.2}$$

Here Λ is the so-called cosmological constant, and quantities that are proportional to it (ΛdV, Λg_{ik}) are called cosmological terms. These field equations obviously satisfy the condition of local Lorentz-invariance and include the equations of motion in the same sense as the equations without Λ do; $T^k{}_{i;k} = 0$, as before.

Einstein initially selected Λ in such a manner as to obtain a stationary cosmological solution with a mean density $T_0{}^0 = \rho c^2 = $ const. that is different from zero; for this, it is necessary that $\Lambda = 8\pi G\rho/3c^2$. After the discovery of the cosmological redshift, Einstein preferred the equations with $\Lambda \equiv 0$.

Both stationary and nonstationary cosmological solutions with $\Lambda \neq 0$ were investigated in detail before 1930; however, until 1967 there were no observational indications as to the necessity or even desirability of introducing Λ. Since 1967, observational data on quasars have suggested that Λ might not actually be zero, but might instead have a value of the order of $\Lambda \approx 10^{-55}$ cm^{-2}. For details, see § 14.1. At present this hypothesis is not at all proved; in fact, it encounters difficulties in explaining the quasar observations. However, in the course of the discussions it has become apparent that the simplest assumption of $\Lambda \equiv 0$, while not refuted, has not been distinctly and uniquely proved. Predictably, difficult and extensive work will be required in years to come, to determine the quantity Λ, or at least to determine limits on it. After a genie is let out of a bottle (i.e., now that the possibility is admitted that $\Lambda \neq 0$), legend has it that the genie can be chased back in only with great difficulty. Even if the specific hypothesis that $\Lambda \approx 10^{-55}\,\mathrm{cm}^{-2}$ is invalidated, we will return not to the simplistic assumption that $\Lambda \equiv 0$, but to a careful $-a < \Lambda < b$, and to a gradual and painful decrease of the limits a and b.

How can the physical meaning of the cosmological constant be understood? Why, in fact, is it interesting for physics as a whole?

One approach was prompted by the dimensions of Λ (cm^{-2}). In this approach one views Λ as the curvature of empty space. But the theory of gravitation links the curvature to the energy, momentum, and pressure of matter. Putting the terms with Λ onto the right side of the field equation, we obtain

$$R_{ik} - \tfrac{1}{2}g_{ik}R = (8\pi G/c^4)T_{ik} + g_{ik}\Lambda \ . \tag{1.9.3}$$

The assumption that $\Lambda \neq 0$ means that empty space creates a gravitational field identical with that in the theory with $\Lambda = 0$, but with matter filling all of spacetime with a mass density $\rho_\Lambda = (c^2\Lambda/8\pi G)$, an energy density $\epsilon_\Lambda = (c^4\Lambda/8\pi G)$, and a pressure $P_\Lambda = -\epsilon_\Lambda$. In this sense one can speak about the energy density and the pressure (stress tensor) of the vacuum.

Notice that our assumptions about ϵ_Λ and P_Λ were formulated in such a manner that the relativistic invariance of the theory was not broken; ϵ_Λ and P_Λ are the same in all coordinate systems moving relative to each other (Lorentz-transformed).

These quantities do not make themselves felt either in elementary-particle experiments or in atomic and molecular physics: the vacuum energy of the vessel in which an experiment takes place plays the role of a constant term which can be canceled in the law of energy conservation.

The only sort of phenomenon in which ϵ_Λ and P_Λ manifest themselves is a gravitational one. In this case ϵ_Λ and P_Λ "work" not only in empty space; they are, as is clear from the above formulae, full and equal members of the field equations even when normal matter is present. Thus, a Cavendish experiment in principle can serve for the discovery and measurement of ϵ_Λ and P_Λ. The attraction of a sphere of lead depends on the sum of the lead density ($11\,\mathrm{g\,cm}^{-3}$), and the vacuum density ($|\rho_\Lambda|$ is less than $10^{-28}\,\mathrm{g}$ cm^{-3}), integrated over the volume of the sphere.

A practical measurement of the influence of P_Λ and ϵ_Λ is not possible, either in laboratory experiments or in observations of planetary motions in the solar system or stellar motions in the Galaxy. Indeed, the mean density of solar-system matter in a sphere with the radius of the Earth's orbit is $\langle \rho \rangle \approx 10^{-7}\,\mathrm{g\,cm}^{-3}$. The mean density of matter in the Galaxy is of the order of 10^{-24} g cm^{-3}. The influence of ρ_Λ will be significant only on the largest of scales—the scale of the whole Universe—i.e., in cosmology.

Let us dwell on the nature of Λ. One can take the viewpoint that a certain Λ and corresponding ρ_Λ, ϵ_Λ, and P_Λ are universal constants that do not require any further explanation. A different viewpoint is also conceivable: assume that in some zeroth-order approximation $\Lambda = \rho_\Lambda = \epsilon_\Lambda = P_\Lambda \equiv 0$. Higher-order values that are different from zero and characterize the vacuum might be derivable from considerations of the theory of elementary particles. There is no such derivation at the present time, any more than there is a theory proving that $\Lambda = 0$.

We consider the second point of view in more detail. The very first attempts to quantize the electromagnetic field led to the paradoxical conclusion of an infinite energy density for the vacuum. The vacuum is defined as the lowest energy state of the system considered, and can be characterized by the Maxwell equations. The particles—photons in this case—represent the elementary excitations of the system. In the analogous quantum-mechanical problem of the motion of the atomic nuclei in a crystal lattice, the situation is similar: the elementary excitations are phonons (sound quanta), and the ground state of the crystal is that which contains no phonons, i.e., which has a temperature of absolute zero. This state is similar to the vacuum.

The ground-state energy in the case of a crystal has a quite certain value which can be measured. The difference in ground-state energies for different isotopes of the same element leads to the dependence of the heat of evaporation on the atomic weight of the isotope. In the simplest variant of field theory the ground-state energy of the vacuum is infinite. However, the theory can be reformulated so that the ground-state energy of the free field becomes identically zero.

In classical Maxwell theory the energy density is equal to $\epsilon = (E^2 + H^2)/8\pi$, where E and H are the electric and magnetic field strengths. As has been emphasized by Berestetsky, Lifshitz, and Pitaevsky (1969), there is no formulation of quantum electrodynamics in which the mean value of E^2 or H^2 is zero in the vacuum (i.e., far from all charges and in the absence of real photons). Consequently, in order to formulate these theories with the help of the usual products of operators, in such a manner that in vacuum $\epsilon = 0$, one must abandon the classical relation between ϵ and the fields.

A second source of the energy of the vacuum arises from the Dirac theory of the electron: the concept of filled levels with negative energy leads literally to an infinite negative value for the energy density. In this case also the theory was soon reformulated to make ϵ identically zero for the vacuum of noninteracting particles. However, it is not at all guaranteed that, when the interaction of the particles is taken into account, the vacuum energy will remain zero. The peculiarity of present-day theory is that the interaction between particles is effective not only when real particles are present to interact, but also when the particles are virtual.

In the above discussion we must remind ourselves that the term "interaction" is not used in the sense of classical physics. There one speaks about the interaction of two colliding bodies, and about the (Coulomb) interaction of a proton and electron. In quantum field theory one speaks about a four-fermion interaction when a neutron breaks up into an electron, a proton, and a neutrino; and one speaks about a photon-electron interaction when an electron produces a photon.

A freely moving electron cannot emit a real photon which is capable of being detected far from the electron. However, one can say that the free electron emits and then reabsorbs photons, and that this causes a change in the properties of the electron (for example, its mass, its magnetic moment, etc.), as has been proved by the Lamb-Rutherford experiment.

Experimental measurement of the change in the electron mass is impossible because there are no experiments which could, even in principle, measure the mass of an electron stripped of its photons. However, the change in the electron's magnetic moment has been corroborated by experiment, with all the precision which is now available.

There are many other processes similar to the above which take place in the vacuum—for example, the creation and annihilation of electron-positron pairs with the absorption and emission of γ-radiation.

The theory of the vacuum state and its properties is not so simple and obvious today as it was about fifty years ago!

Several possible viewpoints may be delineated.

The first is the supposition that the energy of the vacuum is identically zero when fields and interactions are ignored. When they are taken into account, the vacuum energy becomes nonzero; it acquires an additive constant when processes involving real particles are investigated. Particle theory based on this viewpoint faces the problem of calculating all observable processes in a manner which yields answers that do not depend on the unknown—or indeterminate, or even infinite—energy of the vacuum.

It is in this manner that Feynman formulated the problem and devised a successful solution. In his formulation the transition amplitude A_{12} (amplitude for particles plus vacuum in the initial state to change into particles plus vacuum in the final state) is divided by the transition amplitude A_v (vacuum $\rightarrow$ vacuum); and only the ratio A_{12}/A_v represents the real value corresponding to the real particle interaction. Such a method of avoiding the question of the energy of the vacuum is good everywhere except in gravitation theory! The energy density of the vacuum in gravitational problems is, as we mentioned above, a real, measurable quantity!

There is a second, so-called axiomatic, viewpoint on the theory of particles: it is taken as an axiom that the energy density of the vacuum and the corresponding pressure are identically zero.

We have no objection if such a statement is postulated openly as one possibility among others. However, one often meets in the literature statements that such an axiom is necessary; that it and no other is in agreement with relativistic invariance. Such a supposition is quite wrong. It was already noted above that the relation between the pressure and energy density of the vacuum which is provided by the cosmological constant, $P_\Lambda = -\epsilon_\Lambda$, is relativistically invariant.

We shall show below how particle theory can give an order-of-magnitude estimate of ϵ_Λ which is nonzero, while maintaining relativistic invariance.

An error frequently committed arises from introducing a particular normalizing volume V, and examining the energy $E = V\epsilon$. The (three-dimensional) momentum P of the vacuum is obviously zero because there is no preferred direction for it. Energy and momentum comprise the four-dimensional vector (E, P), which in our case, for a given normalizing volume, is $(E, 0)$. Such a combination is clearly noninvariant and will give $P \neq 0$ in other coordinate systems, unless one chooses $E = 0$ (and, therefore. $\epsilon = 0$).

The error lies in choosing a particular volume, since this itself violates relativistic invariance. An unbounded medium, and the vacuum in particular, must be characterized with an energy density which is the T_{00} component of the second-order stress-energy tensor. The tensor as a whole includes components $T_{0a} = T_{a0}$ (where $a = 1, 2, 3$ is a spatial index), which describe the energy flow and simultaneously the momentum density in space, $P_a = T_{0a}V$. The components $T_{\alpha\beta}$ of the stress-energy tensor correspond to the stress which enters into the theory of elasticity. In the case of a gas or fluid (without shear stresses) $T_{\alpha\beta} = \delta_{\alpha\beta}P$.

These well-known truths are repeated here in order to emphasize that the problem lies not in the question of whether the vacuum has an energy-momentum vector, but whether it has a stress-energy tensor. There is no relativistically invariant vector (it is identically zero), but there can perfectly well be a relativistically invariant tensor. It must have the form

$$\text{const.} \times \begin{vmatrix} 1 & 0 & 0 & 0 \\ 0 & -1 & 0 & 0 \\ 0 & 0 & -1 & 0 \\ 0 & 0 & 0 & -1 \end{vmatrix},$$

and it is about precisely such a tensor that one speaks in the case $\Lambda \neq 0$. We cannot rule out such a tensor associated with the vacuum a priori.

The following questions remain:

1. Is there any principle which requires one to set $\Lambda \neq 0$?

1.10 Weak Gravitational Fields

2. Must we regard $\Lambda \neq 0$ as a new, independent, universal constant?

3. Is it possible to calculate Λ (even in order of magnitude) from the other universal constants?

An attempt to answer the third question, without touching on the first two, will be made below. This answer (based only on dimensional analysis and comparisons of orders of magnitude) will probably be useful in the construction of a future theory which is more nearly correct and logically consistent.

The theory of elementary particles yields a quantity with the dimensions of ϵ_Λ: from the fundamental constants of the theory one can construct the energy mc^2, the length $\hbar/mc$, the number density $n = (mc/\hbar)^3$, and the energy density $\epsilon_\Lambda = mc^2(mc/\hbar)^3$. This value of ϵ_Λ certainly is not applicable to gravitation theory; for the electron mass we get ϵ_Λ of the order of 10^{25} ergs cm^{-3}, and for the proton mass we get ϵ_Λ of the order of 10^{38}. Apparently the subconscious aversion of physicists to $\Lambda \neq 0$ has been based on the comparison of such a "theoretical" ϵ_Λ with what is admissible in cosmology: If we can't take a large Λ, then let there be none at all!

Under the influence of suggestions from astronomers, it was pointed out recently (Zel'dovich 1967b) that a sensible ϵ_Λ can be derived by multiplying $mc^2(mc/\hbar)^3$ by the dimensionless quantity $Gm^2/\hbar c$ which is characteristic of gravitation. We can interpret this expression in the following manner: In the vacuum, virtual particles with mass m and average separation $\lambda = \hbar/mc$ are created. Their total proper energy is identically zero, so the gravitational interaction of the neighboring particles determines the energy density of the vacuum:

$$\epsilon_\Lambda = (Gm^2/\lambda)(1/\lambda^3) = Gm^6c^4/\hbar^4 ,$$

corresponding to the relations mentioned above.

The connection between the theory of particles and the cosmological constant is treated in greater detail by Zel'dovich (1968).

There is another side to the problem: Often it is said that $\Lambda \neq 0$ means that gravitons have a nonzero rest mass. But $\Lambda \neq 0$ leads also to the result that, even in the absence of matter, spacetime cannot be flat everywhere. In curved spacetime the very definition of the mass of a graviton is no longer clear. For a thorough discussion of this point, see Treder (1968).

1.10 NEWTON'S LAW AND WEAK GRAVITATIONAL FIELDS

Consider a weak gravitational field. In this case one obviously can construct a reference frame in which the components of the metric tensor differ very little from their Galilean values [denoted by the superscript (0)]:

$$\left.\begin{array}{l} g_{ik} = g^{(0)}{}_{ik} + h_{ik} , \\[2mm] g^{(0)}{}_{\alpha\beta} = -\delta_{\alpha\beta} , \qquad g^{(0)}{}_{0\alpha} = 0 , \qquad g^{(0)}{}_{00} = 1 . \end{array}\right\} \quad (1.10.1)$$

The quantities h_{ik} and their coordinate derivatives will be considered small. In the simplest case the gravitational field is created by slowly moving bodies $(v/c \ll 1)$ and all electromagnetic fields are weak. Then, as has been demonstrated by Einstein (see, e.g., Einstein 1950), coordinates can be selected in which the Einstein field equations take the form (accurate to first order in h_{ik})

$$\Box\, h_{ik} = -\kappa \rho g^{(0)}{}_{ik}\,, \qquad (1.10.2)$$

where $\Box$ is the D'Alembertian operator,

$$\Box = -\partial^2/\partial x^{1^2} - \partial^2/\partial x^{2^2} - \partial^2/\partial x^{3^2} + \partial^2/\partial x^{0^2}\,.$$

The solution of equations (1.10.2), as is well known, can be written in terms of retarded potentials

$$h_{ik} = -\frac{\kappa}{4\pi}\, g^{(0)}{}_{ik}\, \int \left(\frac{\rho dV}{r}\right)_{t-r/c}\,. \qquad (1.10.3)$$

This solution in the non–wave zone—or, put differently, at a distance $r \ll \tau c$, where τ is the characteristic time for mass displacement in the system—has the form

$$h_{ik} = -\frac{\kappa}{4\pi}\, g^{(0)}{}_{ik}\, \int \frac{\rho dV}{r} = -g^{(0)}{}_{ik}\, \frac{2\phi}{c^2}\,, \qquad (1.10.4)$$

where ϕ is the Newtonian gravitational potential. Consequently, expressions (1.10.1) can be rewritten in the form

$$g_{00} = (1 - 2\phi/c^2)\,, \qquad g_{11} = g_{22} = g_{33} = (1 + 2\phi/c^2)\,; \qquad (1.10.5)$$

the remaining g_{ik} vanish. The gravitational force F_a, as given by formula (1.6.1b), is

$$F_a = \partial\phi/\partial x^a\,.$$

The equations of motion (1.5.19) for a slowly moving particle, in the case of the metric (1.10.5), reduce in first approximation to

$$d^2x^a/dt^2 = \partial\phi/\partial x^a\,. \qquad (1.10.6)$$

Equations (1.10.6) are Newton's second law of motion. Thus, equations (1.10.4) and (1.10.6), which follow from Einstein's equations (1.8.1) for a weak field, correspond identically to Newtonian gravitation theory and Newtonian mechanics. Using the metric (1.10.5), one can obtain some first-order corrections to Newtonian theory:

Formula (1.4.8) for the interval of proper time reads

$$d\tau = \frac{1}{c}\,(g_{00})^{1/2}dx^0 = \left(1 - \frac{2\phi}{c^2}\right)^{1/2} dt \approx \left(1 - \frac{\phi}{c^2}\right) dt\,. \qquad (1.10.7)$$

At infinity, $\phi = 0$ and $d\tau_\infty = dt$. Near gravitating masses, the rate of time is slower than it is at infinity:

$$d\tau = \left(1 - \frac{\phi}{c^2}\right) d\tau_\infty\,. \qquad (1.10.8)$$

From formula (1.10.8) it follows, for example, that the vibrations of atomic systems in a gravitational field have lower frequencies (as measured by the clock of a distant observer). Consequently, light quanta emitted by such systems will have lowered frequencies as measured by the distant observer; the quanta will be reddened. This is the famous gravitational redshift, one of the first observational predictions of GTR. We will return to this redshift in a later analysis of precise solutions of the Einstein equations for a strong gravitational field.

If, in writing the equations of motion for the metric (1.10.5), we drop the assumption that the velocity of the test particle is small, then we can easily derive, in place of equation (1.10.6), two other well-known conclusions of GTR which modify Newtonian theory: the relativistic time delay of radar waves, and the deflection of light rays passing near the Sun. If we improve the accuracy of our expression for g_{00} so that it reads $g_{00} = (1 - 2\phi/c^2 + 2\phi^2/c^4)$, then we can derive from the equations of motion (1.5.19) the perihelion shift of Mercury. We shall return to these phenomena in later sections dealing with strong gravitational fields.

Finally, let us drop the assumption that the masses which create the gravitational field have negligible velocities. Then by writing the field equation in a form accurate to higher order (in v/c) than that of equation (1.10.2), we can obtain the first nonvanishing contribution to the spacetime components, h_{0a}, of the metric tensor.

It turns out that, with an appropriate selection of the coordinates, the quantities h_{0a} in lowest approximation are given by the formulae

$$h_{0a} = -\frac{\kappa}{2\pi} \int \frac{\rho v^a}{cr} \, dV \, , \qquad (1.10.9)$$

where v^a are the components of the three-dimensional velocity of the matter. Let us examine the consequences of equations (1.10.9).

In § 1.6, it was demonstrated that if, in a given reference frame, h_{0a} are different from zero (and they cannot be made zero by a transformation of the time coordinate), then the reference frame is rotating; i.e., it experiences Coriolis forces. When the metric g_{ik} differs only slightly from the Galilean metric, and when $\partial g_{ik}/\partial x^0 = 0$, the formulae of § 6 for the angular velocity, Ω, of the reference frame reduce to

$$\Omega = \tfrac{1}{2}c\nabla \times (g_{01}, g_{02}, g_{03}) \, , \qquad (1.10.10)$$

where (g_{01}, g_{02}, g_{03}) denotes a vector with components g_{01}, g_{02}, g_{03}. (Notice that here and henceforth we drop the asterisk from Ω.)

It follows from equations (1.10.9) and (1.10.10) that near a rotating, gravitating body there exists a field of Coriolis forces; i.e., the local inertial reference frames are in rotation relative to the inertial frames at infinity.

Consider, for example, a rotating hollow sphere. In Newtonian theory the gravitational forces do not depend on the motion of matter; consequently, inside the hollow sphere ϕ = constant and there are no gravitational forces. More particularly, because of spherical symmetry, the solution of the Newtonian equation in vacuum $\nabla\phi = 0$ must have the form $\phi = a/r +$ constant. Since there is no singularity at the center, it follows that $a = 0$, ϕ = constant.

Let us next compute the components $h_{0\alpha}$ inside a homogeneous hollow sphere with mass M and radius R, rotating with angular velocity ω. Using equation (1.10.9), we find (the calculations are given in the appendix to this section at the end of chapter 1)

$$h_{10} = (4GM\omega/3c^3R)\ r'\sin\theta\sin\phi\ ,$$

$$h_{20} = -(4GM\omega/3c^3R)\ r'\sin\theta\cos\phi\ , \qquad (1.10.11)$$

$$h_{30} = 0\ .$$

Here, r', θ, and ϕ are spherical coordinates inside the sphere, while the subscripts 1, 2, 3 refer still to the Cartesian system.

By combining equations (1.10.10) and (1.10.11), we obtain for the vector $\boldsymbol{\Omega}$

$$\Omega_1 = 0\ , \qquad \Omega_2 = 0\ ,$$

$$\Omega_3 = -\frac{4}{3}\frac{GM\omega}{c^2R}\ , \qquad \Omega^2 = \Omega_\alpha\Omega^\alpha = \frac{16G^2M^2\omega^2}{9c^4R^2}\ . \qquad (1.10.12)$$

Thus, the rotation of the sphere gives rise to Coriolis forces inside the sphere. This phenomenon is similar to the presence of magnetic forces inside a rotating, charged sphere.

Notice that if an observer located inside the sphere cannot obtain information from outside it, he will not discover any Coriolis forces. Inside the sphere, the precession of gyrocompasses (a system of gyroscopes which delineate constant directions in an inertial frame of reference) is the same at all points, and hence cannot be discovered. Only by comparing his inertial frame with the inertial frame outside the sphere at infinity will the observer discover that his frame is slowly precessing.

A similar effect causes the rotation of inertial frames in the external gravitational field of a rotating body. The formula for this is

$$|\boldsymbol{\Omega}| = (G/c^2)(|K|/r^3)(3\cos^2\theta + 1)^{1/2}\ , \qquad (1.10.13)$$

where K is the total angular momentum of the body.

The presence of Coriolis forces means that an inertial compass (a system of gyroscopes) will remain fixed relative to the distant stars when it is far from all moving masses; but when it is near a rotating body, it will rotate with the above angular velocity, changing its orientation with respect to the distant stars.

The rate of precession of a gyrocompass at the pole of a rotating body ($\theta = 0$) is larger by a factor of 2 than it is at the equator ($\theta = \frac{1}{2}\pi$). At the pole the precession is in the same direction as the rotation of the body, whereas at the equator it is in the opposite direction.

For a uniform sphere rotating with angular velocity ω, formula (1.10.13) can be rewritten in the form

$$|\,\Omega\,| \;=\; \frac{2GM}{5c^2R}\,(3\cos^2\theta + 1)^{1/2}|\omega|\left(\frac{R}{r}\right)^3. \tag{1.10.14}$$

Near ordinary stars and planets, the precession is extremely small (though measurable in principle!). For example, near a pole of the Sun, $\Omega_\odot \approx 5 \times 10^{-12}\,\mathrm{sec}^{-1} \approx 30$ arc sec year^{-1}. On the surface of the Earth, $\Omega_\delta \approx -0.''1$ year^{-1} at the equator, and $0.''2$ year^{-1} at the poles. (Here the direction of rotation of the body is taken to be positive.) Despite the smallness of the effect, Everitt, Fairbank, and Hamilton (1970) are preparing an experiment to measure it, using a superconducting gyroscope in orbit about the Earth.

1.11 The Analogue of the Zeeman Effect in the Gravitational Field of a Rotating Body

It was demonstrated in the preceding section that the gravitational field of a rotating body differs from the field of a nonrotating body, just as in electrodynamics a rotating charged body creates not only an electrostatic field but also a magnetic field. This analogy seems to go even deeper. The components of the gravitational field, h_{0a}, which are analogous to the magnetic field, produce a change in the spectrum of the emitter which is similar to the Zeeman effect (Zel'dovich 1965a).

It will be demonstrated below that the spectral line of an atom with frequency ω_0, emitted from the pole of a rotating body and received by a distant observer above the pole, splits into two components with opposite circular polarization and with frequencies $\omega_0 + \Omega$ and $\omega_0 - \Omega$. Here Ω is the angular velocity of equation (1.10.14).

By contrast with the classical magnetic Zeeman effect, the gravitational effect is universal; i.e., the split of the spectral lines does not depend on the specific properties of the system which emits it. The split is the same for atoms and molecules; it is the same at optical and radio frequencies. For the proof, consider a linear oscillator resting on the pole of a rotating body. Like a Foucault pendulum, the oscillator oscillates always in a fixed direction relative to a local inertial frame, i.e., in a direction which is constant with respect to a local inertial compass. From the viewpoint of a nearby observer, the oscillator emits a plane-polarized wave, which can be interpreted as a superposition of right- and left-hand circularly polarized waves of the same frequency and amplitude.

However, relative to a distant observer, the inertial compass precesses with the angular velocity Ω (cf. § 1.10). Hence, the plane of polarization rotates with this same velocity. Linearly polarized light with a rotating plane of polarization is obviously a superposition of two circularly polarized waves with different frequencies, $\omega_0 \pm \Omega$. Thus, we see that light emitted by a charge, which oscillates in a central field of force at the pole of a rotating body, is received by a distant observer as a superposition of waves with circular polarization, split in frequency. In view of the correspondence principle between quantum theory and classical mechanics, it is obvious that this result will remain valid for any atomic or molecular system.

An alternative description of this phenomenon makes use of the fact that right-hand and left-hand polarized photons (circular polarization) experience different redshifts in the gravitational field. Thus, we are confronted with a specific case of the influence of the angular momentum of a particle (photon) upon its motion in a gravitational field.

It is obvious from the symmetry of the problem that this difference between right- and left-hand polarized photons is entirely due to the rotation of the body which creates the gravitational field. The difference in the frequencies of the photons, which equals 2Ω, does not depend on the undisturbed frequency, ω_0. The rate at which the difference grows as the photons move radially outward is easily calculated. From equation (1.10.14) with $\theta = 0$ (photon at pole) we find

$$|d\Omega/dr| = \tfrac{12}{5}(GM/c^2)(R^2/r^4)\omega_{\text{body}} ; \qquad (1.11.1)$$

and from this we find near the Earth's surface

$$|d\Omega/dr| = \tfrac{12}{5}(GM_\delta/c^2R_\delta^2)\omega_\delta \approx 10^{-23} \text{ Hz cm}^{-1} . \qquad (1.11.2)$$

This frequency split can be compared with the redshift of all photons (right and left) in the fundamental static field of the Earth as measured by Pound and Rebka:

$$(1/\omega)(d\omega/dr) = g/c^2 = GM/R^2c^2 = 10^{-18} \text{ cm}^{-1} . \qquad (1.11.3)$$

For photons with an energy of 14 keV, whose frequency is 4×10^{18} Hz, the redshift is 4 Hz cm^{-1}, so the ratio of the first effect to the second one is $|d\Omega/dr|/|d\omega/dr| \approx \omega_\delta/\omega \approx 10^{-23}$, and the effect of spin (circular polarization) of these hard photons is immeasurably small. The effect of the spin orientation of a proton upon its weight, due to the rotation of the Earth, is of the order of 10^{-28} of its total weight.

1.12 GRAVITATIONAL RADIATION

According to Newtonian theory, the gravitational force of a spherical body decreases as $1/r^2$. In a gravitational field, only relative accelerations are measurable. The relative acceleration (denoted by A) of two neighboring test particles in a gravitational field decreases as $1/r^3$. This is the well-known tidal force. The quadrupole component of the gravitational field of a nonspherical system is of the order of GMR^2/r^4, where M is the mass spread over a distance R that creates the quadrupole moment Q. The relative accelerations of test particles due to this quadrupole component of the field are of the order of $A \approx GMR^2l/r^5$, where l is the distance between the test particles. If the quadrupole moment varies periodically in time (e.g., because of the rotation of a double star), then the relative accelerations change periodically with the same phase. This is the conclusion of Newtonian theory.

The relativistic theory of gravitation asserts that these conclusions pertaining to the quadrupole component are not valid when $r > cT$, where T is the characteristic time of the variation of the quadrupole moment of the source. Beginning at this distance, the relative acceleration of two test bodies decreases as $1/r$. In other words, gravitational waves are generated. In a periodic variation of the quadrupole moment, the phase of the relative acceleration of two test particles is shifted, compared with the phase at the source, by the angle r/cT. From the condition of continuity of A when $r = cT$, we get an estimate for A in the wave zone: $A \approx GMR^2l/(cT)^4r$. At large distances from the source, the relative acceleration of test particles due to the static monopole field of the source (A decreasing as $1/r^3$) is negligibly small compared with the relative acceleration in the quadrupole wave field, which decreases as $1/r$. It is the relative acceleration in the wave that must be measured in order to detect gravitational waves.

There is another aspect of the difference, with respect to gravitational waves, between Newtonian theory and the relativistic theory: A system of two gravitationally interacting particles in orbit around each other does not lose energy in Newtonian theory. The particles in such a Newtonian system move eternally around their center of mass. It is true that such a system can lose energy if there is an inelastic body in its neighborhood. The two-particle system will produce time-varying tidal deformations in such a body. These deformations will heat the body, and the heat will radiate away into space. However, only in the presence of such an energy absorber can there be losses in gravitational energy from the two-particle system. The energy by itself (without such a nearby body) is not radiated and does not go to infinity.

In the relativistic theory, by contrast, the motion of the gravitating particles about each other produces gravitational waves which carry away energy from the system. This fact one usually demonstrates by calculating the flux of gravitational energy associated with the gravitational waves far from the system—i.e., at a distance $r \gg cT$, in the so-called wave zone. However, the energy loss can also be calculated by a close examination of the back reaction of the gravitational field on the radiating system: Burke, Chandrasekhar, Esposito, and Thorne have recently shown, by calculations of the field near a radiating system, how the emission of gravitational waves produces a damping of the system's motion (see § 1.14 for details). The rates of energy loss calculated by the two methods—energy flux in the wave zone, and damping of the system's motion—agree, as of course they must. There can be no doubt that gravitational waves remove energy from the system which emits them.

In an electromagnetic system, charge is conserved, but the dipole moments (magnetic and electric) can change with time. Because of the conservation of charge, the spherically symmetric component of the electric field (which corresponds to Coulomb's law) is also conserved; there exist no longitudinal, spherically symmetric waves by which this component could change in time. The lowest multipoles in electromagnetic waves are electric and the magnetic dipole radiation; and the fields of these waves are transverse.

Consider gravitational waves from the viewpoint of successive approximations. In the absence of gravitational waves, the mass m and the angular momentum K of a system are conserved. Accordingly, in the vacuum surrounding the system, neither the longitudinal, spherically symmetric, static component of the gravitational field (which is proportional to the mass), nor the gravimagnetic, stationary, dipole component (which is proportional to the angular momentum) can have oscillatory changes. Consequently, the lowest multipole of gravitational waves is the quadrupole. However, one should keep in mind the fact that the quadrupole and higher-order waves can carry off mass and angular momentum; and as a result the longitudinal and the dipole stationary components of the field will change.

Let us turn now from this qualitative discussion to a quantitative description of gravitational waves.

In vacuum the metric perturbations, h_{ik}, which describe a weak gravitational field, satisfy the equations

$$\Box \, h_{ik} = 0 \, . \tag{1.12.1}$$

These equations are valid only in a particular type of coordinate system—a system in which[19]

$$\partial/\partial x^k (h^k{}_i - \tfrac{1}{2}\delta^k{}_i h) = 0 \, . \tag{1.12.2}$$

In an arbitrary weak field, such a coordinate system always exists (Hilbert 1917). Equation (1.12.1) is a wave equation similar to the wave equations of electrodynamics. Consequently, nonstationary gravitational fields propagate through space in a manner similar to electromagnetic waves.

By changing coordinates, one can always pass to a system where the new $h_{i'k'}$ will no longer satisfy equation (1.12.1). Obviously, in this case, superimposed on the gravitational excitations are noninertial motions of the reference frame; this is what "spoils" equations (1.12.1).

To separate the propagation of actual gravitational "excitations" in the spacetime curvature from excitations related to arbitrariness of the refer-

19. Here we define $h^k{}_i \equiv h_{mi}g^{(0)mk}$, $h \equiv h^i{}_i$.

ence frame, one must examine the propagation of a quantity which is left unchanged by all small perturbations of the coordinate system ($x^i \rightarrow x^i + \xi^i$; "fake gravitational waves"). Fortunately, the Riemann tensor

$$R_{iklm},$$

which is a small quantity and which produces the relative motions of test particles (geodesic deviation) has this property.

In a coordinate system where equation (1.12.1) is valid, the perturbations in the field h_{ik} propagate in the form of waves with the speed of light c. The perturbation in the Riemann tensor, computed in this coordinate system by the use of h_{ik}, obviously propagates also with speed c. This fact (first suggested by Eddington 1925) is unaffected by coordinate perturbations since they leave R_{iklm} unchanged. Consequently, a "real" gravitational excitation (perturbation of the spacetime curvature) propagates as a wave with the speed of light, as viewed in any reference frame. By contrast, "excitations" in h_{ik}, due to a transformation of coordinates, can be made to propagate at any velocity; these are not perturbations of the curvature, but rather, in a certain sense, mathematical fictions.

On what quantities, then, does the gravitational excitation depend? Consider a region of spacetime which contains a propagating wave h_{ik}. In a small region the wave may be considered as plane-fronted. Orient the x^1-axis along the direction of variation of the field in space (along the direction of propagation of the wave). If we were to calculate now the components of the Riemann tensor, R_{iklm}, we would find that, when the vacuum field equations $R_{ik} = 0$ are satisfied, the components R_{iklm} depend only on the "transverse" components of the metric perturbation, h_{22}, h_{23}, and h_{33}, which are related by

$$(\partial^2 h_{22}/\partial t^2) + (\partial^2 h_{33}/\partial t^2) = 0 . \qquad (1.12.3)$$

Because we are interested only in the time-varying part of the field, we can integrate this equation to obtain $h_{22} = -h_{33}$—as a consequence of which the Riemann tensor depends on only two independent components of h_{ik}: h_{23} and $h_{22} = -h_{33}$.

The scalar C also depends only on these components. It follows that, by using an appropriate choice of coordinates, we can force all components of h_{ik} to vanish except h_{23} and $h_{22} = -h_{33}$. This demonstrates the "transverse" nature of gravitational waves.

Let us emphasize that in a wave (as in any gravitational field) only the relative accelerations of test particles—i.e., only inhomogeneities in the gravitational field—are measurable.

Consider the change with time of the distance between two test particles interacting with the gravitational wave. If $h_{0i} = 0$, then, according to equation (1.6.1a), the lines x^a = const. are geodesics; i.e., test particles initially at rest relative to the coordinate system will always remain at rest. Thus the changes in the distance between the particles will be determined completely by the deformation of the coordinate system itself, i.e., by the time dependence of $h_{\alpha\beta}$.

Taking account of the relation $h_{22} = -h_{33}$, we obtain for the distance between two neighboring particles

$$dl^2 = (dx^1)^2 + (1 + h_{22})(dx^2)^2 + (1 - h_{22})(dx^3)^2 + 2h_{23}dx^2dx^3 . \quad (1.12.4)$$

From this equation it follows that the distance between two particles, whose separation is along the direction of the wave propagation $(dx^1 \neq 0; dx^2 = dx^3 = 0)$, is constant in time. Particles located in a plane perpendicular to the direction of propagation experience a maximum relative acceleration. For such particles, using polar coordinates with the origin at one of them, we can write $(x^2 = r \cos \theta, x^1 = r \sin \theta)$

$$l = r(1 + \tfrac{1}{2}h_{22} \cos 2\theta + \tfrac{1}{2}h_{23} \sin 2\theta) . \quad (1.12.5)$$

The two components of a wave, h_{23} and $h_{22} = -h_{33}$, define two states of polarization. However, by contrast with an electromagnetic wave where the independent polarization states are determined by the oscillating vector of the electric field, in this case the polarization has a tensorial nature. Figure 3 shows, for plane-polarized electromagnetic and gravitational waves, the motions of test charges and test masses.

For the electromagnetic case (left side of Fig. 3) the test charges oscillate relative to an inertial frame. The oscillation is in one direction for one state of polarization (Fig. 3a), and in the perpendicular direction for the other (Fig. 3b).

For the gravitational case (right side of Fig. 3) the test masses oscillate relative to each other. If distributed initially around the circumference of a circle, they will be displaced by the wave to form an oscillating ellipse. For one state of polarization $(h_{23} = 0, h_{22} = -h_{33} \neq 0$ in eq. [1.12.5]; Fig. 3a) the ellipse oscillates in the x^2 and x^3 directions. For the other state $(h_{23} \neq 0, h_{22} = -h_{33} = 0$ in eq. [1.12.5]; Fig. 3b) the oscillating ellipse is inclined at a 45° angle to the x^2 and x^3 directions.

The amplitude equals zero at the moments $t = T/4$ and $t = 3T/4$. Figures 3a and 3b correspond to the case of plane-polarized electromagnetic waves and tensor waves. In a linear theory one can add solutions with arbitrary coefficients. By adding the upper solution (3a) to the lower one shifted by a quarter of a period, we get for the electromagnetic wave a vector which points upward at $t = 0$; to the right at $t = \tfrac{1}{4}T$; downward at $t = \tfrac{1}{2}T$; and to the left at $t = \tfrac{3}{4}T$. The electromagnetic field vector rotates in a clockwise direction. Such a superposition of plane-polarized waves describes a wave field with circular polarization. In a similar manner one can get a solution in which the characteristic ellipse of a tensor wave rotates. This procedure produces gravitational waves with circular polarization, i.e., without principal axes but with a principal direction of rotation. In a number of cases—e.g., gravitational radiation from a double star, see below—gravitational waves carry away not only energy but also angular momentum. In these cases the radiation has circular polarization (at least partially).

The picture (Fig. 3) of the action of tidal gravitational forces on test particles in a gravitational wave completely determines the various possible methods of receiving such waves.

In the first method one examines the periodic component in the distance separating two freely moving bodies.

In the second variant, two bodies are joined by an elastic medium. If this medium prevents the bodies from changing their separation, then it will develop stresses in the process, which can be measured. The system as a

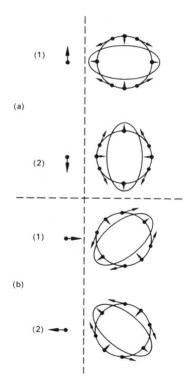

FIG. 3 (adapted from Wheeler 1960).—Displacement of test charges (points on the left) in a polarized electromagnetic wave, and the displacement of test particles (points on the sphere on the right) in a polarized gravitational wave. (*a*) Displacement of particles at opposite phases (1) and (2) of one polarization state; (*b*) displacement of particles at opposite phases (1) and (2) of the second polarization state.

whole should be constructed to have a characteristic frequency equal to the frequency of the waves. Then the waves will produce a resonance (see § 1.15).

Finally, it is possible to detect gravitational waves by means of a rotating pair of bodies or one body rotating about a fixed axis. One can easily see from Figure 3 that a pair of bodies will systematically absorb energy and angular momentum from the waves or put energy and angular momentum into the waves, depending on the relation of their phases, if the angular velocity of their rotation corresponds to half a revolution in one period of the wave.

It is not surprising that a plane-polarized wave (one without angular momentum) can change the angular momentum of a pair of bodies. One can interpret a plane-polarized wave as a superposition of two waves with opposite circular polarizations. One of the circular waves—that which rotates in the same direction as the bodies do—accelerates the bodies, and in the process becomes weaker. The other circular wave does not interact significantly with the bodies. By virtue of this interaction with the pair of bodies, an initially plane wave without angular momentum will have, after passing the bodies, an angular momentum opposite to that which was absorbed by the bodies.

Since the forces and displacements due to a realistic gravitational wave interacting with a realistic body are extremely small, a decisive role in the mechanical reception of gravitational waves is played by noise. The crucial problem is to lift the signal out of the noise (for references, see the end of § 1.13).

It has been suggested that one might detect gravitational waves by their interaction with propagating light and radio waves. However, the effects of this interaction have been grossly exaggerated, particularly in the work of Winterberg. In the case of gravitational waves from double stars, one must take account of their strictly periodic character; such waves are not chaotic. Regions with opposite signs of the metric components, for example h_{22} and h_{33}, are regular and cancel out each other's effects more precisely than in the case of chaotic waves.

In discussing the interaction of light with the thermal gravitational waves of effective temperature about 3° K, one must not use geometric optics.

To compute the intensity of the gravitational radiation emitted by a system of moving bodies, one must write the equations of gravitation with an accuracy of higher order in h_{ik} than was used in § 1.10. We will not go into the computations here (see Landau and Lifshitz 1962; also § 1.14), but will only cite the results. The power dL radiated by the system into a solid angle[20] $d\Omega$ in a direction described by the unit vector n is

20. The symbol Ω for the solid angle must not be confused with the symbol Ω for the angular-velocity vector, which was used in preceding paragraphs.

$$dL = \frac{G}{36\pi c^5}\left[\frac{1}{4}\left(\frac{\partial^3 K_{\alpha\beta}}{\partial t^3}\,n^\alpha n^\beta\right)^2 + \frac{1}{2}\left(\frac{\partial^3 K_{\alpha\beta}}{\partial t^3}\right)^2\right.$$

$$\left. - \left(\frac{\partial^3 K_{\alpha\beta}}{\partial t^3}\right)\left(\frac{\partial^3 K_{\alpha\gamma}}{\partial t^3}\right)n^\beta n^\gamma\right]d\Omega\ . \tag{1.12.6}$$

Here $K_{\alpha\beta}$ is the quadrupole moment of the masses

$$K_{\alpha\beta} = \int \rho(3x^\alpha x^\beta - \delta^\beta{}_\alpha x^\gamma x^\gamma)dV\ . \tag{1.12.7}$$

The total power radiated in all directions is given by

$$-(dE/dt) = (G/45c^5)(\partial^3 K_{\alpha\beta}/\partial t^3)^2\ . \tag{1.12.8}$$

It is explicitly clear from expressions (1.12.6)–(1.12.8) that the gravitational waves have a quadrupole nature. This is the lowest possible multipole order of gravitational radiation. Obviously, dipole radiation cannot occur in the theory of gravitation because the ratio of the "gravitational charge" to the "inertial mass" is the same for all bodies (cf. § 1.1), and the dipole moment therefore always vanishes.

Formulae for the power radiated are given in § 1.13 in forms convenient for astrophysical calculations.

1.13   GRAVITATIONAL RADIATION FROM BINARY STARS

Consider the gravitational radiation from a binary star system with the field assumed to be weak, $\phi \ll c^2$. This problem with an elliptical orbit has been analyzed in detail by Peters and Mathews (1963).

Let the two stars have masses m_1 and m_2. The relative elliptical orbit (i.e., motion of one star with respect to the other) is described by the equation

$$r = \frac{a(1-e^2)}{1+e\cos\psi}\ . \tag{1.13.1}$$

Here a is the semimajor axis of the orbit, e is its eccentricity, ψ is the polar angle, and r is the radius vector. Using equation (1.12.8), and averaging over an entire period of rotation, we obtain the following expression for the total power, L, radiated as gravitational waves:

$$L = -dE/dt = [\tfrac{32}{5}(G/c^5)m_1^2 m_2^2(m_1+m_2)/a^5]f(e)\ . \tag{1.13.2}$$

The dependence on the eccentricity is given by the function

$$f(e) = \frac{1+\tfrac{73}{24}e^2+\tfrac{37}{96}e^4}{(1-e^2)^{7/2}}\ . \tag{1.13.3}$$

A graph of this function is shown in Figure 4. For fixed a the radiation increases with increasing eccentricity. This is because the power radiated, as indicated by equation (1.12.8), is sensitive to variations in the velocity, and hence takes place primarily at the periastron of the orbit.[21] The larger the eccentricity, the closer the stars are at periastron, and the greater are the accelerations and gravitational radiation.

For elliptical motion, the gravitational waves not only exhibit the second harmonic of the orbital frequency, as is the case for the quadrupole radiation from circular motion, but also generate other harmonics.

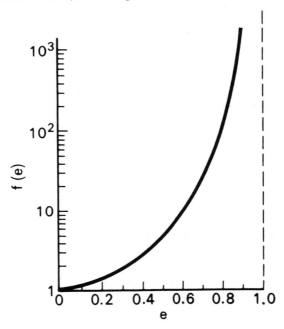

FIG. 4.—The function $f(e)$, the dependence of the power radiated (per unit time, averaged over a period) on the eccentricity e of the orbit, for a fixed semimajor axis a.

Fourier analysis yields the following expression for the total radiation in the nth harmonic of the orbital frequency:

$$L(n) = \tfrac{32}{5}(G^4/c^5)m_1{}^2m_2{}^2(m_1 + m_2)a^{-5}g(n, e) , \qquad (1.13.4)$$

where

$$
\begin{aligned}
g(n, e) = (n^4/32)\{&[J_{n-2}(ne) - 2eJ_{n-1}(ne) + (2/n)J_n(ne) \\
&+ 2eJ_{n+1}(ne) - J_{n+2}(ne)]^2 + (1 - e^2)[J_{n-2}(ne) \\
&- 2J_n(ne) + J_{n+2}(ne)]^2 + (4/3n^2)[J_n(ne)]^2\} .
\end{aligned}
\qquad (1.13.5)
$$

21. The periastron of the relative orbit is the point on the orbit at which the stars are nearest each other.

Here the J_n are Bessel functions. The dependence on n of the function g is shown in Figure 5 for $e = 0.2$, $e = 0.5$, and $e = 0.7$.

For large eccentricities, the radiation is concentrated primarily in the higher harmonics. This is because the radiation, as noted previously, is emitted primarily at the perihelion of the orbit, i.e., during a short portion of the full orbital period.

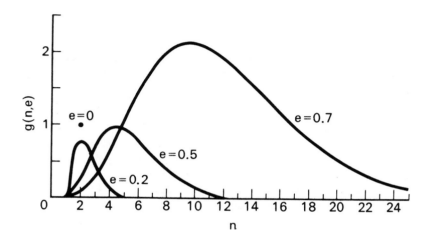

Fig. 5.—Dependence on n, $g(n, e)$, of the relative power in different harmonics for $e = 0$, 0.2, 0.5, and 0.7. The harmonic number n must be an integer. When $e = 0$, all the radiation is concentrated at $n = 2$.

The emission of gravitational waves by binary stars produces changes in the orbit. Let us calculate the rate of change of the semimajor axis a. The total energy of the system is

$$E = -\tfrac{1}{2}Gm_1m_2/a .$$

Consequently,

$$\frac{da}{dt} = \frac{2a^2}{Gm_1m_2}\frac{dE}{dt}, \tag{1.13.6}$$

where dE/dt is determined by formula (1.13.2).

For a circular orbit, using equations (1.13.2) and (1.13.6), we obtain the following formula for the rate of change of the orbital radius (Landau and Lifshitz 1962):

$$da/dt = -\tfrac{64}{5}(G^3/c^5)m_1m_2(m_1 + m_2)a^{-3} . \tag{1.13.7}$$

For highly eccentric orbits de/dt has the form

$$de/dt = (1 - e)(1/a)(da/dt) . \tag{1.13.8}$$

47

Consider next the polarization of the radiation and its directional distribution. We present here the results for motion around a circular orbit ($e = 0$). Formulae for the general case are given by Peters and Mathews (1963).

The directional distributions of the radiation for the two independent states of polarization discussed in the preceding section are given by the formulae

$$dL_1/d\Omega = (1/\pi)(G^4/c^5)m_1^2m_2^2(m_1 + m_2)a^{-5}(1 + \cos^2 \theta)^2 \sin^2 2\phi , \qquad (1.13.9)$$

$$dL_2/d\Omega = (4/\pi)(G^4/c^5)m_1^2m_2^2(m_1 + m_2)a^{-5} \cos^2 \theta \cos^2 2\phi . \qquad (1.13.10)$$

Here θ is the polar angle and ϕ is the difference between the longitudes of the observation point and of the instantaneous separation vector of the stars. The terms $\sin^2 2\phi$ and $\cos^2 2\phi$ determine the time dependence of the directional distribution.

At the poles ($\theta = 0, \pi$), equations (1.13.9) and (1.13.10) describe circularly polarized waves. The tensor of relative acceleration (Fig. 3) rotates with the orbital frequency of the stars; but unlike the stars, it returns to its original orientation after a rotation of 180°. Consequently, the frequency of the radiation is twice the orbital frequency. The circular polarization of the waves is related to the fact that the binary star radiates not only its energy but also its angular momentum. The angular distribution of the total radiation (summed over polarization states and averaged over an orbital period) is given by the formula

$$dL/d\Omega = (G^4/2\pi c^5)m_1^2m_2^2(m_1 + m_2)a^{-5}(1 + 6\cos^2 \theta + \cos^4 \theta) . \qquad (1.13.11)$$

The radiation pattern corresponding to this equation is shown in Figure 6. The radiation is directed mainly toward the poles ($\theta = 0, \pi$).

Let us evaluate the radiated power for real astronomical systems. For the solar system, when we substitute the data for the largest planet, Jupiter, into equation (1.13.2), we get $L = 5 \times 10^{10}$ ergs sec^{-1}. This is roughly a factor 10^{23} smaller than the power radiated by the Sun as light ($L_{\odot} \approx 4 \times 10^{33}$ ergs sec^{-1}).

Astronomers know of binary stars whose radiated power should be considerably greater than this. We present here a table (Table 1), taken from Braginsky (1965), for several binary stars with relatively small orbits.

The total flux of gravitational radiation, which bathes the Earth from all directions, is determined primarily not by a few individual, nearby binary star systems, but by the aggregate of all compact binary systems (systems of the type W UMa) in our Galaxy (Mironovsky 1965b). Let us assume for simplicity that the density, n_*, of binary systems in space is constant out to a distance R from us, and that the gravitational radiation from each system is of the order of L. Then the flux at Earth is

$$F_G = \int_0^R (Ln_*/4\pi r^2)4\pi r^2 dr = Ln_*R ;$$

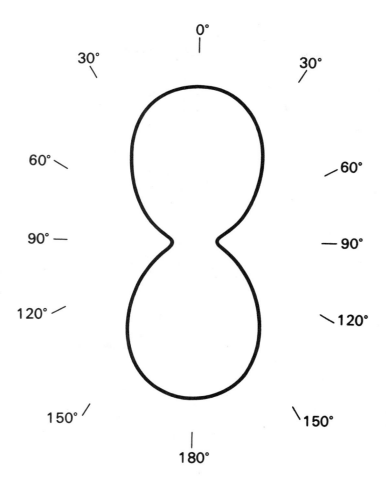

Fig. 6.—Radiation pattern for gravitational waves from a binary star with a circular orbit. Distance from the center of the pattern corresponds to the total intensity averaged over a period in a plane perpendicular to the plane of the stellar orbit.

TABLE 1

The Nature of Gravitational Radiation from Several Double Stars

Star	Period	$m_1/M_\odot$	$m_2/M_\odot$	Distance from Solar System (cm)	dE/dt (ergs sec^{-1})	Flux at Earth (erg sec^{-1} cm^{-2})
UV Leo.........	0^d6	1.36	1.25	2.1×10^{20}	1.8×10^{31}	3.5×10^{-12}
V Pup..........	1^d45	1.66	9.8	1.2×10^{21}	4×10^{31}	2.3×10^{-12}
i Boo..........	0^d268	1.35	0.68	3.8×10^{19}	1.9×10^{30}	1.1×10^{-10}
YY Eri..........	0^d321	0.76	0.50	1.3×10^{20}	2.6×10^{29}	1.3×10^{-12}
SW Lac..........	0^d321	0.97	0.83	2.3×10^{20}	1.1×10^{30}	1.7×10^{-12}
WZ Sge..........	81 min	0.6	0.03	3×10^{20}	3.5×10^{29}	3×10^{-13}

49

i.e., the flux is determined primarily by the distant binary systems. The situation here is the same as it is with the general optical brightness of the sky due to stars.[22] Detailed calculations by Mironovsky (1965b) put the value of F_G at roughly 10^{-9} erg cm^{-2} sec^{-1}. In addition to gravitational waves from binary stars, powerful gravitational radiation should be emitted in certain catastrophic cosmic processes. For more on this, see § 3.11, and also chapter 13. In addition, there should be equilibrium thermal gravitational radiation with a temperature $T \simeq 1°$ K which pervades all of space and is primordial in origin (for more on this, see Vol. II).

In summary, notice that gravitational radiation from celestial bodies can be separated into four types:

1. Radiation from moving bodies in double and multiple systems. This type of radiation was considered in detail above.

2. Radiation from pulsating and rotating stars with deviations from spherical symmetry. The detailed analysis of such systems can be found in the works of K. S. Thorne and his co-workers (for a review, see Thorne 1968).

3. Radiation from stars as they collapse, in a nonspherical manner, through their Schwarzschild radii, and from stars in collision.

4. High-frequency radiation (in the form of gravitons), which is produced by collisions between atoms and particles (e.g., Coulomb scattering of electrons) in the interior of a star. The rate of graviton emission is very small compared not only with the rate of photon emission, but also with the rate of neutrino-antineutrino pair emission (Gandel'man and Pinaev 1959). The power output of the Sun in thermal gravitational radiation is about 10^{15} ergs sec^{-1} (see Mironovsky 1965a), which is 10^{18} times less than the power output in light. For more on gravitons, see § 6 of chapter 2.

We will not go into detail here about the problem of detecting gravitational radiation, except for a brief overview of Weber's recent work in § 14. Those who are interested will find detailed discussions in the book by Weber (1961), in the review by Braginsky (1965), in the paper by Westervelt (1966), and in a series of progress reports by Weber (1967, 1968, 1969).[23]

In order to illustrate the difficulties of detection, we estimate the relative acceleration A produced by a gravitational wave on two test particles separated by a distance l. If the wave is emitted by a double star at a distance

22. The sky brightness at Earth due to the stars of distant galaxies is small because of the cosmological redshift (see Vol. II on cosmology). The radiation flux from the stars of our Galaxy is larger by one and a half orders of magnitude than the radiation flux from stars of other galaxies. For the same reason, the contribution to F_G from binary systems in distant galaxies is insignificant.

23. We note also the very interesting work of Pustovoyt and Gertsenshteyn (1962) and Kopvilem and Nagibarov (1965) about the possibility of coherent emission and detection of gravitational waves, although this work is not directly connected to our discussion.

of 10 pc $= 3 \times 11^{19}$ cm, with stellar masses of the order of the mass of the Sun, and with an 8-hour period of rotation, then from the estimates at the beginning of this section we find

$$A/l \approx 10^{-33} \text{ sec}^{-2} .$$

1.14 GRAVITATIONAL RADIATION DAMPING (BY K. S. THORNE)

Until recently the only way of calculating the energy carried away from a system by gravitational waves was to calculate the energy flux in the wave zone, using a stress-energy "pseudotensor" for the gravitational waves (Landau and Lifshitz 1962, § 100), and to then integrate this flux over a sphere surrounding the source and over time. A similar method was used to calculate the angular momentum carried off by the waves. The back reaction of the radiation on the source was then inferred by demanding that the source lose energy and angular momentum at the same rate as the waves carry it off. The results for binary stars in the last section were calculated by such a procedure.

Skeptics of the reality of gravitational waves (who have declined rapidly in number over the last decade, and hopefully will soon become extinct!) object that the pseudotensor used to measure the stress-energy in the waves has no physical reality. Because it can be set to zero at any event by an appropriate choice of coordinates, the pseudotensor cannot be trusted. Such a critique is usually answered (satisfactorily) by pointing out that the *total* energy and angular momentum radiated are independent of the coordinates and of the choice of pseudotensor. However, in the last two years much better answers have been developed—answers which make absolutely firm the standard conclusions about the energy in a gravitational wave, and about the damping effects of the wave on its source.

The first of these is Isaacson's (1968a, b) construction of a stress-energy tensor (not pseudotensor!) for gravitational waves. Isaacson confines his attention to waves of mean wavelength λ propagating in a smooth "background" spacetime whose mean radius of curvature ℜ is much larger than λ. One can study such a system from two viewpoints: a "fine-grained" one, and a "coarse-grained" one.

The fine-grained viewpoint focuses attention on regions of size not much larger than λ; in such regions the spacetime curvature is dominated by the ripples of the gravitational waves, and the curvature of the background space is hardly noticeable. It is from this fine-grained viewpoint that one studies the propagation of the waves.

The coarse-grained viewpoint concentrates on the large-scale properties of spacetime; it ignores the detailed local features of the waves by averaging all quantities over regions the size of several wavelengths λ. For example,

51

in the curvature invariant, $C = R_{abcd}R^{abcd}$, the coarse-grained viewpoint smoothes out the ripples of the waves but leaves untouched the curvature of the background space; as a consequence, the radius of curvature of the background space is given by

$$\mathfrak{R} \approx \langle C \rangle^{-1/4} \, . \tag{1.14.1}$$

Here the angular brackets represent the coarse-grained averaging over several wavelengths.[24]

Isaacson's analysis expands the metric and curvature tensors in powers of the small dimensionless parameter

$$\lambda/\mathfrak{R} \ll 1 \, . \tag{1.14.2}$$

Notice that this parameter is intimately related to the amplitude of the waves: as we shall see below, the waves themselves help to produce the background curvature. They have an energy of the order of $(c^4/G)(h/\lambda)^2$, where h is their dimensionless amplitude; consequently, through the Einstein field equations, they produce a curvature

$$(\mathfrak{R}_{\text{due to waves}})^{-2} \sim (G/c^4) \times (\text{energy of waves}) \sim (h/\lambda)^2 \, ; \tag{1.14.3}$$

$$h \sim \lambda/\mathfrak{R}_{\text{due to waves}} \lesssim \lambda/\mathfrak{R}_{\text{due to everything}} \, .$$

Hence, in the absence of matter and other fields the expansion in powers of $\lambda/\mathfrak{R}$ is equivalent to an expression in powers of the wave amplitude h.

Isaacson writes the expansion of the metric g_{ij} and of the Einstein curvature tensor $G_{ij} = R_{ij} - g_{ij}R$ in the form

$$g_{ij} = g^{(B)}{}_{ij} + h_{ij} \, ,$$

$$G_{ij} \equiv R_{ij} - \tfrac{1}{2}g_{ij}R = G^{(B)}{}_{ij} + G^{(L)}{}_{ij} + G^{(Q)}{}_{ij} + O(\mathfrak{R}^{-2}[\lambda/\mathfrak{R}]^3) \, . \tag{1.14.4}$$

Here $g^{(B)}{}_{ij}$ and $G^{(B)}{}_{ij}$ are the background metric and curvature, h_{ij} and $G^{(L)}{}_{ij}$ are the "ripply" corrections linear in $\lambda/\mathfrak{R}$, and $G^{(Q)}{}_{ij}$ is the quadratic correction. $G^{(L)}{}_{ij}$ and $G^{(Q)}{}_{ij}$ are calculated in a straightforward manner from the metric (1.14.4) and the standard formulae (1.7.1), (1.7.2) for the Ricci tensor.

In the vacuum region outside all sources the Einstein tensor must vanish, $G_{ij} = 0$. We examine these field equations first from the fine-grained viewpoint and then from the coarse-grained viewpoint. From the fine-grained viewpoint, where the background curvature is not noticeable, the part of the Einstein tensor associated with the ripples must vanish:

$$G^{(L)}{}_{ij} = 0 \, . \tag{1.14.5}$$

---

24. Technically it is a "Brill-Hartle average" (see Isaacson [1968a] and Brill and Hartle [1964]); but in all except the most complex cases it can be any simple-minded type of average.

When written out in terms of the metric perturbation h_{ij}, this equation becomes a curved-space generalization of the wave equation

$$\psi_{ij} \equiv h_{ij} - \tfrac{1}{2} g^{(B)}{}_{ij} h_k{}^k , \qquad (1.14.6a)$$

$$\psi_{ij|a}{}^a + 2R^{(B)}{}_{aijb}\psi^{ab} + R^{(B)}{}_{ia}\psi^a{}_j + R^{(B)}{}_{ja}\psi^a{}_i = 0 . \qquad (1.14.6b)^{25}$$

Here indices are raised and lowered by means of the background metric $g^{(B)}{}_{ab}$, a vertical bar denotes a covariant derivative with respect to the background metric, and the gauge condition

$$\psi^{ij}{}_{|j} = 0 \qquad (1.14.6c)$$

is assumed.

From the coarse-grained viewpoint the vacuum field equations $G_{ij} = 0$ must be averaged over several wavelengths. The average of the background curvature, $\langle G^{(B)}{}_{ij} \rangle$, is just the curvature itself (no ripples!); and the average of the linear perturbations, $\langle G^{(L)}{}_{ij} \rangle$, is identically zero. Consequently, the coarse-grained field equations read

$$G^{(B)}{}_{ij} + \langle G^{(Q)}{}_{ij} \rangle = 0 ,$$

or, equivalently,

$$G^{(B)}{}_{ij} = (8\pi G/c^4) T^{(W)}{}_{ij} , \qquad (1.14.7)$$

where $T^{(W)}{}_{ij}$ is defined by

$$T^{(W)}{}_{ij} = -(c^4/8\pi G)\langle G^{(Q)}{}_{ij} \rangle = (c^4/32\pi G)\langle h_{kl|i} h^{kl}{}_{|j} \rangle . \qquad (1.14.8)$$

In the last expression, which is obtained by averaging the quadratic part of the standard formula for the Einstein tensor (eqs. [1.14.4], [1.7.1], and [1.7.2]), the indices are raised and lowered by means of the background metric, the vertical bar denotes a covariant derivative with respect to the background metric, and the gauge condition (1.14.6c) is assumed.

From the form (1.14.7) of the coarse-grained field equations, we can interpret $T^{(W)}{}_{ij}$ as the stress-energy tensor for the fine-grained waves h_{ij} in the coarse-grained background space. The field equations themselves guarantee that $T^{(W)}{}_{ij}$ satisfies the usual conservation equation for a stress-energy tensor

$$T^{(W)i}{}_{i|j} = 0 , \qquad (1.14.9)$$

where "|" is a covariant derivative in the background space. We must emphasize, however, that the "Isaacson stress-energy tensor" $T^{(W)}{}_{ij}$ is a tensor only in the background spacetime, not in the perturbed spacetime; and that it localizes the energy carried by gravitational waves only from the coarse-grained viewpoint—i.e., it does not say whether the energy is carried by the "crest" of the wave, by its "trough," by its "walls," or by what.

One can verify from formula (1.14.8) that the Isaacson tensor is equal to the spacetime average of the Landau-Lifshitz pseudotensor. (Note that this

---

25. The terms which couple h_{ij} to the background curvature affect the polarization of the waves but do not affect the fact that sharp wave fronts travel with the speed of light.

equality is learned only at the end of the analysis; it was not imposed in the derivation!) Consequently, all results derived with the Landau-Lifshitz pseudotensor (except the fine-grained localization of energy which was always known to be not unique) are formally justified by Isaacson's analysis!

A second way of studying the energy carried off by gravitational waves is to examine the damping produced in their source by the back reaction of the waves. In such an analysis one *does not* calculate the energy in the waves directly by means of the Isaacson tensor or any pseudotensor. Rather, one looks carefully at the influence of the gravitational field on the source, under three different circumstances: (i) when all gravitational waves far from the source are outgoing ("retarded potentials"); (ii) when all distant waves are incoming ("advanced potentials"); and (iii) when all distant waves are standing, i.e., half outgoing and half incoming ("half advanced plus half retarded"). By comparing these three cases one can delineate clearly the effects of the waves on the source. One always finds that for the standing-wave case there is no damping of the source's motion; for the outgoing-wave case there is damping (energy loss); and for the ingoing-wave case there is antidamping, which is quantitatively equal and opposite to the damping of the outgoing-wave case.

Such analyses are well known in electromagnetic theory (see Burke 1969, 1970 for an especially clear exposition). However, in general relativity they have been carried out to definitive completion only recently. In the remainder of this section we shall describe two such general-relativistic calculations.

The first of these is an analysis of nearly Newtonian, slowly moving systems by Burke (1969, 1970), by Chandrasekhar and Esposito (1970), and by Thorne (1969b). (See Burke and Thorne 1969 for a review.) In this analysis one expands the equations of general relativity in powers of the squared velocity of the matter in the source, $(v/c)^2$, which is of the same order of magnitude as the dimensionless Newtonian potential, ϕ/c^2, and as the ratio of pressure to mass-energy density;

$$(v/c)^2 \sim \phi/c^2 \sim p/\rho c^2 \sim 10^{-6} \text{ for the Sun .}$$

At the lowest order, $(v/c)^2$, in the expansion one obtains Newtonian theory. At the next higher order, $(v/c)^4$, one obtains post-Newtonian corrections to Newtonian theory (perihelion shift of Mercury; dragging of inertial frames; relativistic instability in supermassive stars). At the next order, $(v/c)^6$, one finds other, smaller post-Newtonian corrections. And, finally, at the next half-order, $(v/c)^7$, one finds the effects of radiation reaction.

One of the principal results of Burke and Thorne is that the effects of radiation reaction can be understood and calculated quite easily without reference to the post-Newtonian corrections of the intermediate orders [$(v/c)^4$ and $(v/c)^6$]. The key idea is independent of whether one is dealing with general relativity, with electromagnetic theory, or with a scalar-wave theory. Let us illustrate it in the scalar-wave case.

1.14 Gravitational Radiation Damping

Space splits up into a "near zone" ($r \ll \lambda$, where λ is a characteristic wavelength for the waves) and a "radiation zone" ($r \gg \lambda$). The source is assumed to lie entirely in the near zone. In the radiation zone the scalar waves (assumed, for definiteness, to be monochromatic and quadrupolar) will be purely outgoing for realistic problems

$$\psi_{\text{out}} \sim r^{-1}P_2(\cos \theta)e^{i\omega(t-r)}, \quad r \gg \lambda = 2\pi c/\omega . \quad (1.14.10\text{a})$$

However, it is instructive to consider also standing waves

$$\psi_{\text{stand}} \sim r^{-1}P_2(\cos \theta) \cos (\omega r)e^{i\omega t}, \quad r \gg \lambda ; \quad (1.14.10\text{b})$$

as well as ingoing waves

$$\psi_{\text{in}} \sim r^{-1}P_2(\cos \theta)e^{i\omega(t+r)}, \quad r \gg \lambda . \quad (1.14.10\text{c})$$

If the scalar-wave equation has the usual form, $\Box\psi = 0$, then the exact forms of these waves involve spherical Bessel functions in various combinations:

$$\psi_{\text{out}} \sim [n_2(\omega r) + ij_2(\omega r)]P_2(\cos \theta)e^{i\omega t}, \quad (1.14.11\text{a})$$

$$\psi_{\text{stand}} \sim n_2(\omega r) P_2(\cos \theta)e^{i\omega t}, \quad (1.14.11\text{b})$$

$$\psi_{\text{in}} \sim [n_2(\omega r) - ij_2(\omega r)]P_2(\cos \theta)e^{i\omega t} . \quad (1.14.11\text{c})$$

The field, whether it is outgoing, standing, or ingoing, is generated by the source in the near zone. Extrapolated back into the near zone, the fields (1.14.11) take the form

$$\psi_{\text{out}} \sim \{[r^{-3} + \ldots] + \tfrac{1}{3}i\omega^5[r^2 + \ldots]\} P_2(\cos \theta)e^{i\omega t}, \quad (1.14.11\text{a})$$

$$\psi_{\text{stand}} \sim [r^{-3} + \ldots]P_2(\cos \theta)e^{i\omega t}, \quad (1.14.11\text{b})$$

$$\psi_{\text{in}} \sim \{[r^{-3} + \ldots] - \tfrac{1}{3}i\omega^5[r^2 + \ldots]\} P_2(\cos \theta)e^{i\omega t} . \quad (1.14.11\text{c})$$

In each case the dominant ("Newtonian") part of the near-zone field, $\psi_{\text{(Newt)}} \sim r^{-3}P_2(\cos \theta)e^{i\omega t}$, is the same—and it is precisely the kind of near field which should be generated by an oscillating quadrupolar source. In the standing-wave case this dominant part exists alone. However, in the traveling-wave cases there is, in addition, a much smaller part, $\psi \sim i\omega^5 r^2 P_2(\cos \theta)e^{i\omega t}$, which comes from the Bessel function j_2, and which is out of phase with the dominant part (note the factor of i). It is this which leads to radiation reaction forces, so we shall label it $\psi_{\text{(react)}}$:

$$\psi_{\text{(react)}} \sim i\omega^5 r^2 P_2(\cos \theta)e^{i\omega t} . \quad (1.14.12)$$

Notice that inside the source, $\psi_{\text{(react)}}$ is smaller than the Newtonian part of the field, $\psi_{\text{(Newt)}}$, by a factor of $(\omega R)^5 \approx (v/c)^5$, where R is the size of the source and v is its typical internal velocity. The relative phases of $\psi_{\text{(react)}}$ and $\psi_{\text{(Newt)}}$ are such that, in the outgoing-wave case, $\psi_{\text{(react)}}$ produces internal forces in the source which sap energy from it. These are the "radiation reaction forces." For the ingoing-wave case the relative phases are shifted by π, so that $\psi_{\text{(react)}}$ feeds energy into the source.

Electromagnetic theory and general-relativity theory are completely analogous to this scalar-wave case.[26] In the electromagnetic case an analogous argument was used by Lorentz (1915) to calculate radiation reaction for an accelerated classical electron (see also Jackson 1962, § 17.3).

In the general-relativistic case the near-zone Newtonian metric joins onto a standing-wave metric in the radiation zone. If the waves are to be outgoing, the near-zone metric must contain also radiation-reaction terms, which are out of phase with the Newtonian terms, and are smaller by a factor $(v/c)^5$. These near-zone reaction terms, when extended into the radiation zone, combine with the Newtonian terms to produce an outgoing-wave metric. The form of the near-zone reaction terms is fully determined by the outgoing-wave condition.

Burke (1969, 1970), Thorne (1969b), and Chandrasekhar and Esposito (1970) have shown from the first principles of GTR that the effects of the reaction terms on the source, as predicted by GTR, can be described in the language of Newtonian gravity. Put more precisely, *one can incorporate the dominant, general-relativistic, radiation-reaction effects into the Newtonian theory of gravity by merely adding a small piece, $\phi_{(react)}$ to the usual Newtonian potential. The result is the following modified version of Newton's theory:*

Gravity is described by the usual Newtonian potential $\phi(x, t)$, which produces forces on bodies in the usual way

$$F = m\nabla\phi .$$
(1.14.13)

The gravitational potential satisfies the usual source equation

$$\nabla^2\phi = -4\pi G\rho ,$$
(1.14.14)

and as usual it must be nonsingular throughout the system. However, by contrast with the usual Newtonian theory, the boundary condition at $r = \infty$ is not $\phi(\infty) = 0$. Rather, at a particular moment of time t, when the quadrupole moment of the source is (in Cartesian coordinates)

$$K_{\alpha\beta} = \int \rho(3x^\alpha x^\beta - r^2\delta^\alpha_\beta) \, d \text{ volume} ,$$
(1.14.15)

the form of ϕ at large r must be

$$\phi \approx \phi_{(react)} \equiv - \frac{1}{15} \frac{G}{c^5} \left(\frac{d^5 K_{\alpha\beta}}{dt^5}\right) x^\alpha x^\beta .$$
(1.14.16)

26. The nonlinearity of GTR makes the gravitational case different from the scalar-wave case and the electromagnetic case; but those differences are not important for the present discussion. Among the most interesting differences is this: In the post-Newtonian expansion schemes there is a stress tensor

$$t_{\alpha\beta} = \frac{1}{16\pi G} \left[4 \frac{\partial\phi}{\partial x^\alpha} \frac{\partial\phi}{\partial x^\beta} - 2\delta_{\alpha\beta} \left(\frac{\partial\phi}{\partial x^\mu}\right)^2 \right]$$

associated with the Newtonian potential ϕ. In most gauges (e.g., those of Burke and of Chandrasekhar, but not that of Thorne) this gravitational stress generates roughly as much of the gravitational radiation as does the moving matter itself. By contrast, in electromagnetic theory all of the radiation is generated directly by the moving charges.

1.14 Gravitational Radiation Damping

Note that $\phi_{(\text{react})}$ satisfies $\nabla^2\phi_{(\text{react})} = 0$. As a consequence, *everywhere* in space (in the near zone!) the potential of this modified theory is related to that of the usual theory by

$$\phi = \phi_{(\text{usual})} + \phi_{(\text{react})}.$$

The term $m\nabla\phi_{(\text{react})}$ in the force law represents the radiation-reaction force on a particle of mass m. Notice that this force is caused by the behavior of the system as a whole, rather than by the particle's behavior alone! This force saps energy from the system at a time-averaged rate which is readily calculated from the above equations:

$$\left\langle \frac{-dE}{dt} \right\rangle = \left\langle -\int (\rho \nabla \phi_{(\text{react})}) \cdot v \; d \text{ volume} \right\rangle$$

$$= \tfrac{1}{45} \frac{G}{c^5} \sum_{\alpha,\beta} \left\langle \left(\frac{d^3 K_{\alpha\beta}}{dt^3} \right)^2 \right\rangle . \tag{1.14.17}$$

Notice that this is precisely the same rate as the rate at which the gravitational waves carry away energy (eq. [1.12.8]). This equality is significant since the analyses of Burke, Chrandrasekhar and Esposito, and Thorne make no reference to the energy in the waves.

Although our modified Newtonian theory gives, correct to accuracy GM/c^2R, the radiation damping demanded by general-relativity theory, it does *not* give the perihelion shift, the light deflection, or the other non-radiative corrections to Newtonian theory. Because these nonradiative corrections are much larger, instantaneously, than the reaction force, $m\nabla\phi_{(\text{react})}$, the *instantaneous reaction force of the modified Newtonian theory is physically meaningless*. However, its effects, integrated over a long period of time, will come to dominate over the other, neglected, energy-conserving, relativistic corrections. It is these long-term effects that are meaningful—and, indeed, may be crucial in some astrophysical contexts.

The above analysis is not valid for highly relativistic bodies such as neutron stars. However, for the fully relativistic case there is another type of analysis which reveals radiation damping in a clear-cut fashion (Thorne and Campolattaro 1967; Price and Thorne 1969; Thorne 1969a).

Thorne, Campolattaro, and Price (TCP) apply the well-known theory of resonances (as developed, e.g., in quantum mechanics and optics) to the interaction between gravitational waves and a relativistic star. One can distinguish three types of interaction: (i) the scattering of an incoming wave train by the star (a case of little astrophysical interest); (ii) the emission of gravitational waves by an isolated, pulsating star (a case of considerable astrophysical interest); and (iii) the undamped, sinusoidal oscillations of a star coupled to standing gravitational waves, when the whole system, star plus waves, is confined inside a *huge*, perfectly reflecting cavity (a case of absolutely no astrophysical interest).

The scattering problem, the emission problem, and the cavity problem are intimately related: For the scattering problem and the cavity problem there are certain resonant frequencies, ω_0, ω_1, ω_2, . . . , at which the interaction between the star and the waves is particularly strong. For frequencies near one of these, the interaction falls off in the familiar Breit-Wigner manner

$$\begin{pmatrix}\text{Scattering cross-section}\\\text{for scattering problem}\end{pmatrix} \sim \begin{pmatrix}\text{Energy in stellar pulsations di-}\\\text{vided by energy in one wave-}\\\text{length of standing waves for}\\\text{cavity problem}\end{pmatrix}$$

(1.14.18)

$$\sim \frac{1}{(\omega - \omega_n)^2 + (1/\tau_n)^2}.$$

TCP use the familiar techniques of quantum mechanics and optics to show that the resonant frequencies ω_n are the characteristic oscillation frequencies for the isolated, wave-emitting star; and that the half-width, $1/\tau_n$, of each resonance is the rate at which the pulsation amplitude of the isolated star decays, due to radiation reaction. By means of this connection between the three types of problems, they are actually able to derive analytically the radiation-damping formula

$$\text{Pulsation Amplitude} \sim e^{i\omega_n t - t/\tau_n}. \qquad (1.14.19)$$

This result is derived without any consideration whatsoever of the energy carried off by the emitted waves! The derivation focuses its attention, instead, on the direct coupling between the waves and the star. Only afterward, when numerical evaluations are made for particular stars, is it verified that the rate at which the pulsation energy is damped from the star agrees with the rate at which the waves carry away energy.

Thorne (1969a) has used these methods to calculate numerically the characteristic pulsation periods $T_n = 2\pi/\omega_n$ and damping times τ_n for realistic models of neutron stars. For the lowest quadrupole mode of massive ($0.8\,M_\odot \lesssim M \lesssim 2M_\odot$) neutron stars he finds $T_0 \sim 3 \times 10^{-4}$ sec and $\tau_0 \sim$ 0.3 sec. Notice how short this damping time is by astrophysical standards! These numerical results agree to within a factor of $\sim$3 with earlier estimates by Wheeler (1966b) and by Chau (1967), which were based on the linearized theory.

In the case of the radial pulsations of a neutron star one might expect no radiation to be emitted, since GTR does not admit monopole gravitational waves. However, any real star rotates and is deformed by centrifugal forces. These deformations couple the spherical modes of pulsation to the quadrupole modes, causing them to radiate. The predicted radiation-damping time is

$$\tau_{\text{spherical}} \sim e^{-4}\tau_{\text{quadrupole}} \sim (\Omega^2 R^3/GM)^2 \tau_{\text{quadrupole}} \qquad (1.14.20)$$

according to Wheeler (1966*b*). (Here *e* is the star's eccentricity and Ω is its angular velocity.) For the kinds of neutron stars which power the observed pulsars, $e \sim \frac{1}{30}$ to 10^{-3}, so

$$\tau_{\text{spherical}} \sim \text{one week to } 10^4 \text{ years} .$$

Other damping mechanisms are probably more important than this.

1.15 THE DETECTION OF GRAVITATIONAL WAVES (BY K. S. THORNE)

Very recently Weber (1969, 1970) has reported detecting bursts of gravitational radiation from cosmic sources. If Weber's interpretation of his observations is correct, it will have an enormous impact on relativistic astrophysics.

To understand what Weber sees, one must understand his detectors. Each detector is a large, solid, aluminum cylinder, about 2 meters in length and weighing about 1 ton. The cylinder is suspended by wires in vacuum and is mechanically decoupled from its surroundings. Around its middle are attached piezoelectric strain transducers, which couple into electronic circuits that are sensitive to the frequency of the cylinder's fundamental, end-to-end mode of oscillation. The frequency of that mode is

$$\nu_0 = 1660 \text{ Hz} , \tag{1.15.1}$$

and its Q (the number of radians of oscillation required for internal damping to reduce its oscillation energy by a factor of e) is

$$Q \approx 10^5 . \tag{1.15.2}$$

Thus, if the cylinder is struck suddenly by a hammer (or by a burst of gravitational waves), it will "ring" with a period of $T_0 = 1/\nu_0 \approx 6 \times 10^{-4}$ seconds, and with a decay time of $\tau_0 = QT_0/2\pi \approx 10$ seconds.

It is precisely this type of sudden excitation which Weber observes, with no reasonable physical cause except a hypothesized burst of gravitational waves. Moreover, he observes this sudden ringing not in just one cylinder, but simultaneously in two cylinders—one near Chicago, Illinois, and the other near Washington, D.C.; a separation of 1000 kilometers! The precision with which he measures the onset of ringing is 0.44 seconds; and to within this precision the onset is simultaneous.

What are the characteristics of the gravitational waves which could produce this ringing? Clearly, they must have sizable energy at frequencies near $\nu_0 = 1660$ Hz. Moreover, they must come in a burst of duration less than $\tau_0 \approx 10$ seconds, since the observed ringing decays with the natural decay time of the cylinder.

How much energy must the waves deposit in each cylinder? This varies from burst to burst. For the "typical" events seen by Weber—which occur about once each day—the total energy deposited in the fundamental mode of each cylinder is

$$E_{\text{deposited}} \approx kT \approx 4 \times 10^{-14} \text{ erg} . \qquad (1.15.3)$$

Here T is room temperature ($300°$ K); and the figure kT means that the energy in the two cylinders rises simultaneously and suddenly by an amount equal to the mean thermal noise. Such correlations should happen by chance due to random noise less than once each week. Notice that this excitation corresponds to an end-to-end amplitude of oscillation for the cylinder, δL, of

$$\tfrac{1}{2}(10^6 g)(2\pi \times 1.7 \times 10^3/\text{sec})^2 \delta L^2 \approx E_{\text{deposited}} ,$$

$$\delta L \approx 3 \times 10^{-14} \text{ cm} . \qquad (1.15.4)$$

Of course, Weber cannot measure such amplitudes directly; but he *can* and *does* measure the strains that these amplitudes produce in the cylinders!

How much energy must the waves carry with them in order to deposit 4×10^{-14} erg in Weber's cylinders? To calculate this, it is helpful to know that the cross-section of any mechanical oscillator for absorbing gravitational waves *in its bandwidth*, $\Delta\nu_0 \approx \nu_0/Q$, is (see, e.g., Weber 1961, eq. [8.28])

$$\sigma \approx \frac{15\pi^2}{8} \frac{r_g}{\lambda_0} QL^2 . \qquad (1.15.5)$$

Here r_g is the gravitational radius of the oscillator, λ_0 is the wavelength of the radiation to which the oscillator is resonant, L is the oscillator's length, and Q is its "quality factor" (see above). For Weber's cylinders this cross-section is

$$\sigma \approx 3 \times 10^{-19} \text{ cm}^2 . \qquad (1.15.6)$$

It is so small, of course, because the detector's gravitational radius is so small ($r_g \approx 10^{-22}$ cm; $\lambda_0 \approx 10^7$ cm; $r_g/\lambda_0 \approx 10^{-29}$)!

Actually, the above cross-section is relevant for a situation in which a detector at zero temperature is in a steady-state interaction with a long wave train (time of passage $\gg$ detector damping time). Weber's events are very different from this: the temperature and associated noise in the detector are high, and the wave train is short. In this case the theory of the interaction between waves and detector is much more complicated than in the steady-state case (see Braginsky, Zel'dovich, and Rudenko 1970 for detailed discussion). For example, if the burst of waves hits the detector in phase with the thermal oscillations of its fundamental mode, a rather large amount of energy will be deposited:

$$\delta \text{ Energy} \propto \delta(\text{Amplitude}^2)$$

$$\approx 2 \times (\text{Thermal Amplitude}) \times \delta \text{ Amplitude} .$$

Clearly, the larger the thermal amplitude at the moment the burst hits, the greater the energy deposited! However, if the burst hits out of phase with the thermal oscillations, it deposits little energy—and it may even extract energy from the detector!

From studies of the detector response, Weber (private communication, March 1970) believes that he detects only about 10 percent of all bursts of that type which produce about one detector coincidence per day. The other ~90 percent either arrive sufficiently out of phase with one or both detectors or arrive when the thermal amplitude is momentarily so low that they cannot produce significant coincidences.

For those ~10 percent which do produce coincidence events, Weber estimates a cross-section ~20 times higher than equation (1.15.6).

Combining the deposited energy with this cross-section, we see that the total energy per unit area in the detector's bandwidth, which must flow past the Earth in each burst, is

$$(dE/dA)_{\text{at } \nu_0 \text{ in } \Delta\nu_0} \approx 7 \times 10^3 \text{ ergs cm}^{-2} . \tag{1.15.7}$$

Hence, if the source is a distance l_{pc} (measured in parsecs) from Earth, the total energy it emits between ν_0 and $\nu_0 + \Delta\nu_0$ is

$$E_{[\text{emitted at } \nu_0 \text{ in } \Delta\nu_0 = \nu_0/Q = 10^{-5}\nu_0]} \approx 8 \times 10^{41} \, (l_{\text{pc}})^2 \text{ ergs} . \tag{1.15.8}$$

If the bandwidth of the radiation is of the order of ν_0 (a reasonable assumption) and the source is in the central regions of the Galaxy ($l_{\text{pc}} \approx 8.6 \times 10^3$) as Weber's (1970) data suggest, then

$$E_{\text{emitted total}} \approx 5 \times 10^{54} \text{ ergs} \approx 2 \, M_\odot c^2 . \tag{1.15.9}$$

Moreover, bursts with these characteristics occur about ten times per day, according to Weber's estimates. This corresponds to a total rate of mass loss from the central regions of the Galaxy of

$$-\frac{dE}{dt} \approx \frac{6000 \, M_\odot}{\text{year}} \approx \frac{\text{Mass of Galaxy}}{2 \times 10^7 \text{ years}} . \tag{1.15.10}$$

Although these numbers might be incorrect by an order of magnitude, and although there is at least that much uncertainty in our guess at the bandwidth of the radiation, it is clear that theorists will have great difficulty explaining these observations!

APPENDIX TO §1.10

We here calculate the mixed spacetime components of the metric tensor inside a hollow, uniformly rotating sphere. These components are calculated by using formula (1.10.9):

$$h_{0a} = - \frac{\kappa}{2\pi} \int \frac{\rho v^a}{cr} \, dV .$$

(1.10A.1)

Here v^a are the components of the velocity of the sphere in a Cartesian coordinate system, and r is the distance from the volume element dV to the given point. We convert the integral to spherical coordinates. The velocity components become

$$v_x = v^1 = -\omega y = -R\omega \sin \phi \sin \theta ,$$

$$v_y = v^2 = \omega x = R\omega \cos \phi \sin \theta ,$$

$$v_z = v^3 = 0 .$$

Here ω is the angular velocity of rotation, and R is the radius of the sphere. Obviously the component h_{03} vanishes. The quantity r in equation (1.10A.1), when expressed in spherical coordinates, is $r = R(1 + \eta^2 - 2\eta \cos a)^{1/2}$, where $\eta = r'/R$ is the radial coordinate of the point at which we calculate h_{0a}, and where a is the angle between the vector from the center of the sphere to this point and the vector from the center of the sphere to the volume element dV on the sphere. Replacing ρdV by $(M/4\pi) \sin \theta d\phi d\theta$, where M is the mass of the sphere and $M/4\pi$ is the mass per unit steradian on the sphere's surface, we obtain for h_{01} and h_{02}

$$\begin{Bmatrix} h_{01} \\ h_{02} \end{Bmatrix} = - \int \frac{\kappa M \omega}{8\pi^2 c} \frac{\sin^2 \theta \begin{Bmatrix} -\sin \phi \\ \cos \phi \end{Bmatrix}}{(1 + \eta^2 - 2\eta \cos a)^{1/2}} \, d\phi d\theta .$$

(1.10A.2)

We rewrite the denominator of the integrand by first noting that

$$\cos a = \cos \theta \cos \theta_0 + \sin \theta \sin \theta_0 \cos (\phi - \phi_0) ,$$

where θ_0, ϕ_0 and θ, ϕ are the angular coordinates of the field point and of the integration element, respectively. Using this expression, we can write

$$\frac{1}{(1 + \eta^2 - 2\eta \cos a)^{1/2}} = \sum_{n=0}^{\infty} \eta^n P_n(\cos a)$$

$$= \sum_{n=0}^{\infty} \eta^n \Bigg\{ P_n(\cos \theta) P_n(\cos \theta_0)$$

(1.10A.3)

$$+ 2 \sum_{m=1}^{n} \frac{(n - m)!}{(n + m)!} P^m_n(\cos \theta) P^m_n(\cos \theta_0) \cos [m(\phi - \phi_0)] \Bigg\} .$$

Here $P_n(x)$ is the Legendre polynomial and $P^m{}_n(x)$ is the associated Legendre polynomial. For the first term in the braces of equation (1.10A.3), upon multiplying by $\sin \phi$ or $\cos \phi$ and integrating ϕ between 0 and π (cf. expression [1.10A.2]), we obtain zero. The integrals of the remaining terms yield, with the help of

$$\int_0^{2\pi} \sin \phi \cos [m(\phi - \phi_0)]d\phi = \int_0^{2\pi} \tfrac{1}{2}\{\sin [(m + 1)\phi - m\phi_0]$$

$$+ \sin [\phi(1 - m) + m\phi_0]\} \, d\phi = \pi \sin \phi_0 \delta^1{}_m ,$$

$$\int_0^{2\pi} \cos \phi \cos [m(\phi - \phi_0)]d\phi = \int_0^{2\pi} \tfrac{1}{2}\{\cos [(m + 1)\phi - m\phi_0]$$

$$+ \cos [(1 - m)\phi + m\phi_0]\} \, d\phi = \pi \cos \phi_0 \delta^1{}_m ,$$

where $\delta^1{}_m$ is the Kronecker delta symbol,

$$\begin{Bmatrix} h_{01} \\ h_{02} \end{Bmatrix} = \left(\frac{\kappa M \omega}{8\pi c}\right) 2 \int_{-1}^{1} P^1{}_1(x) \sum_{n=0}^{\infty} \frac{(n - 1)!}{(n + 1)!} \, \eta^n P^1{}_n(x) P^1{}_n(\cos \theta_0) \begin{Bmatrix} -\sin \phi_0 \\ \cos \phi_0 \end{Bmatrix} dx$$

$$= \left(\frac{\kappa M \omega}{6\pi c}\right) \eta P^1{}_1(\cos \theta_0) \begin{Bmatrix} -\sin \phi_0 \\ \cos \phi_0 \end{Bmatrix} = -\left(\frac{4GM\omega}{3c^2 R}\right) r' \sin \theta_0 \begin{Bmatrix} -\sin \phi_0 \\ \cos \phi_0 \end{Bmatrix} .$$

This is the expression used in the text (with a small change of notation).

2 INESCAPABILITY OF THE GENERAL THEORY OF RELATIVITY (GTR) AND PROBLEMS IN THE THEORY OF GRAVITATION

2.1 INTRODUCTION

Before turning to processes in strong gravitational fields, let us consider some problems of principle. For coherence of exposition, the conclusions of the previous sections will be restated briefly.

GTR is a theory of gravitation; it describes gravitation as the action of masses upon the properties of space and time; in turn, these properties of space and time affect the motion of bodies and other physical processes. In this respect, the theory of gravitation differs considerably from the theories of other types of interactions, e.g., electromagnetic forces, nuclear forces, and others.

The electromagnetic interaction has been studied in the greatest detail of all. The resemblance between the Coulomb force and Newton's gravitational force is very impressive. Also impressive is the similarity between a planetary system and the electronic structure of an atom, which similarity is due to the resemblance between their laws of interaction.

GTR has an extreme inner beauty and elegance; the construction of GTR required the introduction of only one constant—the constant of gravitation. It has been noted frequently that there exists a marked contrast between the power of the theory and its small number of experimental tests. One should not forget, however, that GTR produces the following: (1) Newton's law of gravitation, itself; (2) a foundation that enables one to apply Newton's law to the interaction of bodies surrounded by infinitely expanding matter; and (3) Friedmann's nonstationary cosmological model, including a prediction of the Hubble redshift for the spectra of distant objects.

Only secondarily do we cite the three famous tests of GTR—the precession of Mercury's perihelion, the deflection of a light beam passing near the Sun, and the variation of the frequency of light in a gravitational field.

2.2 UNIFIED FIELD THEORY, GEOMETRODYNAMICS, AND THE FUNDAMENTAL MASS AND LENGTH

After the creation of GTR, there followed attempts to reformulate electromagnetic theory in a similar way, attempts to construct a geometric theory of the electromagnetic field, and attempts to create a unified field theory which would combine gravitation and electromagnetism. All these attempts failed. The gravitational field acts universally; it imparts equal accelerations to all objects. This universality permits one to describe gravity by a change in the properties of the spacetime through which the objects move. The electromagnetic field does not have such a universality; various bodies and particles have different ratios of charge to mass and experience different accelerations. The electromagnetic field itself creates a gravitational field which is proportional to the square of the field strength. Roughly speaking, the electromagnetic field has energy; and this energy has weight, like any other energy.

It turns out that the equations of GTR for a spacetime which contains an electromagnetic field necessarily force the field to satisfy Maxwell's equations. This result can be compared with the well-known fact that for "conventional" material bodies (point masses or solids), the equations of GTR yield not only a description of the gravitational field but also the equations of motion of these bodies; i.e., they include the equations of Newtonian mechanics. Maxwell's equations are the analogous equations of motion for the electromagnetic field (see Rainich 1924, 1925; Misner and Wheeler 1957).

Let us note another important example of the intrusion of GTR into the theory of other nongravitational fields: from the Lagrangian of a field in curved spacetime, one can obtain the expression for the energy-momentum tensor of the field by varying the metric of spacetime; and it is obtained in an explicitly symmetric form ($T_{ik} = T_{ki}$).

The idea of defining all fields by the spacetime curvatures that they create is known as *geometrodynamics*. Its most articulate exponent is the American physicist John Archibald Wheeler.

In recent years the attraction of this idea in its simplest form has faded somewhat, primarily because of the discovery of new fields and the realization of their diversity. In conventional quantum theory[1] each type of particle is described by a corresponding field: photons, by the electromagnetic field; π- and K-mesons, by fields of the nuclear forces; and similarly for the two types of neutrinos and antineutrinos, etc.

It is clear that with such diversity of fields and particles, a specification of the curvature of spacetime is not sufficient to describe all fields and all particles that fill spacetime. Figuratively speaking, the world is multicolored; and the gravitational field produces only a black and white photograph of it, which cannot do justice to the stormy colors of nature.

Fundamental questions of science are not answered by a majority vote. Nevertheless, the reader must know that the overwhelming majority

---

1. *Editors' note.*—In more recent formulations of quantum field theories the correspondence between number of particles and number of fields is denied and some workers even assume a single "self-interacting" field will describe all particles (including gravitons).

of physicists now consider the program of a unified field theory to be un-realizable in the Einsteinian sense—they feel it is impossible to obtain all the laws of nature from the equations of GTR alone. However, there is a very interesting remark by Wheeler that the origin of the properties of elementary particles and their wide variety may be associated with the wide variety of topological properties of spacetime at small scales.

If, to the constants which are included in GTR (the speed of light, $c = 3 \times 10^{10}$ cm sec^{-1}; and the gravitational constant, $G = 6.67 \times 10^{-8}$ cm^3 sec^{-2} g^{-1}), we also add Planck's constant, $\hbar = 1.05 \times 10^{-27}$ g cm^2 sec^{-1}, then we get a set from which we can construct constants with the dimensions of length $l_g = 1.7 \times 10^{-33}$ cm, of time $t_g = 6 \times 10^{-44}$ sec, and of mass $M = \hbar^{1/2}c^{1/2}G^{-1/2} = 2 \times 10^{-5}$ g. This characteristic mass is enormously larger than the masses of the elementary particles, for example, $m_e = 0.9 \times 10^{-27}$ g, $m_p = 1.6 \times 10^{-24}$ g.

From this, most physicists conclude that the theory of elementary par-ticles, and particularly the theory of their masses, is unrelated to the theory of gravitation and rests on a foundation that is entirely different (and as yet unknown).[2]

2.3   A FLAT-SPACE THEORY OF GRAVITY[3]

Disenchantment with unified field theories has given birth to a new ap-proach to the theory of gravity. This approach can be explained in the following manner. Unquestionably, GTR is a satisfactory theory of gravi-tation. However, is GTR the only possible, the "inescapable," theory of gravitation? Is the concept of spacetime curvature necessary? (At last, we approach the problem that appeared in the title of chapter 2.) Other field theories can be developed within the framework of flat spacetime and New-ton's law is very similar to that of Coulomb. Can't gravitation also be con-sidered as a special field which acts in flat spacetime? What must be the properties of such a field?

It is very instructive to deduce the fundamental properties of such a field from a simple thought experiment. Consider a particle at rest, which creates in the surrounding space a potential $\phi = a/r$. The potential of a particle moving at constant velocity can be obtained from the potential of a particle at rest by performing a Lorentz transformation, i.e., by considering the same field, $\phi = a/r$, from the viewpoint of a comoving observer. Now con-

---

2. Markov (1966) and some other physicists hold that it is gravitation after all which is the underlying foundation of the theory of elementary particles, and that the masses of the particles will eventually be obtained as the value $\hbar^{1/2}c^{1/2}G^{-1/2}$, multiplied by a dimen-sionless coefficient ($\sim 10^{-20}$) which will be derived from the theory. They also suggest that there might be other particles with the fundamental mass $m = \hbar^{1/2}c^{1/2}G^{-1/2}$.

3. Here and below, of course, we work in Minkowski space; the special theory of rela-tivity is neither questioned nor modified.

sider not one particle, but an aggregate of particles contained in a spherical vessel and moving in all possible directions. Assume that all directions are equally probable, so that this aggregate of particles forms a stationary, spherically symmetric system bounded by the walls of the vessel. Let us calculate the total potential of the particles. The resultant total potential depends on the tensor character of ϕ, i.e., on the behavior of ϕ under Lorentz transformations.

If ϕ is a scalar, then it turns out that the field of the aggregate of n moving particles is

$$\phi = n\phi_1(1 - v^2/c^2)^{1/2} ;$$

i.e., ϕ is smaller than for particles at rest. (For calculations, see the appendix to this section at the end of chapter 2.) If ϕ is the zero component of a four-vector, then $\phi = n\phi_1$, so the motion of the particles influences neither their potential nor the fields that they create. Finally, if ϕ is the $(0, 0)$-component[4] of a second-rank tensor ϕ_{ik}, then we obtain

$$\phi = n\phi_1/(1 - v^2/c^2)^{1/2} .$$

(This result can be obtained most simply by considering a nonstationary, spherically symmetric system of particles which all fly radially outward from the origin of coordinates with the same speed.)

It is the second case which corresponds to electromagnetic theory. It is well known that the electrostatic potential ϕ is the zero component of a four-vector $A_\mu = (\phi, A)$; the electric and magnetic fields together form a second-rank tensor $F_{ik} = -F_{ki}$. The electrons in an atom move at high speeds (in heavy atoms these speeds are of the order of the speed of light); however, this does not change their observed charge and does not disrupt the precise balance between the charge of the nucleus and the charge of its electrons.

For the gravitational field it is obviously necessary to select the third variant; i.e., ϕ must be the $(0, 0)$-component of a second-rank tensor. In this case (and in this case only) the gravitational force due to the moving particles, which acts on a test particle at rest, is proportional to the sum of their masses. The fact that the mass of a moving particle increases with increasing velocity, $m = m_0(1 - v^2/c^2)^{-1/2}$, in accordance with the special theory of relativity, has been taken into account.

These results can be stated in yet a different manner. The principle of equivalence requires that all bodies experience the same acceleration in a gravitational field, i.e., that the force acting upon each body be proportional to its mass. Newton's third law, the equality of action and reaction, then requires that the gravitational field produced by a body also be proportional to its mass. The motions of the particles which make up a body increase its mass. Consequently, the motions of the particles must also increase the gravi-

---

4. Under purely spatial rotations, the potential ϕ in all three cases transforms as a three-scalar, and the field $\nabla\phi$ transforms as a three-vector (a vector field of forces).

tational field which they create, by contrast with their electric field which is independent of their speed. Thus, the theory of the gravitational field must differ from the theory of the electromagnetic field (despite their similarity in the static case; cf. the above discussion of the similarity between Coulomb's law and Newton's law).

The gravitational potential in the flat-space theory is a second-rank tensor. Accordingly, the gravitational field has three indices, $\Gamma^i{}_{kl}$. (In the limit of a test particle nearly at rest, only the components $\Gamma^\alpha{}_{00}$ contribute to the force. It is these which we usually identify with the gradient of the Newtonian scalar potential.)

On the basis of this gravitational potential and field, the relativistic theory of gravitation in flat space (RTGFS) can be formulated. Indeed, such crucial observed phenomena as the deflection of a light beam in a gravitational field and the redshift of photons emitted by an oscillator (atom) near a heavy body can be correctly predicted by RTGFS, just as they are predicted by GTR. In other words, the relativistic theory of gravitation in flat space leads to predictions which go beyond the framework of the Newtonian theory of gravitation. Let us reemphasize that the new predictions of RTGFS, as described above, agree fully with the predictions of GTR, because a weak field in GTR can be regarded as a tensor in flat space. It is impossible to choose between RTGFS and GTR on the basis of an investigation of gravimagnetic effects or gravitational waves. Consequently, RTGFS must be considered seriously. There exists an area in which RTGFS, methodologically speaking, is enormously more convenient than GTR: this is the problem of the quantization of the gravitational field, which we shall consider briefly below (see § 2.6).

2.4 Necessity of the Concept of Spacetime Curvature

RTGFS can be a useful substitute for GTR in some problems. But what are the limits of its usefulness? A very explicit and instructive analysis of this question has been given by W. Thirring (1961).

As we have mentioned, RTGFS describes in a quantitatively correct manner the phenomena of the deflection of a light beam in a gravitational field and the redshift of photons emitted near a heavy body.

The first effect is obtained from RTGFS in that in this theory the vacuum Maxwell equations are changed by gravity. This change is embodied in an index of refraction, which depends on the gravitational potential, and which results in a curvature of the light beam. But consider the cost at which the agreement with experiment is obtained: The rate of propagation of electromagnetic waves—the speed of light—no longer equals the fundamental constant $c!$

Consider the second effect, the redshift of spectral lines.

According to RTGFS, in the field of a heavy body there exists a coordinate system in which the body is at rest and in which time flows at the same rate everywhere. The time interval between two events on the surface of the body is identically equal to the time interval between the receipt of two light signals from the events at a distant point in space.[5] The redshift is explained by the fact that the gravitational potential has a real retarding effect upon the vibrations of the oscillator.

Time passes near the body at the same rate as it does near infinity; but the oscillator oscillates more slowly than it would at infinity. Moreover, all other processes are also retarded. No local measurement of the oscillator's frequency can reveal the change in it because a clock located nearby, in the same gravitational potential, is slowed down by the same amount.

Particularly disturbing is the fact that it is not the field which influences the processes; rather, it is the potential, a quantity that is locally unmeasurable.

Change the potential by a constant value, and all observable relationships between the number of vibrations of the oscillator and the number of the oscillations of a pendulum clock, etc., will remain unchanged, even though the frequencies measured in terms of "absolute" time[6] will be different.

Imagine the following experiment: Surround the Earth by a heavy hollow sphere. The potential on the Earth's surface will change, even though all fields remain unchanged. The results of terrestrial experiments will remain unchanged, but in the framework of RTGFS all processes will be slower with respect to absolute time since the absolute timekeeper, by definition, is at infinity.

Thus absolute time in flat space exists theoretically, but it cannot be measured by any experiments performed near the surface of a body. Clearly, this situation resembles somewhat the situation with the ether at the beginning of our century; and it suggests that absolute time, after all, is fictitious. In the cosmological problem, where we are surrounded by infinite space with a constant average density of the matter, absolute time cannot even be defined.

Thirring considers this problem more formally: he demonstrates that there exists a transformation of the potential[7] which leaves all observable quantities unchanged, but which changes the rate of flow of time and the

5. Whatever influence the gravitational field may have upon the propagation of each signal, the propagation time for the two signals in a constant field is the same, and cancels out in the computation of the time interval.

6. The concept of "absolute" time is used here in a sense independent of the gravitational field. This time changes, in passing to a moving reference frame, in accordance with the Lorentz transformation, and in this sense it differs from the Newtonian absolute time.

7. This transformation is similar to a change of gauge in electrodynamics, in which one picks an arbitrary function f and adds df/dt and ∇f to ϕ and A, the scalar and vector potentials. Recall that in RTGFS the potential is a tensor; i.e., it has ten components rather than four as in the electromagnetic case.

rates of clocks as expressed in terms of "absolute" time. In other words, absolute time is not invariant with respect to this transformation; and the change in absolute time can be different at different points.

Thus, a fully developed RTGFS which agrees with GTR in the first corrections to Newtonian theory, in order to explain the universality of the action of gravity, is forced to employ the unphysical hypotheses of an unobservable "absolute" time, and of the influence of unobservable quantities —e.g., the gravitational potential—upon all physical processes.

The concept of spacetime curvature, upon which GTR is based, immediately resolves all difficulties of RTGFS and explains with brilliant simplicity the universality of the action of the gravitational field.

In this light it is clear that, despite the paucity of experimental tests, only GTR, which is endowed with extraordinary elegance, inner beauty, and persuasiveness, can be accepted as the contemporary theory of gravitation.

The specific field equations of GTR, which describe how matter produces the spacetime curvature are, of course, not uniquely forced upon us from among all possible such equations. It would be possible to choose, for example, equations of a higher order than second. The problem of the uniqueness of the equations of GTR in this sense has been debated frequently, starting with the work of Einstein himself. An explicit analysis of the problem can be found in Eddington's work (1925). More recent discussions of the basic assumptions that underlie Einstein's theory are contained in Trautman, Pirani, and Bondi (1965); Ehlers (1965); Hawking (1966*d*); and Trautman (1966). In a certain sense, the equations chosen for GTR are the simplest ones possible. Any modification of the field equations of GTR should be made only on the basis of new and penetrating theoretical and experimental considerations. One modification, the so-called Λ-term, will be discussed in Vol. II (see also § 1.9).

RTGFS can be used as a convenient approximation to GTR only in the case of isolated bodies and weak fields in an infinite, empty space. We have already discussed the inapplicability of RTGFS to the cosmological problem. RTGFS is similarly inapplicable to the problem of a collapsing star near its Schwarzschild radius (see chapter 3).

The truth is that a consistent RTGFS must be a nonlinear theory. This is apparent from the fact that the field far from a two-body system is smaller than the sum of the fields of each body separately, because of the gravitational mass defect. The interaction of the bodies decreases the energy and hence the mass of the system. The nonlinearity can be traced also in gravitational waves, but we shall not pursue this subject here.

In the domain where the nonlinearity is essential, RTGFS loses all its practical convenience and becomes unworkable. Nobody has ever attempted to solve, in terms of RTGFS, the problem of the field of a body near its gravitational (Schwarzschild) radius (cf. § 3.2).

For these reasons, we will not use RTGFS in the rest of this book. The principal advantages and the inescapability of the description of gravitation in terms of spacetime curvature have been demonstrated above.

2.5 ON THE POSSIBILITY OF CALCULATING THE GRAVITATION CONSTANT FROM ELEMENTARY-PARTICLE THEORY

As in Newtonian theory, so also in general relativity the gravitation constant G is considered a universal constant to be determined by experiment. Neither general relativity nor Newtonian theory attempts to express G in terms of other more simple quantities.

However, such an attempt has been made by Sakharov (1967*b*). This attempt is described below. So far this attempt has not led to any concrete results. Sakharov's formula for G contains another unknown quantity. Nonetheless, the novelty of Sakharov's approach to this deep problem of principle by itself justifies our discussing it.

In Newtonian theory G characterizes the force of interaction between particles; a typical quantity is the interaction energy $-Gm_p^2/r$ for two particles, which (for protons) is 10^{37} times less than the electrostatic energy, e^2/r. The relative strengths of the gravitational and electrostatic interactions are characterized by the dimensionless quantities $Gm_p^2/e^2 = 10^{-37}$ and $Gm_p^2/\hbar c = 10^{-39}$.

Fundamental theory should answer the question of why this dimensionless number is so far different from unity. Some theorists attribute it to the influence of the entire Universe on local phenomena; roughly speaking, they say that the smallness of this number is due to the immenseness of the Universe. Such an approach (our attitude toward it is negative!) will be considered later in the light of cosmology. In contrast, Sakharov's starting point is the typical GTR viewpoint, which connects gravity with the concept of spacetime curvature.

The essence of GTR is contained in the expression for the action

$$S = -\Sigma mc \int ds - (c^3/16\pi G) \int R dV , \qquad (2.5.1)$$

where V is the four-dimensional volume. The first term in this equation is a sum over the trajectories of all particles. By varying the trajectories of the particles in a space of fixed metric, we obtain the law of particle motion from the extremal condition on S. In curved space the action associated with a trajectory depends on the curvature, so that the first term in equation (2.5.1) takes into account the influence of the gravitational field on the particles' motion.

By varying the metric in the expression for S we obtain the gravitational field equations. Symbolically,

$$\frac{\delta S}{\delta g^{ik}} = \frac{1}{2c} T_{ik} - \frac{c^3}{16\pi G} (R_{ik} - \tfrac{1}{2} g_{ik} R) = 0 . \qquad (2.5.2)$$

71

The first term in this expression comes from the first term in equation (2.5.1); the second arises from the curvature integral, i.e., from the second term in equation (2.5.1). Roughly speaking, the first term describes the force with which the particles curve space. Such a description is in accordance with the principle of the equality of action and reaction; there is a reciprocity between the action of the curvature on the motion of the particles and the action of the particles on the curvature. This reciprocity manifests itself in the fact that both effects are obtained from one expression, $-mc \int ds$, where ds is calculated in Riemannian spacetime.

The terms containing R and R_{ik} in formulae (2.5.1) and (2.5.2) can be interpreted as describing the elasticity of space—i.e., the "attempt" of space to remain flat, the resistance of space to being curved.

The constant $c^3/(16\pi G)$ associated with the elasticity of the vacuum is the quantity which we wish to calculate. This general-relativistic constant is huge in magnitude (recall that in Newtonian theory we dealt with the inverse of it, which was very small). So that we may deal with a dimensionless quantity, let us think about this elasticity of the vacuum in the following manner: The mass m of an elementary particle spread over its Compton wavelength, h/mc, creates a very small curvature of space because the inelasticity of the vacuum, which resists the curvature, is so large. When we say that the curvature produced is very small, we mean very small when measured in units of $l^{-2} = (h/m_p c)^{-2}$.

Let us remind ourselves that here and in the discussion that follows, the curvature of space and general relativity are considered from a nonquantum, classical, deterministic viewpoint; whereas the motion of elementary particles through this classical curved space is governed by quantum-mechanical laws.

We have not yet overstepped the bounds of conventional general-relativity theory; at most, we have only added decorative words to it.

Now we turn to the goal of Sakharov's discussion, the attempt to calculate the "elasticity" of the vacuum, $c^3/(16\pi G)$.

In connection with the theory of the cosmological constant (cf. § 1.9), we raised the question of whether quantum fluctuations in the vacuum can give to the vacuum a certain energy density and pressure. Sakharov raises the question of how the properties of the vacuum will be changed as a result of a change in the curvature of space. The dependence of the vacuum's energy on the curvature would determine its elasticity, just as the dependence of the energy of a solid on its deformation characterizes its elasticity.

We must emphasize that the connection between the cosmological constant and the vacuum elasticity need not entail any numerical connection or proportionality between them. It is quite conceivable that there is an exact compensation between the contributions of various fields to the vacuum energy (and that, therefore, $\Lambda = 0$); but that, nevertheless, the elasticity is large. The function might equal zero, but its derivative might be nonzero.

The vacuum elasticity depends on the effects of the curvature of space on the quantum-mechanical motion of particles. One can say in this sense that the goal is to obtain the second term in equation (2.5.1) from the curved-space laws of motion, i.e., from the first term in equation (2.5.1), as written out for quantized particles. Recall that we are concerned here with virtual particles which are created alone (e.g., photons) or in pairs (e^+e^-), and with the vacuum—i.e., with space in which there are no real particles.

It follows from the general considerations of relativistic invariance and from dimensional considerations that the correction to the action must depend on the invariant R and must diverge. This means that a finite answer can be obtained only if we suppose that quantum theory breaks down above a certain cutoff momentum p_0.

The correction to the action is $(kp_0{}^2/\hbar)\int R dV$, where k is a dimensionless numerical factor of the order of unity.[8] It is assumed that quantum effects alone completely determine the vacuum elasticity. Put the other way around, Sakharov presumes that the gravitation constant, which we usually find from experiment, can be calculated, at least in principle, from the condition

$$\frac{c^3}{16\pi G} \int R dV \equiv \frac{kp_0{}^2}{\hbar} \int R dV ; \quad G = \frac{k' c^3 L^2}{\hbar} \text{ where } L = \frac{\hbar}{p_0}. \quad (2.5.3)$$

The details of the calculations can be found in the original work of Sakharov. In order to obtain the observed value of the elasticity, one must put the cutoff momentum equal to an enormously large value—a value corresponding to the mass 10^{-5} g. Since we do not know any elementary particle with such a mass, our theory for calculating one quantity, $c^3/(16\pi G)$, containing the unknown G depends upon another unknown quantity, p_0.

The value of p_0 and the corresponding mass 10^{-5} g and length 10^{-33} cm have been discussed for many decades; but now we view them in a logically different context. It has long been conjectured that gravitation should acquire peculiar properties on a length scale of 10^{-33} cm and for a point mass equal to 10^{-5} g. Our new point of view is that the mass of 10^{-5} g or length of 10^{-33} cm are given as something primary, and are the cause of the observed magnitude of the vacuum elasticity, i.e., the observed magnitude of the gravitation constant.

Thus, the formula remains the same as it always was:

$$M^2 c^2/\hbar = c^3/G. \quad (2.5.4)$$

Our new progress lies in the realization that this formula should be read from right to left as a definition of G.

Notice, stepping beyond the bounds of gravitation theory, that Sakharov's ideas can probably be extended to electrodynamics and to the theory of weak interactions (Zel'dovich 1967*a*). One usually writes the action in electrodynamics as

$$S = -mc\int ds - (e/c)\int A_i dx^i - (1/8\pi)\int (H^2 - E^2)dV, \quad (2.5.5)$$

8. We cannot even discuss the exact value of k today because we have no concrete picture of what happens for momenta equal to or larger than p_0. Even the sign of k is not clear.

where the first term is the action for free, charged particles; the second is the action for the interaction between the charged particles and the electromagnetic field; and the third is the action of the free electromagnetic field. Sakharov's train of thought suggests that we take the first two terms as primary and obtain the third (the action of the field) as a result of quantum vacuum corrections, just as we did for the term $\int R\,dV$ in the theory of gravity. In the remarkable work of Landau and Pomeranchuk (1955) one can find justifications for such an approach.

One can introduce a similar viewpoint into the theory of the intermediate charged W-bosons, which characterize the weak interactions according to

$$n = p + \mathrm{W}^-, \qquad \mathrm{W}^- = e^- + \bar{\nu}.$$

In this case the theory leads to the conclusion that the mass of a photon must be zero, in contrast to the mass of a W-boson.

One and the same value of p_0 (under certain assumptions about the spectrum of the fermion masses) gives the correct magnitudes for the gravitation constant, for the charge of an elementary particle, and for the weak-interaction constant.

Returning to gravitation theory, let us emphasize the possible application of the above ideas to the problems of the final state of collapse, of singularities, etc.

It is particularly interesting for astrophysics that, at least in principle, further development of these ideas should give corrections to the gravitation equations, involving the squares of the curvature invariants with coefficients of the order of $\hbar \ln (p_0/mc)$, where m is the mass of an elementary particle. These corrections will be important in regions of space where the curvature is as large as 10^{64} cm^{-2}; and perhaps they will be capable of converting gravitational collapse into expansion near a singularity.

Notice that here we are speaking about nonlinear (in the curvature) corrections to the equations of general relativity, not about the usual nonlinearities (in the deviations of the metric from flat space) which are already contained in the classical equations of general relativity.

From Sakharov's viewpoint, space is classical, not quantized. An alternative idea competes with this viewpoint: perhaps space itself should be quantized, too. Such an idea has been considered in some detail in a number of the works of Wheeler, and from a slightly different point of view in the works of C. DeWitt and B. DeWitt (see, e.g., DeWitt and DeWitt 1964).

2.6 QUANTIZATION OF GRAVITY

The most general considerations indicate that the gravitational field must obey quantum laws. If one does not assume that there is a quantum limit on the precision of gravitational-field measurements (i.e., measurements of the spacetime metric), then one encounters a contradiction with the un-

2.6 Quantization of Gravity

certainty principle for electrons, photons, etc. The quantization of gravity does not at all destroy classical (nonquantum) gravitational field theory; it remains valid as a limiting case of the quantum theory. The correlation is just the same as in the well-known case of nonrelativistic quantum theory and classical mechanics. An even better analogy is classical electromagnetic field theory and quantum electrodynamics.

For both cases (electromagnetism and gravity) the classical theory was worked out historically long before the quantum theory was created. It is obvious that effects due to the quantization of gravity are quantitatively negligible. In a report[9] at the Warsaw conference on gravitation, Feynman (1963) wrote: "There's a certain irrationality to any work in gravitation, so it's hard to explain why you do any of it." That is why we limit ourselves in this book to only a few qualitative remarks and to citations of references where one can study the quantization of gravity in detail.

General relativity using curved spacetime is not similar to electromagnetic theory, in which the field variables (A, E, H) are to be quantized in the arena of flat Minkowski space. In order to quantize general relativity one usually passes to the relativistic gravitation theory in flat space, and regards the small corrections to the metric coefficients, $h_{\mu\nu}$, as field variables. The most important result of quantizing these field variables is the discovery that gravitational waves consists of quanta—"gravitons"—which are just as real as the quanta (photons) of the electromagnetic field.

Gravitons are neutral particles with zero rest mass and spin 2 (in units of Planck's constant $\hbar$); the spin projection on the direction of propagation can take on two values only: $+2$ or -2. Superpositions of waves with these two spin projections define all possible polarization states of a gravitational wave. Gravitons are subject to Bose statistics: states are possible in which many gravitons have identical momenta, i.e., identical frequency and identical wavelength. As is known, such states are well described by the classical (nonquantum) field theory; the greater the number of such gravitons, the more exact the classical description. The gravitational radiation of a binary star need not be described as a flux of gravitons, since its flux is large and the classical theory is applicable; by the same token, we describe the electromagnetic radiation from a radio station by Maxwell's equations, without thinking in terms of quanta.

For the first indications of this result see Rosenfeld (1930); for the first full investigation see Bronstein (1936). For the general theory of fields with higher spins see Pauli and Fierz (1939).

The problem of the division of the quantized field into a static, Coulomb-like part plus a quantum field of gravitons was investigated in the works of

9. This report is of interest not only from the scientific point of view but also as an autobiographical document illustrating Feynman's mode of thought and methods of scientific research.

Gupta (1950, 1952) and in Kibble's review (1965, 1968). For the gravitational interactions of fermions see Kobsarev and Okun (1962).

In more recent work the question of higher-order approximations in the quantum theory of gravity has been considered. In the higher orders the peculiar nonlinearity of the theory becomes clear. Gravitational waves and other gravitational fields have energy and momentum; i.e., they themselves become sources of the gravitational field. It is true that these nonlinearities are small in practice. However, they must be taken into account to make the theory complete and self-consistent. For details see Utiyama (1956), Feynman (1963), and Kibble (1965). About other difficulties see the work of Fadeev (1967) and his report at the Fifth International Conference on General Relativity and Gravitation. We note too the interesting work of Ivanenko and Sokolov (1947), Wheeler (1960), and Vladimirov (1963) concerning the intertransmutations of gravitons and other particles.

High-frequency gravitons, which are similar to gamma quanta, have a very weak interaction with matter because of the smallness of $Gm^2/\hbar c$, where m is the mass of an elementary particle. For this reason, the emission and absorption of gravitons in all processes at the molecular, atomic, and nuclear levels are phenomena that are possible in principle but are nonetheless extremely rare and exotic; hence, they do not play any significant role in nature. For example, in stars the emission of neutrino pairs (a process similar in its effects to the emission of gravitons) is stronger than the emission of gravitons by a factor of 10^{10}.

As an example, consider the "Compton scattering" of a graviton on a particle with zero spin. This process can be thought of as an absorption and reemission of the graviton; in this sense it is a second-order effect. The cross-section as a function of the scattering angle has the form

$$\sigma(\theta) = \text{const.} \left(\frac{\hbar}{mc}\right)^2 \left(\frac{Gm^2}{\hbar c}\right)^2 \frac{1 + 6\cos^2\theta + \cos^4\theta}{1 - \cos^2\theta}.$$

The dimensional factor is $(Gm/c^2)^2 \approx 10^{-100}$ cm^2 for a particle with the proton mass. The cross-section does not depend on Planck's constant. The deflection of the graviton by the gravitational field of the particle (analogous to the 1″.75 deflection of light in the solar field) is the complete and correct source of this effect (Gross and Yakif 1966).

The only exception to the weakness of the gravitational interaction may be near the singularity at the beginning of the "big bang" model of the Universe (cf. Vol. II), where superhigh densities and temperatures exist; there the gravitational interaction may be sufficient to establish thermodynamic equilibrium between gravitons and other forms of matter. In this equilibrium, the energy density of all gravitons should equal the energy density of electromagnetic radiation. This equality is approximately preserved during the later expansion, after the gravitons have "decoupled" from the matter. The existence today of a universal thermal electromagnetic radiation with temperature about 1° K, which resulted from the expansion of the original, hot medium, justifies the assumption that the Universe also contains background thermal gravitational radiation (with a Planck spectrum for the case of a homogeneous, isotropic cosmological model).[10] Its average wavelength is about 0.15 cm, its frequency is 2×10^{11} Hz (cps), and its energy density is roughly 10^{-14} to 10^{-15} erg cm^{-3}. (This problem will be considered in more detail in Vol. II.) The detection of such gravitational radiation is an extremely difficult task. Incidentally, some new ideas have been voiced recently about the molecular generation and detection of gravitational waves in the optical range of wavelengths (Kopvilem and Nagibarov 1965).

In connection with the quantization of gravity, Feynman has made the following interesting remark: "It's clear that the problem we are working on is not the correct problem; the correct problem is: What determines the size of gravitation?" Recall that the preceding section noted Sakharov's attempt to explain why the gravitation constant is so small.

There are several other methods of quantizing gravity: the investigation of the topology of spacetime and its fluctuations (Wheeler 1968), and the evolution of the Universe as a whole studied on the basis of quantum ideas (DeWitt, 1968; Misner 1970). This last question will be discussed briefly in Vol. II in connection with cosmology.

A popular but untrue assertion is that the gravitational interaction is due to an "exchange of gravitons." It is even claimed that all particles emit gravitons and in so doing lose energy (cf. Stanyukovich 1965, and the reply by Zel'dovich and Smorodinsky 1966).

In reality the electromagnetic field, for example, can be divided into longitudinal and transverse components.[11] The transverse components describe propagating waves; their quantization introduces the concept of photons as particles with a certain momentum and energy, and with zero rest mass. The longitudinal components describe the Coulomb interaction; their quantization does not introduce anything new. The Coulomb field is stationary, it does not depend on time; there exist no quanta of a longitudinal field which could escape from the field's source.

Thus, the electrostatic interaction and the emission of photons are two *different* consequences of *one* theory; the electrostatic interaction is not the result of an exchange of free (transverse) quanta.

The difference in character between the longitudinal and the transverse fields clearly illustrates the absurdity of a literal interpretation of the phrase, "Interaction is an exchange of quanta." It is even more important to keep in mind that the emission of waves which carry away energy in the classical

10. See Vol. II regarding the possibility of a substantially different state for the present-day background gravitational radiation.

11. It is assumed that the field is resolved into Fourier components, i.e., into "waves," $a_k e^{ik \cdot x}$. For each k there are two different kinds of a_k: those with a_k parallel to k (longitudinal components), and those with a_k perpendicular to k (transverse components).

theory is of necessity related to an accelerated motion of charge. In the quantum theory, only a system that is in an excited state can emit quanta. In doing so the system makes a direct or cascade transition into its ground state and loses the ability to radiate further.

Meanwhile, the Coulomb field of the charged system is the same in its lower quantum state as in its original state.

This must be kept in mind also for the case of nuclear forces. The theory of scalar mesons (Yukawa's theory—here it is irrelevant that the mesons are pseudoscalar rather than scalar) accounts for two facts: (1) the existence of free mesons with finite mass and zero spin, which are subject to Bose statistics; and (2) the interaction between nucleons.

When there are many mesons, we can use the classical theory of a meson field; it is a scalar field, i.e., it is similar to the temperature field $T(r)$, rather than to the vector field of velocities $v(r)$; and it has no longitudinal or transverse waves.

However, the mesonic theory, too, asserts that only an accelerated or excited nucleon can emit free mesons. A normal nucleon at rest creates about itself a mesonic field which acts upon other nucleons. Inasmuch as the mesonic field is a static one, mesons are not emitted; and the nucleon naturally does not lose mass or energy. This is also obvious from the fact that a static mesonic field decreases exponentially with distance, i.e., as $\exp(-mcr/\hbar)$.

The gravitational interaction between bodies (Newtonian attraction) cannot be construed as a result of graviton emission accompanied by a loss of energy (mass) for two reasons: because of the transverse character of the graviton, and because of energetic considerations. The idea of a gravitational loss of energy can easily be reduced *ad absurdum* by a simple comparison: Why not, in this case, assume also a loss of electrostatic energy or an energy loss due to nuclear forces, stronger by a factor of 10^{40} than the loss due to gravitation?! The time of mass change sometimes claimed for the gravitational losses, 10^{10} years, would be reduced to 10^{-13} sec.

2.7 SCALAR-FIELD GRAVITATION

In concluding this chapter, let us return to the consideration of scalar-field theories of gravity. The generation of the scalar field can be formulated in a very general manner: its source[12] must be a scalar constructed from the second-rank stress-energy tensor. Obviously such a scalar is the trace of the tensor

$$T = Tr(T_\alpha{}^\beta) = g_{\alpha\beta}T^{\alpha\beta}.$$

12. The full solution is always the field due to the source plus a freely propagating wave. As pointed out before, free waves are excluded in the vector and tensor cases if one imposes the condition of spherical symmetry, because for vector and tensor fields free waves are transverse. In the scalar case there are spherical free waves, which can be carefully eliminated to obtain a static solution for a static source.

2.7 Scalar-Field Gravitation

For matter with no anisotropic stresses, at rest in locally pseudo-Euclidean space, the trace is

$$T = T_0{}^0 + T_x{}^x + T_y{}^y + T_z{}^z = \epsilon - 3P .$$

For example, for a nonrelativistic gas

$$\epsilon = n\,m_0 c^2 + n\tfrac{3}{2}RT , \qquad P = nRT , \qquad T = n\,m_0 c^2 - n\tfrac{3}{2}RT .$$

At first glance, a discrimination between the tensor and scalar theories seems easy: take a cold body, weigh it, then heat it at constant volume. In the tensor theory the source strength (the weight) is increased in proportion to ϵ; in the scalar theory the weight is increased as $\epsilon - 3P$.

But this is not the whole story! In a theory of gravitation one must admit that all bodies are field sources. Therefore, one must also take into consideration the vessel which contains the heated body. In a static case, if the body is under pressure, there must be a corresponding negative stress in the walls of the vessel. The overall integral of the pressure and stresses is identically zero; if it were not, there would be accelerated motion.[13] Therefore, in the static case

$$T \neq \epsilon , \qquad \text{but} \qquad \int T dV \equiv \int \epsilon dV ,$$

where the integrals are taken over the whole volume occupied by the matter. Thus, static gravitational experiments cannot discriminate between tensor and scalar gravitation. Only experiments with bodies moving at high velocities can discriminate. This is the basic reason why tests of the Brans-Dicke scalar-tensor theory must rely upon such effects as the deflection of light rays and the advance of the perihelion of Mercury.

Note by K. S. Thorne

The above argument breaks down for bodies whose self-gravity is significant. In that case $\int T dV$ and $\int \epsilon dV$ differ by an amount of the order of the body's gravitational binding energy. A very important consequence of this is the fact that, in scalar-field theories of gravity, the ratio of gravitational mass to inertial mass for a star or planet differs from unity by an amount of the order of (gravitational binding energy)/(rest mass). This means that, in a given external gravitational field, different stars and planets will fall with different accelerations. For example, in the Brans-Dicke theory the Sun should fall with an acceleration which is less by a correction of 10^{-6} than that of a small test body; Jupiter's acceleration should be less by $\sim 10^{-8}$; and Earth's by $\sim 10^{-9}$. These facts and their consequences for solar-system tests of gravitation theories have been pointed out and discussed in a series of papers by Nordvedt (1968a, b; 1970); see also Dicke (1969) and Will (1971).

13. If the heated body is in equilibrium without any vessel, it must have vanishing surface pressure. This is possible only if the body possesses internal attractive forces; it cannot be just an assembly of noninteracting, moving point masses. In this case the stress-energy tensor of the fields producing the attractive forces (electromagnetic field in atomic physics) must be included; and the result is the same as in the text: $\int T dV = \int \epsilon dV$.

APPENDIX TO §2.3

Here we show how the potential of an aggregate of moving particles contained in a spherical vessel is calculated. Consider first the case of a scalar potential ϕ. The source equation for ϕ reads

$$\Box\phi = \frac{\partial^2\phi}{\partial t^2} - \frac{\partial^2\phi}{\partial x^2} - \frac{\partial^2\phi}{\partial y^2} - \frac{\partial^2\phi}{\partial z^2} = a\rho .$$

Suppose that all the particles are at rest. Then $\rho = \Sigma_i \delta(r - r_i)$, where r_i is the location of the ith particle. When there are many particles near each other, one can substitute for the sum of delta functions a smooth function of the coordinates, $\rho(r)$, which is nonzero only in the region where the particles are concentrated (i.e., inside the vessel).

In a stationary situation ($\partial\phi/\partial t = 0$), from the condition

$$\nabla^2\phi = \frac{\partial^2\phi}{\partial x^2} + \frac{\partial^2\phi}{\partial y^2} + \frac{\partial^2\phi}{\partial z^2} = -a\rho ,$$

we find

$$\phi = (a/r)\int \rho dV = aN/r \qquad (2.3A.1)$$

outside the vessel. Here N is the total number of particles.

Next let the particles be in motion. Consider for a moment those particles in the vessel which all have a particular velocity (i.e., a particular speed and a particular direction). Then in a reference frame where these particles are at rest the solution for ϕ has the form (2.3A.1). When we transform to the "laboratory" frame, relative to which the particles are in motion, the values of ϕ and ρ at each event in spacetime remain invariant because ϕ and ρ are scalars: $\phi = \phi'$, $\rho = \rho'$. However, the solution for ϕ, which was stationary in the particles' rest frame, is nonstationary in the laboratory frame (it depends not only on the space coordinates but also on time).

Instead of adding these nonstationary solutions for the potentials ρ due to the particles of various velocities in the vessel, we shall add the strengths of the sources of various velocities. The field due to the summed sources will obviously be stationary because the problem is stationary.

For a given group of particles (with identical velocities) the value of ρ is invariant, but the volume in which the particles are contained is smaller in the laboratory frame than in their rest frame. Consequently, if we let primes refer to the laboratory frame, we have

$$\int \rho' dV' = \int \rho dV' = (1 - \beta^2)^{1/2}\int \rho dV = N(1 - \beta^2)^{1/2} .$$

Let several groups (ρ_1, ρ_2, . . .) of particles inside the vessel move with the same speed $v = \beta c$, but in different directions. The direction of the velocity of each particle changes from time to time as a result of collisions with the walls of the vessel; i.e., particles pass from one group to another so that the average number of particles in each group is conserved. For the aggregate of all particles,

$$\int \rho' dV' = \int \rho_1' dV' + \int \rho_2' dV' + \ldots$$

$$= N_1(1 - \beta^2)^{1/2} + N_2(1 - \beta^2)^{1/2} + \ldots = N(1 - \beta^2)^{1/2}.$$

Note that, although all the particles are in motion, their distribution is on the average constant. Therefore, knowing the source strength $(\int \rho dV)$ in the laboratory frame, we can now construct the stationary solution for our source:

$$\phi'(r) = (a/r)\int \rho dV = (aN/r)(1 - \beta^2)^{1/2} \text{ in laboratory frame}.$$

This result differs from that for particles at rest by the factor $(1 - \beta^2)^{1/2}$, as was asserted in the text.

It is instructive to consider an alternative source of the scalar field ϕ: a system of particles, not contained in a vessel, which all move radially with the same velocity $v = c\beta$, and which all pass through the origin at time $t = 0$. This problem is not stationary. Denote by $r_0(t)$ the distance of the particles from the origin at time t. Then for $t < 0$, $r_0 = -c\beta t$; for $t > 0$, $r_0 = +c\beta t$. At all times $t \neq 0$ the solution for ϕ turns out to have the form

$$\phi = b/r \quad \text{for} \quad r > r_0,$$

$$\phi = (b/r_0)(r/r_0) \quad \text{for} \quad r < r_0.$$

So long as r_0 depends on time, the potential inside the shell of particles also depends on time.

If we rewrite the equation $\Box\phi = a\rho$ in the form $\nabla^2\phi = -a\rho + \partial^2\phi/\partial t^2$, then it becomes clear that the term $\partial^2\phi/\partial t^2$ in the nonstationary solution plays a role analogous to the source in a stationary solution. Inside our shell of particles, $\partial^2\phi/\partial t^2 \neq 0$; consequently, the external field has a source not only on the shell itself but also throughout its interior. As a consequence, the constant b which characterizes the external field strength $(\phi = b/r)$ is *not* equal to $Na(1 - \beta^2)^{1/2}$. The fact that for $r > r_0$ the solution $\phi = b/r$ is independent of time must not be allowed to deceive us; the solution as a whole is nonstationary.

The external potential $\phi(r)$ for a stationary system of particles which are contained forever inside a vessel, $\phi = (aN/r)(1 - \beta^2)^{1/2}$, is different from that for particles which are momentarily inside the vessel (r_0 small) but were previously arbitrarily far outside it. This should not be surprising, since the equation for ϕ is characterized by retardation—the solution depends not on the instantaneous source distribution, but on the retarded ($t' = t - r/c$) distribution. It is this circumstance which enables nonstationary spherical sources to emit spherical scalar waves. We take this opportunity to thank K. S. Thorne, who, while editing this book, noticed that the nonstationary spherical solution did not answer the question of the velocity dependence of ϕ which is posed in the text.

Let us return to the case of interest: a source consisting of moving particles, but a source which, together with its fields, is nonstationary—i.e., particles in a vessel.

When the field ϕ is the time component of a vector, the solution is well known; this is just the electromagnetic case. We shall not analyze it in detail, but shall only make the following remarks. When we transform from the rest frame of a set of charges to a moving frame, we generally find that the vector field acquires nonzero spatial components, A_1, A_2, A_3, from which arises a magnetic field. However, in the spherically symmetric case there is no net magnetic field, even if the individual charges are in motion. Moreover, the external electric field is the same for a system of moving charges confined inside a spherical vessel (stationary problem) as for a shell of charges moving radially (nonstationary problem). However, the vector potential, obtained by adding the potentials of all the individual particles, is different in the two cases. For the nonstationary problem the total vector potential has a nonzero radial component A_r which contributes to the radial electric field: $E = \partial\phi/\partial r - \partial A_r/\partial t$. Although A_r depends on the velocities of the charges, E does not. Consequently, by a change of gauge,

$$\phi \to \phi + \partial f/\partial t, \quad A \to A - \nabla f,$$

we can set A_r to zero, thereby obtaining $\phi = eN/r$ outside the source. This gauge-transformed potential, $A = 0$ and $\phi = eN/r$, for the nonstationary problem is identical with that for the stationary problem—and this was the result quoted in the text.

The fact that the external field for the electromagnetic case is the same, whether the spherical source is stationary or not, is due to the fact that a vector potential (in contrast with a scalar potential) cannot have spherical waves.

Similar remarks apply to the tensor case. The expression for $\phi = h_{00}$ quoted in the text is for a coordinate system in which the components $h_{0\alpha}$ vanish.

3 THE SPHERICALLY SYMMETRIC GRAVITATIONAL FIELD

3.1 INTRODUCTION

Many celestial bodies, as well as some systems of celestial bodies, have spherically symmetric mass distributions. Examples (to high accuracy) are slowly spinning stars, planets, globular clusters of stars, type E0 elliptic galaxies, and spherical clusters of galaxies. The gravitational fields of such bodies are obviously also spherical. Celestial bodies and systems, if they are rotating slowly enough, will also be spherically symmetric when they reach relativistic stages of evolution.

We will analyze in detail spherical gravitational fields. On one hand, our analysis will emphasize direct applications of the theory to celestial bodies and to the cosmological problem; and on the other hand, it will emphasize numerous theoretical problems which can be clarified and solved here, since symmetry simplifies the Einstein equations. Deviations from spherical symmetry will be considered in chapter 4.

The expression for the interval in a spherically symmetric field can be written in the following form:

$$ds^2 = e^{\nu(x^0, x^1)}(dx^0)^2 - e^{\lambda(x^0, x^1)}(dx^1)^2 - e^{\mu(x^0, x^1)}(d\theta^2 + \sin^2 \theta d\phi^2) \ . \quad (3.1.1)$$

Here x^0 is the time coordinate, x^1 is the radial space coordinate, and θ and ϕ are the angular coordinates on a sphere. For convenience we write $g_{00} = e^\nu$, $g_{11} = -e^\lambda$, $g_{22} = -e^\mu$. The mixed components $g_{0\alpha}$ can be set equal to zero, inasmuch as rotation is absent by the assumption of sphericity (see § 1.6). The functions ν, λ, and μ may depend on the time and radial coordinates.

The Einstein field equations for the metric (3.1.1) take the form

$$-(8\pi G/c^4)T_1{}^1 = \tfrac{1}{2}e^{-\lambda}(\tfrac{1}{2}\mu_{,1}{}^2 + \mu_{,1}\nu_{,1})$$
$$- e^{-\nu}(\mu_{,00} - \tfrac{1}{2}\mu_{,0}\nu_{,0} + \tfrac{3}{4}\mu_{,0}{}^2) - e^{-\mu} \ , \quad (3.1.2)$$

$$-(8\pi G/c^4)T_2{}^2 = -(8\pi G/c^4)T_3{}^3 = \tfrac{1}{4}e^{-\lambda}(2\nu_{,11} + \nu_{,1}{}^2 + 2\mu_{,11} + \mu_{,1}{}^2$$
$$- \mu_{,1}\lambda_{,1} - \nu_{,1}\lambda_{,1} + \mu_{,1}\nu_{,1}) + \tfrac{1}{4}e^{-\nu}(\lambda_{,0}\nu_{,0} + \mu_{,0}\nu_{,0} \quad (3.1.3)$$
$$- \lambda_{,0}\mu_{,0} - 2\lambda_{,00} - \lambda_{,0}{}^2 - 2\mu_{,00} - \mu_{,0}{}^2) \ ,$$

$$-(8\pi G/c^4)T_0{}^0 = e^{-\lambda}(\mu_{,11} + \tfrac{3}{4}\mu_{,1}{}^2 - \tfrac{1}{2}\mu_{,1}\lambda_{,1})$$
$$- \tfrac{1}{2}e^{-\nu}(\lambda_{,0}\mu_{,0} + \tfrac{1}{2}\mu_{,0}{}^2) - e^{-\mu} \ , \quad (3.1.4)$$

$$-(8\pi G/c^4)T_0{}^1 = \tfrac{1}{2}e^{-\lambda}(-2\mu_{,01} - \mu_{,0}\mu_{,1} + \lambda_{,0}\mu_{,1} + \nu_{,1}\mu_{,0}) \ . \quad (3.1.5)$$

The remaining equations are identities. Here commas denote partial derivatives; e.g., $\nu_{,0}$ represents $\partial\nu/\partial x^0$, $\lambda_{,01}$ represents $\partial^2\lambda/\partial x^0\partial x^1$, etc.

In a comoving reference system $T_0{}^1 = 0$. In such a system the conservation laws (1.8.5) for the gaseous stress-energy tensor (1.8.2) take the form

$$\lambda_{,0} + 2\mu_{,0} = -2\epsilon_{,0}/(\epsilon + P) , \qquad (3.1.6)$$

$$\nu_{,r} = -2P_{,r}/(\epsilon + P) . \qquad (3.1.7)$$

We return now to the general case of a reference system which might not be comoving. In vacuum, far from the spherical mass, the metric is Euclidean, and the expression for the interval in spherical coordinates becomes

$$ds^2\big|_\infty = (dx^0)^2 - (dx^1)^2 - (x^1)^2 (d\theta^2 + \sin^2\theta d\phi^2) . \qquad (3.1.8)$$

Changes of the coordinates x^0 and x^1

$$x^0 = x^0(x^{0'}, x^{1'}) , \qquad x^1 = x^1(x^{0'}, x^{1'}) \qquad (3.1.9)$$

leave the spherical symmetry unaffected. Using a transformation of this type, we set

$$x^1 = e^{\mu(x^{0'}, x^{1'})/2} , \qquad (3.1.10)$$

where the old coordinates have primes; then we select $x^0 = x^0(x^{0'}, x^{1'})$ so that the metric component g_{10} vanishes (this is always possible). When this is done, the coefficient of the angular differentials becomes $(x^1)^2$—i.e., the same as in expression (3.1.8) for the metric at infinity—and the interval takes on the form

$$ds^2 = e^\nu(dx^0)^2 - e^\lambda(dx^1)^2 - (x^1)^2(d\theta^2 + \sin^2\theta d\phi^2) . \qquad (3.1.11)$$

Understandably, the transformation (3.1.10), which brings the interval to the form (3.1.11), cannot always be performed. Upon performing the transformation (3.1.10) and calculating the coefficients $g_{\mu\nu}$ by using formulae (3.1.9), one may find that in expression (3.1.11) for the interval, the metric component g_{00} is negative and g_{11} is positive.[1] If such is the case, then x^1 is no longer a space coordinate; rather, it is a time coordinate. This means that when all other coordinates (x^0, θ, ϕ) are held constant, the quantity $(g_{11})^{1/2}dx^1$ measures the proper time of a particle which is at rest in the coordinate system. In other words, in the expression

$$ds^2 = g_{AA}(dx^A)^2 + g_{BB}(dx^B)^2 - (x^4)^2(d\theta + \sin^2\theta d\phi^2) ,$$

that coordinate, x^A or x^B, which has a positive coefficient (g_{AA} or g_{BB}) is the time coordinate.

If x^1 turns out to be the time coordinate, then it would be logical to re-index the coordinates—to rename x^1 "x^0" and to rename x^0 "x^1" so that the time coordinate will always be denoted by x^0.

1. The metric components g_{00} and g_{11} cannot have the same sign because of the invariance of the metric's signature $(+ - - -)$.

Then, with the time coordinate appearing as the metric coefficient of the angular differentials, the expression for the interval will read

$$ds^2 = e^\nu (dx^0)^2 - e^\lambda (dx^1)^2 - (x^0)^2 (d\theta^2 + \sin^2\theta d\phi^2) \,. \qquad (3.1.12)$$

Thus, by using a transformation of the type (3.1.10), one can bring the interval for a spherical gravitational field either to the form (3.1.11) (Birkhoff 1923) or to the form (3.1.12) (Novikov 1961, 1962b). Regions of spacetime where the interval is brought to the form (3.1.11) are called *R*-regions, whereas regions where the interval is brought to the form (3.1.12) are called *T*-regions. Obviously, the definitions of *R* and *T* are invariant.

From the definitions of *R*- and *T*-regions and the transformation formulae for the metric, $g_{\mu\nu}$, it is easy to derive conditions that determine the type of region to which any given event belongs. If at the given event in the general coordinate system of equation (3.1.1)

$$e^{\nu-\lambda} > (\mu_{,0}/\mu_{,1})^2 \,, \qquad (3.1.13)$$

then the event is in an *R*-region. If the opposite inequality is satisfied, then the event is in a *T*-region. A more elegant form of this criterion for *R*- and *T*-regions has been pointed out by K. S. Thorne: through the given event passes a sphere of surface area *A* (radius of curvature $[A/4\pi]^{1/2}$). Calculate the gradient of *A*, ∇A, in spacetime. If ∇A is spacelike, the given event is in an *R*-region. If ∇A is timelike, the event is in a *T*-region. If ∇A is lightlike, the event is in neither an *R*-region nor a *T*-region. Far from all masses, where the gravitational field is weak and the metric approaches the asymptotic form (3.1.8), condition (3.1.13) is always satisfied, and we find ourselves in a conventional *R*-region. However, for an exceptionally strong concentration of mass, within distances which are less than "critical,"[2] it turns out that x^1 has no longer the significance of a space coordinate, we are in a *T*-region, and the metric cannot be written in the form (3.1.11). This is discussed in more detail in § 3.13. For the time being we shall assume that the metric has been brought to the form (3.1.11)[3] and that we are in an *R*-region. First we will consider the vacuum gravitational field.

3.2 THE SCHWARZSCHILD GRAVITATIONAL FIELD

The simplest problem of GTR—the problem of the motions of test particles and photons in the strong vacuum gravitational field of a spherical body—exhibits characteristics which are closely linked to the structure of relativistic stars (white dwarfs, neutron stars, supermassive stars), and to the phenomenon of catastrophic stellar contraction (relativistic collapse).

---

2. The critical distance is called the gravitational radius; it depends on the mass, $r_g = 2GM/c^2$. See § 3.2 for further discussion.

3. The transformation (3.1.10) assumes that μ is monotonic in $x^{1'}$. It turns out that if x^1 has the significance of a space coordinate, then μ is always monotonic in $x^{1'}$.

Spherically Symmetric Gravitational Field

The solution of the Einstein equations (3.1.2)–(3.1.5) for the metric (3.1.11) of such a vacuum field $\epsilon \equiv P \equiv 0$ (the Schwarzschild solution[4] [1916]) determines the geometrical properties of space and the rate of flow of time near the body which creates the field. It turns out that this field is always static (even if the matter of the central body performs radial motions while remaining spherically symmetric), and it depends only on the total energy of the body, E (including the rest mass-energy of its particles). [*Editors' note.* —The gravitational self-energy is likewise included.]

The expression for the four-dimensional interval of the Schwarzschild field, expressed in the form (3.1.11), is

$$ds^2 = -(1 - 2GM/c^2r)^{-1} dr^2 - r^2(d\theta^2 + \sin^2 \theta d\phi^2)$$
$$+ (1 - 2GM/c^2r)c^2dt^2 , \tag{3.2.1}$$

where $x^1 = r$, $x^0 = ct$, and $M = $ const. The constant M determines the force of the gravitational field. Far from the body the field is weak and can be described by Newtonian theory. The constant M is the mass which determines that Newtonian field, $\phi = GM/r$. Expression (3.2.1) for ds^2 contains complete information about the gravitational field. Let us recall (cf. § 1.4) how to interpret this expression physically. The sum of the first three terms gives the square of the distance between infinitely close points, dl^2 (with sign reversed), in a spherical coordinate system. A stationary observer located near a massive body can measure the distances within a small region in the conventional manner, by introducing Cartesian coordinates. In these coordinates $dl^2 = dx^2 + dy^2 + dz^2$. If he selects $dz = rd\theta$ and $dy = r \sin \theta d\phi$, then in the flat Euclidean space far from the gravitating body he must find that $dx = dr$. Near the massive body, in its Schwarzschild field, as we see from equation (3.2.1),

$$dx = (1 - 2GM/c^2r)^{-1/2} dr . \tag{3.2.2a}$$

The multiplier in front of dr is different from unity; this reflects the fact that the space geometry is non-Euclidean. From this it follows, for example, that the distance between two close circles drawn about the central body in the same plane is not $(l_2 - l_1)/2\pi$, where l_i are the circumferences of the circles; rather, their separation is

$$\left(\frac{l_2 - l_1}{2\pi}\right)\left(1 - \frac{2GM}{c^2r}\right)^{-1/2} .$$

The last term in expression (3.2.1) is the square of the time interval τ (multiplied by c^2) at a given point

$$\Delta\tau = (1 - 2GM/c^2r)^{1/2} \Delta t . \tag{3.2.2b}$$

Far from the body, when $r \to \infty$, $\Delta\tau = \Delta t$. The closer the observation point is to the gravitating body, the slower is the passage of time; i.e., the time

---

4. On the properties of the Schwarzschild solution, see Landau and Lifshitz 1962.

interval $\Delta\tau$ corresponding to a time interval Δt at infinity becomes smaller and smaller as r decreases. When $r \to 2GM/c^2$, then $\Delta\tau \to 0$.

Let us calculate in a Schwarzschild field the acceleration of a body which is falling freely with a low velocity ($v \ll c$). Using expression (1.6.1b), we can express the acceleration F of a freely falling test particle in the form

$$F = (F_a F^a)^{1/2} = \frac{GM/r^2}{(1 - 2GM/c^2 r)^{1/2}} \cdot \qquad (3.2.3)$$

We see that when $r = 2GM/c^2$, the gravitational force becomes infinite. This indicates that a central body, if it is static, certainly cannot have a radius smaller than $2GM/c^2$. The stationary, nondeforming spherical coordinate system used above is applicable only in regions where $r > 2GM/c^2$. For smaller r the interval (3.1.1) can no longer be brought to the form (3.2.1). The critical radius $r_g = 2GM/c^2$ is called the *gravitational* radius, and the sphere of radius r_g is called the *Schwarzschild sphere*. A nonstatic body can have dimensions smaller than the gravitational radius (see § 13 of chapter 1); however, we shall not discuss this point until later.

At a distance large compared with r_g, the Schwarzschild field is a conventional Newtonian gravitational field with gravitational potential $\phi = GM/r$; the expression for the acceleration, accordingly, is $F = -GM/r^2$. The gravitational radius of the Sun is 2.96 km; the gravitational radius of the Earth is 0.886 cm. Thus, the radii of the Earth and of the Sun are considerably larger than their gravitational radii. It follows that outside the Sun, the Earth, and other normal stars and planets, the gravitational field is Newtonian, to great precision. Inside normal matter the Schwarzschild solution is not applicable.

3.3 THE GRAVITATIONAL FIELD INSIDE A STAR

Consider now the properties of a strong spherical gravitational field inside matter which is at rest. Here the four-dimensional interval takes the form (3.1.11)

$$ds^2 = -e^{\lambda(r)}dr^2 - r^2(d\theta^2 + \sin^2\theta d\phi^2) + e^{\nu(r)}c^2 dt^2 . \qquad (3.3.1)$$

The coefficients $e^{\lambda(r)}$, defining the deviation of the geometry from Euclidean, and $e^{\nu(r)}$, defining the variations in the rate of flow of time, are determined by the distribution of matter

$$e^{-\lambda} = 1 - (8\pi G/c^2 r) \int_0^r \rho r^2 dr , \qquad (3.3.2)$$

$$e^{\nu} = \exp\left\{ \int_r^\infty [(8\pi G/c^4)(\rho c^2 + p)re^\lambda - d\lambda/dr]dr \right\} . \qquad (3.3.3)$$

Recall that ρ is the density of matter, including not only the sum of the rest masses of the particles in a unit volume but also their energies (of motion and interaction,[5] with the gravitational interaction excluded).

Outside the star, in vacuum, expressions (3.3.2) and (3.3.3) yield

$$e^{-\lambda} = 1 - \frac{8\pi G}{c^2 r} \int_0^R \rho r^2 dr , \qquad e^{\nu} = e^{-\lambda} ,$$

where R is the radius of the star's surface, at which $\rho = 0$. These expressions will coincide with the expressions of the preceding section, if the mass is written in the form

$$M = 4\pi \int_0^R \rho r^2 dr . \tag{3.3.4}$$

We recall that because space is non-Euclidean, the volume element is $dV = 4\pi e^{\lambda/2} r^2 dr \neq 4\pi r^2 dr$. Notice that $4\pi r^2 dr$ appears in the integral (3.3.4) rather than dV. As will be demonstrated below, the replacement of dV by $4\pi r^2 dr$ is related to the effect of the energy of the gravitational field upon the mass of the body (see § 10.6).

Another expression for the mass of matter at rest, derived by Tolman (1930), is

$$M = 4\pi \int_0^R (\rho + 3p/c^2) e^{\nu/2 + \lambda/2} r^2 dr . \tag{3.3.5}$$

For the case of matter at rest, expressions (3.3.4) and (3.3.5) yield identical results.

It is apparent from formulae (3.3.2) that the metric coefficients satisfy $e^{\lambda} \geq 1$ and $e^{\nu} < 1$ inside the gravitating mass, just as they do outside it, and that hence the deviation of the interior geometry from Euclidean has the same nature as the deviation of the exterior geometry; in particular, $dV > 4\pi r^2 dr$, and time flows more slowly than it does at infinity.

To obtain formula (3.3.2), we demanded that $e^{\lambda} \to 1$ when $r \to 0$; at this point, the metric becomes Galilean. This, of course, does not mean that space is curved less near $r = 0$ than elsewhere. The truth of the matter is that we are using spherical coordinates so that when $r \to 0$ we are considering only a small region about the center; and in a small region about any given point the metric is Galilean. The space curvature depends on $\lambda_{,11}$ and has dimensions of cm^{-2}; hence, the effects caused by curvature decrease as the square of the linear dimension. Thus when $r \to 0$, the space curvature is not obvious, and $e^{\lambda} \to 1$.[6]

---

5. Naturally, we consider only short-range forces; large-scale electric and magnetic fields are not considered here (see § 13.6).

6. We note that there exist solutions with finite mass and $\rho_c = \infty$. In these solutions, the curvature at the center is infinite, and $e^{\lambda_c} \neq 1$. We shall not discuss these singular solutions at this point.

Actually, the Gaussian curvature (see § 1.4) of space, C_G, at the center of the star is larger than it is at other places. The quantity C_G is given by the formula (see eq. [1.8.13]):

$$C_G = (8\pi/3)(G\rho/c^2) . \qquad (3.3.6)$$

Inasmuch as the density is a maximum at the center of the star, C_G is also a maximum there. Of course, we should not assume on the basis of equation (3.3.6) that space is Euclidean outside the star (where $\rho = 0$), when the gravitational field near the star is strong. Formula (3.3.6) gives only the mean curvature of space averaged over all two-dimensional directions; and this mean curvature actually is zero. However, beyond the star, the Riemannian space curvature (see § 1.4) is nonzero; and, depending on the two-dimensional direction, it may be positive or negative. At the center of the star, all directions are equivalent; there the curvature for any orientation is given by formula (3.3.6), and it is always positive.

The gravitational field obtained by joining the star's interior and exterior solutions has no physical singularities of the type exhibited on the Schwarzschild sphere; everywhere

$$1 \leq e^\lambda < \infty , \quad \text{and} \quad 0 < e^\nu < 1 .$$

3.4 THE RADIAL MOTION OF LIGHT RAYS AND OF ULTRA-RELATIVISTIC PARTICLES

Consider now the radial motion, in a spherical gravitational field, of particles with the fundamental speed c; for example, photons and neutrinos. Since an observer can locally introduce coordinates for which $ds^2 = c^2 dt^2 - dx^2 - dy^2 - dz^2$, and since the speed of light c is independent of reference frame, an equation of motion for a particle with zero rest mass is $ds = 0$ in any coordinate system. Because of spherical symmetry, a particle that starts out moving radially will always move radially. Along such a particle's trajectory θ and ϕ are constant, so the condition $ds = 0$ implies

$$dr/dt = c e^{(\nu-\lambda)/2} . \qquad (3.4.1)$$

Everywhere inside the star $e^{(\nu-\lambda)/2} < 1$. Outside the star, in vacuum, $e^{(\nu-\lambda)/2} = 1 - 2GM/c^2 r < 1$; and this quantity approaches unity as $r \to \infty$. It follows, for example, that for a neutrino emitted from the star's center, the radial coordinate r, as a function of the time t of a distant observer, must have the form shown in Figure 7. The dotted line in this diagram indicates the motion of the neutrino in the absence of a gravitational field.

In the future we will frequently consider the motion of bodies in the immediate vicinity of the gravitational radius of the central body (where the field is especially strong), and even in the limit $r \to r_g$. As noted in § 3.2, a static body cannot have dimensions $r \leq r_g$; and Bondi (1964) has demonstrated that the dimensions of a body in equilibrium cannot be smaller than

$1.6\ r_g$. However, as we will see later (chapter 11), massive and supermassive stars (with masses larger than twice the mass of the Sun) at the end of their evolution may become unstable and contract exponentially fast as measured by a distant observer, eventually reaching the dimensions of the gravitational radius. The external Schwarzschild field in vacuum for such non-equilibrium stars, known as collapsed stars (see chapter 11), reaches down to $r = r_g$. All discussions of the motion of particles in vacuum near r_g are

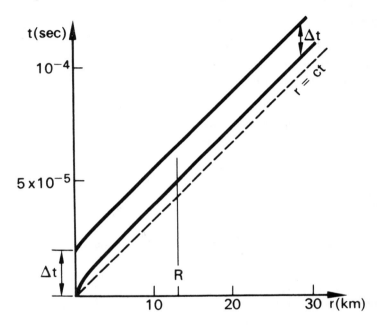

Fig. 7.—Graphs of radius r as a function of the time t, measured by the clock of a distant observer, for two neutrinos emitted at times $t = 0$ and $t = \Delta t$ from the center of a neutron star with mass $0.64\ M_\odot$. R is the star's surface ($R = 6.9\ r_g$).

applicable to such nonequilibrium, unbalanced, contracting objects; in the course of their motion, the particles cannot reach the surface of the body.

But let us return to the problem at hand. Note that in vacuum, as $r \to r_g$,

$$dr/dt = c(1 - 2GM/c^2r) \to 0 \ .$$

Obviously, this does not mean at all that the speed of light approaches zero. The speed measured by an observer at a given point is not dr/dt, but $dx/d\tau$, where dx and $d\tau$ are infinitely small physical distance and time, respectively (see § 3.2). For light, it is always true that $dx/d\tau = c$.

The speed of light measured by using the clock of a distant observer is $dx/dt = c(1 - 2GM/c^2r)^{1/2}$; that is, from his vantage point the light beam near the mass moves more slowly than c. This is the basis for a recent, new experimental test of GTR (Shapiro 1965; Shapiro 1968).

In this test radar signals are reflected off Mercury or Venus just before and just after it passes behind the Sun (Fig. 8). In Newtonian theory, the propagation time of the signals out and back ("delay time") is determined only by the varying distance between Earth and the reflecting planet, and is given graphically by the solid line of Figure 8. According to GTR, the decrease in the speed of the signals near the Sun causes an additional varia-

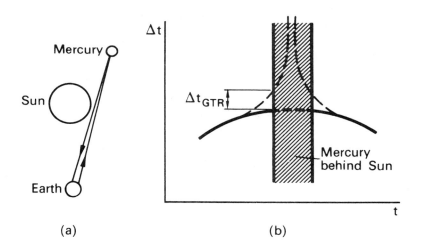

(a)   (b)

FIG. 8.—Radar ranging on the planet Mercury as a test of GTR. (a) schematic diagram; (b) variation of the radar delay time. Vertical axis shows the time interval between emission of the signal and reception of the reflected signal; horizontal axis indicates the moment at which the signal is emitted. Dashed lines show the delay due to effects of GTR. The unobservable period, when Mercury is behind the Sun, is shaded.

tion in the delay time which is measured by the distant observer on Earth.[7] This accounts for the extra delay shown in Figure 8 (*dashed line*). In optimal circumstances the magnitude of the additional delay is of the order of 2×10^{-4} seconds. The actual delay measured by Shapiro (1968) agrees with the prediction of general relativity theory: (measured delay) = (predicted delay) $\times$ (0.9 $\pm$ 0.2). In making his measurement, Shapiro did not have to include corrections for the variation of the speed of radio signals in the plasma of the solar corona, because at their very high frequency (7.84 $\times$ 10^9 Hz) the added delay due to the corona is less than 10^{-6} sec.

7. The trajectory of the beam also changes (the famous deflection by 1″.75 near the edge of the solar disk [see § 6.10]). However, the change in delay time due to this change in the length of the trajectory is one order higher in the small quantity GM/c^2r than the effect described in the text.

Spherically Symmetric Gravitational Field

Notice that the radar delay is closely linked to the gravitational deflection of the ray. To understand this link, imagine a train of plane electromagnetic waves impinging upon the solar system from infinity. Each flat wave front is characterized by a constant value of the phase of the electric field. After the waves have passed the Sun, their wave fronts are no longer flat. They have been curved by the Sun's gravitational field. A single function, that which describes the shape of the wave fronts after passing the Sun, completely determines the results of our two experimental tests of general relativity.

The normal to the wave front defines the direction in which the emitter of the waves (a star, in practice) is seen. The curvature of the surface is accompanied by a deflection of the normal from its original direction, and this we interpret as the gravitational deflection of the ray. By measuring the light deflection for a variety of stars at different angular separations from the Sun, we in effect gain a complete knowledge of the shape of the curved wave fronts produced by plane waves passing the Sun. The curvature of the wave fronts also leads to a change in the arrival time at a fixed point in space. This is the origin of the radar time-delay effect.

Thus, there is an intimate connection between the two effects; they are both determined by the nature of the Sun's influence on plane waves.

Shapiro's method needs no solar eclipse, and it permits one to use all the power of present-day radio-astronomy technology. It is very possible that this method will ultimately yield greater precision than the optical one.[8]

How do the energies of neutrinos and photons—and, hence, also the frequencies of the corresponding waves—change in the gravitational field? Consider the variation of the frequency. Let an emitter at the surface of the star emit two bursts separated by an interval Δt. Since e^λ and e^ν do not de-

8. *Editors' note.*—Within the next one or two years, completely new techniques will replace the standard ones described above for measuring the radar time delay and the light deflection. The time delay will be measured, not by bouncing radar waves off distant planets, but instead by "transponding" radio signals with spacecraft. More particularly, radio signals will be sent out from Earth to a spacecraft on the opposite side of the solar system; the spacecraft will receive the signals and retransmit them to Earth; and a radio telescope on Earth will receive the retransmitted (i.e., "transponded") signals. One will then measure the time delay between transmission and reception back on Earth. Such a technique has been used to track American spacecraft since 1965, and it is now capable of yielding time-delay precision corresponding to a distance uncertainty of the order of 10 meters. The new technique for measuring light deflection is to deal with radio waves from a quasar rather than with light waves from a star. One uses transcontinental interferometry to monitor the separation between the quasars 3C 273 and 3C 279 as 3C 279 passes behind the Sun. The separation is about 8°, and the occultation occurs each October. By using transworld interferometry, it should eventually be possible to measure the deflection in this manner to a precision of about 10^{-3} sec of arc. The first such experiment, tried in October 1968, was a failure; but the second, in October 1969, was a success. However, at the time this book went to press, the precise numerical results of the 1969 experiment were not yet available. *Note added in proof:* Two Caltech groups—Seielstad, Sramek, and Weiler (1970), and Muhleman, Ekers, and Fomalont (1970)—have now reported their results on the 1969 experiment; they are $1.01 \pm .11$ and $1.04^{+.15}_{-.10}$ times the general relativistic prediction, respectively.

pend on t, these will be separated by the same interval, Δt, when they reach a distant observer (cf. Fig. 7). However, in a strong gravitational field the locally measured time interval is

$$\Delta \tau = e^{\nu/2} \, \Delta t \, , \qquad (3.4.2)$$

rather than Δt. Consequently, the frequency of the signal received by the observer, which is proportional to $1/\Delta t$, differs from the frequency of the emitted signal, $\omega_0 = 1/\Delta \tau$, by

$$\omega = \omega_0 \, e^{\nu/2} \, . \qquad (3.4.3)$$

The frequency decreases as the signal leaves the gravitational field, and increases when the signal moves in the opposite direction. Accordingly, the quantum energy $E = \hbar \omega$ also decreases. This phenomenon is known as the *gravitational redshift*. For an observer located on the surface of a star, the emission spectra of atoms look exactly the same as they do in a laboratory on Earth. However, the spectra of the same atoms on the star, as observed at Earth, are shifted to the red by this gravitational phenomenon.[9]

3.5 RADIAL MOTION OF NONRELATIVISTIC PARTICLES

Consider now the radial motion of nonrelativistic particles in vacuum. First we write down the "coordinate" velocity of a freely falling particle in the Schwarzschild field, i.e., the rate of variation of the coordinator r with time t. Using the geodesic equation for the Schwarzschild field (3.2.1), we obtain

$$\frac{dr}{dt} = \left(1 - \frac{r_g}{r} \right) \left[1 - \frac{1 - r_g/r}{1 - r_g/r_0} \right]^{1/2} c \, . \qquad (3.5.1)$$

Here $r_g = 2GM/c^2$ is the gravitational radius of the central mass, and r_0 is the radius at which the fall begins, i.e., at which $dr/dt = 0$. At large distances (r_0 and $r \gg r_g$), formula (3.5.1) takes on the conventional Newtonian form

$$(dr/dt) = \left[\frac{2GM}{rr_0} (r_0 - r) \right]^{1/2} \, .$$

Expression (3.5.1) is the rate of change of the coordinate r as measured by the clock of a distant observer. A local, stationary observer, who is next to the falling body and who measures its velocity, will find it to be

$$\frac{dx}{d\tau} = \frac{1}{1 - r_g/r} \frac{dr}{dt} = \left[1 - \frac{1 - r_g/r}{1 - r_g/r_0} \right]^{1/2} c \, . \qquad (3.5.2)$$

As the body approaches the gravitational radius, $dx/d\tau$ approaches c. However, the velocity dx/dt, according to the time t of a distant observer, is quite different. Using formula (3.5.1), we find that $dx/dt = (1 - r_g/r)^{-1/2} dr/dt \rightarrow 0$

9. A "blue" shift of rays falling upon the Earth from cosmic space is caused by the Earth's gravitational field; but it amounts to only $\Delta \omega/\omega \approx 10^{-9}$, which we will neglect.

as $r \rightarrow r_g$. Naturally, the approach of the velocity dx/dt to zero is caused by retardation of the rate of time flow near r_g (see § 3.4). The velocity $v = dx/d\tau$ is a quantity that has direct physical significance. It is the velocity measured by an observer who is at rest (r, θ, ϕ constant) at the point which the particle is passing. Moreover, it is the velocity that enters into the expressions for the locally measured energy of the particle, $E_{\text{loc}} = mc^2/[1 - (v/c)^2]^{1/2}$, etc. Naturally, when the particle is falling, this velocity is increased continuously by the pull of gravity. The velocity dx/dt, which is determined by the time of the distant observer, does not have such direct significance. Far from the gravitating mass $dx/dt = dx/d\tau = dr/dt$, and for a falling particle dx/dt increases; however, near the mass dx/dt decreases and, as demonstrated previously, it approaches zero when $r \rightarrow r_g$. It has been suggested (McVittie 1956, pp. 85–86) that the decrease of dx/dt at $r \rightarrow r_g$ be interpreted as due to some kind of repulsion at small distances. It seems much more reasonable to us to attribute the decrease entirely to the general-relativistic relationship between the times t and τ described above.

It follows from formula (3.5.2) that during the radial fall of a particle the quantity

$$mc^2 \left[\frac{1 - r_g/r}{1 - v^2/c^2} \right]^{1/2} = mc^2 \left(1 - \frac{r_g}{r_0} \right)^{1/2} = E$$

is conserved. This quantity is the total energy of the particle in the gravitational field.

If $r = r_g$, the integral

$$\Delta t = \int_{r_0}^{r} (dr/dt)^{-1} dr \tag{3.5.3}$$

diverges at its upper limit. Thus, the time t required for a particle to fall to $r = r_g$ is always infinite. Even in the case of light, for which the time of propagation from r_0 to r is obtained by integrating equation (3.4.1) and equals

$$\Delta t = \left(\frac{r_0 - r}{c} \right) + \frac{r_g}{c} \ln \left(\frac{r_0 - r_g}{r - r_g} \right), \tag{3.5.4}$$

the time interval Δt required to reach r_g is infinite; and nothing can move faster than light.

Thus, according to the clock of a distant stationary observer, the time required to reach r_g is always infinite. Any body, regardless of the forces which influence its fall, can approach r_g only asymptotically. What will be the time of fall as read by a clock attached to the falling particle itself? Let us attach a reference frame to the particle. In this frame the clock does not change its position, and hence for it $ds = cdT$, where T is the clock reading. Hence, it follows that $\Delta T = (1/c) \int ds$. However, ds is an invariant quantity which does not change in a transformation to another reference frame; hence, it can be calculated in any arbitrary coordinate system. Computing ds in the Schwarzschild coordinates, we obtain

$$\Delta T = \frac{-1}{c} \int_{r_1}^{r} \left[\frac{1 - r_0/r}{(dr/c\,dt)^2} - \frac{1}{1 - r_0/r} \right]^{1/2} dr . \qquad (3.5.5)$$

Using expression (3.5.1) for dr/dt, we see that equation (3.5.5) converges for any upper limit whatsoever, even for $r = r_0$. In particular, if a particle falls with a parabolic velocity (i.e., $dr/dt = 0$ at infinity), then

$$\Delta T = \frac{2}{3} \left[\frac{r_1}{c} \left(\frac{r_1}{r_0} \right)^{1/2} - \frac{r}{c} \left(\frac{r}{r_0} \right)^{1/2} \right] , \qquad (3.5.6)$$

which is identical with the formula of Newtonian theory if we replace r_0 by its definition, $2GM/c^2$. Here r_1 is the radius of the particle at the beginning of the time interval ΔT. Hence, the time of fall down to r_0, according to a clock on the particle, is finite. That which is infinite on the clock of a distant observer is finite on the clock of a falling observer; can there be a more persuasive illustration of the relativity of the concept of temporal infinity?

There remains only one more clarification to be made. Using equation (3.5.3), we can find $r = r(t)$, i.e., the position of the test particle at time t as measured by the clock of a distant observer. But this, of course, is not the place where this observer *sees* the particle to be at time t; light requires some time Δt to traverse the distance from the particle to the observer. This time is easily computed by using equation (3.5.4). Denote the time of arrival of the light at the observer by t_*:

$$t_* = t + \Delta t . \qquad (3.5.7)$$

As the particle approaches the gravitational radius, $t \to \infty$, $\Delta t \to \infty$, and hence t_* all the more rapidly approaches infinity. Thus, the observer sees that the particle approaches the gravitational radius only asymptotically as an infinite time elapses. By use of the above expressions, it is easy to derive the formula $r = r(t_*)$ for an incident particle; i.e., the law which describes how the observer sees the particle approach the gravitational radius. For $r \to r_0$, the asymptotic form of this formula is

$$r = r_0 + (r_1 - r_0) \exp\left[-c(t_* - t_*^1)/2r_0 \right] . \qquad (3.5.8)$$

Here r_1 is the position of the particle at the moment t_*^1, and

$$(r_1 - r_0) \ll r_0 .$$

Consider now the variation in brightness of a radiation source falling in the Schwarzschild field, as seen by a distant observer. At some given moment let the falling source be near r_0 and be moving with a local velocity $dx/d\tau = v$ along the radius which connects the central body with the distant observer A. Assume that a comoving observer who falls together with the source sees it emit radiation isotropically and with constant intensity. Then the flux density at infinity, I_∞, as measured by observer A is:

$$I_\infty = \text{const.} \left(1 - \frac{r_g}{r}\right)^2 \left[\frac{1 - v^2/c^2}{(1 + v/c)^2}\right]^2. \qquad (3.5.9)$$

Here one factor of $(1 - r_g/r)$ is due to the gravitational redshift, and the second factor of $(1 - r_g/r)$ is due to the curvature of the trajectories of the rays in the gravitational field; one factor of $(1 - v^2/c^2)(1 + v/c)^{-2}$ is due to the Doppler shift, and the second identical factor is due to aberration. It follows from equation (3.5.2) that $1 - v^2/c^2 = 1 - (r_g/r)[(r_0 - r)/(r_0 - r_g)]$, and that as $r \to r_g$,

$$I_\infty = \text{const.} \ (1 - r_g/r)^4. \qquad (3.5.10)$$

The law of variation of r with t_* was formulated in equation (3.5.8). By combining it with equation (3.5.10), we obtain an asymptotic expression showing how the distant observer sees the brightness of a falling source to vary as $r \to r_g$:

$$I_\infty = \text{const.} \times \exp\left[-2c(t_* - t_*^1)/r_g\right]. \qquad (3.5.11)$$

The frequency of a light wave received by the distant observer approaches zero according to a similar law, except that the exponential decay is slower by a factor of 4 ($\omega \propto \exp\left[-c(t_* - t_*^1)/2r_g\right]$).

We shall see further in § 3.10 that, as $r \to r_g$, light rays leaving the source in a certain direction and bending in the gravitational field may circle around the center of attraction for a long time before going off toward the distant observer. These rays create a "halo" around the body (when it is shrinking toward $r = r_g$ and has $(r - r_g)/r_g \ll 1$); the brightness of the "halo" also attenuates exponentially rapidly. In equation (3.5.11) the "halo" is not taken into account. We will return to this problem in § 11.6.

3.6 POTENTIAL CURVES FOR NONRADIAL MOTION

Having clarified the fundamental properties of radial motion, we will consider the general case of nonradial trajectories. The first complete discussion of nonradial trajectories in the Schwarzschild field was by Hagihara (1931). A complete classification of the types of motion is given in the book by Bogorodsky (1962; see also the contributions of Galkin 1961 and of Metzner 1963). An analysis of the stability of motion along circular orbits is given by Kaplan (1949a). The trajectory of a particle's motion always lies in a plane. If that plane is chosen to be $\theta = \frac{1}{2}\pi$, the equations of motion in polar coordinates take the form

$$\left(\frac{dx}{d\tau}\right)^2 = \frac{1}{E^2}\left[E^2 - 1 + (1/r) - (a^2/r^2) + (a^2/r^3)\right], \qquad (3.6.1a)$$

$$\left(\frac{d\phi}{d\tau}\right)^2 = (a^2/E^2r^4)(1 - 1/r). \qquad (3.6.1b)$$

For convenience, these equations are written in dimensionless coordinates: r is the Schwarzschild radial coordinate measured in units of the gravitational radius, $r_g = 2GM/c^2$; $dx = (1 - 1/r)^{-1/2}dr$ is the element of radial distance; τ is time measured by the local observer in units of r_g/c; a is the particle's angular momentum measured in units of mcr_g; E is the energy measured in units of mc^2; and m is the mass of the test particle. The energy includes the rest mass; hence, for a particle at rest at infinity, $E^2 = 1$.[10]

At distances much larger than the gravitational radius (i.e., when $r \gg 1$) and when the energy of motion is small compared with unity ($E - 1 \ll 1$),

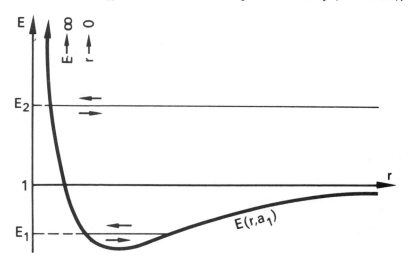

Fig. 9.—The potential curve, $E(r, a_1)$, in Newtonian theory for a particle with fixed angular momentum a_1. The total energy, including the particle's rest mass mc^2, is plotted vertically; $E_1 < 1$ is the horizontal line for finite (elliptical) motion; $E_2 > 1$ is the horizontal line for hyperbolic motion.

we obtain from equations (3.6.1a,b) the Keplerian equations of Newtonian gravitation. More particularly, under such conditions the term a^2/r^3 in equation (3.6.1a) can be neglected, $dx \approx dr$, and $E^2 - 1 \approx 2(E - 1)$. In this case $1/r$ is the gravitational potential and a^2/r^2 is the centrifugal potential. By setting the numerator of equation (3.6.1a) to zero, we obviously obtain the effective radial potential (or "potential curve") for the particle's nonradial motion.

This potential curve of Newtonian theory, $E(r, a_1)$ for fixed a_1, is shown in Figure 9. For every choice of a_1 the potential has a minimum. The quali-

10. The total energy, E, of the particle, which is conserved during the motion, is expressed in terms of velocity, v, and g_{00}, in a form similar to that for radial motion. Actually, from equations (3.6.1a, b) we find $E/mc^2 = [g_{00}/(1 - v^2/c^2)]^{1/2}$. This formula is valid for any static field; see Landau and Lifshitz (1962).

tative properties of the test-particle motion are also shown in Figure 9. The motion takes place at constant energy E_1 and is represented by the horizontal line $E = E_1$. A particle with angular momentum a_1 moves along this horizontal line up to the point where it intersects the potential $E = E(r, a_1)$ (turning point); then it moves in the opposite direction until it again intersects the potential and reverses its direction, thus completing a finite motion in the "potential well." Since the energy in this instance is $E_1 < 1$, and since the energy in GTR includes mc^2 ($mc^2 = 1$ in our units), the particle does not escape to infinity.

If the energy of the particle is $E_2 > 1$ (see Fig. 9), it follows a hyperbola coming in from infinity, reaches a minimum r at the intersection of E with the curve $E = E(r, a_1)$, and then reverses its direction and escapes to infinity. Since the potential curves approach $E = \infty$ as $r \to 0$ (see Fig. 9), for any arbitrarily large energy the incoming particle must whirl around the center of attraction and fly back to infinity, provided, of course, that the particle does not hit the surface of the attracting body. The gravitational capture of one point body by another is impossible in Newtonian theory.

Let us now consider the relativistic theory, i.e., the precise equation (3.6.1a). Here the potential curves have a different form (Fig. 10). Because of the term a^2/r^3, the potential does not rise indefinitely at small r, as it does in Newtonian theory; rather, it curves back downward and reaches zero at the gravitational radius, $r = 1$. Such a curve is shown in Figure 10; it has both a minimum and a maximum.

The motion of a test particle with $E_1 < 1$ in the relativistic potential well (Fig. 10) is similar to the Newtonian motion analyzed above; except that, by contrast with the Newtonian orbit, the relativistic orbit is not a closed curve (see Hagihara's article [1931]). In the Newtonian problem, the period of radial oscillations "accidentally" equals the time for ϕ to change by 2π, which causes the orbit to be closed; in GTR this is not the case. The famous shift of Mercury's perihelion by 42 seconds of arc per century is a manifestation of this.

When $1 < E_2 < E_{max}$ (see Fig. 10) the horizontal line $E_2 = $ const. goes to $r = \infty$ on the right, but on the left it intersects the potential and stops (turning point). In this case the particle comes in from infinity and then returns to infinity in a manner similar to the hyperbolic motion of Newtonian theory.

An important characteristic of the potential curve in the Schwarzschild field is the presence of a maximum. For a particle with $E_3 > E_{max}$, the horizontal line $E = E_3$ does not meet the potential curve. Such a particle reaches the gravitational radius ($r = 1$ in our units) and does not return to infinity. Gravitational capture of the particle occurs. We will discuss in greater detail later this particular characteristic of the relativistic theory.

There is one more peculiar circumstance. If the particle has an energy slightly less than E_{max}, then near the turning point a graph of the right-hand side of equation (3.6.1a) approaches zero with arbitrarily small slope; that is, when r changes by a small amount dr, the particle has enough time to move through an arbitrarily large angle ϕ. This means that near r_{min} the particle may circle the central body many times before it flies back toward infinity. In this case, the orbit near r_{min} does not look like the Newtonian hyperbola at all. When $E = E_{max}$, the particle moves endlessly around the circle $r = r_{E_{max}}$.

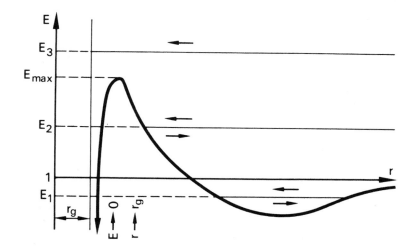

Fig. 10.—The potential curve for particle motion in GTR with fixed angular momentum a_1. $E_1 < 1$ is the horizontal line for finite motion; $1 < E_2 < E_{max}$ is the horizontal line for hyperbolic motion; when $E_3 > E_{max}$, the particle reaches the gravitational radius and never returns to infinity.

3.7   CIRCULAR ORBITS

At an extremum of the potential curve $E(r, a_1)$, $dr \equiv 0$, and the particle moves in a circle with r = constant. Obviously, the circular motion at a minimum of the potential is stable: if the particle's energy E and angular momentum a_1 are perturbed slightly, it will perform finite motion (Fig. 11), corresponding to $E = E_{min} + \delta E_1$, near the bottom of the potential $E = E(r, a_1 + \delta a_1)$. The new trajectory differs very little from the unperturbed circle.

Motion along the circle $r_{E_{max}}$ at the maximum of the potential, E, is unstable: a small perturbation will force the particle either to escape to infinity or to fall toward the gravitational radius.

We have seen that in Newtonian theory the potential curve has a minimum for any a. It follows that in the Newtonian theory there exists a stable circular orbit for any a. The smaller a is, the closer is the orbit to the center of attraction; and when $a \to 0$, we have $r \to 0$. Such is not the case in the Einsteinian theory: on the contrary, there exists a minimum radius for stable circular orbits, and correspondingly a minimum energy for circular motion. This was first pointed out by Hagihara (1931) and later by Kaplan (1949a). To prove this, it is sufficient to plot curves $E = E(r, a)$ for different a (Fig. 12).

We see that when $a < \sqrt{3}$, the curves have no extrema. When $a > \sqrt{3}$, each curve has two extrema—one minimum and one maximum (denoted

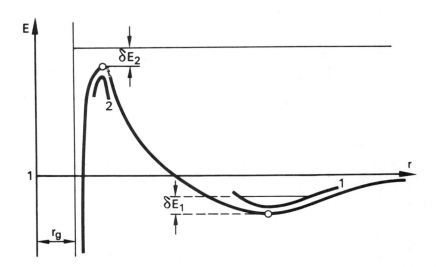

Fig. 11.—Motion along a circular orbit is stable at the minimum of the effective potential and is unstable at the maximum. (1) Potential curve $E = E(r, a_1 + \delta a_1)$; (2) potential curve $E = E(r, a_2 + \delta a_2)$. This figure shows the cases $\delta a_1 > 0$ and $\delta a_2 < 0$.

on Fig. 12 by vertical dashes). The minima correspond to stable orbits and have $r > 3$, and accordingly $\sqrt{(\frac{8}{9})} < E_{\min} < 1$. When a increases from $\sqrt{3}$ to ∞, the coordinate radius of the maximum decreases monotonically from $r = 3$ to $r = \frac{3}{2}$, while the energy $E_{\max}$ increases from $E_{\max} = \sqrt{(\frac{8}{9})} = 0.943$ to $E_{\max} = \infty$.

Thus, the stable circular orbit closest to the center has $r = 3$. The velocity of motion on it is $v_{\mathrm{crit}} = \frac{1}{2}c$, and the corresponding minimum energy is $E_{\mathrm{crit}} = 0.943 \, mc^2$.

Let us recall that for a distant observer all processes in the gravitational field take place with a retardation which is proportional to $\sqrt{g_{00}} =$

$\sqrt{(1 - r_g/r)}$ (see §§ 3.2 and 3.3). This observer will see the motion of the particle along the critical circular orbit with a period of

$$T = \frac{12\pi}{(2/3)^{1/2}} \frac{r_g}{c} .$$

If the particle has a monochromatic emitter with a frequency ω_0, then the light frequency (emitted along or against v) which is received by a distant observer is given by (cf. § 3.5)

$$\omega = \omega_0 \left(1 - \frac{r_g}{r}\right)^{1/2} \left(\frac{1 \pm v/c}{1 \mp v/c}\right)^{1/2} .$$

The first factor after ω_0 produces the time retardation in the gravitational field (gravitational redshift), while the second factor produces the Doppler shift. For a particle moving around the circle at r_{crit} with the plane of its

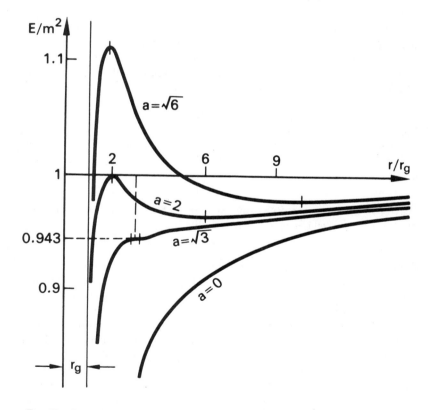

F<small>IG</small>. 12.—Potential curves for different angular momenta a. Curves are labeled by the angular momentum a in units of mcr_g.

orbit passing through the angle of vision of the observer, this formula yields the following: For photons emitted when the particle is moving toward the observer, $\omega = 2^{1/2}\omega_0$ is the blue shift; for photons emitted during motion away from the observer, $\omega = \frac{2}{3}^{1/2}\omega_0$ is the redshift; for a source at rest at this same radius $r_{crit} = 3r_g$, $\omega = \sqrt{(\frac{2}{3})}\omega_0$ is the gravitational redshift.

Closer to the center of attraction, in the interval $\frac{3}{2} < r < 3$, are the unstable circular orbits. The velocity of motion along the smallest of them—the one with $r = \frac{3}{2}$—is the speed of light, $v = c$, which corresponds to infinite energy, $E = \infty$. Closer to the gravitational radius (recall that $r = r_g$ corresponds to $r = 1$ in our units) there are no circular orbits whatsoever; this was first pointed out by Einstein.

3.8 THE MOTION OF A RELATIVISTIC PARTICLE IN A COULOMB FIELD

Let us sidetrack into a different area for a short time and consider the following problem: Analyze the circular motion of a charged particle in a strong Coulomb field. The conclusions obtained from this problem will be useful for understanding the properties of dense stars.

A charged particle in a strong electric field moves at a relativistic velocity. The equation of motion for a charge e in a time-independent field E^* is

$$dp/dt = eE^* .$$

For our problem $E^* = (Z/r^2)n$, where Z is the charge of the central body and n is the unit vector in the radial direction; also, $p = m(1 - v^2/c^2)^{-1/2}v$. Consequently, we obtain for the circular motion of the charge in a Coulomb field

$$m(1 - v^2/c^2)^{-1/2}v^2 = eZ/r . \qquad (3.8.1)$$

When $r \to 0$, then $v \to c$. Rewriting equation (3.8.1) in terms of the angular momentum, $a = m(1 - v^2/c^2)^{-1/2}vr$, we obtain

$$a = eZ/v .$$

It follows from this expression that when the radius of the orbit approaches zero ($r \to 0$) and consequently $v \to c$, the angular momentum does not approach zero as in the nonrelativistic theory, but instead approaches a finite value $a_{min} = eZ/c$.

It seems reasonable that this should be true also for the motion of a relativistic particle in a circular orbit in a Newtonian gravitational field. Such a consideration, however, is inconsistent, because when the velocity of the particle on the circular orbit is comparable with c, there are changes from the Newtonian law of gravitation. However, it would be useful to keep in mind (and this will be applicable also to future discussions) that the inclusion of only special-relativistic effects produces a finite angular momentum for an orbit of zero radius.

3.9 Gravitational Capture of a Nonrelativistic Particle

Thus, in the nonrelativistic theory there exist stable circular orbits with an arbitrary r. When $r \to 0$, the angular momentum a also approaches zero (Fig. 13). In an inconsistent theory, which considers only the effects of STR, the circular orbits can have any radius r. When $r \to 0$, the angular momentum $a \to$ const. (see Fig. 13). In GTR, there exists a minimum radius for a stable circular orbit, r_{min}, with a corresponding minimum angular momentum, a_{min} (see Fig. 13).

3.9 GRAVITATIONAL CAPTURE OF A NONRELATIVISTIC PARTICLE

Consider a case that is important for physical applications—the motion of a body which, at infinity, has a velocity v_∞ that is negligibly small compared with c, and hence for which $E = 1$. We shall trace qualitatively the motion of such a body for various a. On a graph of E versus r this motion is represented by the horizontal line $E = 1$ (Fig. 12). If the angular momen-

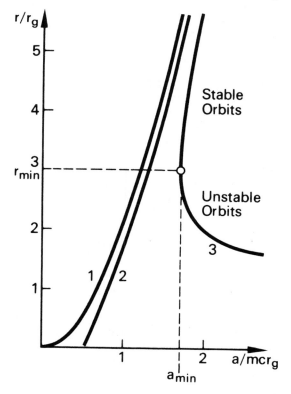

FIG. 13.—Relationship between the radius of the circular orbit, r, and the angular momentum a: (1) in Newtonian theory; (2) in the special theory of relativity; (3) in the general theory of relativity.

tum at infinity is smaller than $a_{crit} = 2$, then the horizontal line $E = 1$ does not meet the potential curve $E = E(r, a)$; there is no turning point; the trajectory of the particle extends into the Schwarzschild sphere.

When $a = a_{crit} = 2$, the trajectory coils itself onto the sphere at $r = 2$. However, if $a > 2$, the body reaches a turning point and then flies back to infinity.

When a is only slightly greater than $a_{crit} = 2$, the particle performs many revolutions near $r = 2$ before flying back. The asymptotic formula for the number of revolutions (Zel'dovich and Novikov 1964b) is

$$N = - \frac{\ln (a - 2)}{2^{3/2}\pi} \cdot$$

Consider now the problem of gravitational capture. According to the Newtonian theory, a particle coming in from infinity, provided it does not hit the surface of the central body, is deflected back toward infinity, and hence gravitational capture is impossible. In the Einsteinian theory, as we have seen, a particle with $a \leq 2$ is captured gravitationally; it does not escape back to infinity. The capture cross-section is determined by the relationship

$$\sigma_E = 4\pi(c/v_\infty)^2 r_g{}^2, \quad v_\infty \ll c. \tag{3.9.1}$$

Compare this capture with the "geometrical capture" of a particle by a gravitating sphere of radius R in Newtonian theory, i.e., the capture when a particle near its periastron encounters the surface of the sphere. In this case, the capture cross-section is

$$\sigma_N = \pi R^2(1 + 2Gm/v_\infty{}^2 R), \tag{3.9.2}$$

where R is the radius of the sphere. Comparing equations (3.9.1) and (3.9.2), we see that in the relativistic case the capture takes place just as it would in Newtonian theory for a central body of radius $R = 4r_g$ and for V_∞ small.

Let us stress the following: In Newtonian theory, capture onto a sphere takes place with an impact against the sphere's surface. In the Schwarzschild field, the captured body approaches the Schwarzschild sphere along a spiraling trajectory after completing a finite number of revolutions, and it asymptotically decelerates its velocity as measured by a distant observer. This approach to the Schwarzschild sphere requires infinite time insofar as the distant observer is concerned, as we saw in detail for radial motion in § 3.4. There is no impact whatsoever in this case. Moreover, the trajectory always approaches the Schwarzschild sphere perpendicularly, along the radial direction (see eqs. [3.6.1a, b]). Therefore, all formulae cited in § 3.4 for a particle that is incident along the radial direction will be asymptotically valid near the Schwarzschild sphere also in the general case of nonzero angular momentum a.[11]

---

11. Naturally, it is assumed that in the relativistic case the central mass has already collapsed, so that the particle does not encounter its surface.

3.10 MOTION OF ULTRARELATIVISTIC PARTICLES AND OF LIGHT RAYS

Consider now the opposite type of particle, one which is ultrarelativistic everywhere, including infinity. Photons and neutrinos are always of this type.

The equation for a particle moving in the Schwarzschild field with the fundamental velocity c is obtained from equations (3.6.1a,b) by the asymptotic transition $v_\infty \rightarrow c$, which corresponds to $E \rightarrow \infty$, $a \rightarrow \infty$. These infinities obviously result from our normalization of the energy to mc^2, and of the angular momentum to mr_0c. Noting that when $E \rightarrow \infty$, a/E approaches l, where l is the impact parameter of the particle's trajectory at infinity, we obtain in the limit $E \rightarrow \infty$:

$$(dx/d\tau)^2 = 1 - l^2/r^2 + l^2r^3 , \qquad (3.10.1a)$$

$$(d\phi/d\tau)^2 = (l^2/r^4)(1 - 1/r) . \qquad (3.10.1b)$$

In flat space the term l^2/r^3 in equation (3.10.1a) and the term $1/r$ in equation (3.10.1b) are absent, and $x = r$. In this case we have uniform motion along a straight line.

The presence of the term l^2/r^3 and the difference between r and x result in the deflection of a light beam which passes near a gravitating mass. For large l (and hence also large $r_{\min}$), this deflection is insignificant. For a beam that grazes the Sun's surface, it amounts to $1".75$. It was this prediction of Einstein, brilliantly confirmed during the total solar eclipse of 1918, which was one of the first experimental proofs of the validity of the general theory of relativity.

For small r, the trajectory of the ray will differ considerably from a straight line. The "turning point" along an orbit, which is given by the function $r_{\min}(l)$, is shown in Figure 14. It follows from this diagram that a ray (or an ultrarelativistic particle) arriving from infinity with an impact parameter $l \leq \frac{3}{2}\sqrt{3} = 2.6$ (recall that distance is measured in units of r_0) does not encounter the turning-point curve, and as a consequence it is gravitationally captured. In this case, as in the case of a nonrelativistic particle, the trajectory approaches the Schwarzschild sphere perpendicularly. The asymptotic formulae near the Schwarzschild sphere, given in § 3.4 for radial motion, are also valid here. Specifically, the time required for a ray to reach the Schwarzschild sphere, as seen by an outside observer, is expanded to infinity.

Thus, the cross-section for the gravitational capture of an ultrarelativistic particle is $\sigma = 27\pi r_0^2/4$.

Notice that a light ray emitted by a source at rest at radius r cannot escape to infinity for all angles of departure. In Figure 15 the rays going inside the shaded cone do not escape to infinity, whereas the rays marked by arrows do. The formula for the half-angle ψ of the capture cone (see Fig. 15) is

$$|\tan \psi| \;=\; \frac{1 - 1/r}{(1/r - 1 + \frac{4}{27}\, r^2)^{1/2}} \,. \qquad (3.10.2)$$

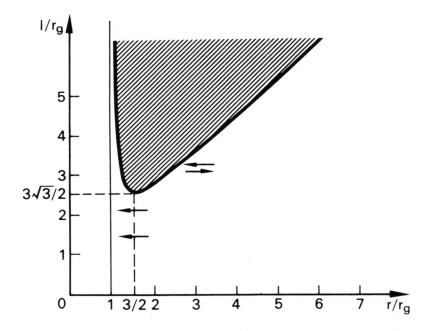

FIG. 14.—Relationship between r_{min} for an ultrarelativistic particle and the impact parameter at infinity, l. Particles with $l/r_g < 3\sqrt{3}/2$ are gravitationally captured. Domain where motion is impossible is shaded.

3.11   PARTICLE MOTION IN THE SCHWARZSCHILD GRAVITATIONAL FIELD, INCLUDING THE EFFECTS OF GRAVITATIONAL RADIATION

The effects of gravitational radiation, even in a weak gravitational field, qualitatively change the nature of the particle motion if the energy of the motion at infinity is small. Thus, for example, a body having very small energy at infinity, will, upon passing near an attracting center and radiating energy as gravitational waves, be left with negative energy; i.e., it will be gravitationally captured. However, such a capture is purely formalistic, since the energy radiated in a weak field is extremely small. From equation (3.11.1) derived below, it follows that for two normal stars of equal mass,

106

which approach each other from infinity to a distance of the order of their dimensions, capture takes place only if their relative speed at infinity is less than approximately 1 cm sec^{-1}.[12] Their maximum separation after "capture" (if the velocity is $v_\infty \ll 1$ cm sec^{-1}) will be $\sim 3 \times 10^{26}$ cm $= 10^8$ pc(!). We recall that the dimension of our Galaxy is four orders of magnitude smaller than this.

Consider now the problem of the radiation of gravitational waves by bodies moving in a Schwarzschild field at distances comparable to r_g, where the radiation is large. The existing theory of gravitational radiation (see

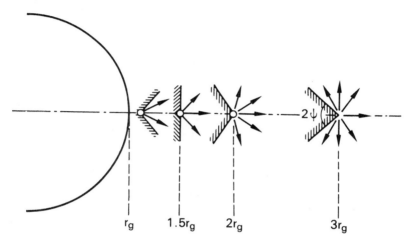

Fig. 15.—Gravitational capture of radiation: rays emitted from each point into the interior of the shaded conical cavity are captured gravitationally.

Landau and Lifshitz 1962; Eddington 1925) is applicable only for processes taking place in weak gravitational fields. This theory was described in detail in §§ 1.12 and 1.13. However, from dimensional considerations it is apparent that the estimates of the orders of magnitude cited in this section must be correct even when motion takes place at distances which are comparable to the gravitational radius of the central body. In this connection, another comment must be made: Just as a charge moving uniformly around a circle with a velocity $v \approx c$ emits primarily higher harmonics, so also the gravitational waves from a body with $v \approx c$ in a strong gravitational field must consist primarily of high harmonics (see the contribution by Pustovoyt and Gertsenshteyn 1962). However, in the gravitational problem, $v \approx c$ is attained only near the gravitational radius, where the amount of radiation is cut down by effects of GTR (gravitational redshift and gravitational capture of

12. In reality, the dispersion in the velocities of stars in the Galaxy near the Sun is of the order of tens of kilometers per second.

radiation); and even when r substantially exceeds r_g, these effects do not influence our order-of-magnitude estimates.

An important aspect of gravitational radiation is the following: When two bodies attracting each other approach to within a distance of the order of their gravitational radii, the total amount of radiated energy must be a function only of their masses, and of G and c. From dimensional considerations, it follows immediately that the small constant G cannot enter the formula at all, and the total energy emitted must be of the order of mc^2 multiplied by a function of the ratio of the masses of the bodies, m/M. If m and M are of the same order, we conclude that the total energy radiated is not small compared with mc^2, where m is the mass of the smaller body (Dyson 1963; Zel'dovich and Novikov 1964b; Fowler 1965). The formulae are cited below. The total radiation is smaller than mc^2 only on account of a dimensionless, numerical coefficient. Consider the effect of the radiation of gravitational waves upon the motion of the mass m (Zel'dovich and Novikov 1964b; see also Smith and Havas 1965). The radiation produces a force on the body, which in turn results in a peculiar radiative gravitational "friction." The frictional force is caused by the interaction of the mass m with its own gravitational field and hence is proportional to m^2, in contrast to the interaction force with an external gravitational field, which is proportional to m. Thus, the change in the motion of the body as a result of its radiation of gravitational waves when $m/M \ll 1$ can be regarded as a small correction to the nonradiative motion in the external field.

In the motion of a nonrelativistic particle m arriving from infinity, the main part of the radiated energy is emitted near the turning point of the trajectory, i.e., near the periastron.

Using the general expression (1.12.8) for the power in the gravitational waves, we can estimate the total energy radiated, ΔE, and the radiation time Δt:

$$\Delta E = \frac{c^2 m^2}{M} \left(\frac{r_g}{r}\right)^{7/2}, \qquad \Delta t = \frac{r^{3/2}}{(2GM)^{1/2}},$$

where r is the radial coordinate of the periastron. The energy lost in radiation results in the gravitational capture of the body by the mass M for values of the angular momentum a which considerably exceed $a = 2$; $a < 2$ is the criterion for capture in the purely mechanical problem of § 3.9.

If the effects of radiation are included, the critical values of the angular momentum and the capture cross-section, a_E and σ_E, depend on the parameter $x = (c/v_\infty)^2 (m/M)$, and are determined as follows:

for $\quad x \gg 10$, $\quad a_E = (2x)^{1/7}$, $\quad \sigma_E = \pi(c/v_\infty)^2 (2x)^{2/7} r_g^2$;

for $\quad x \ll 10$, $\quad a_E = 2 + e^{-20/x}$, $\quad \sigma_E = 4\pi(c/v_\infty)^2 (1 + e^{-20/x}) r_g^2$.

For example, for $v_\infty \approx 10^6$ cm sec^{-1} and $m/M \approx 0.1$ we find that $x \approx 10^8$ and hence $a_E \approx 10$; the capture cross-section σ is larger by a factor of 25 than it would be without the inclusion of the radiation. Of course, these results are only rough approximations, because they are based on the weak-field theory.

As a result of the capture, the body m, upon passing through its periastron, moves away from m no longer to infinity, but to a distance of the order of

$$L \approx \frac{r_g}{2[(m/M)(r_g/r)^{3/5} - (v_\infty{}^2/2c^2)]} \,. \tag{3.11.1}$$

When v_∞ is small and $r = 3r_g$, we obtain $L \approx 600r_g$. As it passes through its periastron a second time, the body emits more energy, and so on. The elongation of the orbit decreases rapidly (see § 1.13).

How does the gravitational radiation affect the circular motion of a particle? This motion is represented by the minima of the curves in Figure 12. As a result of the radiation, the point in the diagram that characterizes the particle's orbit moves from the minimum of one curve inward to the minimum of another, etc. At the beginning, when r is very big, this evolution is very slow (see § 1.13). The radiated power in a circular orbit is determined by equation (1.13.2), which is more conveniently expressed in the form

$$\frac{dE}{dt} = 0.2 \frac{c^5}{G} \left(\frac{m}{M}\right)^2 \left(1 + \frac{m}{M}\right) \left(\frac{r_g}{r}\right)^5 .$$

For ordinary binary stars, the energy loss per year amounts to 10^{-12} of the total energy. When r is small, the evolution rate is considerably greater. For some known close-binary systems, the period of rotation decreases annually by a fraction of 10^{-9}. The circular motion continues down to the last stable orbit with $r_{\text{crit}} = 3r_g$ (see § 3.7). At that point begins the spiraling fall down to the Schwarzschild sphere. The energy of the circling body at the critical radius amounts to 0.943 of its total energy (all rest mass) in an orbit at a large radius. Consequently, the total amount of energy radiated, $\Delta E = 0.06\ mc^2$, does not depend on the mass of the central body. The smaller the ratio m/M, the more revolutions the body must go through in order to emit the energy ΔE and reach r_{crit}.

During one revolution at the critical orbit an energy of $\sim 0.1\ m^2c^2/M$ is emitted. The body passes to a spiraling orbit, which heads toward the Schwarzschild sphere. In this orbit, another $\sim (M/m)^{1/3}$ revolutions are performed. The de-excitation energy emitted in one revolution is always of the same order as when $r = 3r_g$. Thus if $m/M \ll 1$, then upon reaching the critical orbit the body falls to the sphere of gravitational radius, for all practical purposes, without emitting any further energy.

If $m/M \sim 1$, then the number of revolutions the body makes upon reaching the critical orbit is of the order of unity, and the radiated energy is of the same order as it was prior to reaching this orbit. Even though the force of radiative friction is no longer a small correction to the force of the external field, by considerations of dimensionality, symmetry, and the form of the formula for $M \gg m$, we can write an approximate formula for the total energy radiated which is also valid when $m/M \sim 1$:

$$\Delta E_{\text{finite motion}} = a \, \frac{c^2 m M}{m + M} \, , \qquad (3.11.2)$$

where a is of the order of 0.06.

There exists another formula for the total radiated energy from two masses falling toward each other with zero angular momentum (the motion is radial, along a straight line):

$$\Delta E_{\text{radial motion}} = \beta \, \frac{c^2 m^2 M^2}{(m + M)^3} \, . \qquad (3.11.3)$$

Here, $\beta \approx 0.02$; i.e., it is of the same order as a. This formula is applicable for any m/M and is derived from the same considerations as equation (3.11.2). Thus, as a result of gravitational radiation, the system may lose several percent of its rest mass-energy.

3.12 THE R- AND T-REGIONS OF SCHWARZSCHILD SPACETIME

It was demonstrated in § 3.2 that the radius $r = r_g = 2Gm/c^2$ is of crucial significance. The gravitational force F at $r = r_g$ is infinite. Obviously, no static body can have dimensions less than r_g. But what will happen to a dynamically contracting body at $r = r_g$—e.g., a spherical ball of dust particles which collapses under its own gravitational attraction and reaches r_g?

During the process of contraction, the mass M of the ball does not change. Therefore, the dust particles on the surface of the ball simply fall under the influence of a Schwarzschild field with mass M. As we saw in the discussion in § 3.5, the time required to fall to r_g as measured by the falling particles is finite. The ball, in the course of finite proper time, will contract to r_g and must then continue to contract further. However, a nondeforming coordinate system with $g_{22} = -e^\mu = -(x^1)^2$ cannot be used inside the Schwarzschild sphere, because in this region there exist no nondeforming reference frames (Finkelstein 1958; Novikov 1961, 1962b, c). At the Schwarzschild sphere there is no singularity in the geometry of four-dimensional spacetime (i.e., no so-called intrinsic singularity). In particular, the simplest nonvanishing invariant of the spacetime curvature in this case is

$$C = R_{iklm} R^{iklm} = 12 r_g^2 / r^6 \, ,$$

which has no singularity at $r = r_g$.

Let us now follow a contracting ball as its surface falls through the Schwarzschild sphere. To analyze the vacuum gravitational field outside the ball, it is useful to introduce a reference frame made up of freely falling test particles which have zero velocity at spatial infinity (the Lemaître [1933] reference frame; see also Rylov 1961). The motions of such test particles were treated in § 3.5. In contrast to the Schwarzschild reference frame, this reference frame covers the spacetime region inside the Schwarzschild sphere as well as the external region.

The square of the interval in this reference frame takes the form

$$ds^2 = c^2 d\tau^2 - \left[\frac{3}{2}\frac{R - c\tau}{r_g}\right]^{-2/3} dR^2$$

$$- [\tfrac{3}{2}(R - c\tau)]^{4/3} r_g{}^{2/3}(d\theta^2 + \sin^2\theta d\phi^2) \,. \tag{3.12.1a}$$

This reference frame can be obtained from the Schwarzschild frame by setting $r = r_g{}^{1/3}[\tfrac{3}{2}(R - c\tau)]^{2/3}$, and choosing the time coordinate, τ, orthogonal to the radial coordinate, R. The moment τ at which a falling particle at radius R crosses the Schwarzschild sphere is given by

$$\tfrac{3}{2}(R - c\tau) = r_g \,. \tag{3.12.2}$$

(Recall that the freely falling particles remain at rest—R, θ, ϕ constant—in this reference frame.)

Inside the contracting ball, the vacuum solution (3.12.1a) is not applicable; the solution (3.12.1a) must be joined to the ball's interior solution at the ball's surface (Oppenheimer and Snyder 1939; Tolman 1934a). For a contracting dust ball of uniform density, the interior solution has the form (see § 3.13 for details)

$$ds^2 = c^2 d\tau^2 - [\tfrac{3}{2}(R_0 - c\tau)]^{4/3}\frac{r_g{}^{3/2}}{R_0{}^2}[dR^2 + R^2(d\theta^2 + \sin^2\theta d\phi^2)] \,, \tag{3.12.1b}$$

where R_0 is the Lagrangian radius of the ball's surface. Note that $g_{\theta\theta}$ is continuous at the ball's surface; it describes the continuous increase of the radii of spheres about the ball's center with increasing R. However, g_{RR} is discontinuous at the ball's surface, where the matter density is discontinuous. This discontinuity is due to the fact that the change of scale with time inside the ball is different from that in the vacuum exterior.

Let us return to the properties of the vacuum spacetime inside the Schwarzschild sphere.

Using the criteria for R- and T-regions (see § 3.1), we find that the region outside the Schwarzschild sphere, i.e., where

$$\tfrac{3}{2}(R - c\tau) > r_g \,, \tag{3.12.3}$$

is an R-region; but the region inside the Schwarzschild sphere, where the inequality is opposite to inequality (3.12.3), is a T-region.

The spacetime in (R, τ) Lemaître coordinates is shown in Figure 16. The angular coordinates are of no interest because of spherical symmetry; so we will here, and henceforth, use spacetime diagrams with fixed ϕ and θ. Each particle passes $r = r_g$ and then, after finite time, reaches a singularity of spacetime, $r = 0$, where the curvature invariant,

$$C = \frac{12 r_g{}^2}{[\tfrac{3}{2}(R - c\tau)]^6} \,, \tag{3.12.4}$$

becomes infinite. The lines C = const. coincide with the lines $-g_{22} = r^2 = $ const.; i.e., they are the world lines of points on a fixed, spacelike two-sphere. The lines r = const. are shown as dashed lines in Figure 16.

In the T-region the quantity r, defined by $r^2 = -g_{22}$, is not a space coordinate, but has the nature of time (see § 1 of this chapter). For this reason, in the T-region we do not call r a "radius." On the same diagram (Fig. 16) are shown the "light cones" of the radial rays. They are determined by the equations

$$c(d\tau/dR) = \pm (r_0/r)^{1/2}, \quad r = [\tfrac{3}{2}(R - c\tau)]^{2/3}r_0^{1/3}. \quad (3.12.5)$$

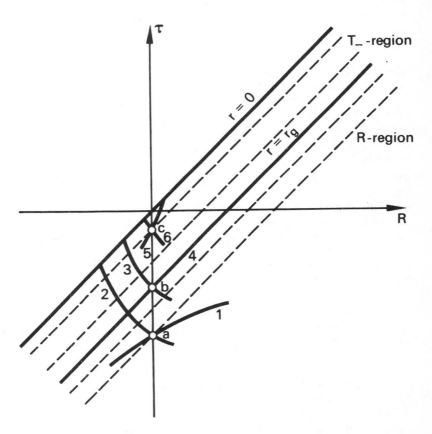

Fig. 16.—Schwarzschild spacetime in Lemaître coordinates with a contracting T-region. *Dashed lines*, lines of constant curvature invariant, C; in the R-region they coincide with r = const. Of the two radial light rays which pass through an event, a, in the R-region, one escapes to infinity (1), while the other propagates to the center (2). For an event, b, on the Schwarzschild sphere, one ray propagates to the center (3), while the other follows the world line of the gravitational radius (4). For an event, c, in the T-region, both rays (5, 6) propagate to the center.

3.12   R- and T-Regions of Spacetime

Outside the Schwarzschild sphere (R-region), the lines r = const. are inside the light cone (timelike), and one of the radial rays goes to the center (toward smaller r) while the other goes to infinity (toward larger r). In the T-domain the lines r = const. are outside the light cone (spacelike), and both rays propagate in the direction of increasing curvature, to the singularity at $r = 0$, $C = \infty$. The world line of any particle must lie inside the light cone. Consequently, any particle in the T-region moves toward the intrinsic singularity $r = 0$, $C = \infty$ and reaches it in finite proper time. Motion outward, from the singularity toward the Schwarzschild sphere, is impossible (Finkelstein 1958). This would amount to motion with speed greater than c.

The motion of the surface of any spherical body can be construed as the motion of a test particle (not free motion, in general, but rather motion under the influence of forces) in the external, Schwarzschild spherical gravitational field. Consequently, the surface of a spherical mass with any arbitrary equation of state (and not just dust with $p = 0$), upon contracting to dimensions smaller than r_0, must continue to contract until, after finite proper time, it is compressed to a point $r = 0$ with $C = \infty$. No internal pressure in the contracting sphere is capable of halting the contraction in the T-region. We will not discuss at this point the stability of the contraction (for that, see Vol. II) or the limits of applicability of GTR at large densities and curvatures C (see § 4.6) or the subsequent fate of the contracting matter (see § 4.6).

It should be stressed that light rays which escape from the surface of the body after it passes $r = r_0$ can never reach an external observer in the R-region. Hence, he cannot acquire any information about processes in the T-region; the regions are separated by an *event horizon*. This phenomenon is known in Russian as *gravitational self-locking* (samozamikaniye).

In the T-region there is an obvious asymmetry in the direction of time flow. In our case, all motions in the T-region are directed toward $r = 0$ where $C = \infty$; there exist no reverse motions. Such a T-region is best referred to as a *contracting* region and denoted by T_-.

The equations of GTR are invariant under time reversal. If we replace τ by $-\tau$ in equations (3.12.1)–(3.12.5), then we will obtain a second type of T-region—an *expanding* T_+-region, which has completely opposite properties (Fig. 17). Here all bodies move only from the singularity $C = \infty$ toward the outside. All light rays escape from inside the Schwarzschild sphere, and no light rays can enter it from the R-region. A vacuum T_+-region can be joined to the internal solution for an expanding sphere, but not to that for a contracting sphere as is the case for a T_--region. A simultaneous coexistence of T_-- and T_+-regions at the same place is obviously impossible.

Thus, the physical extension of the external Schwarzschild spacetime to the region inside the Schwarzschild sphere has a double-valued nature. In one extension the motions of all test particles and light rays are directed inward from the Schwarzschild sphere (T_--region). In the other extension, all motions are directed outward (T_+-region).

113

This ambiguity has been noted on several occasions. However, it is important to emphasize that the choice between the two extensions of the Schwarzschild solution into the T-region is not arbitrary, but is determined physically by the context in which this region occurs (Novikov 1964b, d). If the T-region originates in the compression of a ball to dimensions that are smaller than the gravitational radius—i.e., if the ball's surface, when crossing the Schwarzschild sphere, is moving inward—then, by continuity, the test particles near the surface must also move inward, and a T_--region arises. If, at the very beginning, we demand that the velocity of the ball's matter be directed outward (the ball's size being smaller than r_θ), then the solution inside the ball can be joined only to a vacuum T_+-region.

Consider how the expansion of a ball from inside the Schwarzschild sphere appears to an external observer. During collapse, an observer on the surface of a contracting star crosses the Schwarzschild sphere and reaches the cen-

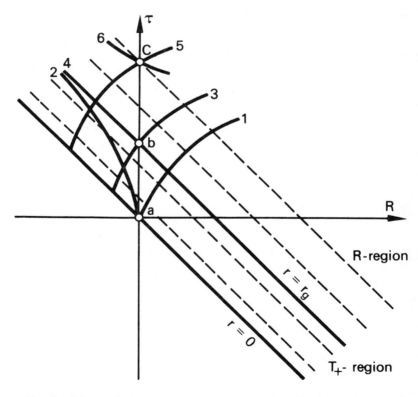

Fig. 17.—Schwarzschild spacetime in Lemaître coordinates with an expanding T_+-region. The radial light rays pass through events a, b, and c.

tral singularity, $C = \infty$, in finite proper time. We will examine this phenomenon with the flow of time reversed. The surface of the star, exploding outward from a point, crosses the Schwarzschild sphere in finite proper time, and continues to expand further. Since the time required to reach the Schwarzschild sphere is infinite as seen by an outside observer for the case of collapse, one might think that the outside observer will also see expansion from the Schwarzschild sphere as an infinitely long process, and will not be able to see what happened prior to the star's escape from under the critical sphere.

In reality, such is not the case. As noted previously, light rays escape freely from the T_+-region, so the picture of expansion as seen by an external observer is not a reversal in time of the contraction picture, but is fundamentally different (Novikov and Ozernoy 1963; Faulkner, Hoyle, and Narlikar 1964). The reasons for this are the following: From the mathematical standpoint the expansion and contraction are precise time reversals of each other. However, in both of these an external observer sees the star by means of outgoing light rays, whereas the time reversal changes outgoing rays into ingoing rays and conversely. The symmetry is broken. The effect of the time dilation of processes during collapse can be explained by the joint action of two phenomena: the retardation of the flow of time in a strong gravitational field, and the (generalized) Doppler effect due to the motion of the star's contracting surface away from the observer. Both phenomena contribute to the time dilation. For the case of expansion, the Doppler effect accelerates processes on the star as seen by an outside observer. It turns out that this effect is stronger than the retardation of processes due to the gravitational field. The external observer does not see the evolution begin from a "frozen" picture when $R = r_g$; rather, he sees the entire process of expansion, beginning from a point.

Let the surface of a ball expand with parabolic velocity (i.e., a velocity which becomes zero at spatial infinity). Figure 18 shows the graph of the variation of the light frequency with time for a ray reaching a distant observer from the center of the ball's visual disk. If the distant observer waits a time $t_\infty \approx 0.28 r_g/c$ after the arrival of the first rays from the initial expansion ($r = 0$, $c = \infty$), he then sees at the center of the disk rays emitted from the ball's surface at the moment when it crossed the Schwarzschild sphere. The frequency of these rays as measured by the distant observer is twice the frequency they had as measured by the emitters. At this moment, the observer sees a disk which has the angular diameter $\phi = 0.43 r_g/r$.

3.13 INTERNAL SOLUTION FOR A NONSTATIC SPHERE

For the interior of a spherical body the Einstein equations (3.1.2)–(3.1.5) cannot be solved analytically, in general, because of complications with pressure, not to mention energy transport, etc. Numerical methods must be used. The first numerical calculations for a contracting relativistic star were performed on computers independently by Podurets (1964*a*) and by May and White (1964, 1966). However, as was demonstrated in the preceding section, the qualitative properties of the motion of the surface of a collapsing star in the vacuum *T*-region do not depend on the equation of state of the matter.

It turns out that several important properties of the solution inside the star do not depend on the equation of state either, and can be obtained by examining the simplest case, $p = 0$ (dust). In this case, the Einstein equations can be integrated analytically (Tolman 1934*a*). We present the solution here (in a form devised by Landau and Lifshitz 1962), along with several conclusions which follow from it. We defer the analysis of other aspects of collapse to the next section of this book and to Vol. II.

The solution is most conveniently presented in comoving (Lagrangian) coordinates. In the absence of pressure, the dust particles move freely, and consequently $g_{00} \equiv 1$ (see § 1.6). The solution has the form

$$ds^2 = c^2 d\tau^2 - r^2(R, \tau)(d\theta^2 + \sin^2\theta \, d\phi^2) - e^\lambda dR^2 \,,$$

$$e^\lambda = \frac{r'^2}{1 + f(R)}\,, \qquad \left(\frac{\partial r}{c\partial\tau}\right)^2 = f(R) + \frac{F(R)}{r}\,, \qquad \frac{8\pi G\rho}{c^2} = \frac{F'(R)}{r'r^2}\,, \tag{3.13.1}$$

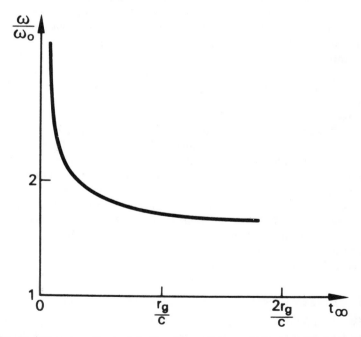

FIG. 18.—Time dependence of the frequency of light reaching a distant observer from the center of the apparent disk of a ball which expands with parabolic velocity.

where the prime indicates partial differentiation with respect to R. The equation for $\partial r/\partial \tau$ is easily integrated. The solution (3.13.1) depends (after the selection of the coordinates) on two arbitrary functions[13] of R, $f = f(R)$ and $F = F(R)$. From the equation for e^λ, it follows that $1 + f > 0$. The functions F and f determine the distribution and the velocity of the matter at an initial moment of time.

An important property of the Tolman solution (3.13.1) is this: Plot the functions f and F, which determine the solution, from the center of the sphere, where $r = 0$, outward along the radius R. The form of each function near the center out to a radius R_0 does not depend on how the function will be plotted beyond the radius R_0.

In other words, the properties of a solution inside the Lagrangian sphere R_0 do not at all depend on the (spherically symmetric!) distribution and motion of the matter outside the sphere. External matter might not exist at all, or it might extend to infinity; but this has no influence upon the matter inside R_0.

This conclusion enables one to ignore the effects of the matter of the expanding Universe[14] on the gravitational field near an isolated body.

If instead of dust we consider matter with nonzero pressure, then this conclusion is changed only as follows: When the solution outside the radius R_0 is modified, the interior will be changed by a perturbation that propagates inward with the speed of sound. In places which this perturbation has not yet reached, the solution as before is unaffected by the change in the external matter. Thus, the external matter in the spherically symmetric case of GTR (and in this case only!) has no gravitational effect upon the internal matter.

As in Newtonian theory, so also in GTR, a spherically symmetric distribution of matter (which moves only radially!) does not create a gravitational field inside a spherical cavity. This is easily demonstrated because in the cavity's vacuum the spherical field can only be a Schwarzschild field (Birkhoff 1923; also see § 3.2); but the field inside the cavity can have no singularities at its center, so it must be a massless Schwarzschild field—i.e., no field at all.

In conclusion, the following remark is in order. In Newtonian theory there exists no field inside a hollow sphere. However, if the potential ϕ is set to zero at infinity, it will be nonzero inside the sphere. Its value there will be

---

13. In reality, upon integrating the equation for $\partial r/\partial \tau$, there emerges a third function of R. However, if the R-coordinate is chosen to have some particular form as a function of proper radial distance at some fixed time τ_0, this third function is not arbitrary.

14. Provided, of course, that the motion of the matter is uniform and isotropic (see Vol. II).

equal to the work per unit mass which is required to remove a particle from the cavity to infinity. In the cavity, ϕ = const. $\neq 0$, and the potential is

$$\phi = \int_{R_1}^{\infty} (1/r)\rho 4\pi r^2 dr , \qquad (3.13.2)$$

where R_1 is the radius of the inner boundary of the matter. The addition of a spherical shell of matter to a sphere which already exists, of course, will not change anything inside; it will not create any field whatsoever. However, it will change the normalization of the potential. If, as before, we demand that $\phi_{\infty} = 0$, then inside the added shell the potential will increase by a constant value given by the integral (3.13.2), where R_1 and ρ now refer to the added shell.

The same applies to the quantity g_{00}, which in the Einstein theory plays a role analogous to the potential ϕ of Newtonian theory. Inside the cavity g_{00} is constant, but it does not equal its value at infinity: g_{00} = const. $\neq (g_{00})_{\infty}$. We will return to this point in § 10.7.

3.14 THE KRUSKAL METRIC

Let us return to the vacuum Schwarzschild solution. We have already shown that the Schwarzschild coordinate system (3.2.1) is applicable only in the R-region; it does not cover the entire spacetime. The Lemaître coordinate system (3.12.1) is applicable both in the R- and T-regions. However, as demonstrated by Kruskal (1960), this system is also incomplete in a certain sense.

To explain this, we will temporarily refrain from joining the vacuum Schwarzschild solution to a solution for the matter which creates the gravitational field; we will extend the vacuum solution throughout its maximum possible world domain. Such a domain in Lemaître coordinates R, τ (with a T_+-region for concreteness; i.e., with the substitution $\tau \rightarrow -\tau$ in eqs. [3.12.1]–[3.12.5]) is the domain shown in Figure 19a—the half-plane to the right of the singularity line $r = 0$. Does this domain cover the entire spacetime, i.e., the history of all moving particles?

Consider a particle falling freely toward the gravitational radius $r = r_g$. Its world line relative to the expanding Lemaître coordinates has the form $(r_g = 1, c = 1)$

$$\text{const.} = R + 2\tau + 4[\tfrac{3}{2}(R + \tau)]^{1/3} + 2 \ln \left| \frac{[\tfrac{3}{2}(R + \tau)]^{1/3} - 1}{[\tfrac{3}{2}(R + \tau)]^{1/3} + 1} \right| . \quad (3.14.1)$$

This world line is shown in Figure 19a. Notice that it asymptotically approaches $r = r_g$. However, we know (see § 3.5) that after finite proper time the particle will reach $r = r_g$, and it must move onward from there. The expanding Lemaître coordinates do not cover those events in the life of the particle which take place after it reaches r_g; consequently, this coordinate system is incomplete. It fully describes the histories of particles that fly outward from under the Schwarzschild sphere, but it does not describe the entire histories of particles that fall inward. Similarly, a contracting Lemaître system (with a T_--region) cannot describe the early history of a particle that flies outward (Fig. 19*b*). Thus, the Lemaître system is incomplete in the sense that it does not include all the events taking place in the vacuum spacetime.

If, instead of considering a spacetime that is everywhere empty, we join the vacuum Lemaître solution to the interior material solution (3.13.1), then the resultant solution, of course, is complete and determines the entire history of every particle. As an example, consider a contracting sphere of homogeneous dust, whose particles had zero velocity at infinity. The spacetime geometry is described by formula (3.12.1a) inside the sphere and by formula (3.12.1b) outside.

If we take into account the matter which creates the spherical field, we must replace Figure 19*b* by Figure 19*c*. The region containing matter is shaded in the figures. At the moment $\tau = 0$ the matter collapses into a point. Focus attention on the world line *a–a'* of a particle which falls into the sphere from infinity, passes through its center, and reemerges out the other side before the sphere crosses its Schwarzschild radius. The world line of Figure 19*c* describes the entire history of this particle—infall, passage through $r = 0$ at time $\tau = \tau_0$, and reemergence out the other side. Similarly, Figure 19*c* describes the entire history of all other particles in spacetime. Thus, the star's spacetime is complete.

The solution for an exploding sphere can be obtained from equations (3.12.1a) and (3.12.1b) by changing the sign of τ. The general picture corresponding to Figure 19*a* will then be replaced by Figure 19*d*. Spacetime is complete here as well.

Of didactic interest is the case of a sphere which does not contract from infinity and does not expand to infinity, but instead has a velocity smaller than parabolic. Such a sphere expands from zero radius to a certain maximum radius and then contracts again. The full spacetime is described (qualitatively; for full detail see below) by Figure 19*e*. The sphere begins to expand from a point at the moment τ_1. Its boundary passes outward through the Schwarzschild sphere at the moment τ_2; it reaches its maximum radius at the moment τ_3; it recontracts through the Schwarzschild sphere at τ_4; and it collapses to the point $r = 0$ at the moment τ_5. This spacetime, like the others, is complete. Spatial infinity in Figure 19*e* is in the right-hand direction. In this case, the solution has both a T_+-region and a T_--region; but, of course, these two regions are in different parts of spacetime.

Now let us return to the more formal approach of considering an everywhere-empty spacetime.

Kruskal (1960) was the first to find a coordinate system that is complete in the sense that it includes all the events that take place in the vacuum

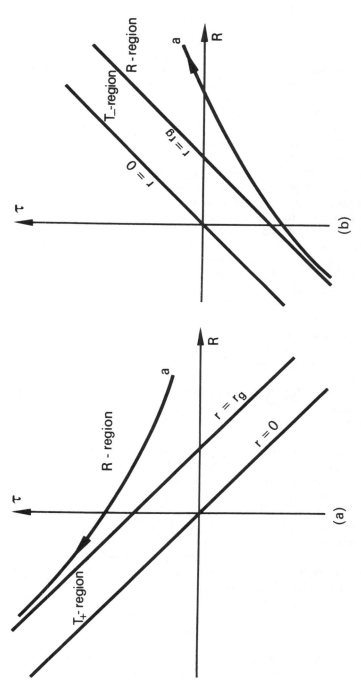

(a)

(b)

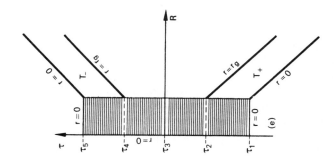

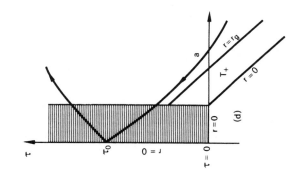

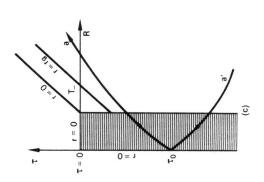

FIG. 19.—(a) World line, a, of a freely falling particle in an expanding Lemaître coordinate system. (b) World line, a, of a particle flying outward with parabolic velocity in a contracting Lemaître coordinate system. (c) World line, a, of a particle that falls radially into and flies through a contracting sphere of dust. (d) World line, a, of a particle that falls into and flies through an expanding sphere of dust. (e) A sphere of dust that expands out of zero volume to a maximum radius, and then recontracts to zero volume.

spacetime. We will present here (with a view to the most recent applications and developments) another coordinate system (Novikov 1963, 1964d), which is just as complete as the Kruskal system but which is generated by a system of freely falling test particles. (By contrast, the particles which map out the Kruskal system do not move freely.)

This system can be derived from Tolman's general solution (3.13.1); in vacuum the last of equations (3.13.1) guarantees that $F = F_0 = $ const. We choose the origin of our time coordinate so that $\dot{r} = 0$ everywhere when $\tau^* = 0$; and we choose the scale of the radial coordinate so that when $\tau^* = 0$, we have $r = r_g(R^{*2} + 1)$.[15] These conditions fully determine the coordinate system. F and f have the following values:

$$F = r_g , \quad f = -1/(R^{*2} + 1) .$$

The solution (3.13.1) can then be put into the following form:

$$ds^2 = c^2 d\tau^{*2} - e^\lambda dR^{*2} - r^2(R^*, \tau^*)(d\theta^2 + \sin^2\theta d\phi^2) , \quad (3.14.2)$$

$$e^\lambda = r'^2(R^{*2} + 1)/R^{*2} , \quad (3.14.3)$$

$$(r^*/r_g c) = -(R^{*2} + 1)\left(-\frac{r^2/r_g^2}{R^{*2} + 1} + \frac{r}{r_g}\right)^{1/2} + (R^{*2} + 1)^{3/2}$$
$$\times \left[\arcsin\left(\frac{r/r_g}{R^{*2} + 1}\right)^{1/2} - \tfrac{1}{2}\pi\right]. \quad (3.14.4)$$

The last equation determines the function $r = r(R^*, \tau^*)$ implicitly. In this coordinate system the singularity $(r = 0)$ and the gravitational radius $(r = r_g)$ are located at

$$r = 0 , \quad (\tau^*/r_g c) = \pm\tfrac{1}{2}\pi(R^{*2} + 1)^{3/2} , \quad (3.14.5)$$

$$r = r_g , \quad (\tau^*/r_g c) = -(R^{*2} + 1)\left(1 - \frac{1}{R^{*2} + 1}\right)^{1/2} + (R^{*2} + 1)^{3/2}$$
$$\times \left[\arcsin\left(\frac{1}{R^{*2} + 1}\right)^{1/2} - \tfrac{1}{2}\pi\right]. \quad (3.14.6)$$

Let us study the vacuum Schwarzschild solution in these coordinates.

Each particle at rest in the coordinate system, $R^* = $ const. $= R^*_0$, begins its history at the intrinsic singularity, $r = 0$, at the moment of time τ^* given by equation (3.14.5) with the minus sign. The particle flies outward through the T_+-region from $r = 0$ to $r = r_g$, where it escapes from under the Schwarzschild sphere at the moment τ^* given by equation (3.14.6); at the moment $\tau^* = 0$ it reaches its maximum radius r, which is given by

$$r = r_g(R^{*2} + 1) ;$$

15. If $\dot{r} = 0$ at a particular event in spacetime, then r cannot be smaller than r_g there, since in a T-region $\dot{r}$ can never be zero (see § 3.12; also Novikov [1964d] for further details).

and it then begins to fall back toward the Schwarzschild sphere. The particle, reaches the Schwarzschild sphere at the moment τ^* given by expression (3.14.6) with the second arcsin, and reaches $r = 0$ at the moment

$$\tau^* = \tfrac{1}{2}\pi(R^{*2} + 1)^{3/2}r_g c \; .$$

The locations of the R- and T-regions in R^* and τ^* coordinates are shown in Figure 20. It is readily apparent that the coordinate system is complete; the world line of every free particle begins and/or ends at the true singularity

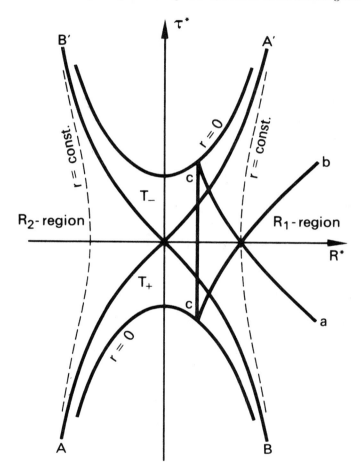

F<small>IG</small>. 20.—Kruskal spacetime. AA' and BB' are the lines $r = r_g$; they are the boundaries of the R- and T-regions. Lines a, b, c are world lines of freely falling particles; a is a particle falling inward from $r = \infty$; b is a particle flying outward from $r = 0$ to $r = \infty$; and c is one of the particles moving with elliptical velocity from whose world line the coordinate system is constructed. *Dashed lines, $r =$ const.*

$r = 0$; and those world lines with just one end at $r = 0$ extend unimpeded toward infinity. The coordinate system covers the entire history of every particle.

By contrast, the expanding Lemaître system covers only part of the space-time shown in Figure 20, namely, the region to the right of the line AA' (or to the left of BB'). Similarly, the contracting system covers the region to the right of BB' (or to the left of AA').

The most amazing aspect of this diagram is the presence of two R-regions, R_1 and R_2. They correspond to two spaces which are Euclidean at infinity, and which are connected by a narrow "throat" ($R^* = 0$), which is the sphere of minimum radius[16] at the given moment $\tau^* = $ const. The radius of the "throat" changes with time, increasing from zero to r_g, and then decreasing back to zero.

The world lines of particles with $r = $ const. in the R_1- and R_2-regions—i.e., of particles at rest in Schwarzschild coordinates—are shown in Figure 20. These lines extend in both proper time and coordinate time from minus infinity to plus infinity.

No signal can reach region R_2 from R_1, and conversely[17] no observer can receive any information from "another" R-region.

Figure 21 shows the radius r as a function of Lagrangian radius R^*, at various moments of time τ^*. The quantity $4\pi r^2$ is the area of a sphere about the center of symmetry. The moment $\tau^* = 0$, when $\dot{r} = 0$, is the moment of maximum expansion of the reference frame. At this moment nowhere in space is there a sphere with $r < r_g$. At the following moment τ^*_1 the system is contracting, and the minimum radius of the throat is less than r_g. At the moment $\tau^*_3 > \frac{1}{2}\pi(r_g/c)$ the spatial section is already disconnected into two parts.[18] In the T_--region of Figure 21, particles and photons move only from right to left. Obviously one cannot get from R_1 into R_2 (and conversely).

The conclusion that there are two Euclidean spaces connected by a narrow throat appears rather strange; at first glance a space with such topology seems to be a mathematical oddity without any physical significance.

Let us right away emphasize that the Kruskal metric does have a definite physical significance (Novikov 1963; Harrison, Thorne, Wakano, and Wheeler 1965). The two Euclidean spaces were obtained because we used the vacuum solution for the gravitational field everywhere and did not perform a join to matter. As we will demonstrate further in Vol. II, when the

---

16. This topology of the extended Schwarzschild solution was first discovered by Flamm (1916) and by Weyl (1917); see also Einstein and Rosen (1937). But in these analyses the two Schwarzschild spaces were joined at the Schwarzschild sphere, and the nonstationary interior region was not analyzed.

17. From T_+ a signal can enter all three other regions, T_-, R_1, and R_2. From R_1 and R_2, one can only reach T_-. From T_- it is impossible to escape, because all other regions are in the absolute past of T_-.

18. For $\tau^* < 0$ the sequence is reversed.

3.14   The Kruskal Metric

solution for the vacuum Kruskal metric is joined to a solution for matter, the presence of a throat does not bring about a second Euclidean space. The *raison d'être* of the T_+- and T_--regions in this case is the same as in the case of Figure 19*e*. As we shall see, a throat occurs in the solution if the boundary of the matter in Figure 19*e* is moved to the left (leaving the sphere's mass unchanged) past the point where the two lines $r = r_g$ cross.

There are contributions in the literature (see, for example, Wheeler 1955; Anderson and Gautreau 1966; Belinfante 1966; Israel 1966) in which the authors attempt to "get rid" of the second R-region without joining the vacuum solution to matter.

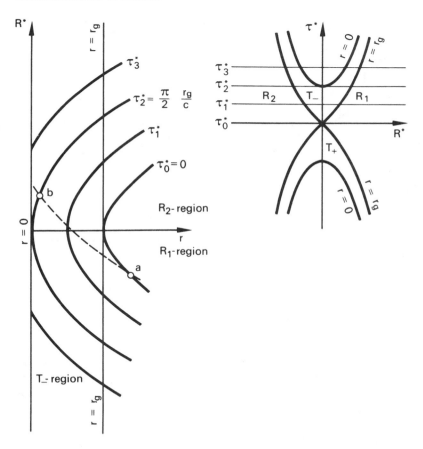

FIG. 21.—Radius r of a sphere as a function of the Lagrangian radius R^*, at various moments of time τ^* ($4\pi r^2$ is the surface area of the sphere). Figure in the upper right shows a series of spatial slices through the spacetime diagram; figure in the lower left shows the radii, $r(R')$, for them. Motion of an inward falling test particle is shown by dashed line *ab*.

For example, Wheeler (1955) has given the following topological interpretation of the Einstein-Rosen (1937) metric, which is somewhat similar to the Kruskal metric; and he has more recently applied it to the Kruskal metric. The throat, or "wormhole" as Wheeler calls it, connects regions of the same physical space which are extremely remote from each other (see Fig. 22 for a two-dimensional analogy of the three-dimensional space at a given moment). Such a topology implies the existence of "truly geometrodynamic objects," which are unknown to physics. Wheeler suggests that such objects have a bearing on the nature of elementary particles and antiparticles and the relationships between them. However, this idea has not

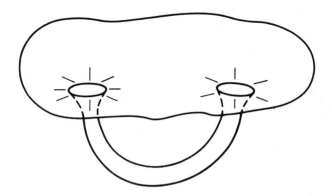

Fig. 22.—Wheeler's topological "wormhole"

yet borne fruit; and there are no macroscopic "geometrodynamic objects" in nature that we know of. Thus, we shall not consider such a possibility further.

Other authors (Anderson and Gautreau 1966; Belinfante 1966; Israel 1966) "get rid" of the second R-region in a more radical manner, namely, by rejecting the principle of causality(!) in the T- and R-regions near r_g. The essence of their suggestion is summarized briefly below.

In Figure 23 consider the point with coordinates (R^*, τ^*) to be identical with the point $(-R^*, -\tau^*)$. What physical conclusions are produced by this identification? Emit photons inward through events D, E, and F from R_1 (R_2 is now the same as R_1!). Upon reaching $r = 0$ at events A, B, C, the photons are "reflected"; from $A' \equiv A$, $B' \equiv B$, and $C' \equiv C$ they proceed toward r_g, where they reenter the R-region and meet themselves in the past at events D, E, and F.

It can be demonstrated that the photons always meet themselves at the radius[19] $r = r_0 \approx 1.28 r_g$. This violates the principle of causality. However,

19. The quantity r_0 is the root of the equation $(r/r_g - 1) \exp (-r/r_g) = 1$.

such violations of causality can only occur in the region $r < r_0$. For events with $r > r_0$ a photon or a particle with finite mass, upon "being reflected" from $r = 0$, returns to the same r at a time after the moment of emission, and there is no violation of causality.

In the T-region this hypothesis is even more "desperate": not only is there no causality, but one cannot uniquely determine the direction of time flow. Thus, a photon from point B' moves left to G'. This means also that it moves from B to G—i.e., "backward" in time with respect to photon EGB. Upon reaching $G' = G$, it cannot enter the R-region (where the direction of time is determined!), so we must postulate that it falls back to $B = B'$, repeating "backward" its own history. A similar situation also takes place for

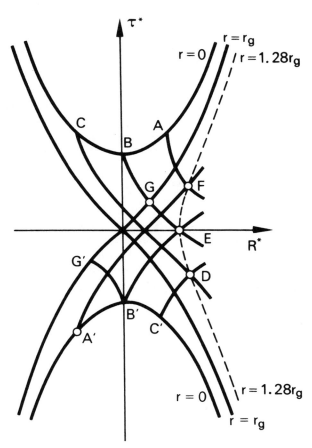

Fig. 23.—World lines of light rays in Kruskal spacetime with the identifications $(R^*, \tau^*) = (-R^*, -\tau^*)$. See text for discussion.

particles with nonzero rest mass. Thus, causality is violated badly in the region with $r < r_g$.

It is impossible to accept such an interpretation of the Kruskal metric for many reasons. *First*, if a light ray passes through the T-region more than once, then causality can be violated at radii $r > r_0 \approx 1.28r_g$, as well as at $r < r_0$. In particular, if the ray emerging from the T-region is reflected back inward at $r < r_0$ (but $r > r_g$), then it can meet the observer who emitted it at $r > r_0$, before the moment of emission, and after the ray's second traversal of the T-region.

Second, this interpretation contradicts the fundamental premise of GTR, the principle of equivalence (see § 1.1). According to this principle, space-time is locally Lorentzian. In a local, free-falling system gravitational accelerations are absent; so in the T-region, as anywhere else, within a small neighborhood of any event one must be able to introduce a locally Lorentz coordinate system in which causality is maintained and the distinction between past and future is determined by a light cone. This is not possible with the identifications of Figure 23.

Third, the symmetry of the representation (Fig. 23) and the possibility of identifying (R^*, τ^*) with $(-R^*, -\tau^*)$ take place only in the total absence of matter. If we were to consider the real problem of the creation of a T-region during the contraction of initially rarefied matter, then the entire domain of Figure 20 to the left of the world line of the contracting sphere's surface is in matter. There the vacuum solution is not applicable; it must be replaced by a material solution (see Vol. II for further details). In this case the identification $(R^*, \tau^*) = (-R^*, -\tau^*)$ is impossible.

We will dwell no longer on this hypothesis. Let it again be emphasized that the Kruskal metric does have a distinct physical meaning. However, we will rest the case at this point, delaying the physical interpretation of the Kruskal metric to Vol. II. This delay is not made so as to intrigue the reader and force him to read another volume. Rather, the problem at hand has a most direct connection to cosmology, so we shall explore it in the volume on cosmology.

4 NONSPHERICAL GRAVITATIONAL FIELDS

4.1 INTRODUCTION

In the preceding chapter we investigated the most important aspects of spherical gravitational fields. Obviously, real astrophysical systems are only approximately spherical. Hence, the question arises as to the extent to which the conclusions of chapter 3 are true in general, and are not due to the assumption of precise spherical symmetry.

To investigate this question, we must study nonspherical gravitational fields. In general, such fields are extremely complex. The motion of nonsymmetrical masses is accompanied by variations in the field, by the emission of gravitational waves, etc. However, some important aspects of this problem can be studied by the investigation of *small* deviations from spherical symmetry. Small deviations can be studied by perturbation techniques, which simplify the investigation considerably.

For astrophysical problems, the deviations from spherical symmetry of greatest interest are those which accompany stellar rotation. Such deviations preserve axial symmetry. Obviously, the retention of axial symmetry permits a considerable simplification in the mathematical calculations and enables one to obtain in some instances exact particular solutions without the assumption of small perturbations from spherical symmetry.

The following three problems, related to deviations from spherical symmetry, are of considerable significance for astrophysics:

First, how will small perturbations in the matter and gravitational field of uniform and isotropic, expanding and contracting matter[1] change with time?

Second, how will the rotation of a superdense star affect its gravitational field?

Third, how will the gravitational contraction of a rotating, flattened star take place when its dimensions decrease to $R \sim r_g = (2GM/c^2)$? Are the qualitative conclusions derived for the spherical case in §§ 4, 5, 9, 10, and 12 of the preceding chapter still valid?

The first problem is of extreme importance for cosmology and will be considered in Vol. II; the other two problems are related to the relativistic

---

1. Obviously, matter which is uniform and expands isotropically can be considered spherically symmetric (see Vol. II).

stages of stellar evolution and to the evolution of stellar systems. In this chapter we will consider the properties of the gravitational field of a rotating star, as well as the properties of the gravitational field of a contracting body which has small deviations from spherical symmetry and is slowly rotating. In Part II of this volume the conclusions of this chapter will be applied to the physics of realistic stars.

We begin with an investigation of stationary solutions. The field of a rotating star in equilibrium is stationary. Moreover, it turns out that numerous conclusions about the variation of the field during the contraction of a nonspherical body can also be obtained from a consideration of stationary solutions. Small perturbations of the Schwarzschild solution were first studied by Regge and Wheeler (1957). In our presentation we will essentially follow the analysis of Doroshkevich, Zel'dovich, and Novikov (1965)—which extends the results of Regge and Wheeler—hereafter omitting references to their paper.

4.2 STATIC FIELDS WITH AXIAL SYMMETRY

The static problem for an axisymmetric field with quadrupole and higher multipoles was solved by Erez and Rosen (1959), who used the method of Weyl (1917, 1919). The interval for a quadrupole field (with an error that appeared in the paper of Erez and Rosen corrected) has the following form:

$$ds^2 = e^{2\psi}dt^2 - m^2 e^{2\gamma - 2\psi}(\lambda^2 - \mu^2)\left(\frac{d\lambda^2}{\lambda^2 - 1} + \frac{d\mu^2}{1 - \mu^2}\right)$$

$$- m^2 e^{-2\psi}(\lambda^2 - 1)(1 - \mu^2)d\phi^2 ,$$

$$\psi = \tfrac{1}{2}\left\{[1 + \tfrac{1}{4}q(3\lambda^2 - 1)(3\mu^2 - 1)]\ln\frac{\lambda - 1}{\lambda + 1} + \tfrac{3}{2}q\lambda(3\mu^2 - 1)\right\} ,$$

$$\gamma = \tfrac{1}{2}(1 + 2q + q^2)\ln\frac{\lambda^2 - 1}{\lambda^2 - \mu^2} - \tfrac{3}{2}q(1 - \mu^2)\left[\lambda\ln\frac{\lambda - 1}{\lambda + 1} + 2\right] \quad (4.2.1)$$

$$+ \tfrac{9}{4}q^2(1 - \mu^2)\left[(\lambda^2 + \mu^2 - 1 - 9\lambda^2\mu^2)\frac{\lambda^2 - 1}{16}\ln^2\frac{\lambda - 1}{\lambda + 1}\right.$$

$$+ (\lambda^2 + 7\mu^2 - \tfrac{5}{3} - 9\mu^2\lambda^2)\frac{\lambda}{4}\ln\frac{\lambda - 1}{\lambda + 1}$$

$$+ \tfrac{1}{4}\lambda^2(1 - 9\mu^2) + \left.(\mu^2 - \tfrac{1}{3})\right] .$$

Here m is the mass of the body which generates the field, and q is a parameter which determines the quadrupole moment. The units of measure are so selected that $c = 1$, $G = 1$. The Schwarzschild field is a particular case of this solution; it corresponds to $q = 0$, in which case the functions ψ and γ become

$$\psi = \tfrac{1}{2}\ln\left[(\lambda - 1)/(\lambda + 1)\right] , \qquad \gamma = \tfrac{1}{2}\ln\left[(\lambda^2 - 1)/(\lambda^2 - \mu^2)\right] .$$

4.2 Static Fields with Axial Symmetry

A transformation to coordinates that are spherical at infinity and that become the usual Schwarzschild coordinates when $q = 0$ can be accomplished by the change of variables[2]

$$\lambda = (r/m) - 1 , \quad \mu = \cos \theta .$$

Let us examine the physical properties of the solution (4.2.1).

Define a *Schwarzschild surface* S_m in a stationary field (i.e., in a field with $\partial g_{ij}/\partial t = 0$) to be any surface on which $g_{00} = 0$. Such a surface has these properties: (1) The gravitational redshift of photons escaping from a stationary source at the surface to infinity is infinitely large, and the energy of the photons is zero; (2) clocks at rest in the field nearer and nearer to S_m tick at rates which approach nearer and nearer to zero compared with clocks at infinity; (3) the gravitational-inertial force F becomes infinite at S_m (cf. § 1.6).

Properties 1, 2, and 3 are closely related to each other. As is well known, in the Schwarzschild field (cf. § 3.2) S_m is a sphere with radius of curve $r = r_g = 2m$ (in units where $G = 1$, $c = 1$). The circumferences of the circles on S_m in the axisymmetric field (4.2.1) at "latitude" $\mu =$ const.—or, equivalently, $\theta =$ const.—are $2\pi \sqrt{(-g_{33})}$. In the spherical case these circles are parallel, and their length is $l \sim \sqrt{(-g_{33})} \sim \sqrt{(1 - \mu^2)}$. In the solution (4.2.1) the Schwarzschild surface $g_{00} = 0$ corresponds to $\lambda = 1$. As $g_{00} \to 0$, the metric component $-g_{33}$ has the asymptotic form

$$-g_{33} = A (1 - \mu^2) g_{00}^{-q(3\mu^2 - 1)/[2 + q(3\mu^2 - 1)]} ,$$

where A is a bounded function of μ ($0 < A < $ const.); for brevity, we will not write out this function. If $q \neq 0$, then as $g_{00} \to 0$ and $q > 0$,

$$-g_{33} \to \infty \quad \text{for} \quad 1 \geq \mu^2 > \tfrac{1}{3} ,$$

$$g_{33} \to 0 \quad \text{for} \quad 0 \leq \mu^2 < \tfrac{1}{3} .$$

For the case $\mu^2 = \tfrac{1}{3}$ in the limit $g_{00} \to 0$, g_{33} remains constant.

Each surface $g_{00} =$ const. is closed. Far from the source of the field these surfaces are spheres (Fig. 24). As g_{00} decreases, the shape of the surface becomes distorted. When $q > 0$, the surface becomes, in the limit $g_{00} \to 0$, two "pears" extending along the polar axis and joined by a cross-connection (see Fig. 24). Each surface with smaller g_{00} is inside the preceding surface with larger g_{00}. Because of the curvature of space, as g_{00} decreases, the areas of the successive embedded surfaces reach a minimum and then begin to increase. When $g_{00} \to 0$, the areas approach infinity and the surfaces themselves approach a limiting two-piece surface, on which $\mu^2 = \tfrac{1}{3}$ everywhere. Each piece extends outward infinitely. Let us emphasize again that this limiting infinite surface $g_{00} = 0$ is inside each surface $g_{00} =$ const. $\neq 0$,

2. In the absence of spherical symmetry, the selection of a convenient coordinate r is to a large extent arbitrary; it could be determined in a manner different from that suggested in the text; for example, by the condition that the surfaces of constant r have $g_{00} =$ const.

131

which has finite area. Similar conclusions are obtained in the case $q < 0$. However, here, as $g_{00} \to 0$, $-g_{33} \to \infty$ for $0 < \mu^2 < \frac{1}{3}$, and $g_{33} \to 0$ for $\mu^2 > \frac{1}{3}$.

It follows that in this field, even when $|q|$ is arbitrarily small, the surface S_m differs considerably in its geometrical properties from the Schwarzschild sphere. As we have emphasized, in the Schwarzschild field at S_m there is no singularity of spacetime, and $C = R^i{}_{klm}R_i{}^{klm} = 12/r_0{}^4 \neq \infty$. In the solution (4.2.1), the scalar C, when $\mu = 0$, has the following asymptotic form: as $g_{00} \to 0$ and $q \to 0$,

$$C = Aq^2 g_{00}^{-1} + 12/r_0{}^4, \quad A = \text{const}.$$

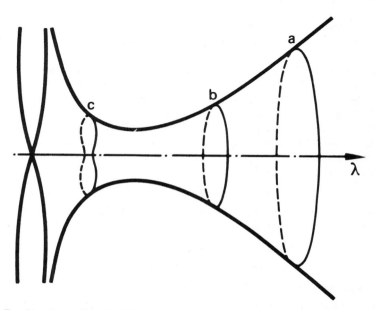

FIG. 24.—A two-dimensional surface whose vertical sections $\lambda = \text{const}$. give the meridional sections of the surfaces $g_{00} = \text{const}$ of the Erez-Rosen solution (1959). Points a, b, c are the poles of successive surfaces which are nested inside each other; the inner surfaces correspond to smaller λ. In the space of the Erez-Rosen solution the points a, b, c lie on a straight line—the axis of symmetry.

We have here the principal divergent term, as well as a term which remains when $q = 0$. Consequently, for an arbitrarily small but nonzero q there are true singularities of spacetime on the surface S_m.

Consider the properties of light rays in the field (4.2.1). Because of the symmetry, light rays which start out radially in the equatorial plane ($\mu = 0$) or along the pole ($\mu^2 = 1$) will continue to propagate radially. The time required to propagate down to a point $\lambda = \lambda_0$ near $S_m(\lambda_0 - 1 \ll 1)$ is

$$t_0 = \text{const}. + \text{const}.' \, (\lambda_0 - 1)^{q^2/8} \quad \text{when } \mu = 0 \text{ (pole)},$$

$$t_0 = \text{const}. + \text{const}.' \, (\lambda_0 - 1)^{-q} \quad \text{when } \mu^2 = 1 \text{ (equator)}.$$

4.2 Static Fields with Axial Symmetry

Thus, by contrast with the Schwarzschild case, the time of light propagation down to S_m is finite for an external observer.[3]

These conclusions, particularly the finite time of light propagation down to S_m, are not a property peculiar only to a quadrupole deviation from spherical symmetry. We shall demonstrate that they are true for any static, axisymmetric solution.

The equations of gravitation for an axisymmetric static field in vacuum can be put into the form (Weyl 1917, 1919)

$$(1/\rho)(\partial/\partial\rho)[\rho(\partial\psi/\partial\rho)] + (\partial^2\psi/\partial z^2) = 0 , \qquad (4.2.2)$$

$$(\partial\gamma/\partial\rho) = \rho[(\partial\psi/\partial\rho)^2 - (\partial\psi/\partial z)^2] , \qquad (4.2.3)$$

$$(\partial\gamma/\partial z) = 2\rho(\partial\psi/\partial\rho)(\partial\psi/\partial z) . \qquad (4.2.4)$$

The coordinates ρ, z are connected to the coordinates λ, μ of the interval (4.2.1) by $\rho = m[(\lambda^2 - 1)(1 - \mu^2)]^{1/2}$, $z = m\lambda\mu$. The equation (4.2.2) for ψ is identical with the equation for the potential in flat space in cylindrical coordinates.

For a source of the type[4] $\sigma = \sigma(z)\delta(\rho)$, where $\delta(\rho)$ is the delta function, the solution to equation (4.2.2) is obviously the potential of a line mass in flat space with linear density $\sigma = \sigma(z)$. Near $g_{00} = 0$ we can write ψ and γ in the following form:[5]

$$\psi = \sigma(z) \ln \rho , \quad \gamma = \sigma^2(z) \ln \rho ,$$

where $\sigma(z)$ is arbitrary. The asymptotic expression for the metric is

$$ds^2 = \rho^{2\sigma}dt^2 - \rho^{2\sigma(\sigma-1)}(d\rho^2 + dz^2) - \rho^{2(1-\sigma)}d\phi^2 .$$

The properties of this metric are similar to the properties which were analyzed above. For example, from the point with coordinates ρ_0, z_0, ϕ_0, moving along the line $z = z_0$, $\phi = \phi_0$ with a velocity that is sufficiently close to the speed of light, we can reach $g_{00} = 0$ in time $t = \rho_0^{[\sigma(z_0)-1]^2} \times [\sigma(z_0) - 1]^{-2}$ as measured by the clock of a distant observer.

The nonspherical problem in vacuum was considered by Regge and Wheeler (1957), who used the method of small perturbations superimposed upon the Schwarzschild solution. It follows from the solution of the equations for small perturbations, as given by Regge and Wheeler, that any static perturbation that dies out at infinity increases indefinitely on approaching the Schwarzschild sphere of the unperturbed spacetime; i.e., it is endowed with the same properties as the exact solutions considered above.

3. An exception is the case $\mu^2 = 1$, $q > 0$, for which the propagation time is a power-law infinity.

4. For a source of this type only at finite distance from the singular surface are the deviations from spherical symmetry small.

5. An exception is the degenerate case of a "point singularity" (see Synge 1960, p. 69, eq. [1]).

To summarize our conclusions: When the perturbations from spherical symmetry are infinitesimally far from S_m, the surface S_m itself differs considerably from the Schwarzschild sphere; it becomes an intrinsic singularity of spacetime; and the time required for light to propagate down to it as measured by an external observer is finite. (See Doroshkevich, Zel'dovich, and Novikov [1965]; and for a more complete discussion see Israel [1967].)

However, can the entire solution with $q \neq 0$, all the way in to $g_{00} = 0$, be realized by a physical system?

In the domain where $g_{00} > 0$ this solution plays the role of the external field of a static body with a quadrupole distribution of mass. However, a static body certainly cannot have $g_{00} = 0$ at its surface, if for no other reason than because the gravitational force will be infinite there. Hence, the entire vacuum solution down to $g_{00} = 0$ cannot be created by a static body. Nor can it arise from a nonstatic body with deviations from spherical symmetry that are small (at the beginning of the contraction), because, as we shall demonstrate in § 4.5, the moment at which the body's boundary crosses S_m is not delineated by any physical phenomenon in the comoving reference frame. At this moment, on the surface of the body, there are no intrinsic spacetime singularities ($C \neq \infty$), whereas they do exist in the static solution with $q \neq 0$. It follows that the solutions investigated above are not the limiting solutions for the collapse of nonspherical bodies.

Thus, it is impossible to apply the solution with $q \neq 0$ for the entire domain down to $g_{00} = 0$, and it is impossible to extend the vacuum field inside the surface $g_{00} = 0$. The occurrence of intrinsic singularities here is completely analogous to their occurrence in the solution for the field of two masses at rest (Synge 1960). In both cases it is due to the fact that the field equations are equations of motion. The occurrence of an intrinsic singularity here means that such a distribution of masses cannot exist. This fact will be a key to our later discussion of the collapse of nonspherical bodies.

The degenerate case of spherical symmetry ($q = 0$) differs in that there is no intrinsic singularity on the Schwarzschild surface $g_{00} = 0$. Radial deformations of the sphere do not change the external field. Therefore, the vacuum solution throughout the domain $g_{00} \geq 0$ can be created by a nonstationary (contracting or expanding) sphere with dimensions that are less than the dimensions of the critical sphere S_m. The vacuum field in this case can be extended inside the sphere S_m (into the T-region). We discussed this point in some detail in the preceding chapter.

4.3 THE EXTERNAL FIELDS OF ROTATING BODIES

We will now consider deviations from spherical symmetry which are related not to variations in the mass distribution which creates the field, but rather to rotation. The effects of rotation in the case of a weak field were analyzed in detail in § 1.10. Here we will consider the role of the rotation when the weak-field condition $\phi \ll c^2$ is not satisfied.

Let us begin with a discussion of an exact particular solution of the Einstein vacuum field equations due to Kerr (1963).

A body whose particles have only rotational motions about an axis of symmetry produces an external field which has g_{03} as its only nonvanishing, off-diagonal metric component, in an appropriate coordinate system (cf. the discussion, later, of the field of a slowly rotating sphere). This follows immediately from considerations of symmetry and of the equivalence of past and future. The Kerr solution can be reduced to such a form by replacing his original coordinate system with a new one (see Boyer and Lindquist 1967). The resultant line element is

$$ds^2 = dt^2 - (r^2 + a^2)\sin^2\theta d\phi^2 - \frac{2mr(dt + a\sin^2\theta d\phi)^2}{r^2 + a^2\cos^2\theta}$$
$$- (r^2 + a^2\cos^2\theta)\left(\frac{dr^2}{r^2 - 2mr + a^2} + d\theta^2\right). \tag{4.3.1}$$

The units are so selected that $G = 1$, $c = 1$. The parameter m is the mass of the body, and $K = -am$ is its total angular momentum. The form of the Kerr solution given here passes over to the usual form of the Schwarzschild solution in the limit $a \to 0$.

The solution (4.3.1) describes the field of a body of mass m with the total angular momentum $K = -am$.

Studies of the Kerr solution (Kerr 1965; Doroshkevich, Zel'dovich, and Novikov 1965; Boyer and Price 1965; Carter 1966a, b, 1968; Boyer and Lindquist 1967) lead to the following conclusions:

The surface $g_{00} = 0$ is *not* a "one-way membrane" in this case; light rays can escape to infinity from beneath it (see Carter 1968; Vishveshwara 1968). The one-way membrane—yes, there is one!—is located inside the surface $g_{00} = 0$. When dealing with the Kerr metric we shall retain the notation S_m for the surface $g_{00} = 0$; we shall call the surface of the one-way membrane the "event horizon"; and we shall denote this event horizon by S_{hor}.

The equations for these surfaces are

$$r_{S_m} = m + (m^2 - a^2\cos^2\theta)^{1/2} \quad \text{for} \quad S_m\,;$$

$$r_{S_{\text{hor}}} = m + (m^2 - a^2)^{1/2} \quad \text{for} \quad S_{\text{hor}}\,.$$

Let us discuss the properties of S_m:

1. When a is arbitrarily small but different from zero, the length L of the circumferences on the surface $g_{00} = \text{const.}$ (these lengths are proportional to $(-g_{33} + g_{03}{}^2/g_{00})^{1/2}$ when $\theta = \text{const.}$ and $g_{00} = \text{const.}$) approach infinity as $g_{00} \to 0$. The asymptotic formula is

$$L = 2\pi a\sin^2\theta/\sqrt{g_{00}}\,.$$

2. Far from the body the precession of a gyroscope is given by the well-known expression (see § 1.10): $\Omega^2 = (c^2 a^2 r_g{}^2/r^6)(1 + 3 \cos^2 \theta)$. Near S_m the precession as measured with local time approaches infinity. The asymptotic formulae are

a) when $\theta = 0$ or π (pole): $\Omega^2 \sim (a/g_{00})^2$;

b) when $\theta = \frac{1}{2}\pi$ (equator): $\Omega^2 \sim (a/g_{00})^2$;

c) when $\theta \neq 0, \frac{1}{2}\pi$, or π: $\Omega^2 \sim (a^6/g_{00}{}^3) \cos^2 \theta \sin^2 \theta(1 - \frac{1}{3}\sin \theta + \sin^2 \theta)$.

The last formula, which gives the expression for the precession near $g_{00} = 0$ at all polar angles with the exception of zero, $\frac{1}{2}\pi$, π, does not pass directly into the expression for the precession on the pole and on the equator, because the entire term of the given order in g_{00} vanishes at $\theta = 0, \frac{1}{2}\pi$, and π.

3. In contrast to irrotational perturbations (cf. § 11.2), the Kerr metric has a curvature scalar $C = R_{iklm}R^{iklm}$ which is nonsingular on S_m. In particular, on the equator of S_m, just as for the Schwarzschild solution at S_m, we have

$$C = 12/r_g{}^4 , \quad r_g = 2Gm/c^2 .$$

Kerr's vacuum gravitational field can be extended through S_m into the region $r < r_{S_m}$. In this region of the solution it is possible to use a coordinate system whose metric coincides with the Kerr metric (4.3.1), but with space and time coordinates interchanged. The singularity of spacetime in the Kerr solution is located at $r = 0$, $\theta = \pi/2$ (by contrast with the Schwarzschild singularity, which is at $r = 0$ for all θ). At the singularity $C = \infty$.

4. A light beam incident on S_m along the pole, and light beams in the plane of the "equator," reach S_m in a logarithmically infinite time of a distant observer. (The clock synchronization here is performed along the trajectory of the beam.)

5. In synchronizing clocks along the equator on the surface $g_{00} = $ const., the difference in time reading ($\Delta\tau = 2\pi g_{03}/\sqrt{g_{00}}$), upon return to the initial point, will approach infinity as $g_{00}{}^{-1/2}$, in the limit $g_{00} \to 0$.

Having studied an exact but special solution for the field of a rotating body, we turn now to approximate general solutions. We shall use the method of small perturbations to study the field of a slowly rotating sphere without assuming that the field is weak. The condition of slow rotation (small angular momentum) is $K \ll mr_g c$.

First we derive the equations for small perturbations of the vacuum Schwarzschild field for a rotating sphere, without assuming that the perturbations are static. (The nonstatic equations will be needed later.)

The sphere may be expanding or contracting radially. It is obvious from symmetry that for slow rotation the only components of the metric perturbation $h_{\mu\nu}$ that are of first order in the angular velocity are h_{03}, h_{13}, and h_{23}. (The perturbations of the diagonal components are of second order.) By using a small transformation of coordinates, it is always possible to annul one of these components; in the transformation $\phi' = \phi + \xi$, the components $h_0{}^3$, $h_1{}^3$, and $h_2{}^3$ acquire the additive changes $\Delta h_0{}^3 = \partial\xi/\partial t$, $\Delta h_1{}^3 = \partial\xi/\partial r$, $\Delta h_2{}^3 = \partial\xi/\partial\theta$. We choose ξ so as to make h_{23} vanish.

The perturbations h_{03} and h_{13} produce the following independent perturbations in the Ricci curvature tensor:

$$\delta R_{23} = -\frac{\partial}{\partial\theta}\left(\frac{\partial}{\partial t}\frac{g_{11}h_{03}}{\sin^2\theta} - \frac{\partial}{\partial r}\frac{g_{00}h_{13}}{\sin^2\theta}\right) = 0\,,$$

$$\delta R_{13} = -\frac{1}{r^2}\left(\sin\theta\frac{\partial}{\partial\theta}\sin^{-1}\theta\frac{\partial h_{13}}{\partial\theta} + 2h_{13}\right)$$

$$- g_{11}\frac{\partial^2 h_{13}}{\partial t^2} + r^2 g_{11}\frac{\partial^2}{\partial r\partial t}\frac{h_{03}}{r^2} = 0\,, \qquad (4.3.2)$$

$$\delta R_{03} = g_{00}\frac{\partial^2 h_{03}}{\partial r^2} + \frac{2}{r}\frac{\partial g_{00}}{\partial r}h_{03}$$

$$- \frac{\sin\theta}{r^2}\frac{\partial}{\partial\theta}\sin^{-4}\theta\frac{\partial h_{03}}{\partial\theta} - g_{00}\frac{\partial}{\partial t}\left(\frac{\partial h_{13}}{\partial r} + \frac{2}{r}h_{13}\right) = 0\,.$$

To find a stationary solution, we set $\partial h_{13}/\partial t = \partial h_{03}/\partial t = 0$. Then the solution to equations (4.3.2) takes on the form

$$h_{13} = \psi(r)r^2\sin^2\theta\,, \qquad h_{03} = (r_g/r)\Sigma_n a_n f_n(r/r_g)P_n{}^1(\cos\theta)\sin\theta\,. \quad (4.3.3)$$

Here, $c = 1, G = 1, \psi(r)$ is arbitrary, $r_g = 2m, a_n = \mathrm{const.}$,

$$f_n(x) = x^3 u_n(x)\int\frac{dx}{x^4 u_n{}^2(x)}\,, \qquad u_n(x) = F(2 + n; 1 - n; 4; x)\,,$$

F is the hypergeometric function, and $P_n{}^1$ is the associated Legendre polynomial (see Gradshteyn and Ryzhik 1965).

Asymptotically, $f_n(x) \sim x^{1-n}$ for $x \gg 1$. If we now perform a small change of coordinates $\phi' = \phi - \int\psi(r)dr$, we obtain $h_{13} = 0$, and the only remaining nonzero component of the perturbation is h_{03}, for which equation (4.3.3) is valid.

The precise form of the field in vacuum (i.e., the values of the coefficients a_n) is determined by matching it to the interior solution at the surface of the body. The junction conditions, which follow from the requirement that the field equations be satisfied on the boundary, require that h_{03} be continuous. For a sphere with rigid-body rotation (but one which is not necessarily stationary; i.e., it may deform radially) this condition, as can be demonstrated, leads to $h_{03} \sim \sin^2\theta$ and $h_{13} \sim \sin^2\theta$ in vacuum. The first of equations (4.3.2) is then satisfied identically, and the solution of the other two equations, after the boundary conditions are imposed, can be reduced to the form

$$h_{03} = +(2K/r)\sin^2\theta \qquad (4.3.4)$$

by a small change of coordinates. Here $K = -am$ is the total angular momentum.

Thus, the external field of such a contracting sphere is constant in time (up to terms of first order in a). Expression (4.3.4) is identical with the one given by Landau and Lifshitz (1962) for a weak field. In reality, it is valid also in a strong field when $a \ll r_g$ (to first order in the dimensionless parameter a/r_g). The expression (4.3.4) for a stationary sphere was independently obtained by Gurovich (1965). For beautiful discussions of its relationship to Mach's Principle see Brill and Cohen (1966), and Cohen and Brill (1968).

In this solution for perturbations of the Schwarzschild field, only terms linear in a were retained in $h_{\mu\nu}$; terms proportional to a^2 and higher powers were neglected. Notice that the surfaces S_m and S_{hor} coincide in this case; they differ only by $O(a^2)$. Those aspects of the Kerr solution at $g_{00} = 0$ which depend on corrections to $g_{\mu\nu}$ which are linear in a are retained by this solution. In particular, here $C_{g_{00}} = 0 = 12/r_g^4 < \infty$.

Hartle and Thorne (1968) have derived the general solution for the exterior gravitational field of a slowly rotating, centrifugally deformed star with a precision of second order in the angular momentum $(a/r_g)^2$ and of first order in the quadrupole moment. Their solution has this curious property: it has no singularity at S_m or S_{hor} if and only if the quadrupole moment Q, the angular momentum K, and the star's mass M satisfy the algebraic identity

$$Q = K^2/M .$$

It is widely speculated that in order for the exact solution to have no real singularities at S_{hor}, every higher multipole moment must be linked to the angular momentum and mass by a relation similar to the above; and it is thought that the only exact solution with such a property (an absence of singularities at the event horizon) is probably the Kerr solution (cf. Hernandez 1967).

Can a solution with a "rotational" deviation from spherical symmetry be realized in nature for the entire domain down to $g_{00} = 0$?

Such a solution cannot be created by a stationary body for the same reason that it cannot in the case of a body with a quadrupole moment. However, for a nonstationary source we cannot apply the same reasoning as we did in § 4.2, because here in the "rotating" Kerr solution the invariant C is finite at $g_{00} = 0$ and at the event horizon, S_{hor}. It will be demonstrated later that a stationary solution with "rotational" perturbations and without a singularity at the event horizon can actually be realized, throughout the domain down to the horizon, as the asymptotic solution for the collapse of a rotating body.

4.4 The Schwarzschild Sphere in an External Quadrupole Field

We conclude our discussion of nonspherical static fields with the following comments.

There exist solutions of the Einstein equations containing Schwarzschild surfaces S_m which do not differ qualitatively from the Schwarzschild surface for the spherical case. However, in these cases the deviations from spherical symmetry must be caused by external fields. For example, if we consider a spherical mass in an external quadrupole field (which increases with distance from the mass m), then an exact solution of the Einstein equations in vacuum has the form (the notation is the same as used in § 2 of this chapter)

$$\psi = \tfrac{1}{2} \ln \left[(\lambda - 1)/(\lambda + 1)\right] + \tfrac{1}{4}q(3\lambda^2 - 1)(3\mu^2 - 1) ,$$

$$\gamma = \tfrac{1}{2} \ln \left[(\lambda^2 - 1)/(\lambda^2 - \mu^2)\right] - 3q\lambda(1 - \mu^2)$$
$$-\tfrac{9}{16}q^2(\lambda^2 - 1)(1 - \mu^2)[9\mu^2\lambda^2 - \lambda^2 - \mu^2 + 1] .$$

The surface $g_{00} = 0$ is determined by the condition $\lambda = 1$. This surface is a Schwarzschild sphere deformed by an external field. The Gaussian curvature of the two-dimensional surface S_m (*not* the Gaussian curvature of the three-dimensional surface created by S_m as it moves through spacetime),

$$C_G{}^* = \frac{1}{4\,m^2} e^q[1 + 3q - 12q\mu^2 - 9q^2\mu^2 + 9q^2\mu^4] ,$$

is different for different μ and is always finite. The physical properties of this S_m are the same as those of the Schwarzschild sphere.

A constant external quadrupole field could be created by distant masses fastened on supports, which prevent them from falling inward. For a limited interval of time the same field could be created approximately also by freely moving distant masses, whose velocities under the influence of reciprocal gravitation are small initially. In this case the field would be almost static.

4.5 THE GRAVITATIONAL CONTRACTION OF A SLOWLY ROTATING BODY WITH SMALL DEVIATIONS FROM SPHERICAL SYMMETRY[6]

Let us briefly review the contraction of a uniform spherical dust cloud of radius r (see § 3.12). For a distant observer, the picture tends to "freeze" as $r \to r_g$ because of time retardation. For this observer, the contracting ball will never have dimensions smaller than r_g. By contrast, an observer on the surface of the contracting ball reaches $R_0 = r_g$ in finite proper time. For him the contraction does not "freeze" at all, but continues through the Schwarzschild sphere into the T-region. The density of the matter in the ball, when $R_0 = r_g$ and the mass is large, has no singularity whatsoever. It can be estimated easily that

$$\rho_g \approx \frac{M}{\tfrac{4}{3}\pi r_g{}^3} \approx 2 \times 10^{16}(M_\odot/M)^2 \text{ g cm}^{-3} ,$$

where $M_\odot \approx 2 \times 11^{33}$ g is the mass of the Sun.

---

6. We will not consider here large-scale magnetic and electric fields surrounding the contracting body. Such fields will be discussed later in § 13.6.

For $M = 10^8 \, M_\odot$, for example, $\rho_g \approx 2$ g cm^{-3}. After the contracting ball passes its gravitational radius, light beams from its surface (see Fig. 25) cannot escape back out through the Schwarzschild sphere and *cannot* reach the distant observer.

If there are small initial perturbations in the density and velocity of the matter inside the body, then they will increase as the contraction proceeds; this was first investigated in detail by Lifshitz (1946) (see Vol. II). However, the surface of the contracting body will at some point cross a surface (event horizon) through which light rays cannot escape to infinity. Although this horizon will differ from the Schwarzschild surface due to the small perturbations, the difference will be so small that we need not make a distinction between the actual horizon and the Schwarzschild surface $r = r_g$. The moment when the star's surface crosses the Schwarzschild surface, $R_0 = r_g$, is in no way special so far as the dynamics of the body's matter are concerned; the density of the matter then is still far from infinite.

Thus, if the initial perturbations are sufficiently small, then at the moment when $R_0 = r_g$ they have not had enough time as yet to become large. The surface of the body in the frame of the comoving observer intersects the Schwarzschild sphere when the perturbations in its matter and in the field around it are still small.

Subsequently, the perturbations in the body increase. But, as demonstrated in § 3.12, an external observer does not receive any information whatsoever from under the Schwarzschild sphere. Consequently, the increasing perturbations inside S_m have no effect whatsoever on the spacetime domain near the Schwarzschild surface, nor on the external domain of the distant observer (R-region). The reader who is inclined to accept this without further explanation may skip the following paragraph. A more formal mathematical proof is included in the appendix to this section.

In fact, disturbances in the gravitational field propagate away from the body at the speed of light. It follows however, from Figure 25 that the trajectories of the rays which escape from the body in the T-region do not approach the Schwarzschild surface. Even trajectories that are fairly large perturbations of null rays do not reach it. This means that the disturbances in the vacuum near the Schwarzschild surface are always small, and that its properties remain unchanged. In particular, no information and no radiation pass through the Schwarzschild sphere to an external observer.[7] Consequently, even when there are disturbances in the collapsing body, the external observer has only a finite interval of the body's evolution accessible to him.

---

7. Let us emphasize once more that, because of the small perturbations, the event horizon actually lies deeper than $g_{00} = 0$ (see § 4.3), so that the properties of the Schwarzschild surface are now smeared out over a narrow region (between the horizon and $g_{00} = 0$). However, that region is so very narrow that we shall simply call it the "Schwarzschild surface."

He can follow the development of the disturbances in the body and in its surrounding field, but only up to the moment when $R_0 = r_g$.

It should be clear by now that the external field of the collapsing body for a distant observer must become stationary in the limit $t \to \infty$; all $\partial/\partial t \to 0$. As a matter of fact, in the reference frame of an external observer, time-dependent perturbations that occur prior to the moment when the surface reaches the sphere r_g must spread out as gravitational waves, and

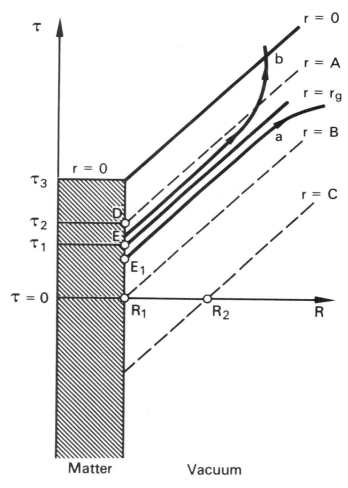

FIG. 25—.Collapse of a spherical dust cloud in a freely falling reference frame (for discussion see appendix to § 4.5). Here a and b are the world lines of light rays. Ray a, emitted from E_1 near time τ_1, propagates near $r = r_g$ for a long time (according to the time of *any* reference frame).

no new disturbances can be received from under the Schwarzschild sphere. Naturally, all these conclusions are valid not only for a collapsing body made of dust but also for matter with nonzero pressure.

Thus, the asymptotic field of a contracting body is stationary, with small deviations from sphericity, when $t \to \infty$.

The rotation of the body brings about an oblateness, i.e., a deviation from sphericity. The perturbations in the external field due to such deviations from sphericity are quantities of second order in the body's angular velocity, as compared with the first-order perturbations of the field due to the rotation itself. The effect of rotation upon the external field was considered in detail in § 4.3.

During the contraction of a rotating ball the external gravitational field does not change with time (up to terms linear in the angular momentum K), and the asymptotic field will be a Schwarzschild field plus a correction caused by the rotation

$$g_{03} = \sin^2 \theta \, (2K/r) \,. \tag{4.5.1}$$

Such a conclusion is not surprising. During the contraction of the ball, the mass M and the angular momentum K are conserved. Hence, the asymptotic field depends on both of these quantities.

What is the nature of the asymptotic field of a contracting body whose nonsphericity is caused not by rotation but, for example, by a nonsymmetrical distribution of mass? The asymptotic field must be stationary. It was demonstrated in § 4.2 that, if corrections to the Schwarzschild field for quadrupole and higher multipole moments (produced by a change in the field's source) are time-independent, then, however small they might be at finite distances from S_m, they become arbitrarily large near S_m and create a true singularity, $C = \infty$, there. On the other hand, as we mentioned at the beginning of this section, in the comoving reference frame of a contracting body with small initial, nonspherical, density perturbations the moment when the body's surface crosses the Schwarzschild surface is not special; it is not accompanied by a true singularity, either in the metric or in the density. Comparison of these results suggests that the quadrupole and higher multipole moments of the external gravitational field must attenuate in the relativistic stages of contraction of a nonsymmetrical body.

Let us attempt to establish the law which governs such attenuation. Consider the approach to the asymptotic field (in the first approximation, and near the Schwarzschild surface) as a passage through a sequence of static states. (About the validity of this assumption, see below.) We shall call this sequence of states "quasistatic" and we shall denote its quadrupole moment by $q(t)$. Examination of the equations for static, axisymmetric, quadrupole perturbations of the Schwarzschild metric indicates that when $g_{00} \to 0$, the components of the metric perturbations that are different from zero take the following form:

$$h_{00} \sim q(1 - r_o/r) \ln (1 - r_o/r) \,,$$
$$h_{11} \sim q(1 - r_o/r) \ln (1 - r_o/r) \,, \qquad\qquad (4.5.2)$$
$$h_{22} \sim h_{33} \sim q \ln (1 - r_o/r) \,,$$

where q is the quadrupole parameter of the perturbation. In these equations multiplicative factors that do not depend on $(1 - r_o/r)$ are omitted.

When the body contracts with $q \neq 0$, all quantities $h_{\mu\nu}$ must remain finite in the comoving frame. Since h_{22} and h_{33} are not affected by a transformation from a comoving coordinate system to the Schwarzschild system, it is obvious that, as $r \to r_o$,

$$q \lesssim 1/\ln (1 - r_o/r) \sim 1/t \,. \qquad\qquad (4.5.3)$$

The last relation in equation (4.5.3) was obtained from the equation (3.5.1) for the law of free fall in a Schwarzschild field. Near the Schwarzschild surface all processes occur at the rate given by equation (4.5.3) (including, e.g., the propagation of gravitational waves). Thus, the quadrupole moment q decays at least as rapidly as $q \sim t^{-1}$. The same applies to higher multipole moments.

More careful analyses of collapse with small nonspherical perturbations (Price 1970; Israel 1970) reveal that the decay of the perturbations is even more rapid than $q \sim t^{-1}$. The body always remains deformed, as seen by a comoving observer. However, as it approaches its Schwarzschild sphere, the body radiates away all nonspherical deformations of its external gravitational field, except the rotational perturbations discussed above. If the gravitational waves did not backscatter off the Schwarzschild curvature of spacetime, the metric perturbations would die exponentially at late times as

$$h_{ij} \sim \exp \left\{ - \frac{1}{\tau} [t - r - r_o \ln (r/r_o - 1)] \right\} \,, \qquad \tau \sim r_o/c \,.$$

However, Price (1970) has shown that, because of backscatter, the outgoing gravitational waves leave behind a "tail." The strength of this tail dies out, at fixed radius, as

$$h_{ij} \sim t^{-(2l+2)} \ln t \,.$$

Here l is the order of the multipole considered ($l = 2$ for quadrupole perturbations, etc.). Interestingly, this decay law for the wave tail is valid for long-wavelength electromagnetic waves, as well as for the gravitational waves which carry off the deformations of the metric.

An important consequence of the above is this: At very late times the asymptotic, external field of a contracting, deformed, slowly rotating body is (to first order) a Schwarzschild field plus stationary "rotational perturbations" in the off-diagonal terms.

An analysis of the motions of test particles and light rays in such a field suggests that the properties of the motion are qualitatively the same as they are in the Schwarzschild field (§§ 3.9 and 3.10). For an external observer, a particle with an impact parameter that is smaller than the critical one is gravitationally captured; upon completing a finite number of spiraling revolutions, it approaches the Schwarzschild surface $g_{00} = 0$ asymptotically as $t \rightarrow \infty$.

The same is true for light rays. No radiation or information can reach an external observer from under the sphere S_m; a gravitational self-closure takes place.

These conclusions are especially important for analyses of the catastrophic contraction of stars, which will be discussed in the next part of this book.

4.6 WHAT HAPPENS TO MATTER AFTER IT FALLS THROUGH THE EVENT HORIZON?

The problem posed in the heading of this section inevitably must be faced. Actually, in the case of strict spherical symmetry, as was demonstrated in § 3.12, the contracting body is necessarily compressed to a point soon after it crosses $r_g = 2Gm/c^2$, so far as the comoving observer is concerned; all of the matter in the sphere is compressed to infinite density. What happens after that? True, the external observer cannot learn anything about it; so far as he is concerned, as $t \rightarrow \infty$ the body's surface approaches r_g, and everything that happens after that is for him in the absolute future (see § 3.12). But what is the final fate of the contracting body, not for an outside observer, but for a comoving observer on the surface of the body?

The relativistic contraction of a spherical body, as noted previously, is unstable. In the course of contraction, perturbations increase without bound as $\rho \rightarrow \infty$ (see Vol. II). Can the evolution of the asymmetries prevent the creation of a singularity in the solution and transform the contraction into expansion after a certain maximum density has been reached? A partial answer to this question has been given by Penrose (1965, 1967) and by Hawking and Penrose (1969). They have shown that if a body contracts to dimensions smaller than r_g, and if other, reasonable conditions are satisfied, then the evolution of a true singularity in the solution is unavoidable. Whether or not *all* of the matter in the body reaches infinite density cannot be established by using the methods of Penrose and Hawking. However, other methods developed by Lifshitz and Khalatnikov (1960a, b; 1963), Lifshitz, Sudakov, and Khalatnikov (1961), and Belinsky and Khalatnikov (1969) have yielded considerable information about the nature of the singularities for a wide class of solutions to the Einstein field equations. These results and the "singularity theorems" of Penrose, Hawking, and Geroch will be discussed in connection with cosmology in Vol. II of this work.

4.6 Through the Event Horizon

The key problem of principle for the fate of contracting matter is not to establish whether, in the course of contraction, the maximum density reached is finite or infinite (although this problem is extremely important), but to determine what happens after that. The body can certainly not reexpand so as to escape from under the Schwarzschild sphere into the R-region of the outside observer who witnessed its compression. For this observer everything that happens after the body reaches r_g is inaccessible.

The problem of the endpoint of the contraction for a comoving observer has not been solved conclusively as yet.

The solution of the problem can be pursued in two directions: First, according to the theorems of Penrose and Hawking, during contraction there is an unavoidable creation of infinite spacetime curvatures (and perhaps also of infinite densities in part of the matter)—or else there is a violation of reasonable conditions on the spacetime structure such as causality. It was demonstrated in chapter 2 that when curvatures exceed the critical value of $C \approx 1/(10^{-33} \text{ cm})^4$, corresponding to ($\rho \approx 10^{93} \text{ g cm}^{-3}$), then GTR is no longer valid; quantum effects must enter the picture. At present there is no quantum theory of strong gravitational fields, so what will happen after such a curvature is reached one cannot say. Some ideas on the subject have been set forth by Harrison, Thorne, Wakano, and Wheeler (1965). Of course, one can assume that, upon reaching the density $\rho \approx 10^{93} \text{ g cm}^{-3}$, the matter cannot reexpand, and spacetime curvatures less than $C = 1/(10^{-93} \text{ cm})^4$ cannot be regained. However, the properties of spacetime under such exotic conditions are totally unknown. This is only one of the possible answers to the question of the final fate of a contracting body.

A study of the relativistic contraction of a charged gravitating sphere (which, of course, is an artificial-model problem) indicates that such a possibility is unlikely for the general case. It suggests, instead, that the contraction of matter might be followed by expansion, but by an expansion in a somewhat unusual sense—expansion into another external space (Novikov 1966a, b; Bardeen 1968a).

Let us examine the qualitative nature of this particular, artificial model. (For details see the papers of Novikov and Bardeen.)

We will investigate the contraction of a uniform sphere of weakly charged dust, $\epsilon^* < mG^{1/2}$, where ϵ^* is the total charge and m is the total mass. The condition $\epsilon^* < mG^{1/2}$ guarantees that the gravitational attraction in the sphere is larger than the electrostatic repulsion. (In the Newtonian stage of contraction the ratio of these forces is constant.) We assume that the matter of the sphere has a low initial density. The charge is uniformly distributed in the sphere and does not become redistributed in the course of contraction.

Consider first the motion of a point on the sphere's surface. This motion can be construed as the motion of a charged test particle in the external gravitational and electric fields of the charged sphere. From time to time

145

particles from the interior of the sphere may drift out across the sphere's surface ("shell crossing"). Subsequently we must regard them as the surface particles, and regard the particles that were formerly on the surface as interior particles. However, our qualitative conclusions about the motion of the surface are not affected by this.

The external gravitational field of any charged sphere is described by the Reissner-Nordstrom metric:

$$ds^2 = (1 - 2Gm/r + G\epsilon^{*2}/r^2)dt^2 - (1 - 2Gm/r + G\epsilon^{*2}/r^2)^{-1}dr^2$$
$$- r^2(d\theta^2 + \sin^2\theta d\phi^2),$$

where we have set the speed of light equal to unity, $c \equiv 1$. The exterior electric field is constant at $r = $ const.: $E = \epsilon^*/4\pi r^2$. As seen by a comoving observer, after finite time the contracting body crosses the Schwarzschild sphere ($g_{00} = 0; r = r_1 = \frac{1}{2}r_0\{1 + [1 - \epsilon^{*2}/Gm^2]^{1/2}\}$), which is also an event horizon. After the body has crossed the Schwarzschild sphere to

$$r < r_1 = \frac{1}{2}r_0[1 + (1 - \epsilon^{*2}/Gm^2)^{1/2}],$$

the gravitational attraction in its interior is replaced by repulsion. In Newtonian theory such a thing would be impossible. From the viewpoint of GTR, a qualitative explanation is that the electric field, which remains constant at a given radius outside the contracting sphere, has a total energy that increases, as the sphere contracts and the exterior region increases in size, until it exceeds the energy inside the sphere. Since the total energy—external plus internal—does not change during contraction, it follows that in the vicinity of a strongly contracted sphere the gravitational field must be equivalent to a negative mass and cause repulsion.

As a result of such gravitational repulsion during strong contraction, as well as of electrostatic repulsion, the surface does not reach $r = 0$; upon reaching a minimum dimension of the order of $r = r_1$, it reexpands. If adjacent layers of matter do not pass through each other, the density does not reach infinity anywhere, except at the center. The replacement of contraction by expansion is nonsimultaneous; it begins at the edge of the sphere and progresses toward the center. The maximum contraction of each layer is $\rho \approx c^6m^4/(\epsilon^*)^6$, where ϵ^* and m are the charge and the mass,[8] respectively, inside the layer. Since m is proportional to ϵ^* in the initial state of low density, the maximum density attained by a layer of interior mass m varies as $\rho_{max} \sim 1/m^2$.

Figure 26 shows the spacetime for this solution, which is similar to Kruskal spacetime, as well as the world line of a particle on the sphere's boundary (*ABCD*). The region inside the sphere is shaded. During the sphere's con-

---

8. The mass m inside a given layer is defined to be the mass, as measured by an external observer, of the configuration obtained by removing all layers outside the given one.

traction, its boundary crosses the Schwarzschild surface[9] $r = r_g$ (point A on Fig. 26), at which point it leaves the external region R'_{ext} and enters the contracting T-region. When $r < r_1$, contraction is replaced by expansion (this takes place in an inner R-region, which is the region $r < r_1$ contained inside the T-region). Upon passing through the expanding T_+-region, the sphere's boundary again crosses the Schwarzschild surface (at point D in Fig. 26), entering into the external region R''_{ext}. This region (cf. Fig. 26) is not the same as the region in which the contraction originated, but is a different one. Relative to the original region, R'_{ext}, it is in the absolute future. The space-at-infinity of this R-region is Euclidean; it is actually "another"

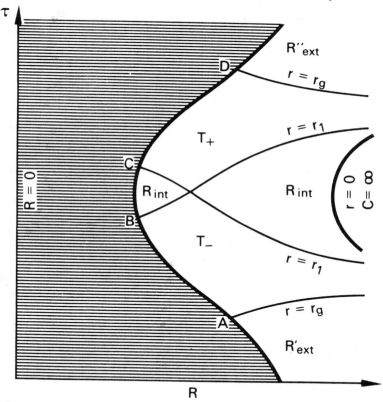

FIG. 26.—Spacetime diagram for a contracting and reexpanding charged sphere. R is a radial space coordinate (which is not Lagrangian inside the matter); τ is a time coordinate; $ABCD$ is the world line of the sphere's boundary. The sphere falls through the Schwarzschild surface ($r = r_g$, at point A) from one external space R'_{ext}; and, upon reexpanding, it enters another external space R''_{ext} (at point D).

9. The value of r_g for a weakly charged sphere is nearly identical with the value $r_g = 2Gm/c^2$ for a neutral sphere.

Euclidean space, different from the one in which the external observer watched the collapse. The real singularity, $r = 0$, in this solution occurs in the vacuum outside the sphere near its point of maximum contraction. Unlike the singularity of the Kruskal metric, this singularity is timelike (see Fig. 26). For this solution, in the limit as the charge approaches zero, the entire interior R-region merges into the singularity at $r = 0$.

The introduction of a charge, which "stratifies" the intrinsic singularity of the Schwarzschild space and permits expansion to follow contraction without the entire matter passing through infinite density, is an artificial trick. However, the result might be indicative of the general fate of a body that contracts inside its Schwarzschild surface; i.e., it might be the answer to the problem posed in the heading of this section. For example, the growth of perturbations during contraction, or processes in the quantum-gravitational domain $\rho \approx 10^{93}$ g cm^{-3}, might convert the contraction into expansion; the expansion, however, would be into another external space!

Thus one might formally join the solutions for the collapse and anti-collapse of a neutral sphere along the line of the intrinsic singularity $r = 0$, keeping in mind that near $r = 0$ the resultant solution does not describe the true nature of the motion and of the field because GTR is not applicable there, but that beyond this narrow domain GTR does govern the true motion. In such a complete solution (involving a single contraction and re-expansion) there exist two spatial domains with Euclidean metric at infinity, which are in the absolute future and past relative to each other (by contrast with the R_1 and R_2 spaces of the Kruskal solution, which are "joined" through a wormhole; see § 3.14, Fig. 20).

We will not dwell in any further detail on these problems, since they are still far from being completely solved. We only wished to direct the reader's attention to the existing possibilities.

Let us reemphasize that these questions do not occur for a Euclidean observer at infinity. No matter what happens inside S_m, he will not find out about it.[10]

And let us reemphasize also that, after the sphere's dimension decreases to $r < r_g$ in finite proper time, its expansion into the external space from which the contraction originated is impossible, even if the passage of matter through infinite density is assumed. The contrary suggestions by Gert-senshteyn (1966*a*, *b*) are fallacious.

10. In a contracting and reexpanding cosmological model, where there is no Euclidean infinity, the situation is somewhat different (Novikov 1964*c*). In this model every "observer" passes into the T-region during the contraction of the model, passes on through the state of infinite spacetime curvature $C = \infty$, moves outward with the reexpanding Universe, and then leaves the T-region. While so doing, the "observer" can see the contraction of the sphere and its reexpansion from under S_m.

APPENDIX TO § 4.5

Let us consider the collapse of a spherical dust cloud. We introduce a comoving reference frame inside the cloud and join it to a freely falling frame outside the cloud's boundary, using the Tolman solution (cf. §§ 3.12 and 3.13). For concreteness we assume that dust particles on the cloud's boundary fall with parabolic velocity, and that the density inside the cloud is uniform.[11] The metric inside the dust is given by the equation which follows equation (3.12.2), whereas the metric outside the dust is the Lemaître metric (see § 3.12) with ds^2 in the form

$$ds^2 = d\tau^2 - \frac{dR^2}{[\frac{3}{2}(R - \tau + \tau_0)]^{2/3}}$$

$$- [\tfrac{3}{2}(R - \tau + \tau_0)]^{4/3}(d\theta^2 + \sin^2\theta d\phi^2) \,.$$

$$(4.5A.1)$$

Here τ is proper time, τ_0 is a constant which fixes the origin of time in this coordinate system, R is the comoving radial coordinate, $c = 1$, and $r_g = 1$. The spacetime of this model is shown in Figure 25. The dashed lines are the lines of constant Schwarzschild radial coordinate, $r = [\frac{3}{2}(R - \tau + \tau_0)]^{2/3} =$ const.

Suppose that this spherical dust cloud is perturbed slightly. Assume that, at the moment $\tau = 0$ (close to the moment τ_1 when the boundary of the cloud crosses the Schwarzschild surface $r = r_g$), the perturbations in the density, the dust velocity, and the metric, $h_\alpha{}^\beta$, are small for all $0 \le r < \infty$. This obviously guarantees that the perturbations are always small at arbitrarily large radii $r =$ const. Then

1. The perturbations $h_\alpha{}^\beta$ will be small in our coordinate system throughout the region $r = [\frac{3}{2}(R - \tau + \tau_0)]^{2/3} > A$, i.e., on the lower right-hand side of the dashed line $r = A$ of Figure 25.

2. A light ray which escapes from the dust after the moment τ_1 will never get out from inside the Schwarzschild surface $r = r_g$ (see Fig. 25). (The limiting moment is actually very slightly later than τ, because of the perturbation of the horizon.)

Let us prove the first statement. It follows from equation (4.5A.1) that that in vacuum the unperturbed metric components $g_{\alpha\beta}$ depend only on $r = [\frac{3}{2}(R - \tau + \tau_0)]^{2/3}$. Therefore, if we consider r and τ as independent variables, rather than R and τ, then the small perturbations in the vacuum metric can be Fourier-analyzed into components of the form $h = e^{i\omega\tau}f(r)$. (The tensor indices α and β will be omitted henceforth.) The function $f(r)$ depends also on θ and ϕ, but this dependence is unimportant for our

---

11. If the collapse begins far from r_g, then near r_g the velocity of the boundary is always close to parabolic. It is not difficult at all to generalize the analysis to the case of a dust cloud with an elliptic or hyperbolic surface velocity and with a radial density gradient in the dust.

considerations, and we shall ignore it. The idea of the proof is to show that the metric perturbations are small along the lines $D-R_1-R_2$ and $r = C$ (cf. Fig. 25); and to use this together with the form of h, $h = e^{i\omega\tau}f(r)$, to show that h is small everywhere within the domain bounded by $r = A$, $r = C$, and $D-R_1-R_2$.

Here is the formal proof: The dust boundary intersects r_g at finite mean density $\langle\rho_c\rangle \approx 2 \times 10^{16} (M/M_\odot)^{-2}$ g cm^{-3}. The solutions of the equations of small perturbations inside the dust[12] demonstrate that h increases without bound only when $\rho \to \infty$, whereas when $\rho = \langle\rho_c\rangle$, it is finite. Thus, up to the moment τ_2 (which is still far removed from τ_3, when $\rho = \infty$) we have $h < \epsilon_1 \ll 1$ throughout the interior of the dust, $R < R_1$.

In a freely falling frame in vacuum there exist solutions which increase without limit at $r = r_g$. However, the above analysis shows that a correct formulation of the Cauchy initial-value problem excludes these solutions so that h is small in the vacuum just outside the surface of the dust for $\tau \leq \tau_2$. Thus, in the vacuum the following arguments can be made:

i) The initial conditions of the collapse are that at $\tau = 0$ and $R > R_1$, $h = f(r) < \epsilon_2$ for some constant $\epsilon_2 \ll 1$.

ii) Because the perturbations on the dust's boundary remain small as it goes through the Schwarzschild surface (see above), for $0 \leq \tau \leq \tau_2$ and $R = R_1$, $h = e^{i\omega\tau}f(r) < \epsilon_3$.

It follows from argument (i) that $f(r) < \epsilon_2$ when $r \geq B = [\frac{3}{2}(R_1 + \tau_0)]^{2/3}$ (see Fig. 25). It follows from argument (ii) that $f(r) < \epsilon_4$, where $\epsilon_4 = \epsilon_3/|e^{i\omega\tau}|_{max}$ for $0 \leq \tau \leq \tau_2$ and $A < r \leq B$ (see Fig. 25).

Thus always

$$f(r) < \epsilon_5 \quad \text{when} \quad r > A , \quad \epsilon_5 = \max(\epsilon_2, \epsilon_4) . \quad (4.5A.2)$$

Since $h < \epsilon_6$ for sufficiently large $r = $ const. $= C$ and arbitrary $\tau > 0$,

$$h_{r=C} = e^{i\omega\tau}f(C) < \epsilon_6 \quad \text{for} \quad \tau > 0 .$$

Thus

$$e^{i\omega\tau} < \epsilon_6/f(C) = \epsilon_7 , \quad \tau > 0 . \quad (4.5A.3)$$

(We have in mind here the absolute value of $e^{i\omega\tau}$.) It follows from equations (4.5A.2) and (4.5A.3) that

$$h = e^{i\omega\tau}f(r) < \max(\epsilon_5, \epsilon_7) = \epsilon_8 , \quad \text{for} \quad r > A , \quad \tau > 0 .$$

This proves statement 1 above. Let us now prove statement 2.

In the unperturbed metric (4.5A.1), for an arbitrary (possibly nonradial)

---

12. The case in which the dust has no boundary is treated in Vol. II. For our dust sphere with boundary, the analysis is modified, and Novikov (1969) has constructed the modified analysis. It demonstrates that with a boundary, as without it, the perturbations of the metric inside the dust remain finite as it collapses through r_g.

ray of light in that part of the T-region where $r < r_g - F$ for some arbitrary constant $F < r_g$, the following inequality[13] is valid:

$$d\tau/dR \geq (-g_{11}/g_{00})^{1/2} > 1 + N .$$

Here $N = \text{const.} > 0$. This inequality states that the slope of the ray is larger by a finite amount than the slope of the line $r = r_g$ in Figure 25.

Above it was demonstrated that for $r > A$, the perturbations in the metric are always small. Obviously, these perturbations cannot change appreciably the value of $d\tau/dR$ along the ray; consequently, the inequality

$$d\tau/dR > 1 + N$$

remains valid. Thus, in the domain $A < r < r_g$, the ray never approaches $r = r_g$ and, of course, cannot cross it. Therefore, we have proved that during collapse with initially small deviations from spherical symmetry, the ray never leaves the T-region.

---

13. We consider a ray for which $(dR/d\tau) > 0$.

II THE EQUATION OF STATE
OF MATTER

5 INTRODUCTION TO PART II

5.1 THE CONCEPT OF PRESSURE; DIFFERENT KINDS OF PRESSURE; THE CASE OF LONG-RANGE INTERACTIONS

In order to analyze the cosmological problem and the theory of stellar evolution, it is necessary to have data on all interactions between the particles which make up bodies. Naturally, the gravitational interaction must be considered separately in this framework.

The contemporary state of the art of the theory of gravitation was described in chapters 1 through 4. In this chapter we will consider short-range forces between particles, and their effects. Regardless of the specific nature of these forces, a common property is the additivity of the interaction energy for a macroscopic system: if the system is divided into macroscopic parts, then the interaction energy between these parts will be negligibly small compared with the energy of each part. Therefore, for short-range forces (and for short-range forces only) we can introduce the concepts of energy density, $\mathfrak{E}$, and specific energy, $E_{sp} = \mathfrak{E}/n$, where n is the density of those particles (baryons) to which we relate the energy. The index "sp" (specific) will be omitted in the subsequent presentation. In a similar manner, we can consider a specific entropy S (per baryon).

Finally, the principal property of short-range interaction in a macroscopic system is the possibility of introducing a pressure P. The pressure is a quantity which permits us to define the interaction force between two parts of a system as an integral over the surface separating the parts

$$\boldsymbol{F} = \int P d\boldsymbol{S} \tag{5.1.1}$$

where P depends only on the state of the matter at this surface $P = P(n, S)$, i.e., on the density n and the entropy S. The fundamental thermodynamic relation has the form

$$dE = -P(n, S)d(n^{-1}) + T(n, S)dS , \tag{5.1.2}$$

where P is the pressure and T is the temperature. The quantity n^{-1} is the volume occupied by one baryon, i.e., the "specific volume." If it is denoted by V, then equation (5.1.2) takes the familiar form $dE = -PdV + TdS$. One consequence of equation (5.1.2) is

$$P = n^2 (\partial E/\partial n)_S . \tag{5.1.3}$$

155

In the nonrelativistic approximation, the mass density $\rho(\text{g cm}^{-3})$ corresponds to the density of the rest mass of baryons $\rho_0 = nm_0$. In this approximation we can introduce the specific energy E_1 per gram of matter and write

$$E_1 = E_1(\rho, S), \qquad P = \rho^2(\partial E_1/\partial \rho)_S. \qquad (5.1.4)$$

However, we will also be confronted with situations in which $P \sim \rho c^2$, $\rho - nm_0 \sim \rho$. In this case the mass density can no longer be considered to be proportional to the density of particles; the nonrelativistic formula (5.1.4) loses its validity; and the relationship between E, ρ, and n is then given by the expressions

$$\mathfrak{E} = En, \qquad \rho = \mathfrak{E}/c^2 = En/c^2, \qquad (5.1.5)$$

along with expression (5.1.3).

In the nonrelativistic theory the energy E is defined with an arbitrary additive constant, whereas in the relativistic theory it is not.

At high temperatures one expects the formation of nucleon-antinucleon pairs. In this situation it is the difference $N - \bar{N}$ which is conserved (the "baryonic charge"). It is to this difference that all thermodynamic quantities (energy, entropy, volume) must be related. For example, the specific energy is the energy per unit of baryonic charge. Only in this sense are equations (5.1.2)–(5.1.5) correct. This will be discussed more fully below. One can also generalize the theory to the case of several types of charge (for example, leptonic charge). When we speak about baryonic charge only, we are tacitly assuming that in our material all other net charges are zero. The case of electric charge is peculiar—see further on in this section.

Of course, we assume that the material considered is in equilibrium (or nearly so). In this local equilibrium the state of the material is fully determined by the conserved quantities. Obviously, without this assumption we could obtain different P or E for given values of S and n by taking an artificial distribution of particle velocities.

But let us return to the fundamental principles of how to treat short-range and long-range forces.

We are discussing commonly known definitions in somewhat more detail than usual because sometimes the concepts of gravitational pressure and gravitational energy density are introduced. Obviously, in the case of gravitation, the force cannot be reduced to an integral over a surface, since the energy of the system is nonadditive. We can consider a system in which the matter is distributed nonuniformly (for example, in the form of individual stars) in such a manner that the scale of the nonuniformities (distance between adjacent stars) is small compared with the dimensions of the entire system. Then we can calculate the actual gravitational energy of the system, W (taking into account the nonuniformity), and the gravitational energy W_0 of a system strictly uniform on a small scale (i.e., without separation into stars, but in the form of a continuous medium) with the same distribution of the mean density.

5.1 The Concept of Pressure

The difference $W - W_0$ is now a local quantity, and it can be expressed as

$$W - W_0 = \int w_1 \langle \rho \rangle dv , \qquad (5.1.6)$$

where w_1 depends only on the local density and the nonuniformity and not on the general size or shape of the system, in contrast to W and W_0. From w_1 we can obtain the quantity P_1,

$$P_1 = \langle \rho^2 \rangle (\partial w_1 / \partial \rho) , \qquad (5.1.7)$$

which plays the role of a gravitational contribution to the pressure. Perhaps we should rephrase this to say that P_1 is the gravitational pressure of the nonuniformities, thus emphasizing their contribution to the expression for force; the gravitational forces computed from the mean density $\langle \rho \rangle$ are not included in P_1; they are determined separately by the gravitational potential ϕ and by the volumetric forces $\rho \nabla \phi$ which correspond to the potential.

In all likelihood, a distinct separation between the volumetric gravitational force and the gravitational contribution to pressure due to nonuniformities is related to certain assumptions regarding the fluctuations. This problem is not yet resolved.[1] The fluctuations are characterized by the concept of spectral density (see Vol. II).

Formally, electric and magnetic fields are a special case. The Coulomb law is identical with Newton's law; hence, at first glance, everything that has been said about gravitation is applicable to the electrostatic interaction. However, a fundamental difference is that in electrostatics there exist charges of both signs, whereas in gravitation all masses have one sign. In astronomy we always deal with electrically neutral systems. Even if the system as a whole is not neutral, in the presence of conductivity free charges exist on the surface of the system, so in a volumetric sense the matter is electrically neutral. Long-range Coulomb forces in electrically neutral matter can be considered by the same procedure used for the analysis of short-range forces; this is because charge neutrality reduces to a negligible size the long-range force due to each single charge. However, charge neutrality occurs only on the average. On an atomic scale, the distribution of the charge is not at all uniform, and this is important. The contributions of electrostatic forces to pressure and to energy are actually quantities which depend on the micro-nonuniformities (on the atomic scale) of the distribution of charge in a medium which is neutral on the average.

On the other hand, it should be remembered that it is the electrostatic forces themselves which cause charge neutrality: it is well known what gigantic fields are generated when there are small deviations of the mean electron density from the proton density in a body of matter. In this way electrostatic forces couple electrons and nucleons and permit us to consider a uniform pressure. Nonetheless, if, as a matter of principle, we were to consider a system with a mean charge density not equal to zero, then we could

1. To familiarize yourself with the statement of the problem, see Shirokov and Fisher (1962).

157

not simplify our concept of pressure. We would have to find the distribution of the potential and the electrical field for the entire macroscopic system, and, along with the local pressure, consider also the volumetric electric force acting upon the charged matter.

The magnetic field generates a volumetric force which acts upon a medium through which electric current flows

$$F = (J \times H)/c \qquad (5.1.8)$$

(in the cgs system). By use of the Maxwell equations which relate the current to the field,

$$\nabla \times H = (4\pi/c)J , \qquad \nabla \cdot H = 0 \qquad (5.1.9)$$

(the field is quasistationary; we neglect $\partial E^*/\partial t$), the expression for the volumetric force can be converted to the form

$$F_a = -\partial T_{a\beta}/\partial x^\beta , \qquad T_{a\beta} = -(1/4\pi)(H_a H_\beta - \tfrac{1}{2}\delta_{a\beta}H^2) , \qquad (5.1.10)$$

i.e., to a form which is similar to the effect of pressure

$$F = -\nabla P . \qquad (5.1.11)$$

The difference is that instead of a scalar quantity (pressure), in the case of a magnetic field we deal with a stress tensor. In the direction of the magnetic field (along the "field lines") there is a tension which is equivalent to a stress with the force per unit area $H^2/8\pi$ dyn cm^{-2}; and in the two perpendicular directions there is a repulsion (or pressure) of the same magnitude.

The same type of analysis can be performed for electrostatic forces (and also for gravitational forces in the case of gravitation theory in flat space). One introduces a corresponding stress tensor, and uses it to rewrite the force acting on a volume in terms of the integral of the stress over its boundary surface.

From this viewpoint the difference between short-range and long-range forces, mentioned above, is diminished in the field theory. The stress is a function of the local fields, just as the pressure is a function of the local density of particles. But the point remains that the local field is determined through field equations by the overall distribution of charges. One of the consequences of this is that the stress tensor of the field is not isotropic, in contrast to the isotropic pressure of matter in equilibrium.

It is possible to describe the effect of the magnetic field by using the pressure concept only when we are dealing with a small-scale chaotic (disordered) field. In this case the averaging of forces over a large surface intersecting the field lines at different angles in different areas of the surface will yield

$$\langle F \rangle = \langle P \rangle S , \qquad \langle P \rangle = \tfrac{1}{24}\langle H^2 \rangle/\pi = \tfrac{1}{3}\langle \mathfrak{E} \rangle , \qquad (5.1.12)$$

where $\langle \mathfrak{E} \rangle$ is the mean density of magnetic energy and S is the surface area.

Thus, in a small-scale, disordered, and on the average isotropic magnetic field, we can discuss the mean energy of the field and the mean pressure of the field. Let us note a possible interpretation of the magnetic energy and the magnetic pressure: Physically these quantities are the result of the motion of charged particles; to be more precise, they are the result of the interaction between moving charged particles. Let us explain this.

If we were to deal with charges of the same sign, then the magnetic interaction due to their motion would be smaller than the electrostatic interaction by a factor of v^2/c^2. However, in an electrically neutral system the electrostatic interaction is zero on the average, whereas the motion of charges of different sign with respect to one another (a current) is still possible.

The motion of charged particles—e.g., electrons—like the motion of any particles, contributes to the energy and to the pressure simply because of the kinetic energy of the particles. Let us express these contributions explicitly:

$$E = \tfrac{1}{2}m_e \langle v^2 \rangle , \quad \mathfrak{E} = \tfrac{1}{2}nm_e \langle v^2 \rangle , \quad P_k = \tfrac{2}{3}\mathfrak{E} = \tfrac{1}{3}nm_e \langle v^2 \rangle . \quad (5.1.13)$$

Here E is the kinetic energy of one particle, while $\mathfrak{E}$ and P_c are the energy density (in ergs per cubic centimeter) and pressure, respectively.

What is the relationship between the magnetic and kinetic energies and the respective pressures? The result depends in a decisive manner on the scale l of those regions in which the electrons undergo correlated motion (i.e., move in the same direction).

Actually, $j \sim nve$, $H \sim lj/c$ (since $|\nabla \times H| \sim H/l$), so

$$P_m \sim H^2 \sim l^2j^2/c^2 \sim l^2n^2v^2e^2/c^2 . \quad (5.1.14)$$

Compare this expression with the kinetic pressure:

$$\frac{P_m}{P_k} \sim \frac{l^2ne^2}{m_ec^2} .$$

Hence, the relationship that we seek is

$$P_m/P_k \sim l^2r_0n , \quad (5.1.15)$$

where

$$r_0 = e^2/m_ec^2 = 2.8 \times 10^{-13} \text{ cm} . \quad (5.1.16)$$

If, for example, we take $n = 6 \times 10^{23}$ cm^{-3} and $l = 1$ cm, we obtain $P_m/P_k \approx 10^{10}$. With an ordered electron motion (even on the scale of 1 cm), the magnetic pressure may play a decisive role. We will return to this problem later, in connection with the theory of quasars and supermassive stars.

Now, after a lengthy introduction, it is time to consider "conventional" pressure. We will begin with an analysis of the pressure of cold matter, where the only variable is density. Then we will turn to the state of matter at high temperatures; special emphasis will be placed on the concept of entropy, which has not been sufficiently popular among astronomers until now. Finally, we will consider the adiabats of hot matter. In this context it will be necessary to investigate in detail the problem of thermodynamic equilibrium and the applications of thermodynamics to systems which are not entirely in equilibrium.

6 COLD MATTER

6.1 CLASSIFICATION INTO DOMAINS

The state of cold matter is determined completely by its density and composition. Moreover, because of the interaction between elementary particles at high density, the composition of matter in thermodynamic equilibrium depends on the density.

The entire range of density variation can be subdivided approximately into the following regions:[1]

1. A density $\rho < 50$ g cm^{-3}, or to put it differently, $P < (1-5) \times 10^6$ kg cm$^{-2} \approx (1-5) \times 10^{12}$ g cm^{-1} sec^{-2}. In this domain there are pronounced individual physical and chemical properties of elements which change from one element to the next in accordance with Mendeleev's periodic law. For pressures up to $(10-20) \times 10^3$ kg cm^{-2}, experimental data obtained by static methods are available for practically all elements. For a number of elements, measurements have been performed up to record pressure ranges of the order of 10^7 kg cm^{-2}, by using explosives. In this area, credit belongs to Al'tshuler (1965) and to the soviet school of research which he created. Methods of numerically calculating pressure from theory were developed by Gandel'man (1962) and Dmitriev (1962); however, because of the great labor involved, the number of calculated curves $P(\rho)$ for various elements is still small.

2. The range 50 g cm$^{-3} < \rho < 500$ g cm^{-3}. In this domain the individual differences between various chemical elements and compounds are already obliterated. At the same time, the theory in which electrons are treated as free particles (see region 3) is still not valid here. The electrostatic fields of atomic nuclei and the interactions of electrons with each other have a substantial influence upon the pressure, which consequently depends not only on the density but also on the atomic number of the nucleus. The difference between the first and second domains is that the dependence of physical properties upon Z in the second domain is smooth and monotonic.

The equations of state for this domain were analyzed in the well-known contribution by Kalitkin (1960), from whose work we will cite graphs and formulae. (The upper limit (500 g cm^{-3}) is for $z = 26$.)

The first and second domains are of little interest to astrophysics (how-

---

1. The boundaries are to some extent arbitrary, since the transitions between the domains are smooth. Moreover, the boundaries depend on the composition; we will cite only typical values of the boundaries.

ever, they are important for the physics of the Earth and of other planets!) because the lower the pressure, the more we must consider the effect of the temperature upon the equation of state at low temperatures; so that even at a relatively low temperature the concept of cold matter is no longer applicable. It should be noted that the equation of state for hot matter at these densities is of utmost astronomical importance; all ordinary stars have central densities precisely in this region.

3. The range 500 g cm$^{-3} < \rho < 10^{11}$ g cm^{-3}. In this rather wide and important domain electrons can be treated as free particles. We may apply the theory of the degenerate electron gas. At the same time, the distance between the nuclei is still larger than the radius of nuclear forces, so that the poorly known interaction between free nucleons (and nuclei) has no effect upon the state of the matter.

This domain is subdivided into a region in which the electrons are non-relativistic ($\rho < 2 \times 10^6$ cm^{-3}) and a region in which most of the electrons have an energy of the order of or larger than $m_e c^2$. The latter region is called the region of relativistic degeneracy ($\rho > 2 \times 10^6$ g cm^{-3}).

Further, when densities are larger than 10^7–10^{10} g cm^{-3} (depending on the composition of the nuclei), high-energy electrons can cause a change in the nuclear composition. Near the edge of the domain $\rho \sim 10^{11}$ g cm^{-3}, a large fraction of the nucleons become free neutrons.

4. The range 10^{11} g cm$^{-3} < \rho < 10^{14}$ g cm^{-3}. In this domain, matter consists predominantly of neutrons; their interaction is substantial in the upper part of this region, near 10^{14}. In other words, the neutrons cannot be treated as separate particles; rather, they form a gigantic nucleus.

5. The range 10^{14} g cm$^{-3} < \rho < 10^{16}$ g cm^{-3}. According to Ambartsum-yan and Saakyan (1963), in this domain there appear, along with neutrons in equilibrium and a small number of protons and electrons, many other types of elementary particles—muons, pions, hyperons.

6. The range 10^{16} g cm$^{-3} < \rho < 10^{93}$ g cm^{-3}. For all practical purposes, this domain has not been explored, either theoretically or experimentally. We can only express some rather general judgments as to the restrictions imposed upon the equation of state by the theory of relativity.

7. A density $\rho > 10^{93}$ g cm^{-3}. At first glance, any classification above $\rho \sim 10^{16}$ g cm^{-3} suggests Averchenko's parody:[2] "The history of the Midianites is lost in the darkness of the ages and is not known; nonetheless, scientists distinguish three distinctly separate periods in it: the first, about which nothing is known; the second, about which one can say the same; and the third, which followed the first two." Actually, the reason for such a classification is related to quantum effects. The density of 10^{93} g cm^{-3} is obtained if we pose the following problem (see chapter 2): Construct a quan-

2. Arcadii T. Averchenko (1881–1925), well-known Russian humorist. The quotation is taken from the beginning of his *Complete World History of Satiricon.*

tity with the dimensions of density from the fundamental constants: $\hbar$, Planck's constant; c, the velocity of light; and G, the gravitational constant. The resultant density has the form

$$\rho_g = \frac{c^5}{\hbar G^2} = 5 \times 10^{93} \text{ g cm}^{-3} .$$ (6.1.1)

When $\rho < \rho_g$, quantum effects have no substantial role in gravitation proper, even though they are essential for describing the behavior of the particles. We may not know the specific dependences of the pressure and energy density on the particle density and the entropy in this regime. We do know, however, that these functions exist; we know the thermodynamic relationships between them; and we know the form of the Einstein equations into which these functions enter. When $\rho < \rho_g$, we can use the concept of continuous space and time.

When $\rho \geq \rho_g$, the situation changes substantially. Quantum gravitational effects become important. But gravitation in the Einstein theory is related to the spacetime metric. Therefore, the spacetime metric per se cannot be treated classically. There exists no specific theory at this time which considers space and time from the viewpoint of quantum theory.

In the spirit of the entire development of physics, it would behoove us to make only one assumption about the domain $\rho \geq \rho_g$: if the density reaches the quantum value ρ_g in a region of space and time which is bounded on all sides, then the integral laws of conservation expressed for states before and after the period $\rho \geq \rho_g$ should not be violated.

However, in the spirit of our discussion of geometrodynamics (chapter 2) we can say the following about the behavior of matter at densities approaching ρ_g. Consider, as an example, an ideal Fermi gas at the "Planck" density, $n = 1/l^3$, where l is the Planck length. Then the Fermi momentum will be $p_f = \hbar/l$, and the corresponding mass (not rest mass, of course; that is negligible for ordinary particles at such high densities!) is equal to the Planck mass of approximately 10^{-5} g. In order of magnitude, the mass-energy density of the matter is just $\rho_{\text{Planck}} \equiv \rho_g$. Now, the energy of gravitational interaction between two such particles at the distance l is equal to their Fermi energy. Consequently, gravitation can no longer be considered a long-range force; it is a necessary part of the equation of state when the density approaches ρ_g. For some ideas about how the equation of state may be changed by this effect (for hot matter, of course), see Sakharov (1966).

Such is the general classification of those domains of pressure and density (which are rather different in their properties and theoretical interpretation) with which one deals in relativistic astrophysics. As to domains 1 and 7, we will limit ourselves to the above comments; the other domains will be considered in detail later.

Let us begin with a consideration of the most important and at the same time the best-understood domain: the third.

6.2 DEGENERATE ELECTRON GAS

For coherence of presentation, let us review well-known formulae, including the numerical values of the coefficients. A clear and coherent exposition of the problem is found in Landau and Lifshitz (1964).

According to the Pauli principle, each quantum level is either occupied by one electron or is vacant. At zero temperature, all levels with energy $E \leq E_F$ are occupied, and all levels with energy $E > E_F$ are empty. The quantity E_F is the boundary energy, also known as the Fermi energy. We can similarly speak about the Fermi momentum p_F. For a given electron motion (given orbital wave function), there exist two levels because the electron spin $\frac{1}{2}\hbar$ may have two projections on an arbitrary axis, namely, $\frac{1}{2}\hbar$ and $-\frac{1}{2}\hbar$.

The number of orbital states of semiclassical motion equals the volume of six-dimensional phase space (space of coordinates and momenta) divided by $(2\pi\hbar)^3$, which is the "volume of one cell."[3] Taking the spin into account, we have

$$dn = 2d^3x \, d^3p/(2\pi\hbar)^3 . \tag{6.2.1}$$

From this we obtain the number of levels in 1 cm[6] with a momentum smaller than p_F (the Fermi momentum), keeping in mind that $\int d^3x = 1$ cm[3] and $\int_{|\vec{p}| < p_F} d^3p = (4\pi/3)p_F^3$:

$$n = 2 \, (4\pi/3)p_F^3/(2\pi\hbar)^3 = p_F^3/3\pi^2\hbar^3 . \tag{6.2.2}$$

It is convenient to express p_F in units of $m_e c$ by introducing a dimensionless parameter $x = p_F/m_e c$ (not to be confused with the coordinates appearing in formula [6.2.1]). We obtain

$$n = (1/3\pi^2)(m_e c/\hbar)^3 x^3 = (1/3\pi^2)(x/l_c)^3 . \tag{6.2.3}$$

Here $\hbar/m_e c = l_c = 3.86 \times 10^{-11}$ cm is the Compton wavelength of the electron. Naturally n, having dimensions of $1/\text{cm}^3$, is proportional to l_c^{-3}. Substituting numerical values, we obtain

$$n = 6 \times 10^{29} x^3 . \tag{6.2.4}$$

Let μ_e denote the molecular weight per electron, i.e., the mean number of nucleons per electron ($\mu_e = 1$ for hydrogen, $\mu_e = 4/2 = 2$ for ^{4}He, $\mu_e = 56/26 = 2.17$ for ^{56}Fe). Then

$$\rho = \mu_e m_0 n = \mu_e(6 \times 10^{29}/6 \times 10^{23})x^3 = \mu_e \times 10^6 x^3 . \tag{6.2.5}$$

---

3. To verify the volume of one cell, compute the number of solutions of the Schrödinger equation in a "box," i.e., in the volume of the parallelopiped with edges of given lengths l_x, l_y, l_z and with the condition $\psi = 0$ on the faces. Verify that the same number of levels will be obtained if the condition $\psi = 0$ (corresponding to an infinite potential) is replaced by the condition of periodicity

$$\psi(x, y, z) = \psi(x + n_x l_x, y + n_y l_y, z + n_z l_z) ,$$

where all n are integers. In this case the levels correspond to plane waves with quantized momentum components.

Here we have used for the mass of the nucleon $m_0 = 1/a$, where $a = 6 \times 10^{23}$ g^{-1} is Avogadro's number. Accordingly, we have

$$x = [(\rho/\mu_e)\, 1.015 \times 10^{-6}]^{1/3} \cong [(\rho/\mu_e) \times 10^{-6}]^{1/3} . \qquad (6.2.6)$$

It follows that when $\rho < \mu_e \times 10^6$ g cm^{-3}, then $x < 1$, i.e., $p_F < m_e c$, and the electrons are nonrelativistic. When

$$\rho > \mu_e \times 10^6 \text{ g cm}^{-3} ,$$

then

$$p_F > m_e c .$$

The precise expression for the total energy of an electron E_F', with a given momentum p_F, which is valid for any (both nonrelativistic and relativistic) value of the momentum is

$$E_F' = [(m_e c^2)^2 + c^2 p_F{}^2]^{1/2} = m_e c^2 (1 + x^2)^{1/2} . \qquad (6.2.7)$$

The limiting cases of this formula are

$$\begin{aligned}
E_F' &= m_e c^2 (1 + \tfrac{1}{2} x^2) & \text{for} \quad x \ll 1 , \\
E_F' &= m_e c^2 [x + (2x)^{-1}] & \text{for} \quad x \gg 1 .
\end{aligned} \qquad (6.2.8)$$

These expressions include the rest energy; i.e., for an electron at rest we have $E_0 = m_e c^2$. For calculations related to nuclear reactions, where energy is expressed in millions of electron volts, we will recall that 1 MeV $= 1.6 \times 10^{-6}$ erg, $m_e c^2 = 0.8 \times 10^{-6}$ erg $= 0.51$ MeV. The question arises: How much energy will be released in the transformation of one electron with energy E_F' into one electron at rest with rest mass-energy $E_0 = m_e c^2$? From formula (6.2.8) we obtain the energy release

$$E_F' - E_0 = \begin{cases} m_e c^2 [(1 + x^2)^{1/2} - 1] = (x^2/2) m_e c^2 & \text{for} \quad x \ll 1 , \\ m_e c^2 [x - 1 + (2x)^{-1}] & \text{for} \quad x \gg 1 . \end{cases} \qquad (6.2.9)$$

Now let us pose the more realistic question about the energy released when compressed matter, with electron energies distributed over the interval from $m_e c^2$ to E_F', is decompressed into conventional low-density gas. In this case electrons pass into a state in which their energy, for all practical purposes,[4] is $E_0 = m_e c^2$. The initial state is characterized by a mean electron energy of

$$\begin{aligned}
\langle E \rangle &= \left[m_e c^2 \int_0^x (1 + x^2)^{1/2} x^2 dx \right] \Big/ \left[\int_0^x x^2 dx \right] \\
&= (3 m_e c^2 / 8 x^3) \{ x(1 + 2x^2)(1 + x^2)^{1/2} \\
&\quad - \ln [x + (1 + x^2)^{1/2}] \} .
\end{aligned} \qquad (6.2.10)$$

---

4. The binding energy of electrons in an atom is small; on the average it amounts to 13.5 eV $= 2.7 \times 10^{-5} m_e c^2$ per electron for H, $7.7 \times 10^{-5} m_e c^2$ for He, and $10^{-3} m_e c^2$ for Fe. Hereafter we neglect this binding energy.

(Note that we use a prime to distinguish energies, e.g., E_F', which include the rest mass-energy; and we omit the prime from energies, e.g., E_F and $\langle E \rangle$, which do not contain $E_0 = m_e c^2$.) Along with this precise expression, the following asymptotic formulae are useful (we solve for $\langle E \rangle - E_0$):

$$\langle E \rangle - E_0 = (3x^2/10)m_e c^2 \qquad \text{for } x \ll 1 ,$$

$$\langle E \rangle - E_0 = m_e c^2 \left(\tfrac{3}{4}x - 1 + \tfrac{3}{4}x^{-1}\right) \qquad \text{for } x \gg 1 . \tag{6.2.11}$$

The pressure of a degenerate electron gas can be expressed in terms of the energy per electron by using the thermodynamic formula

$$P = n^2(d\langle E \rangle/dn) . \tag{6.2.12}$$

In this formula $\langle E \rangle$ is to be differentiated and therefore it is immaterial whether we use $\langle E \rangle$ or $\langle E \rangle - E_0$:

$$\frac{d\langle E \rangle}{dn} = \frac{d\langle E \rangle}{dx}\left(\frac{dn}{dx}\right)^{-1} = \frac{d\langle E \rangle}{dx}\left(\frac{d\ln n}{dx}\right)^{-1} n^{-1} = \tfrac{1}{3}x\frac{d\langle E \rangle}{dx}n^{-1} , \tag{6.2.13}$$

since $n = ax^3$, $\ln n = 3\ln x + \ln a$, and $d\ln n/dx = 3x^{-1}$.

For a gas containing noninteracting particles, the pressure can also be found by analyzing the transfer of momentum through a unit area by particles with a given velocity and momentum

$$P = \int u_z p_z dn .$$

We have selected an area in the (x, y)-plane, i.e., perpendicular to the z-axis. Let θ denote the angle between the momentum vector and the z-axis:

$$u_z = u\cos\theta , \qquad p_z = p\cos\theta ,$$

$$u = c^2 p/E = cx/(1+x^2)^{1/2} , \qquad p = m_e cx ;$$

$$P = \frac{2(m_e c)^4 c}{(2\pi\hbar)^3}\int_0^{2\pi}d\phi\int_0^{\pi}\sin\theta\cos^2\theta d\theta\int_0^{x_F}x(1+x^2)^{-1/2}xx^2 dx$$

$$= \frac{8\pi(m_e c)^4 c}{3(2\pi\hbar)^3}\int_0^{x_F}x^4(1+x^2)^{-1/2}dx \tag{6.2.14}$$

$$= [(m_e c)^4 c/8\pi^2\hbar^3]\{x(1+x^2)^{1/2}[\tfrac{2}{3}x^2 - 1]$$

$$+ \ln[x + (1+x^2)^{1/2}]\} .$$

Both methods naturally yield the same answer:

$$P = (m_e c^2/l_c^3)f(x) , \qquad \text{where} \qquad x = [\rho/(\mu_e \times 10^6)]^{1/3} ,$$

$$f(x) = (1/8\pi^2)\{x(1+x^2)^{1/2}(\tfrac{2}{3}x^2 - 1) + \ln[x + (1+x^2)^{1/2}]\} . \tag{6.2.15}$$

We will again write down the formulae for two limiting cases; the nonrelativistic:

$$P = (m_e c^2/l_c{}^3)(x^5/15\pi^2) = \tfrac{2}{3}(\langle E \rangle - E_0)n$$
$$= (1/15\pi^2)(m_e c^2/l_c{}^3)[\rho/(\mu_e \times 10^6)]^{5/3} = 10^{13}(\rho/\mu_e)^{5/3} ,$$

(6.2.16)

and the ultrarelativistic:

$$P = (m_e c^2/l_c{}^3)(1/12\pi^2)x^4 \cong \tfrac{1}{3}(\langle E \rangle - E_0)n$$
$$\cong (1/12\pi^2)(m_e c^2/l_c{}^3)[\rho/(\mu_e \times 10^6)]^{4/3} = 1.2 \times 10^{15}(\rho/\mu_e)^{4/3} .$$

(6.2.17)

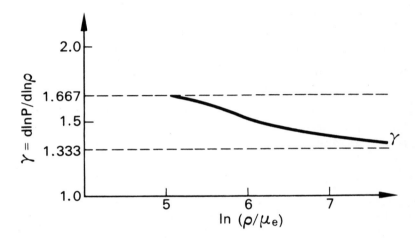

Fig. 27.—Adiabatic index $\gamma = d \ln P/d \ln \rho$ for matter with pressure determined by degenerate electrons and density determined by baryons.

In the last two formulae only the leading term in the pressure (x^4) and in the expression for the energy $(\tfrac{3}{4}x)$ is retained; the subsequent terms are omitted.

Thus, for a degenerate electron gas, the adiabatic index, $d \ln P/d \ln \rho$, is $\tfrac{5}{3}$ at low density and $\tfrac{4}{3}$ at high density.

Figure 27 shows the relationship between the adiabatic index and the density; in particular, the ordinate shows the interval between $\tfrac{4}{3}$ and $\tfrac{5}{3}$, whereas the density[5] on the abscissa is plotted on a logarithmic scale. As we see from the graph, the adiabatic index varies smoothly and monotonically between the indicated limits. (For tables of the adiabatic index, see Schatzman 1958.)

5. To be more precise, it is the quantity ρ/μ_e which is used, so that the graph becomes applicable for all compositions.

6.3 CORRECTIONS IN THE DOMAIN OF HIGH PRESSURES

In the preceding section the electron gas was treated as an ideal gas, without consideration of the reciprocal interaction between electrons, the interaction between the electrons and the nuclei, or the interaction between the nuclei. The interactions we neglected have opposite signs (attractive and repulsive), so that in the first approximation these interactions are mutually suppressing, and hence the approximation of a free-electron gas is justified. This reciprocal compensation derives from the fact that matter in general is electrically neutral. The principal correction is due to the fact that the positive charge is not uniformly distributed in space, but rather is concentrated in individual nuclei. This correction decreases both the energy and the pressure: the repelling nuclei are, on the average, at a larger distance from each other than the mean distance between nuclei and electrons; repulsion is weaker than attraction.

Obviously, at absolute zero temperature (without consideration of quantum effects for the nuclei, which are much heavier than an electron), the nuclei are located in the lattice with the densest packing.[6] We will derive an approximate expression for the energy correction. We replace a cell of the lattice by a sphere of the same volume, $V = 1/N = \frac{4}{3}\pi r_1^3$, where N is the number of nuclei in 1 cm³, and r_1 is the radius of the sphere. The potential of the charge Ze at the center is Ze/r; the interaction energy of Z electrons, uniformly distributed within the volume of the sphere, with the nucleus is

$$E_{n,e} = -\int_0^{r_1} \frac{Ze}{r}\rho_e 4\pi r^2 dr = -\frac{3}{2}\frac{(Ze)^2}{r_1}. \tag{6.3.1}$$

We have used the relation $\rho_e V = -Ze$, where ρ_e is the charge density of electrons in the sphere.

We will now derive the interaction of the electrons with each other. The potential inside a uniformly charged sphere is

$$\phi = \frac{Ze}{r_1}\left(\frac{3}{2} - \frac{1}{2}\frac{r^2}{r_1^2}\right).$$

The interaction energy is

$$E_{ee} = \frac{1}{2}\int \rho_e \phi dV = [(Ze)^2/r_1](\frac{1}{2} \times \frac{3}{2} - \frac{1}{2} \times \frac{1}{2} \times \frac{3}{5}) = \frac{3}{5}(Ze)^2/r_1. \tag{6.3.2}$$

The total electrostatic energy of one cell is

$$E = (-\frac{3}{2} + \frac{3}{5})(Ze)^2/r_1 = -\frac{9}{10}(Ze)^2/r_1. \tag{6.3.3}$$

6. This is the lattice in which, at a given minimum distance between the nuclei, a maximum number density of the nuclei is attained. Obviously, the same lattice also solves the converse problem, namely, the distribution of nuclei with the maximum distance between neighbors for a given number density.

The interaction with electrons and nuclei in other cells can be neglected inasmuch as these cells have no net charge: the external field of a charge of one sign, surrounded by a spherically symmetrical cloud of a neutralizing charge of the opposite sign, is zero. Hence, the error in our calculation is due only to the deviation of the elementary cell of the lattice from spherical shape. A precise calculation indicates that the error of this approximation amounts to less than 0.3 percent of the Coulomb correction.

Another method of calculation, which is theoretically more explicit, consists of the following: First, find the electric field E^* inside the cell consisting of a central nucleus and a distributed negative charge of electrons, which is

$$E^* = (Ze/r^2)[1 - (r/r_1)^3] .$$

Obviously, the field depends on the charge in the domain $r < r_1$, since at $r = r_1$, $E^* = 0$.

Next find the energy of the field, i.e., $\int [(E^*)^2/8\pi] dV$. In calculating, it should be remembered that the energy of matter is chosen to be zero at zero density, i.e., zero energy corresponds to the energy of the nucleus and of electrons which are at a distance far from the nucleus. The zero-density system also has a field:

$$E_0^* = Ze/r^2 .$$

The desired electrostatic energy of contracted matter, relative to the energy at zero density, equals the difference between two integrals

$$\mathfrak{E} = \frac{(Ze)^2}{8\pi} \int_0^{r_1} \frac{1}{r^4} \left[1 - \left(\frac{r}{r_1}\right)^3 \right]^2 4\pi r^2 dr - \frac{(Ze)^2}{8\pi} \int_0^{\infty} \frac{1}{r^4} 4\pi r^2 dr .$$

Rewriting this expression in the following form, we find, in accordance with the previous calculation, the final value

$$\mathfrak{E} = \frac{(Ze)^2}{2} \int_0^{r_1} \frac{1}{r^4} \left[-2\left(\frac{r}{r_1}\right) + \left(\frac{r}{r_1}\right)^6 \right] r^2 dr - \tfrac{1}{2}(Ze)^2 \int_{r_1}^{\infty} \frac{1}{r^4} r^2 dr$$

$$= -\tfrac{9}{10}(Ze)^2/r_1 . \tag{6.3.3'}$$

Let us compare the electrostatic energy with the mean energy of degenerate relativistic electrons. The elementary cell contains one nucleus, and consequently Z electrons. Therefore, the radius of the cell r_1 expressed in terms of the density of electrons is

$$(4\pi/3)r_1^3 = 1/N = Z/n , \quad r_1 = [(3/4\pi)(Z/n)]^{1/3} . \tag{6.3.4}$$

The electrostatic energy referred to one electron and expressed in terms of the electron density is

$$\mathfrak{E}_{el} = -\tfrac{9}{10}(4\pi/3)^{1/3} Z^{2/3} e^2 n^{1/3} . \tag{6.3.5}$$

168

This energy depends upon n in the same manner as does the Fermi energy of a relativistic degenerate electron gas. We recall that in calculating the Fermi energy per electron in the ultrarelativistic limit we found

$$E_F = cp_F = \hbar c (3\pi^2 n)^{1/3} \,.$$

The average electron energy is

$$\langle E \rangle = \tfrac{3}{4} E_F = \hbar c \tfrac{3}{4} (3\pi^2)^{1/3} n^{1/3} \,. \tag{6.3.6}$$

The ratio of the electrostatic energy to the average energy of the degenerate gas is

$$\frac{\mathfrak{E}_{el}}{\langle E \rangle} = -\frac{e^2 Z^{2/3}}{\hbar c} \Big(\frac{4\pi}{3 \times 3\pi^2} \Big)^{1/3} \times \tfrac{4}{3} \times \tfrac{9}{10} = -0.62 \frac{e^2}{\hbar c} Z^{2/3} \tag{6.3.7}$$

$$= -4.56 \times 10^{-3} Z^{2/3} \,.$$

This is the principal correction for the ultrarelativistic gas. Because of the similar dependence on n, the ratio of the correction to the pressure divided by the pressure of the degenerate ideal gas itself is constant. There exist secondary corrections related to the nonuniform distribution of the interacting electrons; and in the ultrarelativistic case, along with the electrostatic interaction, we must also consider the magnetic interaction of electrons. The presence of nuclei somewhat disturbs the density distribution of electrons in space. We quote Salpeter's (1961) final expression for pressure, which also includes other corrections in the ultrarelativistic case ($\rho \gg 10^6$):

$$P/P_0 = 1.00116 - 4.56 \times 10^{-3} Z^{2/3} - 1.78 \times 10^{-5} Z^{4/3} \,. \tag{6.3.8}$$

Here P_0 is the pressure calculated without consideration of the corrections. When $Z = 1$, 6, 12, and 26, this expression yields

$$P/P_0 = 0.9976 \,, \quad 0.9859 \,, \quad 0.9768 \,, \quad \text{and} \quad 0.9598 \,,$$

respectively.

6.4 THE DOMAIN OF MEDIUM DENSITIES, $10^6 > \rho > 500$ g cm^{-3}

It was established above that the principal correction to the equation of state is due to the electrostatic interaction of nuclei and electrons; this correction reduces the pressure, as compared with the pressure of a free, noninteracting, degenerate gas.

We have obtained an expression for the electrostatic energy per electron:

$$\mathfrak{E}_{el} = -\tfrac{9}{10} (4\pi/3)^{1/3} Z^{2/3} e^2 n^{1/3} \,. \tag{6.4.1}$$

In the nonrelativistic domain, the mean energy of a free electron is

$$\langle E - E_0 \rangle = \tfrac{3}{5} m_e c^2 \frac{x^2}{2} = \tfrac{3}{10} m_e c^2 \Big(\frac{n}{m_e^3 c^3 / 3\pi^2 \hbar^3} \Big)^{2/3} \tag{6.4.2}$$

Cold Matter

Let us find the density at which, in this approximation, the pressure is zero. For zero pressure, the equality

$$\langle E - E_0 \rangle = -\tfrac{1}{2}\mathfrak{E}_{\text{el}} \qquad (6.4.3)$$

must be satisfied. This relation, which is obtained from the condition

$$\partial(\langle E - E_0 \rangle + \mathfrak{E}_{\text{el}})/\partial n = 0 , \qquad (6.4.4)$$

is a version of the virial theorem. This approximation determines the "normal density" ρ_0, i.e., the density with which matter is endowed when $P = 0$, which corresponds to

$$n_0^{1/3} = (2^{1/3}/\pi)Z^{2/3}(e^2 m_e/\hbar^2) = (2^{1/3}/\pi)Z^{2/3}(1/a_0) . \qquad (6.4.5)$$

Here $a_0 = e^2 m_e/\hbar^2 = 0.5 \times 10^{-8}$ cm is the Bohr radius, i.e., the radius of the first orbit of the hydrogen atom in the old quantum mechanics of Niels Bohr.

The corresponding density,

$$\rho_0 = \frac{A}{Z}\frac{n}{6 \times 10^{23}} = \frac{2}{\pi^3}\frac{AZ}{6 \times 10^{23}a_0^3} = AZ \qquad (6.4.6)$$

is about unity for hydrogen, 300 for manganese, 1300 for iron, and 16000 for lead. The agreement with experiment is very poor. This is because the computation was conducted by using an unperturbed, uniform distribution of electrons. In reality, however, the interaction with the nuclei changes the distribution of electrons in space considerably at densities much higher than those given above. Therefore, formula (6.2.16) gives only the first two terms of an expansion of the exact dependence of pressure on density. The expansion proceeds with decreasing powers of density—$\rho^{5/3}$ is the first term, and $\rho^{4/3}$ is the second. But when the second term begins to be important, so do the third, fourth, etc.; so that one must start anew, from another viewpoint. An approximate theory was developed in the 1930's, according to which an electron gas with pressure $P = an^{5/3}$ is considered in the electric field of the nuclei and of the electrons themselves. This theory is known as the theory of the *self-consistent* field.

It involves the consideration of the electron distribution inside of a modified elementary cell; for the ordinary cell is substituted a sphere of the same volume, with radius r_1 (see § 6.3). The electron density $n(r)$ depends on the radius r, as does the electrostatic potential $\phi(r)$. The equations have the following form: The Poisson equation, which gives the relationship between ϕ and $n(r)$, reads

$$\nabla^2\phi = -4\pi\zeta(r) = -4\pi[Ze\delta(r) - en(r)] , \qquad (6.4.7)$$

where $e = 4.77 \times 10^{-10}$ (which is a positive number, the absolute value of the electron's charge), and where $\zeta(r)$ is the charge density: $Ze\delta(r)$, where δ is the Dirac delta function, describes the pointlike charge of the nucleus at

170

the center of the cell; and $en(r)$ is the charge of the electrons spread over the cell.

The second equation can be expressed as a condition of mechanical equilibrium for a volume element occupied by electrons, which is acted upon by the electrostatic forces and pressure forces:

$$\zeta(r)E - \nabla P(n) = (-ne)(-\nabla\phi) - [\partial P(n)/\partial r]e_r = 0 . \quad (6.4.8)$$

When the force has a potential and the pressure as a function of density is believed to be known, the equation of equilibrium can be integrated. These assumptions are fulfilled in the case of equation (6.4.8). Therefore, we can rewrite that equation in the following integrated form:

$$ed\phi/dr - (d/dr)\int(1/n)dP = 0 , \quad H(n) - e\phi = K , \quad (6.4.9)$$

where K is a constant, and $H = \int n^{-1}dP = E + P/n$ is the specific enthalpy of the electron gas (enthalpy per electron). (The energy E must not be confused with the electric field E.)

We note that $H(n) = E_F(n)$. Actually, for a nonrelativistic gas $\langle E \rangle - E_0 = \frac{3}{5}(E_F - E_0)$, and $P = n(2/3)(\langle E \rangle - E_0) = n(2/5)(E_F - E_0)$; from which we easily obtain the relationship above between H and E_F with an accuracy up to an unessential additive constant. This relationship between H and E_F is general, and it is also valid in the relativistic domain. Finally, we obtain the equation

$$E_F - e\phi = K' , \quad (6.4.10)$$

where K' is a constant. The significance of this equation is obvious: in equilibrium, the total energy is a minimum; equilibrium in the first approximation must be indifferent to any displacements of a small number of electrons from one place to another. These displacements can be performed by taking an electron at one place (r') and putting it into another place (r''), which yields the condition

$$E_F(r') - e\phi(r') = E_F(r'') - e\phi(r'') . \quad (6.4.11)$$

Here, r' and r'' are arbitrary, and $r' - r''$ need not be small.

Another way to understand equation (6.4.10) involves a shift of all electrons located at a given point by a small distance δr. An analysis of such a displacement leads to the equation of equilibrium in the hydrodynamic form (6.4.8), with a pressure gradient. Obviously, both forms of the condition of equilibrium must be equivalent.

We have considered this problem in such detail because an entirely analogous situation takes place for a star as a whole; we will demonstrate later that the condition of mechanical equilibrium for a star is equivalent to the condition of the constancy of the sum of the chemical potential and of the gravitational potential over the entire star.

But let us return to the microworld. Since the Fermi energy can be expressed simply in terms of the electron density, we obtain

$$An^{2/3} - e\phi = K', \quad n = B(K' + e\phi)^{3/2},$$

and the final equation for the self-consistent field

$$\nabla^2\phi \equiv \frac{1}{r^2}\frac{d}{dr}r^2\frac{d\phi}{dr} = c(K'' + \phi)^{3/2} \quad (6.4.12)$$

—again for the nonrelativistic case and for $0 < r < R$. Here c is a known constant which is expressed simply in terms of $\hbar$, m, and e, whereas $K'' = K'/c$ is a constant without a preassigned value.

The boundary conditions are obtained from the following considerations: At the center of the elementary cell there is a nucleus with charge Ze; hence, when $r \to 0$, $\phi \to Ze/r + \text{const}$. Inside the sphere, there must be Z electrons, so that

$$4\pi \int_0^R n(r)r^2 dr = Z.$$

This condition expresses the charge neutrality of the cell as a whole. From this it follows that (and this can be demonstrated easily in a formal way) when $r = R$, $d\phi/dr = 0$. The second boundary condition, charge neutrality, is satisfied only when K' has a definite value.[7]

The pressure of matter equals the pressure of the electron gas at the cell boundary; i.e., it is known if we know $n(r)$, which in turn depends on $K' + e\phi(R)$. At the boundary, when $r = R$, the electric field is zero, and the cells interact with each other only through the pressure. It can be demonstrated that such an explicit determination of the pressure coincides with the expression for the pressure in terms of a derivative of the energy of the matter as a whole with respect to density.

From dimensional analysis, it follows that the result for the pressure can be expressed in the form

$$P = B\langle n\rangle^{5/3}f\left[\frac{Ze^2/R}{E_F(\langle n\rangle)}\right]. \quad (6.4.13)$$

Here $\langle n\rangle$ is the mean electron density,

$$\langle n\rangle = 6 \times 10^{23}(\rho Z/A) = a(\rho Z/A), \quad a = 6 \times 10^{23}g^{-1};$$

A is the atomic weight; and $B\langle n\rangle^{5/3}$ is the pressure of a degenerate gas of free electrons (without consideration of the variation in density due to the interaction with the nuclei). The expression $B\langle n\rangle^{5/3}$ is given above (eq. [6.2.16]).

The function $f[y]$ is a dimensionless function which describes the influence of the redistribution of the charge upon the pressure. It is clear that this

---

7. We can add an arbitrary constant to ϕ; then K'' decreases by a corresponding value; the sum $K' + e\phi$ is determined uniquely.

function satisfies $f < 1$, since the density of the electrons at the cell boundary must be less than the mean density.

The dimensionless f is a function of the dimensionless ratio of the potential energy to the kinetic energy of one electron,

$$y = \frac{Ze^2/R}{E_F} . \tag{6.4.14}$$

Again, instead of exact values for the energies, we take their characteristic values at a distance R and at a mean density $\langle n \rangle$. We can express y in terms of the density, omitting all dimensionless factors:

$$R = (A/a\rho)^{1/3} ,$$

$$E_F = \frac{p_F{}^2}{2m} \frac{\hbar^2 \langle n \rangle^{2/3}}{m} = \frac{\hbar^2 (aZ\rho/A)^{2/3}}{m} ,$$

$$y = \frac{Ze^2(a\rho/A)^{1/3}m}{\hbar^2(aZ\rho/A)^{2/3}} = \frac{me^2 Z^{1/3} A^{1/3}}{\hbar^2 a^{1/3} \rho^{1/3}} . \tag{6.4.15}$$

The behavior of the function f can be anticipated in two limiting cases: $y \ll 1$ and $y \gg 1$. In the first case, in the limit, $f(0) = 1$; and the next term of the expansion,

$$f(y) = 1 - By ,$$

gives a correction which is proportional to $\rho^{4/3}$ (compare the above with the formulae in § 6.3). The other limiting case is not so obvious. One expects that for a given dimension of the cell and an arbitrarily large nuclear charge Z, the pressure will approach a firmly defined limit. The reason for this can be visualized as follows: We increase the charge of the nucleus and simultaneously increase the number of electrons. The new electrons settle near the nucleus and screen the added charge; the distribution of charge far from the nucleus does not change; it approaches a well-defined limit when $Z \to \infty$. This picture, supported by integration of the equation,[8] leads to the following conclusion: In order for the expression for pressure

$$P = \text{const.} \times Z^{5/3}\rho^{5/3}f[(Z/\rho)^{1/3}]$$

not to depend on Z in the limit $Z \to \infty$, it is necessary that $f(y) = y^{-5}$ when $y \gg 1$. But in this case

$$P = \text{const.} \times \rho^{10/3} \sim 10^{11}(\rho/A)^{10/3}. \tag{6.4.16}$$

It is interesting that such qualitative considerations result in an asymptotic law, at not very high pressure, which is close to $P \sim \rho^3$. The law $P \sim \rho^3$ was derived by Landau and Stanyukovich (1945) for products of explosions. Later, Stanyukovich (1955) demonstrated that this law is very convenient

8. When $Z \to \infty$, the solution has the asymptotic form at $r \ll R$:

$$\phi \sim r^{-4} , \qquad n(r) \sim r^{-6} .$$

for gas-dynamical calculations. Experiments on the compression of iron and a number of other elements within a rather broad range of pressure (up to approximately 10^7 atmospheres) is represented by a similar equation:

$$P = a(\rho^3 - \rho_0^3) . \qquad (6.4.17)$$

The estimates of the equation of state made by using equation (6.4.13) are nonetheless approximate.[9] Still within the framework of an averaged collective description of electrons, it is possible to consider a number of corrections and greatly improve the results. This has been done by Kalitkin (1960), who used contributions by Kirzhnits, Kompaneets, Pavlovsky, and others. Further refinement is possible only by consideration of the individual orbits of electrons and of the specific chemical properties of the elements.

The result obtained by Kalitkin (1960) can be reduced to an expression for pressure in the form of a difference between two terms,

$$P = B(\rho Z/A)^{5/3}f[(ZA/\rho)^{1/3}] - c(\rho Z/A)^{4/3}\phi[(ZA/\rho)^{1/3}] . \qquad (6.4.18)$$

Here the first term coincides with that found previously, and the second term includes the corrections. The functions f and ϕ are defined in Kalitkin (1960). This simple formula describes fairly well, on the average, even such a "delicate" quantity as ρ_0, the density of matter under normal conditions of zero pressure.

We reproduce a graph (Fig. 28) from Kalitkin's work, which compares the calculated and measured ρ_0 as a function of the atomic number Z. Only the oscillations of Figure 28 are beyond the scope of the theory; they are related to the specific chemical properties of the matter. At pressures of the order of 5–10 million atmospheres these oscillations become considerably smoother.

A tabulation for several elements as calculated by Salpeter (1961) is given in Table 2. The table shows the ratio of the total pressure to the pressure of a degenerate gas (eq. [6.2.15]), as a function of the relativity parameter x, for fixed Z.

6.5 Nuclear Processes and Nuclear Interactions: Their Effect upon the Equation of State

In the domain above 10^8 g cm^{-3}, the equation of state depends largely on nuclear processes and interactions. It is possible to formulate precisely the problem of finding the state of minimum energy for a given baryon density and for zero temperature. The energy is the sum of the energies of the nuclei and of the electrons; the latter depends on $Z/A = 1/\mu_e$. For a relativistic Fermi gas, the mean energy of one electron is

$$\langle E \rangle = \tfrac{3}{4}E_F = \tfrac{3}{4}cP_F = \tfrac{3}{4}m_ec^2(\rho/10^6\mu_e)^{1/3} . \qquad (6.5.1)$$

9. The theory of the self-consistent field is described in detail by Gombas (1950).

Accordingly, the energy of the electrons per gram is

$$E_e = \tfrac{3}{4}(m_e c^2/m_P)(\rho/10^6)^{1/3}\mu_e^{-4/3} = 3.75 \times 10^{15}\rho^{1/3}\mu_e^{-4/3} \text{ erg g}^{-1} . \quad (6.5.2)$$

Thus, the energy of the contracted state differs from the nuclear energy of the individual atoms:

$$E = E_n(A, Z) + E_e . \quad (6.5.3)$$

The minimum of E_n corresponds to the stable and most abundant iron isotope, $_{26}^{56}\text{Fe}$.[10] The additional term E_e is of greater significance the higher the

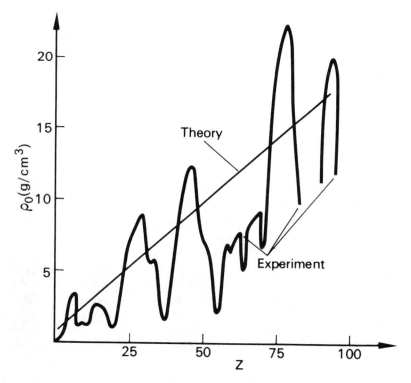

FIG. 28.—Experimental and theoretical (statistical) relationship between the density of uncompressed matter and the nuclear charge.

10. Strictly speaking, the minimum of E_n corresponds to $_{28}^{62}\text{Ni}$ (Kravzov 1965), but the difference $(E_n)_{\text{Fe}} - (E_n)_{\text{Ni}}$ is extremely small,

$$[(E_n)_{\text{Fe}} - (E_n)_{\text{Ni}}]/[(E_n)_{\text{Fe}}] = 4.5 \times 10^{-4} ,$$

and is not very important. At nonzero temperature a more important role in the formation of elements is played by statistical factors.

Editors' note.—A critical parameter, which is more important than statistical factors, is the relative number of neutrons and protons per unit of mass. If weak interactions are density. This additional term displaces the minimum of E toward nuclei with large μ_e.

Let us begin with a very crude picture, neglecting many intermediate steps (see below), and considering only three particular states: iron; a nucleus strongly overloaded with neutrons; and a "gas" of free neutrons.

According to Salpeter (1961), when $E_F = 20.6$ MeV, $\rho = 1.9 \times 10^{11}$ g cm^{-3}, and $P = 3.2 \times 10^{29}$ ergs cm^{-3}, it becomes energetically advantageous to transform iron into strontium. Here, as well as in Tables 3 and 4,

TABLE 2

RATIO OF TOTAL PRESSURE TO THE PRESSURE
OF A DEGENERATE GAS

x	$(2/\mu_e)\rho$(g cm^{-3})	Z 2	6	12	26
0.05	2.44×10^2	0.760	0.564	~0.3	....
0.1	1.95×10^3	0.8802	0.7819	0.6705	~0.5
0.2	1.56×10^4	0.9404	0.8906	0.8341	0.7308
0.3	5.26×10^4	0.9604	0.9266	0.8882	0.8178
0.4	1.25×10^5	0.9705	0.9445	0.9150	0.8607
0.5	2.44×10^5	0.9765	0.9551	0.9308	0.8860
0.6	4.21×10^5	0.9805	0.9620	0.9410	0.9024
0.7	6.68×10^5	0.9833	0.9669	0.9482	0.9138
0.8	1.00×10^6	0.9853	0.9705	0.9535	0.9221
1.0	1.95×10^6	0.9881	0.9752	0.9605	0.9332
1.2	3.37×10^6	0.9898	0.9782	0.9684	0.9401
1.4	5.36×10^6	0.9909	0.9801	0.9677	0.9447
1.6	8.00×10^6	0.9917	0.9814	0.9697	0.9479
1.8	1.14×10^7	0.9922	0.9824	0.9711	0.9511
2.0	1.56×10^7	0.9926	0.9831	0.9721	0.9519
2.5	3.04×10^7	0.9932	0.9842	0.9738	0.9546
3.0	5.26×10^7	0.9935	0.9848	0.9748	0.9562
4.0	1.25×10^8	0.9938	0.9853	0.9757	0.9577
5.0	2.44×10^8	0.9939	0.9856	0.9761	0.9585
7.5	8.22×10^8	0.9939	0.9858	0.9765	0.9592
∞	∞	0.9939	0.9859	0.9768	0.9598

NOTE.—At large Z and small x, when the ratio is less than 0.5, the calculated results are unreliable (upper right-hand corner of the table).

E_F does not include $m_e c^2$. The isotope $^{120}_{38}$Sr is radioactive under normal conditions; however, at the high density considered here, the presence of a Fermi distribution of electrons with a boundary energy $E_F > 20.6$ MeV inhibits the decay of strontium. The transition from iron to strontium with varying density is shown in more detail in Table 3. This table, quoted from Salpeter, shows the Fermi energies (MeV) and the corresponding densities (g cm^{-3}) for transitions at equilibrium conditions between various nuclei (Z, A).

slow compared with evolutionary time scales—as is often the case, especially when densities are not extremely high—then the relative neutron-proton number is determined from the history of the matter.

At densities in excess of $\rho = 10^{11.53}$ g cm^{-3}, it is the state of individual free neutrons that is in equilibrium with strontium. For densities $\lesssim 10^{11.5}$ g cm^{-3} we obtain a theoretical curve $P(\rho)$ for the pressure, which is characterized by a sequence of phase transformations of the first kind (similar to the condensation of steam in water). Figure 29 shows a simplified diagram of the behavior of $P(\rho)$, where only iron, strontium, and free neutrons are considered. The equilibrium between iron and strontium requires a Fermi energy of 20.6 MeV, corresponding to an electron density of 3.9×10^{34} cm^{-3}. By virtue of electrical neutrality, such an electron density corresponds to a matter density of 1.4×10^{11} g cm^{-3} for Fe and 2×10^{11} g cm^{-3} for Sr.

TABLE 3

CRITICAL POINTS FOR THE TRANSITIONS
BETWEEN DIFFERENT NUCLEI

Z	A	E_F	$\log_{10} \rho$
26...............	56	...	...
28...............	62	0.6	7.15
28...............	64	2.5	8.63
28...............	66	3.9	9.15
28...............	68	6.1	9.69
30...............	76	7.0	9.87
30...............	78	8.5	10.13
30...............	80	9.5	10.28
32...............	90	14.8	10.84
38...............	120	20.6	11.28
n...............	1	24.0	11.53

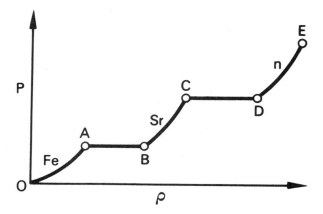

FIG. 29.—Idealized diagram (see Table 2 for details) showing the relationship between the pressure and the density of cold matter.

177

Cold Matter

The equilibrium between strontium and free neutrons at rest requires $E_F = 24$ MeV. This corresponds to a Sr density of 3.2×10^{11} g cm^{-3}.

Now we are ready to give a sketch of the $P(\rho)$ relation. It consists of a segment O–A with pure iron and with $P = a_1\rho^{4/3}$, $a_1 = 4.4 \times 10^{14}$, which ends at $\rho = 1.4 \times 10^{11}$ g cm^{-3}. Along the segment A–B the transformation Fe → Sr occurs.

Points A and B correspond to equal pressure and consequently to equal electron density. The densities of the two elements at points A and B are in a ratio that is inversely proportional to μ_e, so that the density increases during the transformation by a factor of 1.47: $\rho_B = 1.47\rho_A$. Since the nuclear interaction is negligibly small, the distribution of the nuclei has no effect upon the pressure, nor upon the energy; in particular, it is irrelevant whether iron and strontium form two geometrically separated phases, or whether they form a mixed solution. Pure Sr is in equilibrium on the segment B–C, from $\rho = 2 \times 10^{11}$ to $\rho = 3.2 \times 10^{11}$; here $P = a_2\rho^{1/3}$, $a_2 = 0.6\ a_1$.

On the next segment, C–D, we encounter a mixture of Sr and neutrons. Up to a density of the order of $\rho_0 \sim 10^{12}$, it is legitimate to neglect the kinetic energy of the neutrons and their nuclear-interaction energy. In this approximation the electron density, the density of Sr in the mixture, and the pressure (which is due to electrons only) remain constant. The total density grows from C to D due to the growing contribution of free neutrons. There is no sudden change at D. Somewhere near D quantitative changes begin to occur gradually, and the neutron pressure begins to rise smoothly until it reaches the order of magnitude of the electron pressure near D. The subsequent story (at $\rho > \rho_D$) will be told below.

However, this analysis, while theoretically correct (at least in the oversimplified case Fe-Sr-n), is nevertheless an example of excessive rigor or even pedanticism. The truth of the matter is that at low temperature there does not exist a real mechanism which could transform iron into strontium within any reasonable time. Let us reformulate this problem in a somewhat different way: in order to transform $_{26}^{56}$Fe to $_{38}^{120}$Sr within a reasonable time, high temperatures are required; it is necessary to break up the iron nuclei into component parts, i.e., α-particles and neutrons, and reshuffle them in a different way. But at the appropriately high temperature, the equilibrium will also be different. The processes and equilibrium at high temperatures will be considered in the following chapter.

In reality, during the contraction of matter, when a certain threshold E_F is reached, the inverse β-process becomes possible: an electron is captured by a nucleus and one of the protons of the nucleus is transformed into a neutron:

$$(Z, A) + e^- \to (Z - 1, A) .$$

This process is called the neutronization of matter. It takes place at the rate of weak interactions.

During the compression of iron, when the level $E_F = 3.7$ MeV, $\rho = 1.15 \times 10^9$ g cm^{-3}, is reached, it becomes possible to transform iron into $^{56}_{25}$Mn. However, the transformation threshold of odd-odd $^{56}_{25}$Mn into even-even $^{56}_{24}$Cr is smaller: it is 1.6 MeV. Hence, each nucleus of $^{56}_{25}$Mn will be transformed immediately into $^{56}_{24}$Cr.[11]

Thus, during compression to a density which exceeds the critical density $\rho_{cr} = 1.15 \times 10^9$ g cm^{-3} by a very small amount, iron begins to transform into chromium. During this process, the pressure is

$$1.2 \times 10^{23}(E/m_ec^2)^4 = 3.3 \times 10^{25} \text{ ergs cm}^{-3}.$$

The density increases by the ratio of $26/24 = 1.08$. We will consider the rate of this reaction below. In order of magnitude, when P_c is exceeded by 1 percent, a time of the order of 10^6 sec is required for the reaction to take place.

The two-step nature of the reaction (Fe $\rightarrow$ Mn, slow; Mn $\rightarrow$ Cr, rapid, with an excess of energy) causes an irreversibility of the process. If matter, which had been transformed by compression into $^{56}_{24}$Cr, were to expand slowly, then the decay of chromium would begin only after E_F had dropped below 1.6 MeV; that is, at density $\rho = 7.6 \times 10^7$ g cm^{-3} and pressure 1.2×10^{25} ergs cm^{-3}. Each decay Cr $\rightarrow$ Mn $+ e^- + \nu$ will be followed immediately by Mn $\rightarrow$ Fe $+ e^- + \nu$, since the second process, in essence, is an above-threshold process; that is, the energy of decay is 3.7 MeV, and $E_F = 1.6$ MeV. In the pressure interval from $1.2 \times 10^{23}(1.6/0.511)^4 = 1.15 \times 10^{25}$ ergs cm^{-3} to $1.2 \times 10^{23}(3.7/0.511)^4 = 3.3 \times 10^{26}$ ergs cm^{-3}, the matter may contain mixtures of iron and chromium in any proportion, depending on its preceding history.

The same phenomena occur in the transitions Cr $\rightarrow$ Va $\rightarrow$ Ti, etc.

We assumed above that in the course of thermonuclear reactions at high temperatures and not very high densities a complete thermodynamic equilibrium was attained, which was followed by cooling; the equilibrium state of matter upon cooling is ^{56}Fe. After this, contraction and the transformations

$$\text{Fe} \rightarrow \text{Mn} \rightarrow \text{Cr}$$

take place.

In principle the contraction of matter, which has not been exposed to a high temperature, or which has experienced insufficiently high temperature and consists not of iron but of other elements, is possible. Table 4 gives the neutronization thresholds for various nuclei.

Of interest are the especially low transformation threshold of ^{3}He, and the exceptional stability of ^{4}He. The reaction threshold for hydrogen is such that at this threshold density, even in cold matter, the process $p + p \rightarrow$ D $+ e^+ + \nu$ or $p + p + e^- \rightarrow$ D $+ \nu$ takes place rather rapidly. Such

11. Theoretically, when $E_F = 2.65$ MeV a dual process Fe $+ 2e^- \rightarrow$ Cr $+ 2\nu$ is possible. However, the probability of such a process is negligible, and hence we will not consider it here.

processes taking place at low temperature are called *pycnonuclear;* for calculations of the rate of the first process see Zel'dovich (1957) and Wildhack (1940). These processes do not have a precise threshold, but their rates are strongly influenced by the density. At the critical density for $H + e^- \rightarrow n + \nu$, the time of both pycnonuclear reactions is of the same order: 10^{10} seconds.

With further increase of the pressure above the level which corresponds to $E_F \sim 20$ MeV, a qualitatively new phenomenon takes place. At lower pressure, there occurred transformations of some nuclei into others, but all those nuclei had the same mass number A, and were also stable with respect to nuclear forces. At higher pressures, nuclei are obtained which decay with an emission of free neutrons. A specific example of such a situation is included

TABLE 4

NEUTRONIZATION THRESHOLDS

Nucleus	E_F (MeV)	ρ_c(g cm^{-3})	P_c(erg cm^{-3})	Reaction Chain Product at Threshold
H.........	0.78	10^7	5.5×10^{23}	$H \rightarrow n, n+H = D$
$^3_2$He.......	0.018	2.8×10^4	1.8×10^{17}	$^3He \rightarrow T$
$^4_2$He.......	20.6	1.32×10^{11}	3.2×10^{29}	$^4He \rightarrow T+n, e^-+T=3n$
$^{12}_6$C........	13.4	4.45×10^{10}	5.7×10^{28}	$^{12}C \rightarrow ^{11}B$
$^{28}_{14}$Si........	4.6	1.92×10^9	7.8×10^{26}	$^{28}_{14}Si \rightarrow ^{28}_{13}Al$
$^{56}_{26}$Fe........	3.7	1.15×10^9	3.3×10^{26}	$Fe \rightarrow Mn, Mn \rightarrow Cr$

in Table 3; in the course of the reaction of $e^- + {}^4$He one might expect the formation of ^{4}H. However, the nucleus ^{4}H is not stable; it decays to a neutron plus a tritium nucleus. In this case, the neutronization threshold of T (9.5 MeV) is lower than the threshold of ^{4}He, so that as a result of an initial reaction of $e^- + {}^4$He, only free neutrons are left in the course of subsequent reactions. Neutrons do not attach individually to ^{4}He because the ^{5}He nucleus is unstable.[12]

A similar phenomenon will also take place in any other chain. For example, when $A = 56$ after $Fe \rightarrow Mn \rightarrow Cr$ with increasing E_F, the process will continue; and approximately when $Z = 12$ (i.e., for the exotic nucleus $^{56}_{12}$Mg), when $E_F \cong 20$–22 MeV, the emission of free neutrons will begin and the reactions

$$e^- + {}^{56}_{12}Mg \rightarrow {}^{53}_{11}Na + 3n + \nu , \quad e^- + {}^{53}_{11}Na \rightarrow {}^{50}_{10}Ne + 3n + \nu ,$$

12. It is not impossible that, at very high density, ^{6}He and ^{8}He are formed by pairwise attachment of neutrons $^4He + 2n \rightarrow {}^6He + \gamma$, $^6He + 2n \rightarrow {}^8He + \gamma$. The threshold for $e^- + {}^6He \rightarrow T + 3n + \nu$ is somewhat higher than the threshold for the ^{4}He reaction (E_F is larger by 3 percent and P_c is larger by 12 percent). For ^{8}He E_F amounts to 24 MeV. The very existence of ^{8}He was predicted by one of the authors (Zel'dovich 1960); it has now been confirmed and its binding energy has been measured. From this E_F is calculated.

will take place. Thus, in the course of contraction, we can outline three domains: (*a*) the domain of stable nuclei; (*b*) the domain of β-radioactive nuclei originating from a nuclear reaction with electrons, during which the protons inside the nucleus are transformed into neutrons, the resultant nuclei being stabilized by the presence of electrons; and (*c*) the domain of a neutron gas mixed with β-radioactive nuclei which are not able to attach free neutrons.[13] The β-decay energy of such nuclei is of the order of 24 MeV. It follows that in equilibrium with those nuclei there are enough electrons to provide $E_F = 24$ MeV, i.e., $n_e = 6 \times 10^{29}(48)^3 = 7 \times 10^{34}$ cm^{-3}.

Under these conditions,

$$P = 6.6 \times 10^{29} \text{ dyn cm}^{-2}.$$

Assuming $Z/A = 1/4$ for the nuclei (e.g., ^{8}He), we find their mass density to be

$$\rho = n \times (A/Z) \times (1/a) = 4 \times 7 \times 10^{34}/6 \times 10^{23} = 4 \times 10^{11} \text{ g cm}^{-3}.$$

Consequently, at an overall density ρ, which exceeds 4×10^{11} g cm^{-3} and corresponds to a pressure of 6.6×10^{29} dyn cm^{-2}, a neutron gas with $\rho_n = (\rho - 4 \times 10^{11})$ g cm^{-3} is in equilibrium with electrons of density $n_e = 7 \times 10^{34}$ cm^{-3}, and with neutron-rich nuclei[14] of density $\rho_{\text{nuclei}} = 4 \times 10^{11}$ g cm^{-3}.

There exist statements in the literature to the effect that matter in such conditions ($4 \times 10^{11} < \rho < 3 \times 10^{13}$) may be treated as a uniform mixture of neutrons and protons. Actually, so long as the overall density of the matter is less than nuclear density, the protons undoubtedly are not distributed uniformly, but rather are agglomerated with an appropriate number of neutrons into that type of nucleus which yields the maximum binding energy Q_p per proton for an arbitrary number of neutrons. In the nucleus ^{4}He, $Q_p = 14$ MeV; in the stable ^{12}C, $Q_p = 15.3$ MeV, but in ^{16}C, $Q_p = 18.5$ MeV; in the radioactive nucleus ^{38}S, $Q_p = 20$ MeV; in ^{49}Cu, $Q_p = 21$ MeV. These values are not the maximum possible; the maximum is unknown because it lies in a regime of neutron-rich nuclei that has not been explored experimentally. A theoretical extrapolation, with account taken of nuclear forces, gives $Q_p = 30$ MeV in ^{72}Ti, and $Q_p = 36.5$ MeV in ^{22}C.

6.6 The Properties of a Neutron Gas

In first approximation, we will consider free neutrons as an ideal gas; i.e., for the time being we will disregard the interaction of neutrons with the

---

13. A typical nucleus which is not able to absorb individual neutrons is ^{4}He; nuclei which are not able to absorb any number of neutrons are T and ^{8}He.

14. This is true up to a density of the order of 3×10^{13}, where the Fermi energy and the interaction of free neutrons begin to be important.

remaining nuclei and with each other. However, we will keep in mind that neutrons obey the Fermi statistics.

For a neutron gas, the Fermi momentum is $p_F = m_e c (\rho_n / 10^6)^{1/3}$, where $\rho_n = \rho - 4 \times 10^{11}$. At nuclear density, $\rho_n = 2 \times 10^{14}$, we obtain $p_F = 450\, m_e c = \frac{1}{4} m_n c$, where m_n is the mass of the neutron. Thus $E_F = p_m^2/2M = \frac{1}{32}\, m_n c^2 = 30$ MeV, which means that the neutrons still can be regarded as approximately nonrelativistic when $\rho_n \approx 2 \times 10^{14}$ g cm^{-3}. When $\rho < 2 \times 10^{14}$ g cm^{-3}, the neutron pressure is that of a nonrelativistic, ideal Fermi gas of neutrons

$$P_n = \tfrac{2}{3}\langle E_n \rangle = \tfrac{2}{3} \times \tfrac{3}{5} \times E_F n = 5.5 \times 10^9 \rho_n^{5/3} . \qquad (6.6.1)$$

The total pressure equals the sum of the pressure of the electron gas with $E_F = 24$ MeV and of the neutron Fermi gas. Thus,

$$P = 6.6 \times 10^{29} + 5.5 \times 10^9 (\rho - 4 \times 10^{11})^{5/3} . \qquad (6.6.2)$$

It is instructive to determine the density at which the pressures of the neutrons and of the electrons are equal. We obtain from equation (6.6.2)

$$\rho_n = 1.1 \times 10^{12} \text{ g cm}^{-3} , \qquad \rho = 1.5 \times 10^{12} \text{ g cm}^{-3} .$$

At nuclear density, $\rho = 2 \times 10^{14}$ g cm^{-3}, the neutron pressure is 3.5×10^{33} dyn cm^{-2}, which is greater than the electron pressure by a factor of 5000; the density of neutrons (per cubic centimeter) is greater than the electron density by approximately a factor 2000, and is greater than the density of the nuclei present in matter at this density by a factor of 500.

However, when the Fermi energy of the neutrons is 30 MeV, the condition that determines the Fermi energy of the electrons has already changed. Actually, during the reaction $e^- + p \rightarrow n + \nu$, we must not only consume the binding energy per proton in the nuclei,[15] which equals 24 MeV (see § 6.5 above), but we must also place the neutron just formed on top of the Fermi distribution of neutrons, consuming another 30 MeV (previously, when this energy was less than 24 MeV, we neglected it; it varies as $\rho_n^{2/3}$). Therefore, it could be expected that the Fermi energy of the electrons will increase up to 54 MeV. Accordingly, we will obtain for the density of nuclei, not $\rho/\mu_e = 4 \times 10^{11}$ g cm^{-3}, but 5×10^{12} g cm^{-3}, or 2.5 percent ρ_n; the electron density (per cubic centimeter) will be 0.6 percent of the neutron density; and the electron pressure will be 1.6×10^{31} (0.7 percent of the total pressure), all at a total density of

$$\rho \cong \rho_n = 2 \times 10^{14} \text{ g cm}^{-3} .$$

---

15. It must be understood, therefore, that we are not speaking of free protons! More exactly, we should write $e^- + (Z, A) \rightleftharpoons (Z - 1, A) + \nu$ or $e^- + (Z, A) \rightleftharpoons (Z - 1, A') + (A - A')n + \nu$—the transmutation of a proton inside a nucleus, perhaps accompanied by the emission of one or more free neutrons. The inverse process is β-decay "in flight": a neutron which is not initially bound in a nucleus undergoes β-decay into a bound proton, $n + (Z, A) \rightarrow (Z + 1, A + 1) + e^- + \bar{\nu}$, with energy release including the binding energy of the proton.

The numerical relationships and formulae just given should be treated as order-of-magnitude estimates because the interaction of the neutrons was not considered. Direct experimental studies of neutron-neutron scattering cannot be accomplished at the present time. An estimate of the neutron interaction can be obtained from reactions of the type $\pi^- + D \rightarrow n + n$, $\pi^- + D \rightarrow n + n + \pi^0$, and $\gamma + D \rightarrow n + n + \pi^+$, as well as from the interaction between neutrons and nuclei, if the contribution of the protons which are present in the nuclei is subtracted. But there exists a more powerful, though indirect, approach to the problem.

Investigations of the properties of nuclei have resulted in the formulation of the principle of isotopic invariance (abbreviated PII; also called charge independence of nuclear forces). For purposes of this discussion, PII is formulated as follows: Neutrons and protons in the same state experience the same nuclear interactions.

The concept of a quantum-mechanical state includes the energy, the orbital angular momentum L, and the spin angular momentum S of the system. In particular, the nuclear interactions $p + p$ (two protons with each other), $p + n$ (interaction between a proton and a neutron), and $n + n$ (interaction between two neutrons) are the same in equivalent states with $L = 0$ and $S = 0$, and with the same energy. It is necessary to make two stipulations here: (1) The nuclear interaction is the same for all nucleons, but the electromagnetic interaction does not satisfy PII and is different; this difference is easily taken into account. (2) A proton and a neutron in the context of the Pauli principle are different particles and thus, for example, for the system $p + n$ the condition $L = 0$, $S = 1$ (parallel spins) is allowed, whereas for $p + p$ and for $n + n$ this condition is not possible. Therefore, it is not surprising that $p + n$ can form a bound deuteron D (in the state with $L = 0$, $S = 1$) whereas $p + p$ and $n + n$ do not form such a bound system. Experimental investigations of $p + p$ and $p + n$ systems show that there is no bound state with $L = 0$, $S = 0$ in these pairs: in particular, it is well known that there is no $2p = {}^2\mathrm{He}$ nucleus and no excited, bound state of a deuteron with zero spin. From this we can conclude that there is also no bound state of two neutrons.

In scattering studies of neutrons on protons it is possible to delineate two processes: the scattering of a neutron on a proton with parallel spins (so that the total spin $S = 1$), and the scattering with antiparallel spins (total $S = 0$). The second process confirms the absence of any bound state for $S = 0$ and gives information about the interaction potential. It can be approximated by a well of depth $\sim$20 MeV and range $r_0 \approx 10^{-13}$ cm; for these values there is just barely no bound state. The absence of bound states in a system with attractive interaction is a typical quantum-mechanical effect. Qualitatively it is a consequence of the Heisenberg uncertainty principle. In a loose system (with large mean distance between the neutrons), the probability of neutron interaction is small; in a dense system, the kinetic energy of the neutrons increases too much, so there exists no state with a negative total energy. Experimentally, though with less certainty, it is found that a bound state n^4 (tetraneutron) does not exist (see, for example, the latest summary by Argan *et al.* [1965], and the review by Baz', Gol'dansky, and Zel'dovich [1965]).

However, the neutron-neutron attraction must decrease the pressure of a neutron gas substantially (by several factors of 2) as compared with the pressure of an ideal Fermi gas.[16] In the region of density less than nuclear density, up to approximately 5.2×10^{13} g cm^{-3}, only the interaction of neutron pairs in the state with $L = 0$ is important. The conditions of the problem are rather simple; nonetheless, the problem of the state of a neutron gas has not yet been solved precisely even in this formulation.

From dimensional arguments,[17] it follows that in a wide domain, $10^{11} < \rho < 2 \times 10^{13}$,

$$P_n = \beta \times 5.5 \times 10^9 \rho_n^{5/3} \, ,$$

where β is a constant numerical coefficient ($\beta < 1$) which is unknown (it is not even certain that $\beta > 0$). However, one does know that β is independent of density.

With further increase of the density, the more complex nature of nuclear forces becomes apparent. In addition to collisions of neutron pairs with $L = 0$, at higher densities collisions with $L = 1, 2, \ldots$, are also important. Low density ($\rho < 2 \times 10^{13}$ g cm^{-3}) corresponds to a small energy of the colliding neutrons $E_{\mathrm{F}} < 10$ MeV; here one could consider the interactions with $L = 0$ only, by use of the so-called virtual level with energy $E_\nu = +60$ KeV which characterizes low-energy scattering. (Neglecting E as compared with E_{F}, we could conclude that $P \sim \rho_n^{5/3}$.) However, when $L = 0$, $E_{\mathrm{F}} > 10$ MeV, one cannot limit oneself to this approximation.

At small distances (which occur in high-energy scattering of neutrons and protons) the attraction is replaced by repulsion, as is also shown by scattering experiments, but at high energies. Finally, at high density the interaction cannot necessarily be reduced to a combination of paired interactions. It is not clear whether data on nucleon-nucleon scattering are sufficient for the development of a theory. An experiment involving the collision of several high-velocity particles or involving a direct determination of the pressure, is obviously impossible. A quantitative theory of the equation of state at nuclear density and at higher density does not exist; and its development is impossible until a definitive theory of strong interactions has been devised. The expressions for the pressure which are cited in the literature are not reliable, and authors exaggerate their accuracy.

16. For more detailed discussion on the subject of a neutron gas, see Brueckner, Gammel, and Kubis (1960).

17. See n. 16 above.

6.7 Density Greater than Nuclear

Having warned the reader adequately as to the necessity of healthy skepticism with respect to the accuracy of the formulae, we will, for reference purposes, cite the formula used by Cameron (1959a):

$$P = 5.3 \times 10^9 \rho_0^{5/3} + 1.6 \times 10^{-5} \rho_0^{8/3}$$
$$- 1.4 \times 10^5 \rho_0^2, \qquad \rho_0 = m_n n. \tag{6.6.3}$$

The corresponding velocity of sound is

$$(a_{\text{sound}})^2 = \left(\frac{\partial P}{\partial \rho}\right)_s$$

$$= \frac{8.8 \times 10^9 \rho_0^{2/3} + 4.3 \times 10^{-5} \rho_0^{5/3} - 2.8 \times 10^5 \rho_0}{c^2 + 1.3 \times 10^{11} \rho_0^{2/3} + 2.6 \times 10^{-5} \rho_0^{5/3} - 2.8 \times 10^5 \rho_0}. \tag{6.6.4}$$

More complex expressions have been developed by Salpeter (1961).

In formulae (6.6.3) and (6.6.4), ρ_0 is the density of rest mass; i.e., it is a quantity which is proportional to the baryon density (in this case, when $\rho < 10^{15}$ g cm^{-3}, the baryons are neutrons). The total mass density also includes the energy:

$$\rho = \rho_0 \left(1 + \frac{E_1}{c^2}\right) = \rho_0 \left(1 + \frac{1}{c^2} \int_0^{\rho_0} \frac{P}{\rho_0^2} \, d\rho_0\right). \tag{6.6.5}$$

The relationship between ρ and ρ_0 will be described in more detail in § 6.7.

Notice that the expression (6.6.4) leads to a velocity of sound greater than the velocity of light when $\rho > 3.7 \times 10^{15}$ g cm^{-3}. Consequently, the formula (6.6.3) is certainly incorrect when $\rho > 3.7 \times 10^{15}$ g cm^{-3}. In a subsequent paper Tsuruta and Cameron (1966) have given a new formula, which takes into account the partial transformation of neutrons into other particles.

6.7 DENSITY GREATER THAN NUCLEAR

Consider the relationship between pressure and density in the density domain $\rho > 10^{14}$ g cm^{-3}. The upper boundary of the domain considered is $\rho \sim 10^{93}$ g cm^{-3}. In this domain there exist no experimental data which could provide a foundation for the desired relation. Therefore, the content of the following section is an elucidation of the limits within which the pressure is contained, i.e., of those restrictions which are imposed by general physical laws. First of all we must define our terminology in this particular context.

Let n denote the density (number per cubic centimeter) of baryons. If the system includes both baryons (n_1) and antibaryons ($\bar{n}_1$), then $n = n_1 - \bar{n}_1$; and in this sense it might be more precise to consider n to be the baryon charge density.[18]

---

18. This situation (the presence of antibaryons) occurs at superhigh temperatures.

Cold Matter

In the nonrelativistic domain, the density is $\rho = nm_0$, where m_0 is the mass of one baryon.[19] The work done during compression (per baryon), ϵ_1, is comparable to $m_0 c^2$ by the time the matter has reached $\rho > 10^{14}$ g cm^{-3}; consequently, the density ρ is greater than nm_0. We will henceforth refer to $nm_0 = \rho_0$ as the "density of rest mass." Obviously, $\rho = n(m_0 + \epsilon_1/c^2)$ rather than ρ_0 appears in the equations of hydrodynamics: the quantity $dm = \rho dV$ is used to define the mass of a volume element. To be more precise, the energy density ϵ, which is the 0–0 component of the stress-energy tensor T_{ab}, is $\epsilon = T_0{}^0 = \rho c^2$. The pressure is given by the diagonal space components, $T_1{}^1 = T_2{}^2 = T_3{}^3 = P$. All other components are zero in the rest frame of matter in equilibrium. The equations of hydrodynamics and gravitation involve both ϵ and P, but not ρ_0: the inertial mass is $c^{-2}(\epsilon + P)dV = (\rho + P/c^2)dV$; the gravitating mass density is $c^{-2}(\epsilon + 3P) = \rho + 3P/c^2$.

However, in the equations of thermodynamics, quantities related to a unit rest mass are of basic importance. Consider a thought-experiment which is performed with a given quantity of matter enclosed in an impenetrable vessel; the rest mass in the vessel $\rho_0 V$ is conserved, whereas the mass ρV changes in accordance with the work performed. From the changes in ρV we can determine the specific energy E and the specific entropy S per unit rest mass.[20] We shall include in E the rest energy, so that $E = E_{nr} + c^2$, where E_{nr} is the excess of the energy over the rest mass-energy; E_{nr} is what is called simply "energy" in nonrelativistic physics. Therefore, we have

$$\rho = \rho_0 E/c^2 . \qquad (6.7.1)$$

At constant entropy, and in particular when $S = 0$,

$$P = -\frac{\partial E}{\partial V} = -\frac{\partial E}{\partial(1/\rho_0)} = \rho_0{}^2 \frac{\partial E}{\partial \rho_0} = \rho_0{}^2 c^2 \frac{\partial(\rho/\rho_0)}{\partial \rho_0} . \qquad (6.7.2)$$

The velocity of sound a_{sound} (see, for example, Landau and Lifshitz 1954) is

$$(a_{\text{sound}})^2 = dP/\partial\rho . \qquad (6.7.3)$$

Since there exist two different densities, ρ and ρ_0, we can discuss two adiabatic indices, depending on whether the pressure P is considered as a function of ρ or of ρ_0.

Let us illustrate these thermodynamic relations for the example of a power law (which may be asymptotically correct at high densities). Let

$$\rho = a\rho_0{}^b . \qquad (6.7.4)$$

Then, by using formulae (6.7.2) and (6.7.4), we find

$$P = (b - 1)c^2 a\rho_0{}^b = (b - 1)c^2\rho . \qquad (6.7.5)$$

---

19. To be more precise, m_0 can be determined as the mass per one baryon in the lowest energy state when $\rho = T = 0$, i.e., as one fifty-sixth of the mass of the nucleus ^{56}Fe.

20. Sometimes it is convenient to relate all quantities to one baryon or, equivalently, to a unit of baryon charge. In the case of cold matter, $S = 0$.

Thus, for the asymptotic law $d \ln P/d \ln \rho_0 = b$ corresponding to any adiabatic index b with respect to rest-mass density ρ_0, one obtains an adiabatic index, $d \ln P/d \ln \rho$, of unity for the relationship between pressure and density ρ.

We find further that

$$(a_{\text{sound}})^2 = (b - 1)c^2 . \qquad (6.7.6)$$

Thus, it appears that the relativistic requirement $(a_{\text{sound}})^2 \leq c^2$ brings us to the restriction $b < 2$ (see § 6.12).

6.8 An Ideal Neutron Gas at Superhigh Density

In their pioneering work Oppenheimer and Volkov (1938) considered a neutron gas in the first approximation, i.e., on the assumption that there exists no interaction between neutrons whatsoever. Neutrons are Fermi particles; thus, the entire theory of a degenerate neutron gas is similar to the theory of a degenerate electron gas. This similarity occurs because we use the particle density n, or equivalently the rest-mass density ρ_0, as an independent variable. In particular, the relationship between the Fermi momentum and n coincides precisely for electrons and neutrons:

$$n = 2 \times \frac{4\pi}{3} \frac{p_{\text{F}}^3}{(2\pi\hbar)^3} = \frac{p_{\text{F}}^3}{3\pi^2\hbar^3} . \qquad (6.8.1)$$

Let us find the characteristic rest density at which a transition from a nonrelativistic to a relativistic neutron gas occurs; the condition[21]

$$p_{\text{F}} = m_0 c$$

gives

$$(\rho_0)_c = \frac{m_0^4 c^3}{3\pi^2\hbar^3} = 5.25 \times 10^{15} \text{ g cm}^{-3} ,$$

which is much greater than the density of atomic nuclei ($\sim 2 \times 10^{14}$). Denote the dimensionless density of rest mass by

$$\rho_0/(\rho_0)_c = \chi , \qquad (6.8.2)$$

and introduce the dimensionless Fermi momentum,

$$t = P_{\text{F}}/m_0 c = \chi^{1/3} . \qquad (6.8.3)$$

Keeping in mind the relativistic expression for the neutron energy

$$E_n = (m_0^2 c^4 + c^2 p^2)^{1/2} = m_0 c^2 [1 + (p/m_0 c)^2]^{1/2} ,$$

---

21. We do not differentiate here between m_0 (see n. 19, this chapter) and the neutron mass. The difference (which is about 1 percent) is negligibly small compared with the influence of other simplifying assumptions.

we can easily derive expressions for the density ρ and pressure P:

$$\rho = 3(\rho_0)_c \int_0^t (1 + q^2)^{1/2} q^2 dq \tag{6.8.4}$$

$$= \tfrac{3}{8}(\rho_0)_c \{(2t^2 + 1)t(t^2 + 1)^{1/2} - \ln [t + (t^2 + 1)^{1/2}]\} \,,$$

$$P = (\rho_0)_c c^2 \int q^4 dq/(1 + q^2)^{1/2}$$

$$= \tfrac{3}{8}(\rho_0)_c c^2 \{t(1 + t^2)[\tfrac{2}{3}t^2 - 1] + \ln [1 + (t^2 + 1)^{1/2}]\} \,, \tag{6.8.5}$$

$$a_{\text{sound}} = 3^{-1/2} ct(t^2 + 1)^{-1/2} \,. \tag{6.8.6}$$

In particular when $\chi = t = 1$, i.e., at the critical rest density $\rho_0 = (\rho_0)_c$, we have $\rho = 1.26(\rho_0)_c$, $P = 0.154(\rho_0)_c c^2$.

It is easy to find the asymptotic behavior when $\chi \to \infty$:

$$\rho = \tfrac{3}{4}(\rho_0)_c \chi^{4/3} \,, \qquad P = \tfrac{1}{4}(\rho_0)_c c^2 \chi^{4/3} = \tfrac{1}{3}\rho c^2 \,, \qquad a_{\text{sound}} = 3^{-1/2} c \,. \tag{6.8.7}$$

Thus, within the limits of $\rho \gg \rho_0$, the rest mass can be neglected. The ideal ultrarelativistic Fermi gas has a velocity of sound which asymptotically approaches

$$3^{-1/2} c = 0.58c \,.$$

This result is reasonable, since nearly all particles move with the velocity of light—in all directions, however. Getting somewhat ahead of ourselves (for the equation of state at high temperature, see chapter 8), we note that the relationships

$$P = \tfrac{1}{3}\rho c^2 = \tfrac{1}{3}\epsilon \,, \qquad a_{\text{sound}} = 3^{-1/2} c \,, \tag{6.8.8}$$

hold asymptotically for any ultrarelativistic gas; for particles of zero rest mass, i.e., for neutrinos and for photons, this relationship is exact.

6.9 Ideal Gas with the Inclusion of Reciprocal Transformations between Particles

Above we have considered an ideal gas which consists of neutrons only. This was based on "neutronization" calculations, i.e., calculations of the inverse β-process:

$$e^- + p = n + \nu \,.$$

We recall that at nuclear density, the calculation brought us to the conclusion that the equilibrium concentration of electrons and protons amounts to less than 1 percent. This results from the fact that at nuclear density the electrons are relativistic, so their energy is considerably larger than the energy that corresponds to their rest mass, and at the same time the neutrons and protons are not yet relativistic, so their energy, for practical purposes, does not differ from the energy that corresponds to their rest mass.

However, when neutrons, too, become relativistic, their energy and associated chemical potential increase, and the equilibrium again shifts to the left in the reaction above, i.e., in the direction of increasing numbers of electrons and protons. The interaction between neutrons, particularly their repulsion at small distances (i.e., at densities greater than nuclear), also affects the neutron chemical potential. It is generally thought that repulsion at small distances is a common property for every pair of baryons (n-n, p-n, Λ-n, etc.). In this case the chemical potentials corresponding to each baryon type must counterbalance each other in the equation pertinent to each reaction that transforms one type of baryon into another. (These equations are discussed below.)

Above we have laid particular emphasis on the existence of nuclei (but not free protons) sprinkled about in the neutron gas, when the gas density is smaller than nuclear densities. In the opposite case of a high neutron density, it is possible that the protons are dissolved uniformly in the dense neutron liquid. For a static star, neutrinos and antineutrionos may be considered to escape freely. Consequently, their concentration and Fermi energy are zero. Hence, the condition of equilibrium for the reaction $e^- + p \rightarrow n + \nu$ is the balance of chemical potentials

$$\mu_p + \mu_e = \mu_n . \tag{6.9.1}$$

For a cold Fermi gas, the chemical potential coincides with the Fermi energy.

The density of rest mass and the baryon density n are determined by the sum of the concentrations of protons and neutrons. We denote

$$n_p = an , \quad n_n = (1 - a)n , \tag{6.9.2}$$

and we impose the condition of charge neutrality, i.e.,

$$n_{e^-} = n_p = an .$$

The rest masses must be included in the Fermi energies; in the ultrarelativistic case, the rest mass can be neglected (at small densities, it would be prudent to neglect the mass of the electron and the difference of the rest masses of the proton and of the neutron, but not m_n itself). The equation (6.9.1) for the chemical potentials yields (in the ultrarelativistic limit)

$$m_0 c^2 (a\chi)^{1/3} + m_0 c^2 (a\chi)^{1/3} = m_0 c^2 [(1 - a)\chi]^{1/3} . \tag{6.9.3}$$

Hence, it follows that

$$8a = 1 - a , \quad a = \tfrac{1}{9} , \quad 1 - a = \tfrac{8}{9} ,$$

so that we have in equilibrium 88.9 percent neutrons and 11.1 percent protons, plus the electrons whose number density equals the number density of protons. Thus, during a monotonic increase of density, the fraction of all baryons in the form proton plus electron passes through a deep minimum. Actually, at very low pressures, the number of neutrons and protons in equilibrium is approximately equal, so $a \sim 50$ percent[22]; while at nuclear

---

22. To be more precise, for iron $^{56}_{26}\text{Fe}$, $a = \tfrac{26}{56} = 46.3$ percent, $1 - a = \tfrac{30}{56} = 53.7$ percent.

density (as demonstrated in §§ 6.5 and 6.6) $a \sim 10^{-3}$. In accordance with the general principles of thermodynamics (LeChatelier-Brown), the transformation of some of the neutrons back into protons and electrons somewhat reduces the pressure at a given χ (relative to the pressure of a pure neutron gas). Denoting the pressure of pure neutrons by P_n, we find

$$P = [a^{4/3} + a^{4/3} + (1 - a)^{4/3}]P_n = 0.96 P_n \, .$$

The asymptotic law relating pressure to rest-mass density changes only by the multiplicative factor 0.96 in the ultrarelativistic limit; the power in the exponent, and the expression for the sound velocity, $a_{\mathrm{sound}} = 3^{-1/2} c$, do not change. In the pressure the effect of interactions (repulsion) must be included; as has already been shown, it increases the sound velocity.

Ambartsumyan and Saakyan (1960) were the first to consider in extensive detail the situation which occurs due to the fact that modern physics has considerably expanded the assortment of elementary particles. Along with the neutrons, protons, and electrons, we must consider muons ($\mu^+, {}^0\mu^-$), pions (π^+, π^0, π^-), and kaons ($K^0, \bar{K}^0, K^+, K^-$), as well as various baryons ($\Lambda, \Sigma^+, \Sigma^0, \Sigma^-, \Xi^0, \Xi^-$). The general principles of expressing the conditions of equilibrium are the same as in the elementary example which we discussed above. It can be demonstrated, however, that there is no need to write out the relationships between the chemical potentials (Fermi energies) which correspond to all possible reactions that can transform the particles. The transformations of the particles satisfy two laws: the conservation of baryon charge[23] and the conservation of electric charge. The conservation of neutrino charge in this problem is of no significance, since it is assumed that all neutrinos and antineutrinos freely escape from the system.

For any arbitrary particle r, let

$$\mu_r = a q_{\mathrm{er}} + b q_{\mathrm{br}} \, , \qquad (6.9.4)$$

where q_{er} is the electric charge and q_{br} is the baryon charge of the particle.

The coefficients a and b are identical for all the different types of particles which are included in the system. It is easy to demonstrate that in this case, for any reaction, e.g.,

$$r_1 + r_2 = r_3 + r_4 \, ,$$

for which the laws of conservation

$$q_{e1} + q_{e2} = q_{e3} + q_{e4} \, , \qquad q_{b1} + q_{b2} = q_{b3} + q_{b4} \, , \qquad (6.9.5)$$

are satisfied, the equilibrium condition

$$\mu_1 + \mu_2 = \mu_3 + \mu_4 \qquad (6.9.6)$$

is also satisfied when equation (6.9.4) is satisfied.

Knowing the chemical potential of a given type of particle, we can easily

---

23. If all the enumerated types of particles are considered, the conservation of baryon charge means constancy of the sum of the numbers $n, p, \Lambda, \ldots, \Xi$ (the concentration of each particle is denoted by its letter).

find its concentration. Thus the total composition of an equilibrium mixture consisting of a large number of components is expressed by two parameters only, a and b.

The condition of charge neutrality determines the value of a. It is easy to demonstrate that by increasing a, we increase the concentration of positively charged particles and decrease the concentration of negative particles. For each b, there exists a value of a, $a_0(b)$, which gives charge neutrality. The baryon density depends monotonically on the one parameter b, and thus by interpolation it is easy to find an equilibrium composition for any ρ_0.

Note two peculiarities of the solution at zero temperature:

1. For fermions, when $\mu_r < m_r c^2$, the corresponding particles (r) are totally absent in equilibrium;

2. For bosons (π- and K-mesons), under the assumption of absolutely no interaction, the concentration is also zero when $\mu_r < m_r c^2$. When bosons are present in a concentration $r > 0$, then $\mu_r \equiv m_r c^2$, and m does not depend on the concentration r. This follows from the condition that bosons (without interaction!) are all in the ground state, i.e., at rest.

It follows from the first property that, as ρ_0 increases, particular types of particles are created at certain threshold densities. From the second property it follows that in an equilibrium system, neutral mesons are totally absent: when $q_{cr} = 0$ and $q_{br} = 0$ we see that $\mu_r = 0$, and consequently $r = 0$.

It follows also that in equilibrium the existence of π^- precludes the presence of K^- and of other, heavier, negative mesons. Also forbidden are positive mesons, positrons, and antibaryons.

Notice that the conservation of strangeness (or of the so-called hypercharge) must not be imposed in the calculations; in reality, strangeness is conserved in strong interactions, which take place during the collisions of particles in times of 10^{-20} seconds and smaller. However, in times of $\sim 10^{-8}$ seconds (which are small compared with the scale of nonstationary phenomena in stars) processes involving a change of strangeness have enough time to occur. An equilibrium composition depending on ρ_0, found by Ambartsumyan and Saakyan (1963), is shown in Figure 30.

Similar calculations have been performed with other sets of particles (Cameron 1959a; Tsuruta and Cameron 1965). The general properties of the resultant pressure-density relation differ very little from those for the simplified systems (n; npe^-) considered above. For a review see Saakyan (1968).

Recently calculations have been performed which take more accurate account of nuclear forces (Nemeth 1968). The main result is that, as a result of the neutron-proton attraction, the concentration of protons at a given density is greater than thought previously—for example, the proton concentration is approximately 4 percent of the neutron concentration at $\rho \sim 3 \times 10^{14}$ g cm^{-3}. Negative muons also begin to appear earlier, at $\rho \sim 1.5 \times 10^{14}$ g cm^{-3}. At low densities the calculations cited above are not correct: obviously, at densities much smaller than nuclear the matter

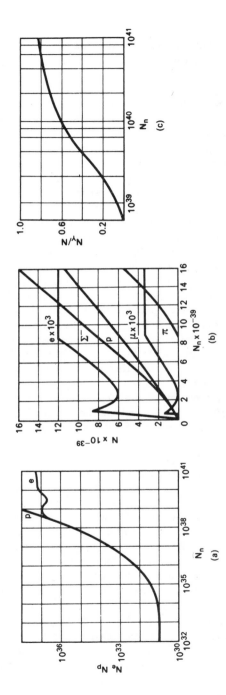

Fig. 30.—(a) Curve showing the proton and electron densities as functions of the neutron density, N_n. (b) Densities of protons, Σ^--hyperons, μ^- and π^--mesons, and electrons as functions of the neutron density. The densities of electrons and mesons are multiplied by 10^3. (c) Dependence of the fraction of all baryons that are hyperons upon the number density of all baryons. All data are taken from Ambartsumyan and Saakyan (1963).

192

is no longer a uniform solution of protons (or even pairs of protons in a neutron fluid); rather, it is a mixture of neutron-rich, complex nuclei and free neutrons.

When one considers a definite, finite number of types of particles without interaction, one finds that P = const. $\rho_0^{4/3}$ and $a_{\text{sound}} \rightarrow 3^{-1/2} c$ asymptotically. The pressure and energy density are smaller than they are for a neutron gas only by the ratio $N^{-1/3}$, where N is the number of types of baryons, Assume that the interaction, particularly the repulsion at small distances. depends only on the baryonic charge. In this case the numbers of particles are not changed, but the overall pressure is increased by an interaction-dependent amount. The results mentioned above (P = const. $\rho_0^{4/3}$ and the magnitude of the constant) are valid only for the kinetic part of the pressure in this case.

6.10 ARE ALL "ELEMENTARY" PARTICLES REALLY ELEMENTARY?

The consideration of eight, eighteen, or even larger numbers of baryons and resonances (see, for example, reviews by A. Rosenfel'd *et al.* [1965], and Zel'dovich [1965d]) as independent types of elementary particles appears to be unnatural to any physicist who uses his common sense. Until recently, resonances were considered as excited states of "conventional" baryons. A more radical viewpoint holds that all baryons (conventional, as well as resonances) are complex systems which consist of three even more fundamental particles, called quarks.

From either viewpoint, at superhigh densities, when the volume per particle becomes less than the intrinsic volume of a particle (which is composite and has its own spatial structure), the particles certainly cannot be considered elementary.

We can assume, for example, that when the total kinetic (Fermi) energy of degenerate baryons per one quark becomes considerably larger than the binding energy of quarks,[24] compressed matter can be considered a quark gas.

In such a case, we obtain asymptotically an equal number of quarks of all three types, $n_1 = n_2 = n_3$.[25] Here each number equals n, the baryon density, since one baryon consists of three quarks. Hence, the pressure in this model at a given rest-mass density is 3 times greater than the pressure of an ideal neutron gas (but less than the pressure calculated for a mixture of various resonances).

At first glance, this result contradicts the general Le Chatelier-Brown principle; in the preceding section, it was emphasized that when one permits transformations of the particles, the pressure surely must be below the pres-

---

24. We are discussing the binding between three quarks which form a baryon.

25. Quark (n_1) has charge $+\frac{2}{3}$ and strangeness 0 (p); (n_2) has charge $-\frac{1}{3}$ and strangeness 0 (n); (n_3) has charge $-\frac{1}{3}$ and strangeness -1 (λ).

sure of an ideal neutron gas P_n. Isn't there an inner contradiction, or an error in the quark model with $P_q = 3P_n$? The problem of the asymptotic behavior of the pressure of composite particles has been considered by one of the authors (Zel'dovich 1965d).

It appears that if the particles (in this case, neutrons) are composed of quarks, then they cannot be regarded as noninteracting particles; they repel. This repulsion is a result of the Pauli principle for quarks.

An ideal neutron gas of high density in this case is not a physically admissible state, and therefore there is no violation of the Le Chatelier-Brown principle.

Since about 1967, the idea of physical quarks with fractional charges $(+\frac{2}{3}e, -\frac{1}{3}e, -\frac{1}{3}e)$ has been fading slowly. Experimental searches have not confirmed the existence of quarks either in cosmic rays (Kasha and Stefanski 1968; Bondarev et al. 1968; Garmire, Leong, and Sreekanton 1968), in accelerator experiments (Bellamy, Hofstadter, and Lakin 1968; Foss et al. 1967), or in small amounts at rest in ordinary matter (Braginsky et al. 1968a, b; Stover, Moran, and Trishka 1967), or in the solar photosphere (Leacock, Beavers, and Daub 1968).

Detailed theoretical investigations of the possible structures of baryons as composites of quarks have also led to difficulties (Braginsky et al. 1968a, b). Cosmology predicts the existence of primordial, relict quarks—if quarks exist at all, of course (see Vol. II). Taken all together, the evidence against fractionally charged quarks is strong.

There are more complicated schemes with nine different types of integer-charged quarks which are much more difficult to prove or disprove. The thermodynamics of quarks should be considered as an illustration of principles rather than as a necessary description of reality.

6.11 THE ELECTROMAGNETIC INTERACTION OF PARTICLES

Can anything be said about the equation of state in the presence of an interaction between the particles? Yes, in one very important case, namely, for an electromagnetic interaction. The energy density of the field is $(E^{*2} + H^2)/8\pi$, where E^* is the intensity of the electric field and H is the intensity of the magnetic field.

In the general case the forces are determined by the Maxwell stress tensor, and in ordered fields the forces are strongly anisotropic (the tension along the electric field, $E^{*2}/8\pi$, and the repulsion in the two perpendicular directions, $E^{*2}/8\pi$, are analogous to those for the magnetic field).

However, the sum of the normal stresses upon three perpendicular areas always is identically equal to $(E^{*2} + H^2)/8\pi$; i.e., it equals the energy density. This result does not depend on the nature of the fields, i.e., static or variable; nor does it depend on whether the fields are located in a domain which includes charges and currents, or whether the fields (specifically the field of an electromagnetic wave) are in vacuum. When can we discuss a pressure for an electromagnetic field? To do so it is necessary that, in averaging in time or over macroscopic areas, all directions be equivalent. Then the sum of the stresses is $3P$; consequently, for an electromagnetic field $\epsilon = 3P$, always, where ϵ is the volumetric density of energy. We have already noted that this same result holds for an ultrarelativistic gas, and in particular for a collection of light quanta—i.e., speaking classically, for a collection of electromagnetic waves.

However, the result obtained above is richer: in fact, the electromagnetic field is not entirely represented by a field of waves. This is already apparent from the fact that electromagnetic waves are transverse. The Coulomb field of a charge, for example, is longitudinal and cannot be reduced to free electromagnetic waves.

The pressure and energy density of a system of charged particles need not necessarily be computed as derivatives of the energy. In matter of a given density one can mentally construct a plane and find the flow of momentum normal to its surface, i.e., the pressure; this operation can be performed both in classical mechanics and within the framework of quantum theory. It is not necessary to perform the theoretical calculations here; this approach is introduced only to justify splitting the pressure of the interacting particles into the "kinetic" pressure of moving noninteracting particles plus the pressure of the electromagnetic field, which accounts for the interaction:

$$P = P_p + P_f . \qquad (6.11.1)[26]$$

In the same manner, the interaction energy can be described as the energy of the field. Hence, the energy density can also be rendered in the form of a sum

$$\epsilon = \epsilon_p + \epsilon_f . \qquad (6.11.2)$$

For the particles,

$$P_p \leq \tfrac{1}{3}\epsilon_p ; \qquad (6.11.3)$$

the equality sign refers to particles with zero rest mass. For the field,

$$P_f = \tfrac{1}{3}\epsilon_f . \qquad (6.11.4)$$

Hence,

$$P = P_p + P_f \leq \tfrac{1}{3}(\epsilon_p + \epsilon_f) , \qquad \epsilon - 3P \geq 0 . \quad (6.11.5)$$

Such is the general conclusion for electromagnetically interacting particles, and for the electromagnetic field.

Here a subtle point should be noted. An isolated, charged particle has an electromagnetic field of its own. One can calculate the volume integrals of its field energy density and stress. For a point charge this calculation is plagued by the divergence of the field near $r = 0$; thus, it is feasible only

26. Here, the indices are p for particles and f for field.

for a smeared-out charge. But we shall not dwell on this part of the problem. What we wish to emphasize is that the field energy of a single particle is already included in its experimentally measured rest-mass energy (mc^2); and the field stresses are counterbalanced by other, nonelectromagnetic forces inside the particle.

It follows that a naïve calculation of the contributions of the electromagnetic field to the energy and stresses in an interacting gas would lead us to an error. By taking the total field E and squaring it, we would include energies already counted in the particle rest masses; in the equations

$$E = \sum_i E_i , \quad E^2/8\pi = \sum_i E_i{}^2/8\pi + \sum_{i \neq k} E_i \cdot E_k/8\pi$$

the sum $\Sigma E_i{}^2$ is actually a part of the energy of the noninteracting particles. It is only the cross terms which should be included in ϵ_f and P_f.

Now, this remark does not destroy the fundamental equality $\epsilon_f = 3P_f$; one can easily show that it holds not only for $(E_{\text{total}})^2$ but also for $(E_{\text{total}})^2 - \Sigma_i E_i{}^2$ (and also for the corresponding magnetic contributions). But after subtraction we can no longer claim that ϵ_f and P_f are positive; and, of course, they are negative in a nonrelativistic, neutral gas.

Let us return to equations (6.11.5), and particularly to the inequality $\epsilon - 3P \geq 0$. The range of applicability of this inequality is extremely wide. It can be applied to an ideal Fermi gas of neutral particles. It also remains applicable for the electromagnetic interactions of charged particles of a cold Fermi gas (incidentally, in this instance the interaction yields only a very small correction of the order of $e^2/\hbar c = \frac{1}{137}$, or $\frac{1}{137}Z^{2/3}$ in the presence of heavy nuclei—see above). But this inequality is also applicable to a hot gas, where the energy density of the photons is larger than the energy density of conventional particles (particles which have a rest mass). It also remains in force at temperatures high enough to create particle-antiparticle pairs (electrons and positrons).

Finally, the inequality (6.11.5) is also applicable to a magnetoturbulent medium, where more or less chaotic electric currents create chaotic magnetic fields, and interact with those fields. In our derivation we did not have to apply the condition of freezing-in of the magnetic field or other considerations as to the conductivity of the medium: the lifetime, i.e., the future evolution, of the given state depends on the conductivity and freezing-in; however, the pressure and the energy of the state at a given moment do not depend on these factors.

For magnetoturbulence there must be a scale of currents and fields on which there exist ordered motions of tremendous numbers of charged particles; only in such a case will the magnetic pressure exceed the kinetic pressure. On the other hand, the scale of currents and fields must be small compared with the dimensions of the entire system so that, by averaging over the perpendicular areas, we can discuss a mean pressure. In a system resembling a magnetic dipole, there is no "pressure" which follows Pascal's law; there exist only anisotropic stresses.

For the same reasons, we can discuss the pressure and the density of the energy only for an electrically neutral system. In equilibrium all uncompensated charge moves away to the surface of the system, or to infinity in the case of uniform, infinite matter. The general problem regarding the differing interpretation of short-range and long-range forces was described at the beginning of Part II (§ 5.1).

In the general-relativistic theory of the energy-momentum tensor, the inequality $\epsilon - 3P \geq 0$ is rendered in an especially compact and elegant form.

In the local Euclidean comoving frame of reference, provided Pascal's law is satisfied,

$$\epsilon = T_0{}^0 \; ; \quad -P = T_1{}^1 = T_2{}^2 = T_3{}^3 \, , \quad \text{i.e.,} \quad T_\alpha{}^\beta = -P\delta_\alpha{}^\beta \, , \quad (6.11.6)$$

where $\delta_\alpha{}^\beta$ is the Kronecker delta symbol. Contracting this stress-energy tensor, we find

$$T = T_\alpha{}^\alpha = T_0{}^0 + T_1{}^1 + T_2{}^2 + T_3{}^3 = \epsilon - 3P \geq 0 \, . \quad (6.11.7)$$

The actual validity of this inequality for a wide range of systems, and its compact, generally covariant formulation suggest the assumption that $T = \epsilon - 3P \geq 0$ is a general law of nature. It was once assumed that it would be possible in the future to derive a general proof relating not just to the electromagnetic interaction but to any type of interaction.

This biased view was refuted by Zel'dovich (1961), whose argument is briefly summarized in the following section.

Notice that the asymptotic relation $\epsilon = 3P$ is elegantly rendered in the form $T = T_\alpha{}^\alpha = 0$. However, the expression derived for the velocity of sound, $a_{\text{sound}} = c/\sqrt{3}$, is less elegant. It is difficult to conceive that relativistic considerations would lead to anything other than the condition $a_{\text{sound}} \leq c$.

6.12 A RIGOROUS LIMIT UPON THE EQUATION OF STATE?

Consider a system of particles interacting with a vector field. This vector field is analogous to the electromagnetic field, with one difference only: One adds to the Lagrangian density a term proportional to $A_k A^k$, where A_k is the vector potential (four-vector ϕ, A_a):

$$L = -(1/16\pi)F_{ik}F^{ik} - (1/8\pi)\mu_*{}^2 A_k A^k \, . \quad (6.12.1)$$

The remaining equations are the usual ones. Such a change in the Lagrangian produces a change in the nature of the solutions of the field equations. In particular, the solutions that describe waves propagating in vacuum have a relationship between wavelength and frequency which is different from standard electromagnetic theory.

In the expression

$$A_k = a_k e^{i\omega t + ikr} \tag{6.12.2}$$

the equations yield

$$\omega^2 = c^2 k^2 + \mu_*^2 c^2 . \tag{6.12.3}$$

A relation of the form (6.12.3) corresponds, in quantum theory, to a field whose quanta ("vector mesons") have a rest mass of

$$m = \hbar\mu_*/c . \tag{6.12.4}$$

In particular, vector mesons can be at rest; electromagnetic quanta with zero rest mass, by contrast with vector mesons, always move at the velocity of light.

For a given wavenumber k, the wave has three independent solutions corresponding to three types of polarization (by contrast with two types for the electromagnetic field). These three solutions correspond to the quantum representation of a vector meson as a particle with spin 1 (angular momentum $\hbar$). The projection of this spin upon an arbitrary axis can take three values: 1, 0, -1.

For a system of free waves of such a field, we obtain

$$\epsilon > 3P ,$$

as must be expected from the quantum representation of the motions of the particles (vector mesons) with nonzero rest mass.

However, these solutions are not exhaustive with respect to the content of the theory: This is already apparent from the fact that there are four components of the potential, whereas free waves have only three independent solutions. It is necessary to consider also the static field of the charge and the interaction of the charges. In the electromagnetic problem

$$\phi = e/r, \text{ (interaction energy)} \equiv \epsilon_{12} = e_1 e_2/r_{12} . \tag{6.12.5}$$

In this case, however, we obtain the Yukawa potential

$$\phi = g e^{-\mu r}/r , \quad \epsilon_{12} = g_1 g_2 e^{-\mu r_{12}}/r_{12} , \tag{6.12.6}$$

where g plays the role of the charge and determines the interaction between the particle and its vector-meson field. The essence of the problem is in this change of the law of interaction.

Because of the exponential term, a macroscopic system can exist with constant charge density. Let the dimensions of this system, $R \gg 1/\mu$, be larger than the characteristic length of attenuation. Then, in spite of the charge, the external field of the system is negligible, and we can consider the energy density and the pressure of the charged system to be functions of the charge density.

Let us calculate the energy by simply summing the interactions of all pairs of particles, which we assume to be uniformly distributed. We are assuming that the mean distance between particles, $n^{-1/3}$, is less than the Yukawa interaction radius, μ^{-1}, and we will substitute integration for summation. The interaction energy of the particles in a volume V is

$$E_V = \tfrac{1}{2}n^2g^2 \int_V \int_V (e^{-\mu r_{12}}/r_{12})dV_1dV_2 = 2\pi g^2 n^2 V/\mu^2 . \qquad (6.12.7)$$

If the charges (particles) have mass m_0 and are at rest, then we obtain the energy density and from it the density ρ:

$$\rho = nm_0 + 2\pi g^2 n^2/\mu^2 c^2 = \rho_0 + b\rho_0^2 , \qquad (6.12.8)$$

where

$$b = 2\pi g^2/m_0^2\mu^2 c^2 . $$

From this, we find the pressure

$$P = c^2 b\rho_0^2 , \qquad (6.12.9)$$

so that asymptotically

$$P \to c^2\rho = \epsilon . \qquad (6.12.10)$$

Accordingly, for the velocity of sound we obtain

$$(a_{\text{sound}})^2 = (dP/d\rho) = 2b\rho_0/(1 + 2b\rho_0)c^2 , \qquad (6.12.11)$$

and asymptotically $a_{\text{sound}} \to c$ as $\rho \to \infty$. For the sake of simplicity we have considered charges at rest. If the charged particles are fermions, then at high density they will become relativistic, and their contribution to the density and pressure will be $n^{4/3}$. However, the motion of particles at a given density does not change the potential that they create; this is a well-known fact in electromagnetic theory,[27] and it does not change in the theory of vector mesons. Therefore, the motion of the particles does not change the term proportional to n^2 in the energy density and pressure. For sufficiently large n, this term will always dominate. A more detailed and formal proof of these results is given by Zel'dovich (1961). He shows that these same results follow from relativistically invariant field theory beginning with a Lagrangian. It should be pointed out that only a vector field gives repulsion between particles and attraction between a particle and an antiparticle. Scalar and tensor fields give attraction between particles.

This result has this theoretical significance: It demonstrates the possibility of developing a consistent relativistic theory of an interaction with $\epsilon \to P$, so that the inequality (6.11.7) is violated and $\epsilon - 3P < 0$. Thus, the assumption that the inequality $\epsilon - 3P \geq 0$ is a consequence of the theory of relativity is refuted; this inequality should be regarded as the property of a definite class—but not the most general class—of systems. Another question remains which is considerably more difficult: What actually takes

27. Example: the eighty-two electrons of an atom of lead, some of which move at a velocity of the order of $\tfrac{1}{2}c$, cancel precisely the charge of eighty-two protons contained in the nucleus (see chapter 2).

Cold Matter

place in superdense matter with baryons at high density; what is the inequality for them? To this question, we cannot offer any definite answer at this point. In nature, not everything permitted by the theory of relativity takes place.

It has been demonstrated experimentally that there exist neutral vector mesons ρ^0 and ϕ^0 with masses of about the mass of a nucleon.[28] Their interaction with nucleons corresponds approximately to the above assumptions; but it has not been investigated at too high a momentum.

The asymptotic form of the interaction which we require depends on whether ω and ϕ can be constructed as field quanta, or whether ω and ϕ themselves are compound particles, put together from quarks and antiquarks or from other stuff. In the latter case, at large energies the interactions of ω and ϕ with baryons will change. This presents both a new problem and new possibilities: What interaction is it that holds together quarks and antiquarks in mesons and baryons? Isn't this interaction, as yet entirely unknown, an appropriate candidate for the role of a vector field? This enumeration of questions without answers could be continued. There is only one conclusion which can be made from all that has been said: In a domain where we know nothing specific, the only restriction imposed by general principles is that $a_{\text{sound}} \leq c$.

There are no other restrictions (and specifically not $\epsilon \geq 3P$) in the general case. Thus, the result of the investigation is not a positive statement, but the destruction of a prejudice.

Tsuruta and Cameron (1966) give a smooth curve joining the results of calculations which take into account baryon interactions, to the asymptotic law $p = \epsilon$. Their curve can be approximated by the equations

$$P/c^2 = \rho_0^2/r, \quad \rho = \rho_0 + \rho_0^2/r, \quad P/c^2 = r[\rho/r + \tfrac{1}{2} - \tfrac{1}{2}(1 + 4\rho/r)^{1/2}].$$

Here $\rho_0 = nm_p$ is the rest-mass density and n is the number density of baryons. The constant r is

$$r \equiv 5 \times 10^{15} \text{ g cm}^{-3}.$$

These equations hold for $\rho > 10^{13}$ (without any known upper limit) with an error of the same order as the uncertainty in the equation of state.

---

28. A unitary scalar which interacts in the same way with any baryon, independent of its strangeness, must be constructed as a linear combination of ω and ϕ.

7 PROPERTIES OF MATTER AT HIGH TEMPERATURES

7.1 PHYSICAL CONDITIONS IN ORDINARY STARS

Nearly the entire scope of astrophysics deals with matter at high temperatures. Even when we considered cold matter, it was merely assumed that the temperature was insufficiently high to change the properties of matter substantially. At densities of the order of 10^6 g cm^{-6} this means that $T < 10^9$ ° K, and at a density of 10^{14} g cm^{-3} it means $T < 10^{11}$ ° K; by earthly standards, these temperatures are rather high. It is not surprising that white dwarfs with densities of $\sim 10^6$ g cm^{-3} have high surface temperatures, $T \approx 10^4$ ° K, and that the neutron stars (with densities of 10^{14} g cm^{-3}) could have such high surface temperatures, $T \approx 10^6$ ° K, that they must emit X-rays. Nevertheless, the mechanical equilibrium of such dense objects is determined by the properties of cold matter and does not depend on their temperatures, since the temperatures of their interiors are below the indicated limits.

Main-sequence stars (i.e., stars that are uniform in their chemical composition and in which hydrogen-burning nuclear reactions take place), which make up most of the stellar population, have central temperatures of the order of 10^7 °–10^8 ° K at densities of 0.1–100 g cm^{-3}. In this case one deals with hot stars; their equilibrium is fully determined by the thermal pressure. This is seen in the fact that the density of the matter in such stars is low. For a stellar mass of $M < 100\ M_\odot$, the principal contributions to the pressure and to the energy density are due to the plasma, which is made up of atomic nuclei and electrons that form a monatomic, nonrelativistic gas with an adiabatic index of $\frac{5}{3}$.

In the outer layers the temperature is lower, and on the surfaces of the stars it goes down to 50000°–3000° K. Recently objects of an unknown nature have been discovered which have surface temperatures of the order of 700°–1000° K. Therefore, in the external shells of these stars, the ionization is not at all complete; along with free electrons, the gas contains ions which partially retain their electron shells. At the very surface of the atmosphere of a cool star there exist neutral atoms, molecules of chemical compounds, and even molten or solidified particles of heat-resistant substances (metal oxides, perhaps).

Thermal-radiation equilibrium plays a decisive role in determining the energy flux in a star, even if the energy density of the radiation is small compared with that of the plasma. The thermal conductivity of hot matter determines the transfer of radiative energy.[1]

At high temperatures the paths of photons depend on scattering by free electrons (Compton effect); at lower temperatures, the predominant factors in determining the thermal conductivity are absorption and emission of photons by electrons in the fields of the nuclei. These processes occur in all variants: a free electron $(E > 0)$ before and after interaction with the photon; change of the electron from a bound $(E < 0)$ to a free $(E > 0)$ state during the absorption, and from a free to a bound state during the emission of the photon; and electron transfer from one bound state $(E < 0)$ to another bound state $(E < 0)$.

All the fundamental aspects of ordinary stars have been worked out in detail. Calculations of nuclear reactions in stars, of heat fluxes, and even of the characteristics of the observed spectra are performed with great accuracy at the present time. Also well known are all the material parameters relating to this domain (the cross-sections of the nuclear reactions, the opacity of the matter, etc.). Actually, this area is not part of relativistic astrophysics. Hence, for its treatment we will refer the reader to well-known books on astrophysics (see bibliography in chapter 9).

A number of problems related to the ionization, heat conductivity, and hydrodynamics of a hot gas are described in Zel'dovich and Rayzer (1966). We will note here only one effect which has emerged within the last two years and which has not yet been reflected adequately in the literature. We are referring to the influence of the bound-bound electron transition, in which an isolated atom or ion contributes to an emission or an absorption spectrum consisting of individual lines. It had been assumed that the role of individual lines in the opacity of matter was insignificant. However, heavy elements give a very large number of lines, and interactions with electrons and other ions widen those lines. It turns out that at the densities and temperatures of ordinary stellar matter the contribution of the lines to opacity may decrease the thermal conductivity by a factor of 2. The fundamental calculation of thermal conductivity was made by Cox, Stewart, and Eilers (1965). Calculations of the change in the structures of main-sequence stars because of increased opacity caused by the absorption lines have been performed by Imshennik and Nadezhin (1967).

7.2 High Temperatures

Of interest for relativistic astrophysics are the higher temperatures which on one hand bring about the instability of a star and on the other hand arise during catastrophic stages of stellar evolution. The early, prestellar period

---

1. In certain stellar domains (depending on the mass of the star and the stage of its evolution) the energy transport is produced by convection.

of the evolution of the Universe presumably was also characterized by rather high temperatures.

In the high-temperature domain, the most typical process is the creation of particles. First of all (as the temperature increases) photons are created. At temperatures approaching $m_e c^2/k \approx 6 \times 10^9 \, ^\circ$K, the creation of electron-positron pairs becomes important. At higher temperatures, the creation of other, heavier neutral particles is also possible (e.g., π^0, K^0, $\bar{K}^0$), as well as charged ($\mu^\pm$, $\pi^\pm$, $K^\pm$) elementary particles extending on up to nucleon-antinucleon pairs ($\bar{n}$, $\bar{p}$, n, p), and also particles which can be considered excited states of nucleons ("strange" particles Λ, Σ, ... ; resonances Δ, Σ^*, ...). Obviously, in this domain of very high densities of strongly interacting particles, the quantitative predictions are rather uncertain; in fact, a complete list of particles or states which should be examined is not even known.

The problem of the equilibrium concentration of neutrinos and antineutrinos has a different character. It has been demonstrated in experiments within the past few years that there exist two types of neutrino and two types of antineutrino, i.e., the so-called electron neutrinos $\nu_e \bar{\nu}_e$, and the muon neutrinos $\nu_\mu \bar{\nu}_\mu$. Since the rest masses of these particles are zero, their equilibrium concentration and the corresponding energy density and pressure are rather close to the values that are applicable to quanta of electromagnetic radiation.

Therefore, a system consisting of a definite number of various particles may convert to a system of the same composition, plus an arbitrary number of quanta. However, it is impossible to create a neutrino or antineutrino only, without changing the composition of the rest of the system. From the viewpoint of the laws of conservation, only pairs $\nu_e \bar{\nu}_e$ or $\nu_\mu \bar{\nu}_\mu$ can occur or vanish without hindrance. Finally, according to contemporary views, neutrinos are helical particles: this means that for a given direction of the momentum, the neutrino spin can only be antiparallel to that momentum (the antineutrino spin, accordingly, must be parallel to its momentum). Therefore, the energy, the pressure, and the number of neutrino-antineutrino pairs are twice as small as the same quantities for ultrarelativistic electron-positron pairs, when $kT > m_e c^2$.

There exist substantial differences between photons and neutrinos: photons are fully neutral, whereas neutrinos are neutral only in the sense of electric and baryonic charge. There exists also the concept of leptonic charges (electron and muon) with two laws of conservation:

$$n(e^-) + n(\nu_e) - n(e^+) - n(\bar{\nu}_e) = \text{const.},$$

$$n(\mu^-) + n(\nu_\mu) - n(\mu^+) - n(\bar{\nu}_\mu) = \text{const.}$$

The main characteristic of the neutrino is the weakness of its interaction with all other particles, as well as with other neutrinos. Very special conditions are required in order to maintain thermodynamic equilibrium between neutrinos and other particles. These conditions can occur only in an unbounded, uniform distribution of matter, i.e., in a "big bang" model of the Universe. In this case the large mean free path of the neutrino in space has no significance because, to replace the neutrinos which escape from a given volume, we have an equal number of neutrinos arriving from adjacent volumes. At early stages of the cosmological expansion, when the temperature is above 2–3×10^{10} ° K for $\nu_e \bar{\nu}_e$ (or $\sim 10^{11}$ ° K for $\nu_\mu \bar{\nu}_\mu$), the rate at which equilibrium is established is high compared with the rate of variation of density and other quantities. In such a situation we actually deal with a complete thermodynamic equilibrium of all particles, including neutrinos. A more detailed discussion will be included in Vol. II (Cosmology).

During normal stellar evolution, and even at the beginning of the catastrophic collapse or explosion of a star, the time required for a neutrino to escape from the star is rather small compared with all other time scales. Therefore, the instantaneous concentration of neutrinos at each moment is negligibly small compared with an equilibrium concentration, or compared with the concentration of photons.

Therefore, as a rule, when considering the state of matter in a star we have to assume a state of incomplete equilibrium without neutrinos.[2] Such processes as the creation and annihilation of pairs, $e^+ + e \rightleftarrows 2\gamma$, and the creation and absorption of photons, $e + Z \rightleftarrows e + Z + \gamma$ (where Z is a nucleus), always occur quickly; for this reason, in a star there is always complete local equilibrium among the thermal energies of the nuclei and the electrons, and among the energies of photons and positrons. (*Editors' note:* There is one important exception—the stellar atmosphere.)

The production of neutrinos and antineutrinos, which escape immediately from the star, must be considered one of the factors responsible for the evolution of the star. There is a theoretical difference between the creation of electron-positron pairs and the creation of neutrino-antineutrino pairs.

Both processes are accompanied by an energy consumption and reduction of pressure (compared with the pressure which would have existed without the creation of those pairs). However, e^+e^- do not leave the system. In equilibrium at a given T and ρ, they are created and annihilated at the same rate. Therefore, they do not use up energy. The purchase of e^+e^- pairs is an investment; it is a one-time expenditure, not a constant drain as, for example, buying everyday food and necessities. If in the course of contraction e^+e^- are created, then during expansion they disappear again giving back their energy to the total energy supply.

When e^+e^- pairs are considered, the explicit forms of the functions

---

2. On this subject see the contributions by Tsuruta and Cameron (1965) and by Imshennik, Nadezhin, and Pinaev (1966). More recent work, in greater detail, is that of Imshennik and Chechetkin (1969), which will be discussed in some detail in § 3.

$E(T, \rho)$, $S(T, \rho)$, and $P(\rho, s)$ change; however, the equation $dS/dt = 0$ is retained.

The creation of neutrino-antineutrino pairs, which irrevocably leave the system, leads to $dS/dt = -\omega(T, \rho) \neq 0$ (a number of papers have dealt with the calculation of ω; see below).

Finally, it should be pointed out that in some catastrophic events such as supernova explosions, the ν's and $\bar{\nu}$'s are unable to leave the system; this has been shown by Arnett (1967, 1968). It is not only the short time scale of the explosion that prevents neutrino escape; it is also, and more important-ly, the very high values of the matter density, the electron temperature, and the neutrino energy.[3]

As shown in Arnett's calculations, during the formation of a neutron star when a density and temperature of $\rho \sim 10^{14}$ g cm^{-3}, $T \sim 10^{11}$ ° K are reached by a mass of the order of 1 $M_{\odot}$ (so that the radius is $R \sim 10^6$ cm), the rates of escape of ν_e and $\bar{\nu}_e$ are small; inside the star an equilibrium is reached which includes the neutrino concentrations. In that equilibrium the number of ν_e's and $\bar{\nu}_e$'s is constrained by the value of the leptonic charge, $l = \nu_e - \bar{\nu}_e + e^- - e^+ = (e^- - e^+)_0$.

At $T \sim 10^{11} \approx 10$ MeV, muon neutrinos are important. Their birthrate is small: in a first-order process the birth of a ν_μ is necessarily associated with the birth of a μ^+ or the disappearance of a μ^-. Therefore, there is a multiplicative factor of exp $(-m_\mu c^2/kT)$ in the rates of all processes in-volving ν_μ and $\bar{\nu}_\mu$. But muon neutrinos can escape easily from the neutron star, so they are the most important factor in its cooling immediately after it is created.

After electron neutrinos and muon neutrinos one should consider the creation and escape of gravitons. Their birthrate is even smaller than that of muon-neutrino pairs; correspondingly, their scattering and absorption in matter is also smaller, and their escape is easier. We do not know at present of any astronomical object or process, observed or theoretically predicted, for which the emission of high-frequency gravitons would be important.

7.3 VARIOUS TYPES OF EQUILIBRIUM

The most explicit and simple case of equilibrium is one in which all pro-cesses of reciprocal transformation (creation and annihilation of particles) take place quickly compared with the escape rate of the particles, and com-pared with the rate of variation of thermodynamic quantities (density or pressure, entropy or temperature). In this case total thermodynamic equi-librium is attained.

The state of a system in total thermodynamic equilibrium is determined

---

3. The scattering cross-section of ν on e is proportional to E_*^2, where E_* is the energy in the center-of-mass frame. If $E > m_e c^2$ and $kT > m_e c^2$, then $\langle E_*^2 \rangle \sim EkT$.

by a complete set of parameters which are conserved during the particle reactions. It is necessary to assign a baryonic charge and leptonic charges to the system. The electric charge, as a rule, is zero: on account of the long-range interaction of electrostatic forces, even a small excess of particles with one sign of electric charge leads to immense electric fields; this excessive charge rapidly escapes to the surface of the system and vanishes.

However, such total thermodynamic equilibrium is not always attained. There exists one other very important case where we can also discuss the equilibrium and apply formulae of statistics and thermodynamics. This is the case in which some of the processes take place very rapidly and the others take place very slowly. A good example of this is gas at temperatures of 10^6 ° -10^7 ° K and at densities between 10^{-3} and 10^7 g cm^{-3}, and with dimensions l greater than $10^{-12}T^{7/4}\rho^{-3/2}$ cm; i.e., $l > 10^5$ cm for $\rho = 10^{-3}$ and $T = 10^7$, or for $\rho = 10^7$ and $T = 10^6$.

In such gas, a Maxwellian distribution of electron and ion energies is rapidly established. There is also a quickly established equilibrium of ionization and equilibrium with respect to the electromagnetic radiation. (The dimensions were assumed large enough that the escape of radiation is slow.) On the other hand, the rate of creation of pairs $\nu_e \bar{\nu}_e$ is small, and the rate of creation of $\nu_\mu \bar{\nu}_\mu$ is extremely small. The rate of nuclear reactions is also rather insignificant. Therefore, in this case we can apply the concept of equilibrium, but of a limited equilibrium. First, it is an equilibrium without neutrinos. But that is not all; second, this equilibrium corresponds to the maximum entropy at the given energy and given number of nuclei of various types. To assign a limited equilibrium, we need to assign a larger number of parameters than those required in total equilibrium: we must assign separately the concentration of hydrogen, helium, etc., and not just one total concentration of baryons. This obviously follows from the fact that, at a temperature when nuclear reactions are for practical purposes negligible, not only the total number of baryons but also the number of nuclei of each individual type is conserved.

A fortunate circumstance is the exponential dependence of the rates of nuclear reactions on temperature. These reactions are produced by strong interactions; i.e., they amount to a regrouping of protons and neutrons. An example of such a process is the decay $^{56}_{26}\text{Fe} \rightarrow 13\ ^4\text{He} + 4n$. Because of the strong temperature dependence, the entire thermodynamic domain may be separated into a region where the process does not take place at all (limited equilibrium), and a region where the given process takes place rapidly and reaches equilibrium (this equilibrium being more complete and less restricted). When all "strong" regrouping processes are fast, we must assign concentrations of protons and neutrons (free as well as bound in nuclei) as independent variables.

In equilibrium, the concentrations of all different types of nuclei (charge

7.3 Various Types of Equilibrium

Z, atomic weight A) are determined by the equilibrium conditions, which can be written down by using chemical potentials:

$$\mu(Z, A) = Z\mu(p) + (A - Z)\mu(n) . \qquad (7.3.1)$$

Here the chemical potentials $\mu(p)$ and $\mu(n)$ are for free protons and neutrons. When the nuclei are not degenerate, the equilibrium condition has the form of the conventional law of mass action, which is used extensively in physical chemistry

$$[Z, A] = \text{const.} \ [p]^Z[n]^{A-Z} . \qquad (7.3.2)$$

Here brackets denote the respective concentrations.

However, such a picture is frequently too idealized. The true story of the kinetics and equilibrium of nuclear reactions in hot plasmas in stars is complex, because the separation of reactions into fast and slow processes does not really coincide with their classification into processes which depend on strong interactions (i.e., which only regroup the protons and neutrons) and processes which depend on weak interactions (i.e., where transformations of protons into neutrons and conversely take place).

Strong-interaction processes take place slowly when they require collisions between charged particles and the temperature is not very high. The rates of weak-interaction processes vary widely. The rate is especially small when the particles must interact during the time of collision, as, for example, in the p-p reaction

$$p + p \rightarrow D + e^+ + \nu .$$

A spontaneous β-decay, on the other hand, is characterized by rates which vary strongly and depend primarily on the energy released during the decay. Let us cite three examples: For the decay of tritium, $T \rightarrow {}^3\text{He} + e^- + \bar{\nu}$, the energy released is 18 keV and the half-life is 12 years. For neutron decay $n \rightarrow p + e^- + \bar{\nu}$, the energy released (if the rest mass of the electron is subtracted) is 0.8 MeV and the half-life is 11 minutes. For helium 8, ${}^8\text{He} \rightarrow {}^8\text{Li} + e^- + \bar{\nu}$, the half-life is 0.01 second.

At very high temperatures, when many electron-positron pairs are in equilibrium, the time required to establish equilibrium of weak interactions is reduced; it becomes less than the time of spontaneous decay of a neutron because new processes can occur:

$$e^+ + n \rightarrow p + \bar{\nu} , \quad e^- + p \rightarrow n + \nu ,$$

and similar processes with complex nuclei. (Such processes are called *URCA reactions* because, just as the URCA casino in Rio de Janeiro was an excellent sink for money, so also the ν and $\bar{\nu}$ of these reactions are an excellent sink for a hot star's energy.)

Here there is a substantial difference between the situation in cosmology and in stars.

In cosmology, as noted previously, the neutrinos do not escape anywhere.

This means that besides the processes mentioned above, their inverses also take place, i.e.,

$$\nu + p \rightarrow n + e^+ , \quad \nu + n \rightarrow p + e^- ;$$

and total thermodynamic equilibrium is developed.

In stars, ν and $\bar{\nu}$ escape freely and therefore a total, detailed equilibrium is impossible.[4] Recently Imshennik, Nadezhin, and Pinaev (1966) have sought a stationary solution which satisfies the following conditions:

$$\frac{d[p]}{dt} = -\alpha[p][e^-] + \beta[n][e^+] + \gamma[n] = 0 ,$$

$$\frac{d[n]}{dt} = \alpha[p][e^-] - \beta[n][e^+] - \gamma[n] = 0 , \tag{7.3.3}$$

where α and β are the rate constants of the corresponding processes, and γ is the rate of decay of a free neutron. The ratio between the concentrations $[p]$ and $[n]$ in the stationary state appears to be close to the ratio which is obtained in equilibrium with equal quantities of ν and $\bar{\nu}$, i.e., with the condition that the chemical potential of the neutrino is zero. The aim of these calculations and of others which followed (Imshennik, Nadezhin, and Pinaev 1967) was to calculate the rate of energy loss by neutrino emission in the URCA process. In a later paper Nadezhin and Chechetkin (1969) calculated the rate of energy loss by neutrinos due to an URCA process in which complicated nuclei participate. (The first calculations of energy-loss rates due to the URCA process were those of Tsuruta and Cameron 1965.)

Yet another paper (Chechetkin 1969) considered not only neutrons and protons but also more than fifty nuclear species with the purpose of finding the composition in a stationary state with escaping neutrinos. This work should be compared with that of Clifford and Tayler (1965), which considered the equilibrium of strong-interaction processes at a given proton:neutron ratio. Chechetkin finds, for every temperature and density, the $p:n$ ratio which is produced by weak-interaction processes.

Thus, a new type of equation of state is arrived at: energy, pressure, composition, entropy, etc., are found at given ρ and T in the case of escaping neutrinos. This is an open system (due to the escape of $\nu\bar{\nu}$); therefore, one cannot deal with thermodynamic relations. To find the effective adiabatic index (important for the star's stability), one must consider compression or expansion with weak interactions frozen. This procedure is not quite exact, because the rate of energy loss depends upon the same type of process as governs the establishment of the stationary state. The approximation is good for low temperatures, but it breaks down when catastrophic collapse begins.

It is important that Chechetkin's calculations give a more narrow in-

4. This is true until, in the course of collapse, the density attains nuclear values (10^{14}) and the temperature exceeds 10^{11}. In these extreme conditions, as we noted, the core of a star is opaque to electron neutrinos, and the energy loss occurs mostly through muon neutrinos.

stability region ($\gamma < \frac{4}{3}$) than did the thermodynamic calculations made earlier with the assumption $\mu_\nu \equiv 0$ (see § 7.6).

Now we can turn to another question, one connected not with neutrinos but with the properties of strongly interacting particles: Is there an upper limit on the temperature?

The list of strongly interacting particles has increased rather rapidly within the past few years. Along with the so-called strange baryons (Λ, Σ, Ξ, Ω) and mesons (K, $\bar{K}$) with lifetimes of the order of 10^{-8}-10^{-10} sec, this list for various reasons also includes the so-called resonances with lifetimes of 10^{-18}-10^{-22} sec. The number of these resonances by now is approaching 200, and it continues to grow. On the other hand, the theory has been advanced that all this multiplicity of particles may be construed as combinations of three types of quarks and three types of antiquarks, hypothetical sub-elementary particles.[5] Hypotheses about the structures and interaction of elementary particles have decisive influence upon the relationship between temperature and energy density at superhigh temperatures ($kT \sim Mc^2$, where M is the mass of a baryon).

One extreme hypothesis that has been advanced claims that all particles and resonances are statistically independent. Moreover, it is assumed that at superhigh temperatures the interactions of particles can be neglected. If the number of different kinds of particles is great, we can treat the number of particles in a given mass interval between m and $m + \Delta m$ as a smooth function $\Delta N = f(m)\Delta m$. The same can be said about the average statistical weight ($2s + 1$, where s is the spin) of the particles, $g(m)$.

The energy density in a neutral gas is then equal to (cf. § 7.2)

$$\epsilon = \int F(m, T)f(m)g(m)dm ,$$

$$F(m, T) = 4\pi c \int_0^\infty P^3 \{\exp [c(m^2c^2 + p^2)^{1/2}/kT] \pm 1\}^{-1}dp . \qquad (7.3.4)$$

It was shown in Zel'dovich (1969) that if $f(m)g(m) = am^n$, the answer is

$$\epsilon = \text{const. } T^{4+n+1} . \qquad (7.3.5)$$

But if $f(m)g(m)$ is a growing exponential, $f(m)g(m) = b \exp (mc^2/\theta)$, the temperature has an absolute maximum: $\epsilon \to \infty$ as $T \to \theta$ like

$$\epsilon = \text{const. } \ln \left(\frac{\text{const.}}{\theta - T}\right) . \qquad (7.3.6)$$

The idea that there is an absolute upper limit on temperature was suggested previously by Hagedorn (1965). On the basis of experiments with high-energy scattering he calculated $T_{\max} = 1.5 \times 10^{12} = 150$ MeV, which of course is very low.

On the other hand, in the quark hypothesis, it is natural to assume that

5. For a popular exposition of the quark hypothesis, see Zel'dovich (1965d). See also the remarks about the recent state of quark theory on page 193.

when $kT \geq M_q c^2$ (M_q is the mass of the quark), we will have in equilibrium only the quarks, antiquarks, and leptons; there will be no conventional strongly interacting particles, just as at normal conditions of high temperatures there are neither atoms nor molecules but only nuclei and electrons. Presumably, M_q is larger than the mass of the proton by a factor of 5 or 10.

In this hypothesis, asymptotically,

$$\epsilon = a_1 a T^4 , \tag{7.3.7}$$

where the contribution of quarks to a is $3 \times 2.75 \approx 8$. Finally, with the inclusion of μ^+, ν_e, ν_μ, e^+, and γ, we obtain $a \sim 18$.

The last formula (7.3.7) appears to be more plausible for ultrahigh temperatures than the strange relationship (7.3.5). But where then is the error in reasoning that led to formula (7.3.5)? Multiple types of particles do indeed exist.

Let us begin with a simple example. A hydrogen atom unquestionably exists as a bound state. Its mass differs little from the mass of a proton. Let us view a hydrogen atom as a particle. Considering this matter formally, we are bound to arrive at the conclusion (obviously incorrect) that at a temperature of $kT \sim m_p c^2$, when there are many protons and antiprotons, there also must be many atoms of hydrogen (p, e^-) and antihydrogen ($\bar{p}$, e^+).

But what is the formal cause of the error? The truth of the matter is that we did not take into account the interaction between the hydrogen atom and the gigantic number of electrons, positrons, protons, and antiprotons which are present in a high-temperature plasma. Nor did we consider the volume of a hydrogen atom. Individual atoms can exist only when the volume $4a_0^3$ (where a_0 is the Bohr radius) does not include any other particles. On the other hand, our reasoning regarded atoms as particles at densities considerably larger than $1/(4a_0^3)$.

By analogy, it is obvious that the error which led to formula (7.3.5) was contained in the assumption of the existence of different types of particles, mesons, baryons, resonances; our assumption, that a gas can be considered ideal when the density of these particles is high, was incorrect.

One should expect that including the interactions of different particles in the analysis will cause it to yield the same results as those that come from the quark model. The repulsion of particles will restrict their production and will decrease the heat capacity of the system.

For an analogous problem in the theory of cold matter at superhigh densities this idea (the universal repulsion of baryons) was developed by Zel'dovich (1959). Thus the existence of a tremendous number of elementary particles does not obviously bring us to an upper limit on temperature. An attempt to establish an upper temperature limit for the domain where quantum effects are combined with the effects of GTR has been made by Sakharov (1966). We refer those who are interested to his contribution.

8 THERMODYNAMIC QUANTITIES AT HIGH TEMPERATURES

8.1 NEUTRAL GAS, PLASMA; IONIZATION EQUILIBRIUM

We will now, without derivation, cite the fundamental formulae which are found in textbooks. Energy will be expressed in ergs, temperature in degrees Kelvin, and mass in grams. To be more precise (this will become important in subsequent discussions), the thermodynamic quantities will refer to 1 gram of mass at rest.

For a nonrelativistic, monatomic gas

$$P = RT\rho/\mu, \qquad E = \tfrac{3}{2}RT/\mu + \text{const.}, \qquad (8.1.1)$$

where ρ is the density of rest mass, and μ is the molecular weight, $\mu = 1$ (and not 1.008) for hydrogen, $\mu = 4$ for ^{4}He. The gas constant is $R = 8.31 \times 10^7$ ergs (g $^\circ$ K)$^{-1}$.

The general expression for specific entropy (entropy per gram) is

$$S = -(R/\mu) \ln (\rho/\mu m) + \tfrac{3}{2}(R/\mu) \ln (kT) + (C + \tfrac{5}{2})(R/\mu), \qquad (8.1.2)$$

where C is the so-called chemical constant, $C = \ln [g(m/2\pi\hbar^2)^{3/2}]$.

The numerical value of C for atomic hydrogen is $C = 20.204$, and for ^{4}He it is $C = 22.824$.

In specifying the value of C for hydrogen, we have considered that the basic state is fourfold degenerate due to the proton and the electron spins. It is assumed that $kT \gg \Delta E$, where ΔE is the difference in energy between the state with parallel spins, $\mathfrak{F} = 1$, and the state with antiparallel spins, $\mathfrak{F} = 0$; thus ΔE corresponds to a radio line of wavelength $\lambda = 21$ cm.

The same formulae apply for a fully ionized gas, the difference being that the effective molecular weight is determined by all the particles (nuclei and electrons)

$$\mu = \frac{\mu_0}{1 + Z/A}, \qquad (8.1.3)$$

where μ_0 refers to a neutral gas: $\mu = 0.5$ for hydrogen and $\mu = \tfrac{4}{3}$ for ^{4}He at full ionization. The entropy constants are: $C/\mu = 41.794$ for hydrogen, and $C/\mu = 17.1$ for helium.

The equation for ionization equilibrium (the so-called Saha equation) may be expressed in terms of the concentrations of the particles, i.e., number of particles per cubic centimeter:

$$\frac{n_1 n_e}{n_0} = \frac{(2\pi m_e kT)^{3/2}}{(2\pi\hbar)^3} \frac{g_1 g_e}{g_0} \exp(-I/kT) . \qquad (8.1.4)$$

The equations for the second ionization ($n_2 n_e/n_1$), third ionization, etc., have similar forms. Here g_1 is the statistical weight of the ion, g_e is 2 for an electron, and g_0 is the statistical weight of the neutral atom. Hence, we obtain numerically for hydrogen

$$\frac{n_1 n_e}{n_0} = \left(\frac{m_e kT}{2\pi\hbar^2}\right)^{3/2} \exp(-13.6 \text{ eV}/kT)$$

$$= 2.4 \times 10^{15} T^{3/2} \exp(-1.58 \times 10^5/T),$$

and for helium (first ionization! n_1 = [He II])

$$\frac{n_1 n_e}{n_0} = 4\left(\frac{m_e kT}{2\pi\hbar^2}\right)^{3/2} \exp(-24.6 \text{ eV}/T)$$

$$= 9.6 \times 10^{15} T^{3/2} \exp(-2.85 \times 10^5/T).$$

Table 5 summarizes the implications of these equations for hydrogen.

TABLE 5

TEMPERATURE AT WHICH HYDROGEN IS HALF-IONIZED

PARAMETER	ρ: 10^{-24}	10^{-16}	10^{-8}	10^{-4}	10^{-2}
$n_e = n_1 = n_0$	0.3	3×10^7	3×10^{15}	3×10^{19}	3×10^{21}
T (° K)	3.2×10^3	5×10^3	11×10^3	27×10^3	61×10^3
kT/I	2×10^{-2}	3.2×10^{-2}	7×10^{-2}	1.7×10^{-1}	3.9×10^{-1}

Even though one might expect the heat energy (kT) be of the order of the ionization energy I at the stage of half-ionization, the quantity kT/I in reality is considerably smaller than unity due to a statistical factor (cf. Table 5).

As for a neutral, monatomic gas, so also for a completely ionized gas the adiabatic index γ is $\frac{5}{3}$; i.e.,

$$(\partial \ln P/\partial \ln \rho)_S = \tfrac{5}{3} .$$

However, in a domain where the ionization is neither zero nor complete and where consequently the degree of ionization depends on density and pressure, the adiabatic index is considerably smaller than $\frac{5}{3}$ (although $1 \leq \gamma \leq \frac{5}{3}$).

In the ($\log_{10} P$, $\log_{10} \rho$)-plane the curves S = const. have the form shown in Figure 31, with two asymptotes. The asymptotes satisfy the equation $\log P = \frac{5}{3} \log \rho + \text{const.}$; the left asymptote A is for the neutral gas, whereas B is for the fully ionized gas.

The vertical distance Δ between the two parallel asymptotes differs for different values of S. For the theory of the hydrostatic equilibrium of matter in a gravitating star, the intersection of the curve $P(\rho)$ for $S = \text{constant}$

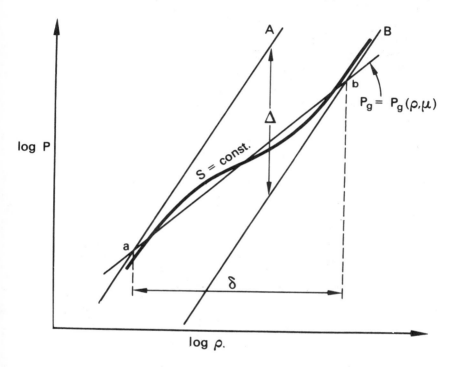

Fig. 31.—Isentropes in the ($\log P$, $\log \rho$)-plane for hydrogen (schematic). Line A is asymptote $S = \text{const.}$ before ionization; line B is asymptote $S = \text{const.}$ after ionization.

with the curve $P_g = 0.36 \, GM^{2/3}\rho^{4/3}$ is significant.[1] In logarithmic coordinates the latter curve is a straight line with slope smaller than the slope of the asymptotes of the adiabat (the slopes $\gamma = (d \log P/d \log \rho)$ are $\frac{4}{3}$ and $\frac{5}{3}$, respectively). For a given stellar mass M this curve intersects the asymptotes at points a and b. For a certain interval of masses M, the $P(\rho)$ curve

1. The expression P_g gives the gravitational force due to half the mass of a star, M, acting on the other half through a distance of the order of the radius of the star, R, divided by R^2: $(GM^2/R^2)/R^2 = GM^2/R^4$. Taking $R \sim M^{1/3}\rho^{-1/3}$, we obtain an expression of the type given in the text. The numerical coefficient is obtained from stellar-model calculations. The equality of $P(\rho)$ and P_g is demanded by the balance between the forces which expand the star [the plasma pressure $P(\rho)$] and the forces which contract the star (the gravitational "pressure" P_g); see chapter 10.

intersects the adiabat three times; this implies the existence of two different stable equilibrium configurations for the given mass, the configurations at the outer crossing points. (The middle crossing point corresponds to an unstable configuration.)[2]

The difference in density between these two configurations corresponds approximately to the distance δ along the horizontal between b and a (see Fig. 31). It is easy to demonstrate the $\delta = 3\Delta$. Consequently, the ratio of the densities of the two configurations is $\rho_b/\rho_a = 10^\delta$, and the ratio of their pressure is $P_b/P_a = 10^{1.33\delta}$, since δ relates to a graph with a logarithmic scale.

8.2 THE THERMODYNAMICS OF RADIATION

The expression for the energy density of radiation has the form

$$\epsilon = \frac{2}{(2\pi\hbar)^3} \int_0^\infty \frac{cp \times 4\pi p^2 dp}{\exp (cp/kT) - 1} , \qquad (8.2.1)$$

where p is the momentum of a photon. Upon integration we obtain

$$\epsilon = \frac{\pi^2 k^4}{15 \hbar^3 c^3} T^4 = aT^4 . \qquad (8.2.2)$$

The numerical expression, ϵ (ergs cm^{-3}) in terms of T, in different units is

$$\epsilon = 7.57 \times 10^{-15} T^4 (^\circ K) = 7.57 \times 10^{21} T_9^4 = 1.37 \times T^4 \text{ (eV)} ,$$

where $T_9 = T/10^9 \,^\circ K$. The corresponding pressure is

$$P = \tfrac{1}{3}\epsilon .$$

The entropy per unit volume is

$$S_v = \tfrac{4}{3} (\epsilon/T) = \tfrac{4}{3} aT^3 \qquad (8.2.3)$$

(the dimensions of the entropy are ergs (cm^3 deg)$^{-1}$ so that S_v depends on the units in which the temperature is expressed).

In the technical literature, it is conventional to denote by σ the coefficient of the expression for the energy flux (and not the energy density) of an equilibrium radiation field. Thus

$$q \text{ (ergs cm}^{-2} \text{ sec}^{-1}) = \sigma T^4 = \frac{c}{4} \epsilon = \frac{c}{4} aT^4 , \qquad (8.2.4)$$

$$\sigma = 5.7 \times 10^{-5} \text{ erg (cm}^2 \text{ sec } ^\circ K^4)^{-1} .$$

Below, we will consider plasma which is in equilibrium with radiation. Over a wide range of temperatures and densities, the energy and the entropy

---

2. These considerations indicate the theoretical possibility of the existence of two configurations. This situation is considered in detail later (see Part III).

of the atoms, ions, and electrons of which the plasma is composed can be neglected. However, the density of the rest mass is determined fully by the density of the plasma alone. Let us first derive the thermodynamic relations which hold in this range of T and ρ, and then determine precisely what conditions on T and ρ must be satisfied in order for these relations to be valid. We obtain for specific energy and entropy (per gram)

$$E = aT^4/\rho , \qquad S = \tfrac{4}{3}aT^3/\rho .$$ (8.2.5)

For adiabatic compression, $S = $ const., we find

$$T = \left(\frac{3S\rho}{4a}\right)^{1/3}, \qquad P = \frac{3^{1/3}S^{4/3}}{4^{4/3}a^{1/3}}\,\rho^{4/3} .$$ (8.2.6)

Thus, the adiabatic index is $\gamma = \tfrac{4}{3}$ for matter whose principal contribution to the energy is due to radiation.

The number of photons per unit volume is obtained from the expression

$$n_\gamma = \frac{2}{(2\pi\hbar)^3} \int_0^\infty \frac{4\pi p^2 dp}{\exp{(cp/kT)} - 1}$$ (8.2.7)

$$= 0.244(kT^0/\hbar c)^3 = 20T^3(^\circ \text{K}) ,$$

from which the mean energy per photon is

$$\langle\hbar\omega\rangle = \epsilon/N = 2.8kT .$$

The number of photons per gram of plasma is proportional to T^3/ρ; i.e., it is proportional to the specific entropy. This circumstance is discussed extensively in § 8.5.

Let us compare the energy and pressure of the radiation with the energy and pressure of the plasma. The ratio of the plasma pressure to the total pressure is conventionally denoted by β:

$$\beta = \frac{RT\rho/\mu}{\tfrac{1}{3}aT^4 + RT\rho/\mu} .$$ (8.2.8)

For fully ionized hydrogen with $\beta = \tfrac{1}{2}$, we obtain

$$T^3/\rho = (3R)/(\mu a) = 6.6 \times 10^{22} .$$

Thus, the relationship between density and temperature, $T_{1/2}$, when $\beta = \tfrac{1}{2}$, is expressed by

$$T_{1/2} = 4 \times 10^7 \rho^{1/3} .$$

It is convenient to transform the expression for β to the form

$$\beta = \frac{1}{1 + (T^3\mu a)/(3R\rho)} = \frac{1}{1 + (T/T_{1/2})^3} .$$ (8.2.9)

When $\beta = 0.5$, the plasma pressure equals the radiation pressure; but the radiation energy is larger than the plasma energy by a factor of 2. The

215

additional mass density which, according to the principle of equivalence, is related to the energy $\Delta\rho = \epsilon/c^2$, amounts to a fraction of the rest-mass density of the plasma given by

$$\frac{\Delta\rho}{\rho_0} = \frac{\epsilon}{\rho_0 c^2} = \frac{1.5aT^4}{\rho_0 c^2} = 10^{-35}\frac{T^4}{\rho_0} = \left(\frac{2T}{10^9\rho_0^{1/4}}\right)^4. \tag{8.2.10}$$

Thus, there exists a wide domain in which the radiation energy is comparable to, or even greater than, the plasma energy; but in which the mass density of the radiation is small compared with the rest-mass density of the plasma. At greater temperatures one must take into account the increase of the total mass density, $\rho_t = \rho_0 + \Delta\rho$, over the rest-mass density ρ_0. However, in typical relativistic stellar models the difference $\Delta\rho$ is only one of several relativistic corrections which are all of the same order of magnitude.

TABLE 6

DEPENDENCE OF β AND γ ON THE MASS OF A HOMOGENEOUS STAR OF HYDROGEN ($\mu = 0.5$) OR OF IRON ($\mu = 56/27$)

				$M*$				
0	5.7×10^{34}	1.0×10^{35}	4.1×10^{35}	1.4×10^{36}	3.2×10^{36}	1.4×10^{37}	∞
(0)	(3.3×10^{32})	(5.9×10^{33})	(2.4×10^{34})	(7.9×10^{34})	(1.9×10^{35})	(7.8×10^{35})	(∞)
β...... 1	0.9	0.8	0.5	0.3	0.2	0.1	0
γ...... $\frac{5}{3}$	4.687/3	4.533/3	4.278/3	4.158/3	4.103/3	4.05/3	4/3

* Numbers outside parentheses are for $\mu = 0.5$; numbers inside parentheses are for $\mu = 56/27$.

Notice that for each given value of β there exists a definite ratio T^3/ρ and consequently a definite value of

$$\frac{P}{\rho^{4/3}} = \frac{1}{3}\frac{aT^4}{\rho^{4/3}} + \frac{RT}{\mu\rho^{1/3}} = \frac{R}{\mu}\left(\frac{3R}{\mu a}\right)^{1/3}\left[\left(\frac{T}{T_{1/2}}\right)^4 + \frac{T}{T_{1/2}}\right]$$
$$= \frac{R}{\mu}\left(\frac{3R}{\mu a}\right)^{1/3}\left[\left(\frac{1-\beta}{\mu}\right)^{4/3} + \left(\frac{1-\beta}{\beta}\right)^{1/3}\right]. \tag{8.2.11}$$

As noted previously (see the discussion connected with Fig. 31), the ratio $P/\rho^{4/3}$ depends directly upon the mass of the star. This will be discussed in greater detail in § 10.2; but even at this early stage in the book it is useful to summarize (Table 6) the roles of radiation pressure and plasma pressure in stars of different masses.

The last line of Table 6 gives the values of the adiabatic index γ determined from the isentropic relationship between pressure and density. Obviously, γ depends only on β. It is easy to derive the formula for a fully ionized plasma

$$\left(\frac{\partial\ln P}{\partial\ln\rho}\right)_s = \gamma = \frac{4}{3} + \frac{1}{3}\beta\frac{4-3\beta}{8-7\beta}, \tag{8.2.12}$$

which was used to calculate the values of γ listed in the table. There exist in the literature other expressions for the effective index γ as a function of temperature, density, etc., but those expressions will not be used here.

8.3 PAIRS AND NEUTRINOS

The expressions for the equilibrium number densities of electrons n_- and positrons n_+ in the form of integrals follow:

$$n_- = \frac{2}{(2\pi\hbar)^3} \int_0^\infty \frac{4\pi p^2 dp}{\exp\left[(E_e - \mu)/kT\right] + 1}, \qquad (8.3.1)$$

$$n_+ = \frac{2}{(2\pi\hbar)^3} \int_0^\infty \frac{4\pi p^2 dp}{\exp\left[(E_e + \mu)/kT\right] + 1}. \qquad (8.3.2)$$

Here $E_e = (m_e c^2 + c^2 p^2)^{1/2}$, and μ is the chemical potential of the electrons. The fact that the sum of the chemical potentials of positrons and of electrons in equilibrium is zero has been taken into account. The value of μ is determined from the condition of charge-neutrality for the gas, $n_- - n_+ = \Sigma Z_i n_i$, where n_i is the number density of the ith nucleus and Z_i is its charge expressed in units of the electron charge.

The energy per unit volume is determined by similar integrals:

$$
\begin{aligned}
E_- &= \frac{2}{(2\pi\hbar)^3} \int_0^\infty \frac{E_e 4\pi p^2 dp}{\exp\left[(E_e - \mu)/kT\right] + 1}, \\[2mm]
E_+ &= \frac{2}{(2\pi\hbar)^3} \int_0^\infty \frac{E_e 4\pi p^2 dp}{\exp\left[(E_e + \mu)/kT\right] + 1}.
\end{aligned}
\qquad (8.3.3)
$$

However, here we make a stipulation: Generally, the energy of cold matter is considered to be zero; but equation (8.3.3) includes the energy of the rest mass of those electrons which were contained in the cold matter. Their number is $\Sigma Z_i n_i = n_- - n_+$. Hence, the total energy of a unit volume in this system is

$$E = E_- + E_+ - m_e c^2 (n_- - n_+). \qquad (8.3.4)$$

(Here E is the excess of energy of the hot plasma due to its pairs—but excluding the energy of photons—over the energy of a cold plasma.)

Finally, the expression for the entropy per unit volume of the particles is

$$S_- = k\frac{2}{(2\pi\hbar)^3} \int \ln\left[\exp\left(\frac{E_e - \mu}{kT}\right) + 1\right] \frac{4\pi p^2 dp}{\exp\left[(E_e - \mu)/kT\right] + 1}, \qquad (8.3.5)$$

and similarly (with a change in the sign of μ) for S_+.

These general expressions are considerably simplified in limiting cases, which we will consider now.

The first limiting case involves a nondegenerate nonrelativistic gas of

Thermodynamic Quantities

electrons and positrons. A necessary condition for this is $\mu < m_e c^2$; this occurs when

$$m_e c^2 \, (\rho/10^6)^{1/3} < kT < m_e c^2 \, . \tag{8.3.6}$$

In this case two simplifications are possible: We can neglect the one in the denominator of equations (8.3.1) and (8.3.2) as compared with exp $[(E_e \pm \mu)/kT]$, and we can write the energy expression in the nonrelativistic form

$$E_e = m_e c^2 + (p^2/2m_e) \, . \tag{8.3.7}$$

Introducing the notation $\mu' = \mu - m_e c^2$, we obtain

$$
\begin{aligned}
n_- &= \exp\left(\frac{\mu'}{kT}\right) \frac{2(2\pi m_e kT)^{3/2}}{(2\pi\hbar)^3} \, , \\
n_+ &= \exp\left(-\frac{\mu' + 2m_e c^2}{kT}\right) \frac{2(2\pi m_e kT)^{3/2}}{(2\pi\hbar)^3} \, ,
\end{aligned}
\tag{8.3.8}
$$

and the condition of equilibrium in the form of the law of mass action:

$$n_- n_+ = \exp\left(-\frac{2m_e c^2}{kT}\right) \frac{(m_e kT)^3}{2\pi^3 \hbar^6} \, . \tag{8.3.9}$$

Equation (8.3.9) and the condition (charge neutrality) which yields $n_- - n_+$ are two equations for the two quantities n_- and n_+. In the same approximation, if $kT/m_e c^2$ terms in the distribution are ignored, there are increments

$$\Delta E = 2n_+ m_e c^2 \, , \quad \Delta S = -\frac{\Delta E}{T}$$

in the energy and entropy density due to the creation of pairs.

The second limiting case corresponds to a charge symmetry expressed by $\mu \approx 0$, $n_+ \approx n_-$. This case, at low plasma density, is attained at low temperature, so that there exists a domain where the two limiting cases overlap. The greater the density of the plasma, the higher the temperature must be to produce the second limiting case. When $\mu = 0$, all quantities per unit volume depend on one parameter, $x = kT/m_e c^2$; for example,

$$n_+ = n_- = \frac{1}{\pi^2}\left(\frac{m_e c}{\hbar}\right)^3 \int_0^\infty \frac{z^2 dz}{\exp\left[(1+z^2)^{1/2}/x\right] + 1} \, . \tag{8.3.10}$$

Similar expressions are also obtained for other quantities. Convenient asymptotic formulae for small and large x are:

$$
\left.
\begin{aligned}
n_+ &= n_- = \frac{1}{\pi^2}\left(\frac{m_e c}{\hbar}\right)^3 e^{-1/x} x^{3/2} \, , \\
E_+ &= 2n_+ m_e c^2 \, , \quad S_+ = \frac{E_+}{T} \, , \quad P_+ = \frac{kT}{m_e c^2} E_+
\end{aligned}
\right\} \quad x \ll 1 \, , \tag{8.3.11}
$$

$$
\left.
\begin{aligned}
n_+ &= n_- = \frac{1}{\pi^2}\left(\frac{m_e c}{\hbar}\right)^3 2x^3 \left(1 - \frac{1}{2^3} + \frac{1}{3^3} - \dots\right) \\
E_+ &= \tfrac{7}{4} a T^4 \, , \quad P_+ = \tfrac{1}{3} E_+ \, , \quad S_+ = \tfrac{7}{3} a T^3 \, ,
\end{aligned}
\right\} \quad x \gg 1 \, , \tag{8.3.12}
$$

where a is the constant which relates to photons (see the preceding section).[3]

Let us consider a gas consisting of plasma and radiation at high temperature. By reasoning similar to that used in the preceding section, we can neglect the energy and the pressure of the plasma, but we express all quantities in terms of a unit mass of plasma. We can also neglect the density of the initial electrons, and obtain as a result the charge-symmetrical limit.

In two limiting cases, we obtain simple situations: when $x \ll 1$, only the radiation is important:

$$E = aT^4/\rho, \quad P = \tfrac{1}{3} aT^4, \quad S = \tfrac{4}{3} aT^3/\rho \; ; \qquad (8.3.13)$$

when $x \gg 1$, then, in addition to radiation, the relativistic positron-electron gas contributes:

$$E = (aT^4 + \tfrac{7}{4} aT^4) = \tfrac{11}{4} aT^4, \quad P = \tfrac{11}{12}aT^4, \quad S = \tfrac{11}{3}aT^3/\rho. \qquad (8.3.14)$$

From these formulae it is easy to find the expressions $P(\rho, S)$ for both cases. The curve $P(\rho, S = \text{const.})$ in logarithmic coordinates has the shape shown in Figure 32, a (not to scale).

TABLE 7

DEPENDENCE OF γ ON TEMPERATURE

				T (keV)					
50	70	100	125	150	200	300	500	700	∞
1.320	1.267	1.221	1.234	1.259	1.301	1.336	1.345	1.342	$\tfrac{4}{3}$

γ ......

The slopes of the left and right asymptotes (shown by dotted lines) are the same, and correspond to $\gamma = \tfrac{4}{3}$.

It is easily demonstrated that the right asymptote is below the left one by the amount $\tfrac{1}{3} \ln (\tfrac{11}{4})$. Thus, in the intermediate interval, γ must be smaller than $\tfrac{4}{3}$ on the average, and it can be demonstrated that

$$\int_{-\infty}^{+\infty} (\gamma - \tfrac{4}{3})d \ln \rho = -\tfrac{1}{3} \ln \tfrac{11}{4}.$$

G. V. Pinaeva (1964) has considered in detail the dependence of the adiabatic index on temperature. We cite in Table 7 and Figure 32, b, her values of $\gamma = (\partial \ln P/\partial \ln \rho)_S$.

The adiabatic index γ is considerably smaller than $\tfrac{4}{3}$ in the domain where pure radiation changes to radiation with e^+e^- pairs as T increases. When $T \gtrsim 300$ keV, γ becomes larger than $\tfrac{4}{3}$. This happens because the number of e^+e^- pairs grow slowly with isentropic contraction, and the Fermi gas of the pairs always has $\gamma > \tfrac{4}{3}$.

3. Here E_+, S_+, P_+ are quantities relating to pairs; i.e., they include the contributions of both positrons and electrons.

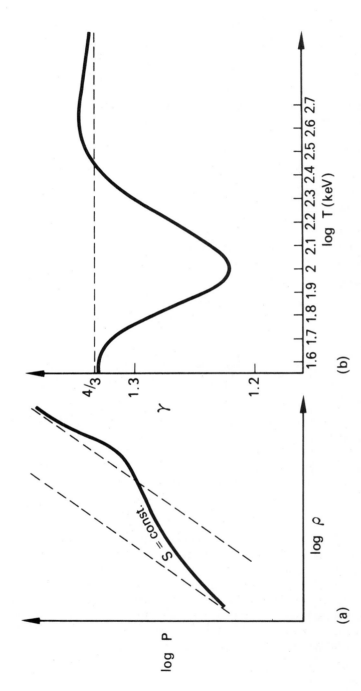

FIG. 32.—(a) Schematic representation of the relationship between pressure P and density ρ when plasma pressure can be neglected, and in the temperature domain where $e^{+}e^{-}$ pairs exist in equilibrium. (b) Relationship between $\gamma = d \ln P/d \ln \rho$ and T for the same conditions as are shown in (a).

8.4 Dissociation of Nuclei

The significance of the foregoing derives from the fact that the decrease of γ below the critical value $\gamma = \frac{4}{3}$ produces an instability in stars (see § 10.1). Moreover, in the cosmology of the "big bang" model of the Universe, this thermodynamic regime is encountered during the adiabatic expansion of the gas (which initially consists of pairs and radiation) to lower temperatures, where the pairs no longer exist.

The expressions for the concentration and the thermodynamic parameters of neutrinos are identical with the expressions for ultrarelativistic electrons (in which $m_e c^2/kT \to 0$) with only one correction: Instead of the statistical weight of 2 for the electron which corresponds to spin $\frac{1}{2}$, we must use for neutrinos, which are helical particles, the statistical weight 1. Hence, in equilibrium we have

$$\epsilon_{\nu_e} = \epsilon_{\bar{\nu}_e} = \epsilon_{\nu_\mu} = \epsilon_{\bar{\nu}_\mu} = \tfrac{7}{16}aT^4 , \qquad (8.3.15)$$

$$\epsilon_{\nu_e} + \epsilon_{\bar{\nu}_e} = \epsilon_{\nu_\mu} + \epsilon_{\bar{\nu}_\mu} = \tfrac{7}{8}aT^4 . \qquad (8.3.16)$$

8.4 The Dissociation of Nuclei

In the preceding section we considered the thermodynamics of radiation, neglecting the role of the plasma. This approximation is valid at low matter densities.

At high plasma densities (10^2–10^9 g cm^{-3}) and high temperatures (up to several times 10^{10} ° K), nuclear processes play a decisive role in the thermodynamic behavior of matter.

At low temperatures the equilibrium of nuclear processes corresponds to putting all the nucleons into the nucleus with the least energy, i.e., with the largest mass defect.[4] This nucleus is the most abundant (and this is not accidental) isotope of iron, $^{56}_{26}$Fe.[5] However, as was first pointed out by Hoyle and Fowler, at high temperatures one should expect a dissociation ^{56}Fe $\to$ 13 ^{4}He $+ 4n$. The energy required for this process is 124.4 MeV. The ratio of this energy to the increase in the number of particles $\Delta N = 13 + 4 - 1 = 16$ appears in the thermodynamic formulae; it is $Q_1 = 124.4/16 \simeq 7.7$ MeV per new particle. As for the ionization of an atom, this dissociation takes place at temperatures considerably smaller than Q_1/k; i.e., at $T = aQ_1/k$, where the dimensionless factor a is considerably smaller than unity.

Table 8 (lines 2 and 3) shows the temperatures for 50 percent dissociation of iron ($T_{1/2}$), computed for several values of density. The calculations were performed by using the obvious formula of the Saha type, i.e., all the nuclei of Fe and ^{4}He were assumed nondegenerate and nonrelativistic; the

---

4. To be quite precise, the sum of the mass of the nucleus and of the electrons which neutralize it must be a minimum; it is this sum, which differs little from the mass of a neutral atom, that is cited in mass-spectrometric tables.

5. But cf. note 10, chapter 6.

first excited levels of the nucleus ^{56}Fe were taken into consideration along with the ground state; the nuclei of ^{4}He and n have no excited levels.

The dissociation ^{4}He $\rightleftarrows 2p + 2n$ follows. The energy of this process is 28.4 MeV, yielding $Q_1 = \frac{1}{3}$ (28.4) = 9.5 MeV per new particle. Thus, the Q_1 for dissociation of He is larger than the same quantity for dissociation of Fe. The two processes are separated to some extent in temperature: There exists a domain (even though it is not wide) in which iron has dissociated to form He, but He has not yet dissociated.

Lines four and five of Table 8 show $T_{1/2}$ and a for He.

All these calculations are very rough. On one hand, an attempt should be made to consider the entire set of nuclei, stable and unstable, which occur in equilibrium at high temperature. On the other hand, one must examine all the links of the kinematics of the processes which lead to equilibrium.

TABLE 8

THE TEMPERATURE FOR 50 PERCENT DISSOCIATION OF
Fe INTO He (IN UNITS OF 10^9 ° K)

	ρ			
	1	10^3	10^6	10^9
Fe:				
$(T_9)_{1/2}$	3.43	4.1	5.79	9.57
$a = kT/Q_1$	3.83×10^{-2}	4.59×10^{-2}	6.48×10^{-2}	10.7×10^{-2}
He:				
$(T_9)_{1/2}$	4.2	5.49	8	15.2
$a = kT/Q_1$	3.85×10^{-2}	4.98×10^{-2}	7.26×10^{-2}	1.37×10^{-1}

The fundamental process is photodissociation of nuclei by the photons of the equilibrium radiation field. At temperatures at which the dissociation is appreciable, $T \sim T_{1/2}$, the density of photons with energy greater than Q_1 is sufficiently large that the detachment of neutrons from iron (like the inverse process of attachment of neutrons) takes place quickly.

The barrier for ^{4}He nuclei (α-particles) is not large in nuclei with medium atomic weight. The indirect argument in favor of the rapid establishment of equilibrium is that we are considering temperatures which are *higher* than those at which the synthesis of iron took place.

For the case of the dissociation of helium there is a distinct barrier: at a mean energy per particle of $Q_1 = 9.5$ MeV, the first step in the dissociation, ^{4}He $\rightarrow$ ^{3}He $+ n$ or ^{4}He $\rightarrow$ T $+ p$, requires an energy of 20 to 21 MeV. In general, the problem of the rate at which equilibrium is established requires additional research.

Recently, Imshennik and Nadezhin (1965) made a detailed study of the thermodynamics of plasma, taking into account the possible transformation $p \rightleftarrows n$, i.e., including weak-interaction processes along with the assumption of the creation of pairs. Since these calculations were performed with the theory of stars in mind, the chemical potential of the neutrino was assumed to be zero. Imshennik, Nadezhin and Pinaev (1966) demonstrated that calculations based on this set of assumptions are sufficiently accurate.

The principal result of these calculations is the determination of the adiabatic index γ for a wide domain of densities and temperatures. We cite from their work Figure 33, *a*, which shows the density-temperature plane (the density is plotted on a logarithmic scale). Of greatest interest is the domain where the adiabatic index is less than the critical value $\gamma = \frac{4}{3}$, which—as mentioned in § 8.3—determines the limit of stability of stars. (See also § 10.1.)

It is apparent from Figure 33 that this unstable domain is bounded both on the left and the right (*shaded curves*).

The dissociation of iron, the subsequent dissociation of helium, and the creation of pairs reduce γ to a value less than critical ($\gamma = \frac{4}{3}$) when $T \sim (2–6) \times 10^9$ (depending on the density) at the left boundary.[6] However, after the dissociation is complete and the relativistic pairs produce a gas with $\gamma = \frac{4}{3}$, the large number of nonrelativistic particles, i.e., nucleons n and p, again increase γ to a value higher than $\frac{4}{3}$, for $T > (10–20) \times 10^9$.

Figure 33, *a*, shows the following curves: (1) $x_{Fe} = x_{He}$ is the line on which iron is half-dissociated; (2) $x_{He} = x_p + x_n$ is the line of half-dissociation of He; (3) $x_+ = \frac{1}{2}x_-$ is the line on which the number of e^+e^- pairs equals the number of initial electrons, so that $e^- = e_0^- + e^-_p = 2e^-_p = 2e^+_p$. The diagram clearly shows that the bottom left projection of the shaded curve is close to this third line; this means that when $\rho \leq 10^6$ g cm^{-3} and $T \sim 4 \times 10^9$, the transition of γ through $\frac{4}{3}$ is caused specifically by the creation of pairs.

At higher densities the dissociation of iron is the most fundamental effect. However, a return to $\gamma > \frac{4}{3}$ in the course of increasing temperature is accomplished only after the He dissociation is completed.

Most obvious is the existence of two minima of γ, $\gamma_{min_1} = 0.98$ and $\gamma_{min_2} = 1.06$, which are due to the joint action of pairs and of the dissociation of iron in the first case, and to the joint action of pairs and of the dissociation of He in the second case. The degeneracy of electrons begins to play a significant role in the upper left corner at high density: The curve $P_e = 2P_e^{deg}$ indicates the region in which the pressure of the electrons becomes doubled due to degeneracy; in this domain, pairs are of no significance.

The degeneracy of electrons is essential also for the equilibrium between neutrons and protons; as one sees from the curve $x_n = x_p$ of Figure 33, *a*,

---

6. It follows from this that the onset of instability in a star (when γ becomes smaller than $\frac{4}{3}$ in the course of slow contraction and heating) may produce, for a certain range of conditions, rapid contraction down to a new stable state with $\gamma > \frac{4}{3}$ (to the right of the right boundary), which would be a state of higher temperature and density. In reality, the situation is more complex than this, as we shall see in Part III of this book.

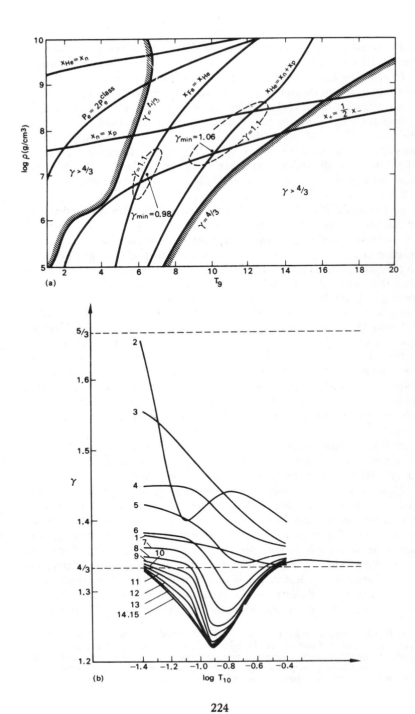

224

where $n = p$, this equilibrium depends essentially on the density, rather than the temperature.

Figure 33, b, taken from Bisnovatyi-Kogan and Kazhdan (1966), indicates the behavior of γ along the isentropes for various values of S. A comparison of this diagram with Figure 32, b, shows the influence of the presence of matter upon the adiabatic index γ. In the domain of the parameters shown here (which is important for the theory of the equilibrium of large stars) the effects of the dissociation of iron and helium, which are essential at high temperatures, are not included. The thermodynamic quantities for such temperatures are given in Imshennik and Nadezhin (1965).

Calculations of this type (along with data on the rate of energy loss due to neutrinos, the rates of nuclear reactions, the opacity of matter, etc.) represent the real foundation of the theory of stellar evolution in its present form. However, as was pointed out in § 7.2, a more exact description of hot matter in stars is obtained by using the kinetic equations for escaping neutrinos rather than thermodynamic equilibrium with zero chemical potential for the neutrinos.

The new calculations of Chechetkin and Imshennik (see § 7.3) lead to a much smaller region of instability ($\gamma < \frac{4}{3}$).

8.5 DENSE MATTER AT LOW TEMPERATURES

The properties of matter at zero temperature were described in chapter 6. What are the changes in pressure and energy density at low temperature when thermal effects just begin to be noticeable? The radiation produces an energy density of $\epsilon = aT^4$. The thermal energy density of noninteracting Fermi particles is $\epsilon_{\text{th}} = \text{const.}\ T^2 \cong mkT\ (kT/E_{\text{F}})$. For the thermal pressure we have $P_{\text{th}} = \frac{2}{3}\ \epsilon_{\text{th}}$ in the nonrelativistic case and $P_{\text{th}} = \frac{1}{3}\ \epsilon_{\text{th}}$ in the relativistic case. The exact values of these contributions are of importance for calculations of the cooling of neutron stars.

As has been pointed out by Ginzburg (1968, 1969) and his colleagues, the low-temperature behavior may be strongly affected by the superfluidity and superconductivity which might occur in dense matter at temperatures far above the familiar 4°–20° K. The heat capacity of superfluid matter is exponentially small; therefore its cooling time is drastically diminished.

The real significance of these remarks for astrophysics is not yet clear, so we restrict ourselves to only mentioning them and giving references.

FIG. 33.—(a) (ρ, T)-diagram for hot, dense matter (for notation and discussion see text). (b) Relationship between γ and T for pure iron, with entropy S held fixed. Isentropes are plotted for the following values of $S_{10} = [S \text{ ergs } (\text{g}^\circ\text{ K})^{-1}]/10^{10}$: (1) $S_{10} = 0.003981$; (2) $S_{10} = 0.01$; (3) $S_{10} = 0.01585$; (4) $S_{10} = 0.02512$; (5) $S_{10} = 0.03981$; (6) $S_{10} = 0.0631$; (7) $S_{10} = 0.1$; (8) $S_{10} = 0.1585$; (9) $S_{10} = 0.2512$; (10) $S_{10} = 0.3981$; (11) $S_{10} = 0.631$; (12) $S_{10} = 1.0$; (13) $S_{10} = 2.512$; (14) $S_{10} = 10.0$; (15) $S_{10} = 15.85$. Temperature is shown in terms of $T_{10} = T^\circ\text{ K}/10^{10}$.

Thermodynamic Quantities

8.6 DIMENSIONLESS ENTROPY

To conclude this part of the discussion dealing with the equation of state, we will derive convenient formulae which express the entropy in dimensionless units. In the classical theory, entropy is determined differentially by $dS = dQ/T$ up to an additive constant; it has the dimensionality of cal (g deg)$^{-1}$. Quantum theory determines the absolute value of the entropy. There, $S = k \ln W$, where k is the Boltzmann constant and W is the statistical weight of the state. The use of special thermal units involves a certain arbitrariness; in a rational system, T is measured in energy units, so that $k \equiv 1$. We will find the entropy per nucleon, $S_1 = \ln W_1$ in this system. If the system consists only of nucleons (the total number of nucleons is N, and their number density is n), then W_1 is the mean number of cells of phase space per nucleon; i.e., it is the ratio of the number of quantum levels Γ available to the nucleons, divided by the number of nucleons in a nonrelativistic gas

$$\Gamma = \frac{\langle p \rangle^3 V}{(2\pi\hbar^3)}, \quad \text{where} \quad \langle p \rangle = (3Tm)^{1/2}, \quad V = \text{(volume of sample)};$$

$$W_1 = \frac{\Gamma}{N} = 3^{3/2} m^{3/2} T^{3/2} n^{-1} \hbar^{-3} = 1.8 \times 10^{79} m^{3/2} T^{3/2} n^{-1}.$$

For ionized hydrogen with the density $\rho = 10^{-29}$ g cm^{-3} ($n = 6 \times 10^{-6}$ cm^{-3}), when $T = 10^6\,^\circ$ K, including the contribution of electrons (as well as numerical factors which were omitted above), we find $W_1 = 10^{34}$, and $S_1 = \ln W = 78$. In a star,

$$\rho = 1 \text{ g cm}^{-3} \ (n = 6 \times 10^{23} \text{ cm}^{-3}), \quad T = 10^8, \quad W_1 = 10^8, \quad S_1 = 18.4.$$

Consider plasma with overwhelming radiation. The energy per unit volume and the corresponding entropy are

$$\epsilon = aT^4, \quad S = \tfrac{4}{3}aT^3, \quad a = \frac{8\pi^5 k^4}{15 c^3 h^3} = 7.57 \times 10^{-15} \text{ erg (cm}^3\,^\circ \text{ K}^4)^{-1}.$$

The dimensionless entropy per nucleon is

$$S_1 = S/kn = 72.5 \ (T^3/n),$$

where T is in degrees Kelvin. For comparison, the number of photons per unit volume is

$$n_\gamma = \frac{\epsilon}{\langle h\nu \rangle} \approx \frac{\epsilon}{2.7kT}, \quad n_\gamma = \frac{8\pi}{(2\pi\hbar)^3} \int_0^\infty \frac{p^2 dp}{\exp(pc/kT) - 1}$$

(the second formula for n_γ is exact). Here p is the photon momentum, and the number of photons per nucleon is

$$N_\gamma = n_\gamma/n = 20.4 \ (T^3/n) = S_1/3.7.$$

8.6 Dimensionless Entropy

A similar calculation for an equilibrium spectrum of neutrinos and anti-neutrinos (with chemical potential $\mu = 0$ and with account taken of the helical nature of the neutrino) yields

$$\epsilon = \tfrac{7}{8} a T^4 , \quad S = \tfrac{7}{6} a T^4 ;$$

the dimensionless entropy is

$$S_2 = \frac{S}{kn} = 64 \frac{T^3}{n} , \quad n_{\nu\bar\nu} \approx \frac{\epsilon}{6.22 kT} ;$$

$$n_\nu = n_{\bar\nu} = \frac{4\pi}{(2\pi\hbar)^3} \int_0^\infty \frac{p^2 dp}{\exp(cp/kT) - 1} ;$$

and the number of neutrinos per nucleon is

$$N_{\nu\bar\nu} = \frac{n_\nu + n_{\bar\nu}}{n} = \frac{15.2 \, T^3}{n} = \frac{S_2}{4.2} .$$

The combined entropy in dimensionless units is expressed by the formula

$$S \approx 4(N_\gamma + N_\nu + N_{\bar\nu} + \ldots) . \tag{8.6.1}$$

This formula is derived as follows. (We note that a mistake in the first Russian edition was pointed out by K. S. Thorne.) We wish to calculate the entropy of N indistinguishable massless particles ($N \gg 1$) which occupy R states. If the particles were distinguishable, the number of dispositions for them would be R^N; but since they are not, the number of different dispositions is $W = R^N/N!$. Therefore, the entropy per particle (recall: there are N particles per nucleon) is

$$S_1 = \frac{1}{N} \ln\left(\frac{R^N}{N!}\right) \approx \frac{1}{N} \ln\left[\frac{R^N}{(N/e)^N}\right] \approx 1 + \ln(R/N) .$$

The main contribution is given by states for which the occupation number N/R is small (this was already tacitly assumed during the calculation of W); therefore, for $\mu = 0$ we have

$$N/R = n = e^{-(E/kT)} , \quad S_1 = 1 + E/kT .$$

In this approximation ($n \ll 1$) Bose-Einstein and Fermi-Dirac statistics are the same. Now, in the expression $1 + E/kT$ for the entropy per nucleon of a class of relativistic particles (photons or neutrinos), we must average over the whole distribution. After doing so, we obtain $S_1 = 1 + \langle E \rangle/kT \cong 4$ because $\langle E \rangle/kT = 3$ in our approximation; i.e.,

$$\frac{\int (E/kT) \exp(-E/kT) E^2 dE}{\int \exp(-E/kT) E^2 dE} = \frac{3!}{2!} = 3$$

after neglecting the factor of 1 in the expression $(e^{E/kT} \pm 1)^{-1}$. Adding up the entropy per baryon of the various types of relativistic particles, we obtain the formula (8.6.1) given above.

Thermodynamic Quantities

8.7 GENERAL THERMODYNAMIC RELATIONS FOR TRULY NEUTRAL MATTER

Let us consider truly neutral matter, i.e., matter with baryonic charge, leptonic charge, and electric charge all equal to zero.

The ground state of such matter is obviously the vacuum without any particles at all; it is the state reached at zero temperature. At $T \neq 0$ massless particles (photons, neutrinos, etc.) will be present in equilibrium at first; and as the temperature rises, particles with $mc^2 \gtrsim kT$ will also appear. Obviously, particles and antiparticles exist in equal numbers in the equilibrium state.

In this section we wish to point out the general relation which is valid for such matter even if interactions between all the particles (and antiparticles, of course) are considered.

In conventional thermodynamics the state of a sample of matter is determined by $Z + 1$ independent variables, where Z is the number of independent species of matter present. In the case of high temperature Z is the number of conserved quantities. Due to the long-range character of the electrostatic interactions, we exclude electric charge in enumerating these conserved quantities. Hence, we have $Z = 3$ corresponding to one baryonic and two leptonic charges. Thus, the thermodynamic properties of high-temperature matter depend on three charge densities and one thermal quantity, e.g., the entropy density or the temperature. However, in a truly neutral substance only one independent variable survives; for concreteness we take it to be the entropy density S.

In the familiar expression

$$dE = -PdV + TdS \qquad (8.7.1)$$

every quantity is referred to a unit of "matter," i.e., to a unit of baryonic charge. What is the implication of this fundamental equation for a truly neutral substance? We can obtain it by assuming a small baryonic charge density ($n \to 0$ in the limit), which does not influence the properties of the substance; it merely serves to give a frame of reference, i.e., to determine a definite volume $V = 1/n$.

The energy density is $\epsilon = nE$, the entropy density is $s = nS$, and in saying that the baryonic charge density has no influence we mean that

$$\epsilon = \epsilon(s) , \quad E = (1/n)\epsilon \, (nS) . \qquad (8.7.2)$$

This is a restriction of the more general relation $E = \psi(S, n)$. Inserting this equation for E into equation (8.7.1), we obtain

$$dE = -\frac{1}{n^2}\epsilon dn + \frac{S}{n^2}\epsilon' dn + \epsilon' dS = \frac{p}{n^2} dn + TdS .$$

Here a prime denotes differentiation relative to $s = nS$. From this relation we see that

$$\epsilon' = d\epsilon/ds = T \; ; \quad -\epsilon + s(d\epsilon/ds) = P \; . \tag{8.7.3}$$

These basic equations do not contain n explicitly, so they are valid also in the limiting case $n = 0$. For noninteracting particles with zero rest mass, $\epsilon \sim T^4$, $s \sim T^3$, $\epsilon \sim S^{4/3}$, $d\epsilon/ds = \frac{4}{3} (\epsilon/s)$, so equation (8.7.3) gives the familiar result $P = -\epsilon + \frac{4}{3} \epsilon = \frac{1}{3} \epsilon$.

When interactions and nonzero rest masses are important, the interdependence of the thermodynamic quantities can be more complicated. However, formula (8.7.3) remains valid and still gives useful information.

For example, if $\epsilon = as^n$, then $T = n(\epsilon/s) = nas^{n-1}$, so that $\epsilon \sim T^{n/(n-1)} = T^k$, $P = (n - 1)\epsilon = \epsilon/(k - 1)$. Some of these formulae are right even if the power-law relation between ϵ and s holds only over a small interval. To be precise, let us define n by $n = d \ln \epsilon/d \ln s$ at $s = s_0$ and $\epsilon = \epsilon(s_0) = \epsilon_0$; but let us allow $d^2 \ln \epsilon/d (\ln s)^2 \neq 0$. In this case, near $s = s_0$ the difference between the explicit $\epsilon(s)$ and the approximate formula $\epsilon = as^n$ with $a = \epsilon(s_0)s_0^{-n}$ is of the order of $(s - s_0)^2$. The difference between the exact $T(s)$ and $T = nas^{n-1}$ is of the order of $(s - s_0)$. At $s = s_0$ the relation $P = (n - 1)\epsilon$ holds; but if we define $k = d \ln \epsilon/d \ln T$, the relations $n/(n - 1) = k$ and $P = \epsilon/(k - 1)$ are not true. (The statement to the contrary in the paper of Zel'dovich [1969] is erroneous.)

Zel'dovich (1969) has discussed the influence on thermodynamics of interactions and of various assumptions about the spectrum of rest masses. Since no firm, conclusive results for the actual behavior of matter were reached, we merely give the reference; previous work is also cited therein.

III RELATIVISTIC STAGES OF EVOLUTION OF COSMIC OBJECTS

We now turn our attention to celestial bodies for which the effects of the general theory of relativity produce qualitative changes as compared with Newtonian theory.

Before 1967, astronomers had discovered no objects for which GTR effects were significant. Theory predicted the existence of neutron stars and of collapsed stars, but no such stars had ever been observed. Since then, the discovery of pulsars and their interpretation as neutron stars has solved one-half of the puzzle. Still undiscovered are collapsed stars, whose properties—unlike those of neutron stars—are wholly dependent on general relativity. But let us proceed from the beginning.

The theory of the structures of stars in slow evolution has been developed in detail and agrees well with the observational data. The distributions of temperature and density in the Sun and other stars containing an adequate supply of hydrogen have been computed in full. These stars form the main sequence of the Hertzsprung-Russell diagram. Their luminosities (total rate of energy release), radii, spectra, and evolution have also been computed. The relationships obtained agree with observational data. This remarkable accomplishment of the past two decades supports the basic premises of the theory, which relate to the properties of matter at the temperatures found in stellar interiors and to the rates of nuclear reactions under such conditions.

Since the theory is correct as applied to the hydrostatic states of stars, one must also take seriously the predictions of the theory which relate to their final fates. The general direction of the evolution is toward consumption of nuclear fuel and a gradual increase of temperature and density in the center of the star.

It is difficult, for purely technical reasons (even by using electronic computers), to perform a detailed computation of the later stages of stellar evolution; consequently, this has not yet been done. Therefore, the final state of a star is sought not by tracing the entire evolution in detail, but by using a different approach. One assumes that all nuclear fuel has been consumed (otherwise the reactions would continue) and that the temperature has dropped to zero (otherwise energy would have to be radiated away); and one calculates the distributions of such matter which satisfy the condition of hydrostatic equilibrium.

For stars with a mass less than 1.2 solar masses (1.2 $M_\odot$), the answer is well known: an equilibrium state is obtained in which the electron shells are crushed, but the nuclei are still at relatively large distances from each other. The pressure of the degenerate electron gas balances the force of gravitation; such stars are called white dwarfs.[1] The observations confirm this theoretical prediction.

When the mass of the star is larger than 1.2 $M_\odot$, but smaller than the critical value $M_{cr} \approx 2\ M_\odot$, the equilibrium state is that of a neutron star. The matter is compressed to a density which is of the same order as the density of an atomic nucleus (10^{14} g cm^{-3}).[2] The radius of the star is of the order of 10 km; the gravitational potential is of the order of 0.1 c^2.

The magnetic field and rotation of a neutron star are not very important for its bulk properties such as mean density, gravitational potential, and composition. Also unimportant for bulk properties are superconductivity, superfluidity, and solidification of matter in different parts of the neutron star.

However, the magnetic field and rotation lead to very peculiar pulsed radio emissions accompanied, in young pulsars, by optical and X-ray emission and by the acceleration of relativistic particles and injection of them into the surrounding nebula. Superconductivity, superfluidity, and solidification may also be important for some of the observed peculiarities of pulsars. No body except a neutron star can withstand such rapid rotation as that corresponding to the interval between pulses of a pulsar. The examples of the Crab and Vela pulsars confirm the idea that neutron stars are created by supernova explosions. Pulsars are probably strong emitters of gravitational waves (see, e.g., Ostriker and Gunn 1969; Shklovsky 1969b).

Obviously, for such stars one must consider those changes in the law of gravitation which follow from GTR.

What does GTR contribute to the problem of the fate of a star? When the mass is less than M_{cr}, only quantitative changes occur. But the existence of a maximum critical mass, M_{cr}, is a result of GTR; and it appears that a critical mass exists for any conceivable equation of state which is consistent with relativity theory. When the mass is greater than critical, there exists no equilibrium configuration! The final stage of evolution must be unlimited contraction.

---

1. This term is historical. As noted in chapter 6, physicists may refer to matter as "cold" when its temperature does not affect the equation of state. White dwarfs have a surface temperature of $\sim 10^4$ ° K; however, they may be considered "cold" in this sense. At temperatures of 10^4 ° K, these stars have a white color; hence the designation. For the history of the discovery of white dwarfs, see Schatzman (1958). Recently it has been shown that at high densities the heat capacity falls, and the cooling time also decreases. Probably old dwarfs with masses near the Chandrasekhar limit are now red or even black. Some of them, perhaps, remain undiscovered. For a review see Greenstein (1969).

2. A neutron star can have a mass smaller than 1.2 $M_\odot$ (but it must be larger than ~ 0.1 $M_\odot$). However, in the interval 1.2 $M_\odot < M \lesssim 2\ M_\odot$, the neutron star is the unique equilibrium state for cold matter (see below).

In this context, consideration of the general theory of relativity leads to a conclusion which, at first glance, appears to be paradoxical: any distant observer sees an asymptotic approach of the contracting star to a certain state. This state is not an equilibrium state; it may be referred to as a "frozen" state. In reality, there is no paradox; it is just that the conclusion of the theory is unexpected and unusual.

The time interval between two given events is not the same as viewed by different observers; there is no Newtonian universal time. For an observer in the star the contraction does not grind to a halt. It is just the law of transformation of time intervals which makes a distant observer see the contraction halt. Perhaps one could call this phenomenon "relativistic retardation of the passage of time": a given interval of internal time requires an increased amount of time as measured by a distant observer; internal clocks are retarded from the point of view of a distant observer. The relativistic retardation of the passage of time simultaneously implies an approach of photon frequencies to zero (we refer to the photons that are received by a distant observer). A gravitational self-closure of the star occurs. It stops the emission of energy to an outside observer and it stops the outward flow of information.

Thus, the theory predicts three types of celestial bodies at the final stage of stellar evolution: (1) white dwarfs, (2) neutron stars, (3) "frozen" stars; and which type of body a given star can become depends upon its mass.

There is a conflict between the theory and the observations in that no frozen stars have thus far been discovered. Thus, those bodies in which GTR plays the most decisive role are just the bodies that have not yet been found. The problem of the existence of such bodies is of great importance for cosmology, also; the general curvature of space on a large scale depends on the density of all types of matter, including neutron stars and frozen stars. The first approximate estimate by Hoyle, Fowler, Burbidge, and Burbidge (1964) was that the total mass of frozen stars might be comparable to the mass of all visible stars. However, this result depends largely on the simplifying assumptions that were made.

What then are the possible ways of resolving the conflict between theory and observation? On one hand, it is necessary to consider the difficulty of observing frozen stars. What are their properties? How do they manifest themselves by interactions with other stars and with the interstellar medium which contains dust, gas, and magnetic fields? Is it possible that there exist many frozen stars but that they are difficult to observe? (It should be remembered that theoretical predictions of the properties of neutron stars played no role in the discovery of pulsars.)

On the other hand, one must analyze the assumptions which lead us to conclude that these are the final states of stellar evolution; in particular, one must examine all possible methods of mass loss, the role of rotation of the star, and the role of its magnetic field. In the course of the evolution of a massive star the increase in density is accompanied by increasing temperature. At a certain point, the star approaches the onset of instability, after which a catastrophic contraction takes place. However, the stellar matter may still contain a reserve of nuclear energy.[3] Release of this energy might halt the contraction and lead to an explosion of the star. It should be kept in mind, however, that the observed frequency of stellar explosions is many times smaller than the number of stars which are initially too massive to conclude their evolution by becoming white dwarfs; in other words, the observational data seem to indicate that not all stars escape the transformation into a neutron or frozen state by explosion. The very existence of white dwarfs with $M < M_\odot$ confirms this idea: the evolution of a star with $M < M_\odot$ into the white-dwarf state requires a time which exceeds not only the age of the Galaxy but the age of the Universe itself. Such dwarfs must be remnants of stars with greater mass which evolved more rapidly and subsequently lost some of their mass. It is unknown what portion of its mass may be lost by a star as a result of quasistatic, hydrodynamic emission of matter from its surface.

A new method of investigating the final states of stellar evolution appears to have arisen in the second half of 1969—the experimental study of cosmic gravitational radiation. Weber (1969) claims to have discovered pulses of such radiation at high frequencies. The total energy flow in these pulses is tremendous—greater than or of the order of 10^5 ergs cm^{-2} (see § 1.15). Only frozen stars (in the process of formation or in collision) can give such an output of gravitational radiation without so much accompanying electromagnetic radiation that observers would have seen it. The amount of information about Weber's observations is not yet complete enough to make a decisive judgment; but the bright perspective of a new type of investigation is before us.

The disparity between the conclusions of the theory and the observations has existed for a long time. However, the discovery of quasars put the entire situation into a somewhat different perspective. In the theory of quasars an important role must be played, on one hand, by small relativistic corrections to the laws of gravity, and on the other hand, by macroscopic motions and magnetic fields—i.e., by factors which do not play an important role in the theory of the structure of normal stars.

With the discovery of quasars, dozens of theoreticians turned their attention to general-relativistic aspects of the theory of the equilibrium and contraction of stars; and astronomers suddenly remembered the classical, 1938–1939 work of Oppenheimer, Volkoff, and Snyder. Simultaneously, following a natural psychological law, the assumption occurred that the two riddles were related—i.e., that the fate of normal stars and the nature of

---

3. This reserve essentially depends on whether or not convective mixing of the various layers of the star has occurred.

quasars might have a common answer. At the present time we feel that this is not so, even though both problems have overlapping aspects, some of which are related to GTR.

In this part of the book, we will consider the aforementioned problems.

The exposition begins with a very brief review of the data on stellar equilibrium and stellar stability. Then the theory of the equilibrium states of stars consisting of inert matter (including supermassive stars) is described. Next, we consider the evolution of a star, and its transition to the final state. Naturally, our discussion does not include a detailed analysis of the slow evolution of the star; this is covered in appropriate monographs (see the end of this chapter). We consider only the general nature of the evolution, its rate, and its stability. These are important in order to determine the conditions which lead to a stage where GTR effects are essential; and conversely, they also help us to determine the means by which a star may avoid the relativistic stage.

Next we consider the catastrophic collapse of a star which has become unstable in the course of its slow evolution, we examine the outbursts of supernovae and other processes which accompany the collapse, and we consider the final stage of stellar collapse with $M > 2 M_\odot$, i.e., the relativistic collapse which produces a frozen star. Following this, we discuss the properties of the evolution of a supermassive star and the role of rotation in its evolution. These problems are important for the theory of quasars. Then we examine processes which must take place in the vicinities of relativistic objects and analyze methods for the discovery of such stars. Finally, we analyze the theory of star clusters and the possible physical nature of quasars.

The problems discussed have been described previously in reviews by the authors (Zel'dovich and Novikov 1964a, 1965, 1966a; also Novikov and Zel'dovich 1965).

Following is a brief list of some monographs and papers which cover the problems that are discussed in this part of the book.

The classical theory of stellar structure and nonrelativistic stages of stellar evolution are discussed in monographs by Chandrasekhar (1939), Burbidge and Burbidge (1958), Schwarzschild (1958), and Frank-Kamenetsky (1959); also in the collective work *Stellar Structure* edited by Aller and McLaughlin (1965); in *Hdb. d. Phys.*, Vol. 51, 1958; in Baade (1963); in Iben (1967); in the text by Martynov (1965); in the text by Clayton (1968); and in the popular book by Kaplan (1963).

The fundamental conclusions drawn from the work of Oppenheimer and Volkoff and of Oppenheimer and Snyder are presented in the texts by Landau and Lifshitz (1962, 1964) and in the lecture notes by Thorne (1967). Problems of the theory of superdense configurations of cold matter have been considered by Ambartsumyan and Saakyan (1963); in our reviews listed above; and also by Harrison, Thorne, Wakano, and Wheeler (1965).

Theoretical problems of the influence of small effects of GTR upon the stability of a star in a near-critical state have been detailed by Kaplan (1949*b*), W. A. Fowler (1964*a*, *b*), Chandrasekhar (1964*a*, *b*, 1965), Thorne (1967), and in our reviews.

Problems of neutrino radiation have been discussed in reviews by V. S. Pinaev (1963*a*) and Chiu (1964). The latter report also touches upon other problems of relativistic astrophysics.

The theory of the explosion of a supernova has been discussed by Fowler and Hoyle (1964), and the astronomical aspects have been described in a monograph by Shklovsky (1966).

A number of new contributions on the theory of the explosion of supernovae have appeared recently; see § 11.4. A review of the problem of supermassive stars has been given by Wagoner (1969); and reviews of the problem of quasars are available by E. M. Burbidge (1967), by E. M. and G. Burbidge (1967*b*), and by M. Schmidt (1969).

Naturally, this list is not and does not claim to be complete.

10 THE EQUILIBRIUM AND STABILITY OF STARS

10.1 THE EQUILIBRIUM AND STABILITY OF A STAR AS A WHOLE

In its normal state, a star is a gas sphere which is in hydrostatic and thermal equilibrium. The hydrostatic equilibrium is provided by the equality of the gravitational force and the pressure force acting upon each mass element. Thermal equilibrium indicates an equality between the energy produced in each mass element in the star and the energy radiated from the surface of each mass element. If either condition of equilibrium were not satisfied, then the star's structure would begin to change, and the star would cease to be a stationary object (changes produced by thermal non-equilibrium would usually be quite slow).

Let us assume that the pressure does not exactly compensate gravity, so that under the influence of a noncompensated force the matter acquires an acceleration comparable to the acceleration of free fall, $g = GM/R^2$. Then an element of matter will be displaced a distance of the order of the radius of the star, R, during the following time interval (called the hydrodynamic time scale):

$$t_H \approx (R/g)^{1/2} = (GM/R^3)^{-1/2} . \tag{10.1.1}$$

Here the subscript H stands for hydrodynamic.

Substituting the values for the Sun, $M = M_\odot = 2 \times 10^{33}$ g and $R = R_\odot = 7 \times 11^{10}$ cm, into this equation, we obtain $t_H \approx 10^3$ sec. For a star to exist in a stationary state, it must satisfy the condition of hydrostatic equilibrium, which has the form

$$- dP/dr = GM(r)\rho/r^2 . \tag{10.1.2}$$

The left-hand side is the pressure force acting on a unit volume, while the right-hand side is the gravitational force of attraction of that volume by the mass $M(r)$ contained inside the sphere of radius r. In this section we will analyze problems of equilibrium and stability using order-of-magnitude estimates which involve a mean density $\langle \rho \rangle$ and pressure $\langle P \rangle$. While such a procedure is approximate, it does reveal very explicitly the physical essence of the problems. The precise theory of the stability of an equilibrium stellar model will be discussed in § 10.5, and in subsequent sections. This exact

239

theory confirms the approximate estimates. Using mean values for the whole star and using the condition of hydrostatic equilibrium, we can relate the mean density and pressure to the star's mass and radius:

$$M \approx \tfrac{4}{3}\pi R^3 \langle \rho \rangle , \qquad \frac{\langle P \rangle}{R} = \langle \rho \rangle \frac{GM}{R^2} . \qquad (10.1.3)$$

By using these formulae one can easily estimate the pressure in a star, and then by using the equation of state of an ideal gas one can estimate the temperature of the interior.[1] Substituting into equations (10.1.3) the values for the Sun, we find $\langle P_\odot \rangle \approx 10^{16}$ dyn cm^{-2}, $\langle T_\odot \rangle \approx 10^{7}$ ° K.

If we use these estimates, we arrive at a rather important conclusion: The characteristic time for hydrodynamic processes in a star is considerably smaller than the corresponding times for thermal processes, and for the consumption of nuclear fuel. For the Sun, for example, the characteristic time for thermal process is determined by the condition

$$t_{T\odot} \approx E_{T\odot}/L_\odot \approx (3kT_\odot/m)\,(M_\odot/L_\odot) \approx 3 \times 10^7 \text{ years} ,$$

where $E_{T\odot}$ is the thermal energy of the Sun, m is the mass of the proton, $L_\odot$ is the luminosity of the Sun, and $M_\odot$ is the mass of the Sun ($L_\odot = 4 \times 10^{33}$ ergs sec^{-1}). Thus, the Sun, without any nuclear sources of energy, could exist for 30000000 years. By the same token, the time for burning of nuclear fuel is

$$t_{N\odot} \approx E_{N\odot}/L_\odot \approx (0.01\ c^2)(0.1\ M_\odot)/L_\odot = 10^{10} \text{ years} .$$

Here, $E_{N\odot}$ is the reserve of nuclear energy in the Sun, $0.01\ c^2$ is the maximum energy released by nuclear reactions per unit mass, and $0.1\ M_\odot$ is the mass of the core where the temperature is sufficiently high for nuclear reactions.

We see that $t_H \ll t_T$ and $t_H \ll t_N$. The estimates obtained suggest that finding the configurations of density and pressure which satisfy hydrostatic equilibrium is the primary task in the theory of stars.

Evolution involving thermal and nuclear processes, or involving loss and accretion of matter, produces changes in the equilibrium configuration. Reaching the limit of the existence of such configurations leads to a disruption of the equilibrium and to catastrophic phenomena.

For an analysis of the hydrostatic equilibrium, we use the energy method. A detailed description of the method and calculations is included in the appendix to this section, p. 317 (see also Harrison, Thorne, Wakano, and

---

1. Formulae (10.1.3) allow us another approach to find t_H. This approach consists of an estimate of the time required for a sound wave to travel a distance of the order of the radius of the star. The velocity of sound is $v_{\text{sound}} = (dP/d\rho)^{1/2}$. Using for an estimate the averaged equations of equilibrium (10.1.3), we find that

$$v_{\text{sound}} = (dP/d\rho)^{1/2} \approx (\langle P \rangle / \langle \rho \rangle)^{1/2} = (GM/R)^{1/2} ,$$

whence follows equation (10.1.1) for t_H.

10.1 Stability of a Star as a Whole

Wheeler 1965, and contributions by Chandrasekhar 1964a, b, 1965). In the text, we will formulate only the general conclusions.

The condition of hydrostatic equilibrium is equivalent to the condition of an extremum of the total energy of the star for a given number of conserved elementary particles (baryons), and a given distribution of specific entropy. The equation for the pressure gradient is the Euler equation of the variational problem of finding the extremum of the energy by varying the distribution of the matter. This assertion is valid in classical Newtonian theory as well as in the general theory of relativity. Therefore, it is natural to construct a theory of equilibrium configurations in which one analyzes their energy with respect to the free parameters (mass distribution). A minimum of the energy corresponds to stable equilibrium, and a maximum of the energy to unstable equilibrium; in the energy approach, the exploration of stability does not require additional computations. By contrast, the more direct approach of analyzing the solutions of the differential equation of hydrostatic equilibrium does not permit a determination of stability; there one must also investigate the linearized equations of small perturbations.

Especially, it is necessary to emphasize the role of entropy: its role is specified by the thermodynamic relationship $P = - [\partial E_1/\partial(1/\rho)]_S$, where E_1 is the specific energy and S is the specific entropy (all quantities per unit rest mass). This relationship permits us to establish the connection between the energy of the star, which includes E_1, and the equation of equilibrium, which includes P. Therefore, the theory involves E_1 explicitly as a function of ρ and S, rather than of ρ and temperature T.

To illustrate the general situation, we will first roughly approximate the entire star by a mean density $\langle \rho \rangle$ and a mean energy per gram of matter $\langle E_1 \rangle$. The total energy of the star of mass M can be expressed in the form

$$E = \int_V E_1 \rho dV - G \int_V (m\rho/r) dV . \qquad (10.1.4)$$

The first term is the internal energy, the second term is the gravitational energy, and m is the mass inside the sphere of radius r. Using the mean values, we rewrite equation (10.1.4) as follows:

$$E = \langle E_1 \rangle M - a_1 GM^2/R , \quad a_1 = \text{const.} ;$$

and by expressing R in terms of $\langle \rho \rangle$, where $\langle \rho \rangle = 3M/4\pi R^3$, we find

$$E = \langle E_1 \rangle M - a_2 GM^{5/3} \langle \rho \rangle^{1/3} , \quad a_2 = \text{const.} \qquad (10.1.5)$$

The equation of state of an ideal gas can be expressed in the form

$$E_1 = K(S)\rho^{\gamma-1} + L(X) ,$$

where $K(S)$ is a function of the entropy of the gas and of its chemical composition, $L(X)$ is a function only of the chemical composition, γ is the adiabatic index, $(d \ln P / d \ln \rho)_{S=\text{const.}}$, and P is the pressure. Recall that for a monatomic, nonrelativistic, ideal gas $\gamma = \frac{5}{3}$. In the theory of stars, the so-called polytropic index n is frequently used rather than γ; $\gamma \equiv 1 + 1/n$.

Using our ideal-gas formula for E_1, we can rewrite equation (10.1.5) in the final form

$$E = C_1 M \langle \rho \rangle^{\gamma-1} + C_2 M - C_3 M^{5/3} \langle \rho \rangle^{1/3} \,, \tag{10.1.6}$$

where C_1, C_2, and C_3 are constants at fixed entropy. The additive constant in the energy is obviously not needed for the solution of the variational problem.

If $\gamma > \frac{4}{3}$, then curve E, as a function of $\langle \rho \rangle$, has a minimum. It corresponds to the position of the stable equilibrium of the star.

If $\gamma < \frac{4}{3}$, then the curve $E(\langle \rho \rangle)$ cannot have a minimum and the star has no stable equilibrium state. In this case, the curve has a maximum which corresponds to an unstable equilibrium.

Finally, when $\gamma = \frac{4}{3}$ and $C_1 = C_3$, the energy of the star does not depend at all on the mean density; i.e., a neutral equilibrium of the star occurs at any density. This neutral equilibrium exists only with respect to contraction of the star as a whole. It can be demonstrated that the star is stable with respect to any deformation of its density distribution.

We will now find the relationship between the density of an equilibrium star and its mass. For this, we set to zero the derivative of E with respect to ρ:

$$dE/d\langle \rho \rangle \,|_{M=\text{const.}\,,\ S=\text{const}} = 0\,.$$

From this we obtain

$$M = \text{const.} \, \langle \rho \rangle^{(\gamma-4/3)(3/2)} \,. \tag{10.1.7}$$

This formula shows that the sign of $(\partial \rho / \partial M)_{S=\text{const.}}$ for an equilibrium configuration is the same as the sign of the difference $(\gamma - \frac{4}{3})$. The result can be formulated as follows: When the star is stable, then $\partial \langle \rho \rangle / \partial M > 0$, and when the star is unstable, then $\partial \langle \rho \rangle / \partial M < 0$. The calculation of $\partial \langle \rho \rangle / \partial M$ involves a comparison of two models of the star both of which consist of matter that has the same equation of state and the same entropy but in which the masses differ by δM.

This criterion is natural: in a stable state, an addition of mass causes contraction and an increased pressure, which compensates the increased gravitational force. The analysis of an exact (not averaged) equation of state leads us to the stability criterion $(\partial \rho_c / \partial M)_S > 0$, where ρ_c is the central density of the star (see Zel'dovich 1963a).

The process of energy radiation by a star into surrounding space is governed by the diffusion of the radiation from the interior to the surface.[2]

---

2. Under certain conditions, the energy flux is transported not by radiation, but by convection; but this does not change the essence of these considerations.

The flux of energy at the surface, L, depends on the distribution of T and on the opacity of the star:

$$L = 4\pi r^2 D \ (dE/dr) \, , \qquad (10.1.8)$$

where D is the diffusion coefficient, and E is the energy density of radiation, $E = aT^4$.

Can a star possibly lose stability due to the strong coupling that exists between the temperature and the processes of energy release by nuclear reactions? This coupling for small intervals of T is expressed by the relationship

$$A_{nuc} = A_0\rho T^\nu \, , \qquad (10.1.9)$$

where A_0 and ν are constants. For the proton-proton reaction, for example, $\nu = 4.5$ in the temperature range $(0.9$–$1.3) \times 10^7$ ° K; for the carbon cycle, $\nu = 20$ when $T = (1.2$–$1.6) \times 10^7$ ° K.

The energy density of radiation included in expression (10.1.8) is proportional to the fourth power of the temperature, $E \propto T^4$. Usually, the quantity ν in equation (10.1.9) is larger than 4, $\nu > 4$; and, as a consequence, the energy release has a stronger dependence on T than the heat transfer. It would appear that an accidental small increase of nuclear energy release over the energy loss by radiation would bring about an increase of T, and consequently a drastic increase of A_{nuc}: the perturbation would grow. A phenomenon would occur which is analogous to a thermal explosion in chemical systems. Actually, this is incorrect. We have emphasized on several occasions that hydrodynamic processes in a star take place considerably faster than thermal processes. Therefore, an increase of the energy release will bring about a departure from equilibrium: the internal pressure will exceed the gravitational force. This will lead to an expansion of the star, and ρ will decrease. Substituting into expression (10.1.3) the equation of state $P = (\Re/\mu)T\rho$, we find an expression for T (here $\Re$ is the gas constant):

$$\langle T \rangle = (\tfrac{4}{3}\pi)^{1/3} \frac{\langle\mu\rangle}{\Re} GM^{2/3}\langle\rho\rangle^{1/3} \, . \qquad (10.1.10)$$

We see that the decrease of ρ results in a decrease of T,[3] and hence also of A_{nuc}; consequently, the perturbation will not grow.

When the nuclear energy release exceeds the heat loss by radiation, the star is pushed toward the opposite condition and equilibrium is restored. Thus, the star regulates the output of the sources of nuclear energy, bringing them into correspondence with the energy radiated from the surface.

When the energy release exceeds the heat transfer, the temperature of the star goes down. In this sense, we can discuss a negative heat capacity

---

3. In massive stars, most of the pressure is radiation pressure, so $P = \tfrac{1}{3}aT^4$. In this case, too, a decrease of ρ is accompanied by a decrease of T.

of the star. This heat capacity differs from the heat capacity at constant pressure or that at constant volume, which are conventionally used in physics. In this case, the heat capacity is determined under the condition of stellar equilibrium, with the effects of gravity included.[4]

This assertion can be formulated differently. We multiply the equation of equilibrium (10.1.2) by $4\pi r^3 dr$, integrate dr over the entire radius of the star, and perform an integration by parts on the left side of the equation. As a result, we obtain

$$\int_0^R 3P4\pi r^2 dr = \int_0^R \rho \left[GM(r)/r\right] 4\pi r^2 \, dr , \qquad (10.1.11)$$

where R is the radius of the star. If we now use the equation of state of an ideal gas, then in the left integral of equation (10.1.11) we can substitute $3P = 2E_1\rho$, thereby revealing that the integral is twice the thermal energy of the star. Now, the integral on the right side is minus the gravitational energy, so equation (10.1.11) is equivalent to

$$2E_T = -U . \qquad (10.1.12)$$

This relationship is the virial theorem for normal stars. The total energy can now be expressed in the form

$$E = E_T + U = -E_T . \qquad (10.1.13)$$

The last expression indicates that when energy is added to a star, its thermal energy is reduced, and conversely, radiation of energy brings about an increase in both thermal energy and temperature.

In a stationary state, the release of nuclear energy in a star precisely compensates the radiated energy losses. However, the decreased concentration of nuclear fuel eventually brings about a disruption of the balance; and the energy losses begin to exceed the release of energy by some amount, however small. This brings about an increase of the temperature, which becomes stabilized at such a level as to provide the needed rate of release of nuclear energy at decreased concentration of nuclear fuel, or to provide a transition to the combustion of another fuel with a higher burning temperature (for example, from hydrogen to helium). These changes constitute a slow evolution of the star with gradual exhaustion of the reserves of nuclear energy.

In accordance with the negative heat capacity of the star as a whole, the gradual increase of temperature is accompanied by a decrease of entropy. To show this, we express the equation of state first in terms of tem-

---

4. We must stipulate here that everything which has been said refers only to normal stars. At the end of stellar evolution, in the so-called white-dwarf state, a star may have a positive heat capacity (see § 11.1). In this case, degenerate electrons play a significant role in the gas pressure.

perature and second in terms of entropy; finally, we express P in terms of M and ρ by using the averaged equilibrium conditions (10.1.3):

$$P = \text{const. } T\rho, \qquad (10.1.14)$$

$$P = \text{const. } \exp{(C_1 S)}\rho^{5/3}, \qquad (10.1.15)$$

$$P = \text{const. } \rho^{4/3}. \qquad (10.1.16)$$

From equations (10.1.14) and (10.1.16) it follows that

$$T = \text{const. } \rho^{1/3}; \qquad (10.1.17)$$

and from equations (10.1.15) and (10.1.16), it follows that

$$\exp{(C_1 S)} = \text{const. } \rho^{-1/3}. \qquad (10.1.18)$$

Thus, it follows from equations (10.1.17) and (10.1.18) that an increase in density results in an increase of T and a decrease of S.

Another remark is in order. The hydrodynamic time t_H is smaller than the thermal time t_T by approximately twelve orders of magnitude. However, the thermal time t_T is only approximately 300 times shorter than the nuclear time t_N. It is the nuclear time that determines the slow evolution. In view of the relatively small difference between these times, the conditions of thermal equilibrium at some periods of the life of the star may become disrupted by rapid evolution. This actually occurs, for example, when the transition from hydrogen burning to helium burning occurs, i.e., when the structure of the star changes drastically.

In this section, we have delineated only the general picture and the general relationships. Several exceptions, which may bring about an instability of the star, will be discussed later.

10.2 GENERAL ASPECTS OF THE THEORY OF STELLAR EQUILIBRIUM

In this section, as above, we shall describe stellar matter by its mean density and mean temperature, $\langle \rho \rangle$ and $\langle T \rangle$. The chemical composition of the matter will be regarded as determined by the nuclear reactions at previous stages of the evolution. If the reactions are rapid, then we shall consider the composition to be in equilibrium (the relationship of the number of protons, neutrons, and various nuclei will correspond to thermodynamic equilibrium at the given density and temperature).

Thus, we deal with three quantities (for example, M, ρ, T; the angular brackets for the average are omitted in this section), between which a relation is established by the condition of hydrostatic equilibrium of the star (eq. [10.1.3]). Thus, for example, for a given M and ρ, one can find a temperature T [here $T = T(M, \rho)$] which provides equilibrium. One relationship between three variables indicates that in the density-mass plane (Fig. 34) each point describes a unique stellar configuration, and accordingly each point has its own temperature. However, it is obvious that in the (ρ, M)-plane there exist certain restrictions. A certain cold pressure $P = P(\rho, 0)$ corresponds to the lowest temperature $T = 0$; in order to create such pressure, it is necessary to have a finite mass $M(\rho, 0)$. Below this line, there exist no stellar configurations. The peculiar shape of the line $T = 0$ is caused by differing equilibrium conditions of matter in different domains of density (see chapter 6, and §§ 10.8 and 10.9).

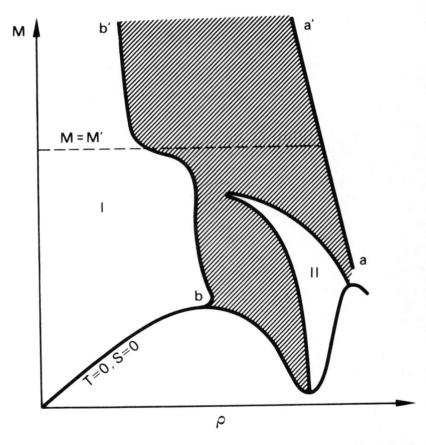

Fig. 34.—A (ρ, M)-diagram for stellar configurations (not to scale). Line $T = 0, S = 0$ corresponds to cold stars. Below this line no equilibrium configurations exist; a–a' corresponds to a line of equilibrium configurations determined by GTR. There exist no equilibrium configurations to the right of this line. Domain of unstable-equilibrium configurations is shaded. Domains I and II correspond to stable-equilibrium configurations. Horizontal line $M = M'$ separates the lower domain, where plasma pressure dominates, from the upper domain, where radiation pressure dominates.

10.2 General Aspects of Stellar Equilibrium

Further, we know (see chapter 3) that, because of GTR effects, static configurations certainly do not exist when the radius of the body is close to the Schwarzschild radius, $r_g = 2GM/c^2$. From this we obtain the restriction $\rho < 3M/4\pi r_g^3$, i.e., $\rho < 2 \times 10^{16}(M/M_\odot)^{-2}$, i.e., $M < 10^8 \ M_\odot\rho^{-1/2}$. This line is shown on Figure 34 (line a–a').

Thus solutions exist only inside the domain that is bounded by the ordinate and by two lines which intersect[5] in the region $\rho \approx 2 \times 10^{16}$ g cm^{-3}, $M \approx M_\odot$. Inside this domain, at some distance from the boundary lines, one can (a) use the Newtonian theory of gravity, since the point is located well to the left of the line a–a' which is determined by GTR, and (b) disregard the pressure of degenerate gas, since the point is well above the line $T = 0$. It follows from procedure a that in order of magnitude the pressure can be calculated by using formula (10.1.3). We render this expression in the form

$$P = k_1 G M^{2/3} \rho^{4/3}. \tag{10.2.1}$$

The dimensionless coefficient k_1 depends on the method of averaging and can be approximately set to $k_1 = 0.4$ (see §10.8).

According to procedure b we can consider our gas to be ideal, including radiation pressure:

$$P = \Re T\rho/\mu + \tfrac{1}{3}aT^4 . \tag{10.2.2}$$

The condition of hydrostatic equilibrium yields:

$$\Re T\rho/\mu + \tfrac{1}{3}aT^4 = k_1 G M^{2/3}\rho^{4/3} , \tag{10.2.3}$$

$$(\Re/\mu)(T/\rho^{1/3}) + \tfrac{1}{3}a(T/\rho^{1/3})^4 = k_1 G M^{2/3} . \tag{10.2.4}$$

From this we obtain a remarkable result which was first developed by Eddington: the mass M is uniquely related to the parameter $T/\rho^{1/3}$. But the ratio of the radiation pressure to the plasma pressure depends on the same parameter (for a constant molecular weight μ and a constant coefficient a in the expression $E = aT^4$). By equating these two pressures

$$\Re T\rho/\mu = \tfrac{1}{3}aT^4$$

we obtain the mass of the star in which radiation pressure and gas pressure are equal:

$$M' = \left[\frac{2}{k_1 G} \frac{\Re}{\mu} \frac{T}{\rho^{1/3}} \right]^{3/2} \approx \frac{50 M_\odot}{\mu^2} .$$

Recall that μ is the average mass of a free particle in units of the proton mass. For hydrogen, $\mu = \tfrac{1}{2}$, $M' = 200 \ M_\odot$; for iron, $\mu = 2$, $M' = 12 \ M_\odot$.

In the (ρ, M)-plane, one can draw the horizontal line $M = M'$ separating the lower domain, where plasma pressure (ideal gas consisting of elec-

5. In the precise theory, of course, the line of the static solutions with $T = 0$ and the line of the boundary of static solutions cannot have common points; however, such a refinement at this point is immaterial.

trons and nuclei) dominates, from the upper domain, $M > M'$, where radiation pressure dominates. In these domains, the laws are different; for example, both the isotherms (lines of constant temperature, T = const.) and the adiabats (lines of constant entropy, S = const.) have different shapes in the different domains.

The following important consideration is related to the stability of the hydrostatic equilibrium. Not every solution corresponding to a point in the admissible domain of the (ρ, M)-plane is stable. As we demonstrated in § 10.1 on the curve $M = M(\rho, S$ = const.), the regions with positive derivative $\partial M / \partial \rho|_S > 0$ are stable, and those parts of the curve where $\partial M / \partial \rho|_S < 0$ are unstable. Inasmuch as when $T = 0$, we have also $S = 0$, it is easy to verify that two stable and two unstable domains alternate on the bottom curve. With increasing temperature (and consequently increasing M for fixed ρ), the second stable domain II soon vanishes (see Fig. 34).

In the domain where degeneracy plays no role but where $M < M'$, the equilibrium of a monatomic ideal gas is stable up to temperatures at which energy-consuming nuclear reactions take place. On the other hand, when $M > M'$, the gas whose pressure is essentially determined by radiation has only a small margin of stability. Even small corrections for GTR (as well as for electron-positron pair creation) disturb the stability of the equilibrium. This explains the strong bend of the boundary of the stable domain near the horizontal line $M = M'$ of Figure 34. (The domain of unstable configurations is shaded.)

The following sections are devoted to an explanation of the entire picture, and to a calculation of the boundaries of these domains.

Obviously, unstable configurations do not occur in nature. In a stable configuration, a small perturbation causes oscillations about the equilibrium state (which attenuate due to energy dissipation if there are no processes that drive the oscillations). An unstable configuration differs in that small perturbations grow exponentially with time, since the theory is linear. A small contraction causes an increase of the force of gravity which exceeds the increased pressure, and the contraction accelerates. And, in the same way, a small expansion causes a reduction of the force of gravity and further exponential growth of the expansion.

However, in the course of its evolution, a star does not immediately enter the depths of the instability domain. A star is initially a stable object; its evolution begins in the domain of stability. Prior to reaching the domain of instability, the star must cross the boundary between the domains. It can be demonstrated (see below) that on the boundary of stability the linear theory is not sufficient, and it is always a catastrophic contraction that occurs, rather than an expansion which would return the star into the stable domain.

Let us make one more essential remark. Far from the boundary at which instability sets in, the rate of variation of the entropy of the star is considerably smaller than the rate at which hydrodynamic equilibrium is established (as we pointed out in § 10.1). At the onset of instability, these rates become comparable. Therefore, in the domain of stability, when approaching its boundary, one must use equations of hydrodynamics to compute the evolution (but only in the immediate vicinity of the boundary).

One look at Figure 34 is sufficient to reveal the significance of mechanical instability for the theory of stellar evolution. Let us now proceed to a more detailed description of this picture.

10.3 ANALYTIC THEORY OF POLYTROPIC GAS SPHERES (LANE-EMDEN THEORY)

The Newtonian theory of hydrostatic equilibrium in the particular case

$$P = K\rho^\gamma = K\rho^{1+1/n} , \qquad (10.3.1)$$

where $n \equiv 1/(\gamma - 1)$, is quite simple and is of considerable importance for astrophysics (Emden 1907; the essential conclusions of the theory have been described by Krat 1950). There is a far-reaching mathematical similarity between bodies with given values of n but different total masses M and constants K. The hydrostatic structure depends, as we shall see, on a single dimensionless function of one dimensionless variable; for example,

$$\rho/\rho_c = \psi(m/M) , \qquad (10.3.2)$$

where M is the total mass of the star, ρ_c is the central density, and ρ is the density of the shell inside which a mass m is contained. The function ψ depends only on the polytropic index n.

The early authors were impressed by the fact that a diatomic gas has $\gamma = \frac{7}{5}, n = \frac{5}{2}$, and that a monatomic ideal gas has $\gamma = \frac{5}{3}, n = \frac{3}{2}$, at a given, radially independent entropy $S = $ const. The modern viewpoint is that the polytropic law is never realized exactly, but that polytropic theory still gives good approximations in the absence of accurate numerical computer calculations. The polytropic theory also yields insight into some qualitative features of stellar theory. Even a highbrow general relativist should know the basic elements of this theory.

To exhibit the relationship between polytropic models of different M and K, we introduce dimensionless variables θ and ξ. By tradition θ is related to the density and pressure by

$$\rho = \lambda\theta^n , \quad P = K\rho^{1+1/n} = K\lambda^{1+1/n}\theta^{n+1} , \qquad (10.3.3)$$

where λ is the central density, $\rho_c = \lambda$, so that $\theta = 1$ at the center of the star. The variable ξ is related to the radial coordinate r by

$$r = a\xi ; \quad a = \left[\frac{(n+1)K}{4\pi G}\lambda^{1/n-1}\right]^{1/2} . \qquad (10.3.4)$$

In terms of ordinary variables the equations of hydrostatic equilibrium read

$$\frac{dP}{dr} = -\frac{Gm(r)}{r^2}\rho , \quad \frac{dm}{dr} = 4\pi r^2 \rho , \quad \frac{1}{r^2}\frac{d}{dr}\left(\frac{r^2}{\rho}\frac{dP}{dr}\right) = -4\pi G\rho . \quad (10.3.5)$$

When translated into the new variables (eqs. [10.3.3]–[10.3.5]), these equations take the form

$$\frac{1}{\xi^2}\frac{d}{d\xi}\xi^2\frac{d\theta}{d\xi} \equiv \Delta_\xi \theta = -\theta^n ; \quad (10.3.6)$$

this is the so-called Lane-Emden equation. As noted above, the constants a and λ are so adjusted that the boundary condition at the star's center is

$$\theta = 1 , \quad d\theta/d\xi = 0 \text{ at } \xi = 0 . \quad (10.3.7)$$

Equations (10.3.6) and (10.3.7) taken together are sufficient to carry the integration, for a given n, from the center of the star to its surface. The

TABLE 9

POLYTROPIC PARAMETERS

n	ξ_1	μ_1	ρ_c/ρ	H_1
0	2.45	4.30	1.00	0.817
0.5	2.75	3.79	1.84	0.643
1.0	3.14	3.14	3.29	0.554
1.5	3.65	2.71	5.99	0.488
2.0	4.35	2.41	11.4	0.439
2.5	5.36	2.19	23.4	0.396
3	6.90	2.02	54.2	0.364
4	14.98	1.80	622	0.315
5	∞	1.73	∞	0.270

* SOURCE.—From Chandrasekhar (1939).

solution obtained is a decreasing function, $\theta_n(\xi)$ of radius, ξ. At a particular value of ξ the function θ drops to zero: $\theta_n(\xi_1) = 0$; $\xi_1 = \xi_1(n)$. This value of ξ corresponds, of course, to the surface of the star.

An important quantity is the integral

$$\int_0^{\xi_1} \theta^n \xi^2 d\xi$$

From equation (10.3.6) we see that

$$\mu_1 \equiv -\xi_1^2 \left(\frac{d\theta_n}{d\xi}\right)_{\xi=\xi_1} = \int_0^{\xi_1} \theta^n \xi^2 d\xi . \quad (10.3.8)$$

The values of ξ_1, of the integral (10.3.8), and of several other quantities are given for various values of n in Table 9. The transformation back to physical variables has the form

$$r/R = \xi/\xi_1 ; \quad \rho/\rho_c = \theta^n ; \quad \langle\rho\rangle/\rho_c = 3\mu_1/\xi_1^3 . \quad (10.3.9)$$

Important conclusions can be drawn from these formulae even without any knowledge of the numerical values of μ_1, ξ_1, etc.

Let the mass M of a given model and the properties of its gas (n, K) be given, and ask what its density and radius must be in order for it to be in equilibrium. By combining expression (10.3.9) for M with Eq. (10.3.4) you will find

$$\rho_c = \lambda = \left\{ \frac{M}{\mu_1(n)} \left[\frac{4\pi G^3}{K^3(n+1)^3} \right]^{1/2} \right\}^{2n/(3-n)} \qquad (10.3.10)$$

That the index $n = 3$, $\gamma = 1 + 1/n = \frac{4}{3}$ is critical one can see directly from this equation: for $n < 3$, $\gamma > \frac{4}{3}$, ρ_c grows with increasing M, which is the normal behavior for a stable star in equilibrium. When additional mass is added to its surface, the star is compressed, and its density and pressure increase so that there is a restoring force to support the added mass. The reader should go through the similar argument for $n > 3$, $\gamma < \frac{4}{3}$ and obtain a feeling for the instability in that case.

By combining the above equations, we obtain the following expression for the radius of the star as a function of M, K, n:

$$R = M^{-(n-1)/(3-n)} K^{(3-7n)/[2(3-n)]} G^{(7n-3)/[2(3-n)]}$$

$$\times (4\pi)^{3(n+1)/[2(3-n)]} (n+1)^{(3-7n)/[2(3-n)]} \xi_1(n) [\mu_1(n)]^{-2n/(3-n)} . \qquad (10.3.11)$$

Expressions for the internal energy, the gravitational energy, and the total energy of a polytropic star can also be given. Their derivation involves the virial theorem; we omit that derivation and give only the definitions and results (for a detailed derivation see the appendix to § 10.5, p. 319):

$$E_{\text{int}} = \int \frac{1}{\gamma - 1} \frac{P}{\rho} dm = nMK\rho_c{}^{n-1}\mu_1{}^{-1} \int \theta^{n+1}\xi^2 d\xi = \frac{GM^2}{R} \frac{n}{5-n} ,$$

$$E_{\text{grav}} = -G \int \frac{m \, dm}{r} = -\frac{GM^2}{R} \frac{3}{5-n} , \qquad (10.3.12)$$

$$E_{\text{tot}} = E_{\text{int}} + E_{\text{grav}} = -\frac{GM^2}{R} \frac{3-n}{5-n} .$$

The formulae given above confirm and make more precise the considerations presented in the last part of § 10.1—considerations which, there, were derived from a simplified approach.

A useful expression for the central pressure of a polytropic star is

$$P_c = H_1 GM^{2/3} \rho_c{}^{4/3} \qquad (10.3.13)$$

where H_1 is a function of the index n, and is given in the table. Notice that the dependence of central pressure on the mass M and central density ρ_c is independent of n.

Three cases in which the Lane-Emden equation can be integrated in terms of elementary functions should be mentioned:

1) The case $n = 0$ corresponds to an incompressible liquid

$$\left.\begin{array}{l} \rho = \rho_c = \text{const.} , \qquad p = \text{const.} , \\ \theta_0 = P_c(1 - \tfrac{1}{6}\xi^2) \end{array}\right\} . \tag{10.3.14}$$

2) The case $n = 1$, i.e., $P = K\rho^2$, makes the Lane-Emden equation linear; the solution is

$$\theta_1 = \frac{1}{\xi} \sin \xi ; \qquad \xi_1 = \pi . \tag{10.3.15}$$

In this case the radius R is independent of the mass M, as one sees from equation (10.3.11).

3) For $n = 5$, the solution is

$$\theta_5 = (1 + \tfrac{1}{3}\xi^2)^{-1/2} , \tag{10.3.16}$$

which does not vanish for any finite ξ. Consequently, the radius R for $n = 5$ is infinite. The peculiarity of the case $n = 5$ is also seen from the denominator $(5 - n)$ in equations (10.3.12).

The case $n = 3$ does not have an analytic solution in terms of known functions, but it is singular for the formulae giving ρ_c and R as functions of the mass M. In this case, for a given K there is a solution only for a single, particular value of the mass $M = M_3$. For this M, ρ_c and R can have any values whatsoever, but they are related by

$$\rho_c \frac{4\pi}{3} R^3(\langle \rho \rangle / \rho_c)_3 = M_3 . \tag{10.3.17}$$

The function $\psi(m/M)$ (see eq. [10.3.2]) is presented for the important cases $\gamma = \tfrac{4}{3}$ and $\gamma = \tfrac{5}{3}$ in Figure 35.

10.4 THE ADIABATIC AND POLYTROPIC INDICES

In the last section we assumed an equation of state of the form $P = K(S)\rho^{1+1/n}$, for which the energy is $E = nK(S)\rho^{1/n}$ and the entropy per baryon is constant throughout the star. One can imagine instead a star with variable entropy, with a power law for the adiabatic pressure-density relation, $P = K(S)\rho^{\gamma_1}$, and with a distribution of S carefully designed so that $K(S) = K_2\rho^{\gamma_2-\gamma_1}$ throughout the star. In this case the pressure and density in different places are linked by the polytropic power law

$$P = K_2\rho^{\gamma_2} . \tag{10.4.1}$$

In this case the two indices γ_1 and γ_2 should be distinguished from each other. The index γ_1 gives the dependence of P on ρ when a sample of matter is compressed and expanded without inflow or outflow of heat. The other index, γ_2, characterizes the distribution of pressure and density with radius in the star. Clearly, the radial dependence of ρ on r depends on γ_2.

The first index is important for stability considerations. It is $\gamma_1 = \frac{4}{3}$ which is critical for the stability against radial perturbations. The Emden solutions for polytropes with $\gamma_2 < \frac{4}{3}$ (i.e., $n > 3$) are meaningful so long as the adiabatic index γ_1 is greater than $\frac{4}{3}$. Obviously, whenever γ_2 and γ_1 differ, the entropy is variable. In the case $\gamma_1 \neq \gamma_2$, $S \neq$ const., formulae (10.3.12) must be revised; they become

$$E_{\text{grav}} = -\frac{GM^2}{R}\frac{3}{5 - n_2}, \qquad E_{\text{int}} = -E_{\text{grav}}\frac{n_1}{3} = \frac{GM^2}{R}\frac{n_1}{5 - n_2},$$

$$E_{\text{tot}} = -\frac{GM^2}{R}\frac{3 - n_1}{5 - n_2}. \tag{10.4.2}$$

The adiabatic index depends upon the properties of the gas, as has already been explained in the chapter about equations of state: it is $\gamma_1 = \frac{5}{3}$ for a nonrelativistic gas, degenerate $(S = 0)$ or nondegenerate $(S \neq 0)$; it is $\gamma_1 = \frac{4}{3}$ for a relativistic gas, including both the case of a relativistic, degenerate electron gas and the case of a gas dominated by radiation pressure.

The polytropic index depends upon the distribution of entropy with radius in the star. That distribution is governed by the processes of heat flow and nuclear energy generation. The pressure and density distribution can follow an exact polytropic power law only by chance. However, in a variety of situations they can be close to a polytropic law.

There are important restrictions on the possible values of γ_1 and γ_2. As we have discussed already, the condition for stability against compression or expansion of the star is $\gamma_1 > \frac{4}{3}$. This same restriction follows from the requirement that the total energy be negative—i.e., that the star be gravitationally bound.

It can be shown that only stars with $\gamma_1 > \gamma_2$ are stable against convective motions. Obviously, in this case $K(S)$ and S increase with increasing radius; K and S have their minimum values at the star's center. Roughly speaking, this situation corresponds to a light gas in the stellar atmosphere, lying on top of a heavy gas in its core. In the opposite case, $\gamma_1 < \gamma_2$, the entropy is a maximum at the center.

To evaluate the stability, one must not merely compare the actual density ρ in different layers of the star—such a comparison is sufficient only for an incompressible fluid. For a gas, we must instead compare the density of the matter at a particular radius (r_1) with the density that a sample of gas would have if it were moved from its initial, equilibrium position (r_2) to the radius r_1, and were expanded adiabatically to the pressure P_1 during the move. This comparison of densities yields the well-known Schwarzschild criterion for instability against convection, $dS/dr < 0$.

In the formal theory of small perturbations $dS/dr < 0$ causes the equilibrium configuration to be unstable against nonradial perturbations in which the material sinks downward in one part of the star and rises toward the surface in another. For a given entropy distribution this mixing process will end when the matter with minimal S is at the center and $\gamma_1 > \gamma_2$. The reasoning is easily generalized to the case in which not only the entropy but also the chemical composition is nonuniform.

If heat flow by conduction or radiative diffusion generates a temperature distribution corresponding to $dS/dr < 0$ in some layers of a star, convection will set in and will replace the normal heat-flow mechanism, causing $dS/dr = 0$. Therefore in convective stars one expects to have $\gamma_1 = \gamma_2$.

We conclude with the warning that the theory of polytropic gas spheres is an extreme idealization. The actual nature of Newtonian nonrelativistic stars is treated comprehensively in many books (see, e.g., Schwarzschild 1958).

10.5 THE ENERGY APPROACH TO THE THEORY OF EQUILIB- RIUM FOR A STAR CONSISTING OF MATTER WITH γ CLOSE TO $\frac{4}{3}$

The discussion of the relationship between the energy of a star and one parameter (mean or central density, or radius) given in §§ 10.1 and 10.2 is only an approximate, illustrative one because in reality one must consider the energy not as a function of one parameter but as a functional, $F[\rho(m)]$. However, there is a rather important case in which the energy approach with a single parameter becomes asymptotically precise. This is the case of matter which has an adiabatic index close to $\gamma = \frac{4}{3}$ (polytropic index $n = 3$) throughout the star. This case is important because, as we saw above, it is the value of $\gamma = \frac{4}{3}$ that is critical for the transition from stability to instability. There exist specific examples where the equation of state of matter in stars has an adiabatic index close to $\frac{4}{3}$. One such example is the white dwarfs. In the future, we will be greatly concerned with the parameters of critical states. Therefore, the case $(\gamma - \frac{4}{3}) \ll 1$ will be investigated in considerable detail. In this investigation the solution of the Newtonian equation of hydrostatic equilibrium for a state with $\gamma = \frac{4}{3}$ will be used as a zero-order approximation.

For each $\gamma = $ const. there exists (as we have seen in § 10.3) a unique shape of the stellar density distribution, i.e., a certain relationship between the dimensionless density (expressed in units of the central density) and the dimensionless mass, m/M:

$$\rho = \rho_c \psi(m/M) .$$

We will be concerned with ψ for $\gamma = \frac{4}{3} (n = 3)$. This function is shown in Figure 35. Deviations of the thermodynamic equation of state from the corresponding $n = 3$ equation of state (i.e., from $P = b(S)\rho^{4/3}$) can be considered as small corrections.

Let us make one more essential remark. Effects of the general theory of relativity become decisive when the gravitational potential ϕ becomes of the order of c^2. For this it is necessary that the dimension of a body, R, be comparable to r_g. It might seem that in all cases in which $R \gg r_g$, the effects of GTR cannot have a qualitative influence upon the structure of a celestial body and upon its evolution. However, for a star with $(\gamma - \frac{4}{3}) \ll 1$, which

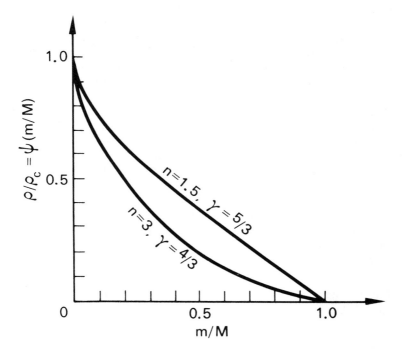

FIG. 35.—The Emden function $\rho/\rho_c = \psi(m/M)$ for the polytropic index $n = 3$ ($\gamma = \frac{4}{3}$), curve $n = 1.5$ ($\gamma = \frac{5}{3}$) is plotted for comparison. The asymptotic behavior of the curves is $d\rho/dm \to \infty$ when $m/M \to 1$ (although it does not show up on the coarse scale of this figure).

is at the boundary of stable equilibrium, this is not true. In this case, a small correction for GTR is sufficient to change the stability of the equilibrium.

According to an analysis by Kaplan (1949b) (see also Kaplan and Lupanov 1965; Fowler 1964a, b; Chandrasekhar 1964a, b, 1965; and Zel'dovich and Novikov 1965), in this case even small effects of GTR result in qualitative changes of the picture. The effects of GTR can be construed as corrections to the Newtonian theory with $n = 3$ taken as the zero-order approximation. All corrections for the deviation of the equation of state from $P = b(S)\rho^{4/3}$

as well as for GTR are computed by using the $n = 3$ polytropic distribution, and hence are functions of one parameter—the central density ρ_c.

Corrections in the equation of state and corrections due to GTR of the order of $a = \Delta E_1/E_1 \ll 1$ cause changes in the density function ψ which are of the same order, a. However, because of the extremal properties of ψ as a solution in the zero approximation, any variation in ψ of the order of a causes a variation in the energy of the order of a^2. (The first variational derivative of the total energy with respect to the function ψ is zero.)

Therefore, a computation of the corrections that uses the unperturbed Emden function ψ for the density distribution gives an exactly correct first term (of the order of a) for the expansion of the energy in powers of a.

In this sense, we can discuss the asymptotically precise theory (with an error of order a^2) of stellar equilibrium when $(\gamma - \frac{4}{3}) \sim a$.

The further removed we become from the critical state $\gamma = \frac{4}{3}$, the less precise (quantitatively) the expressions obtained become. Qualitatively, all conclusions of the single-parameter theory (with the zero-order approximation of $\gamma = \frac{4}{3}$) remain valid, and even the quantitative estimates do not change substantially. As an illustration, the Emden function ψ for $\gamma = \frac{5}{3}$ is shown in Figure 35 (equation of state of a nondegenerate, monatomic, ideal gas). Note that it does not differ too much from ψ for $\gamma = \frac{4}{3}$.

As emphasized before, for our purposes critical states are especially important, and for states far from critical, approximate estimates are quite adequate. For this reason we will use the single-parameter method in the future.

10.6 RELATIVISTIC EQUATIONS OF STELLAR EQUILIBRIUM

Before going on, we must formulate the GTR equations for stellar equilibrium. We shall consider a spherically symmetric distribution of mass in hydrostatic equilibrium, assuming that the effects of rotation and ordered magnetic fields are negligible. The star will be regarded as in an absolutely static state; in reality, of course, any star slowly evolves; but the velocities and accelerations of all particles are small, as is the flux of heat through the star. (Estimates of these quantities will be given later.) Assuming that there are no other bodies in the neighborhood, we shall find the static, spherically symmetric solution for the star's gravitational field—i.e., for its metric.

We shall look for a metric of the form

$$ds^2 = e^\nu c^2 dt^2 - e^\lambda dr^2 - r^2(d\theta^2 + \sin^2 \theta \, d\phi^2) \, . \tag{10.6.1}$$

For a static, equilibrium situation we can put

$$\lambda_{,t} = \lambda_{,tt} = \nu_{,t} = T_0{}^1 = 0 \, . \tag{10.6.2}$$

Since the star's fluid is at rest in our coordinate system, Pascal's law is valid:

$$T_1{}^1 = T_2{}^2 = T_3{}^3 = -P \, , \qquad T_0{}^0 = \epsilon = \rho c^2 \, . \tag{10.6.3}$$

Inserting equations (10.6.1)–(10.6.3) into the Einstein field equations, we obtain

$$\frac{\kappa P}{c^2} = e^{-\lambda}\left(\frac{\nu'}{r} + \frac{1}{r^2}\right) - \frac{1}{r^2}, \tag{10.6.4}$$

$$\frac{\kappa P}{c^2} = \tfrac{1}{2}e^{-\lambda}\left(\nu'' + \frac{\nu'^2}{2} + \frac{\nu' - \lambda'}{r} - \frac{\nu'\lambda'}{2}\right), \tag{10.6.5}$$

$$\frac{\kappa\epsilon}{c^2} = -e^{-\lambda}\left(\frac{1}{r^2} - \frac{\lambda'}{r}\right) + \frac{1}{r^2}. \tag{10.6.6}$$

Equation (10.6.6) can be integrated independently of the others. Defining the quantity m by

$$m(r) = 4\pi \int_0^r \rho r^2 dr, \qquad m' = 4\pi\epsilon c^{-2}r^2, \tag{10.6.7}$$

and using the boundary condition $\lambda(0) = 0$, we obtain

$$e^{-\lambda} = 1 - \frac{\kappa m(r)}{4\pi c^2 r} = 1 - \frac{2Gm(r)}{c^2 r}. \tag{10.6.8}[6]$$

This equation is equally valid inside the star $(r < R;\ \rho > 0)$ and outside the star. Obviously, at any point r_1 outside the star,

$$m(r_1) = 4\pi \int_0^{r_1} \rho r^2 dr = 4\pi \int_0^R \rho r^2 dr = m(R) = M, \tag{10.6.9}$$

where R is the star's radius. One can see from the expression for the metric at large R that M is the star's total mass—i.e., it is the mass which governs the Keplerian motions of distant planets.

By using Eq. (10.6.4) we can express ν' in terms of P, λ, r; then by differentiating that relation with respect to r we can obtain an expression for ν'' in terms of P', P, λ, λ', and r. By combining this equation with relations (10.6.4)–(10.6.6) and (10.6.8), we can eliminate ν'', ν', λ', and λ, and thereby obtain an equation for dP/dr in terms of P, ρ, m, r:

$$\frac{dP}{dr} = -\frac{G(\rho + P/c^2)(m + 4\pi Pr^3/c^2)}{r^2(1 - 2Gm/c^2 r)}. \tag{10.6.10}[7]$$

This equation is the GTR generalization of the nonrelativistic equation of hydrostatic equilibrium:

$$\frac{dP}{dr} = -\frac{Gm}{r^2}\rho. \tag{10.6.11}$$

Our derivation of the equation of hydrostatic equilibrium (10.6.10) from the Einstein field equations for the metric is a particular case of the fact that the field equations themselves contain the equations of motion of the

6. For a discussion of the singular solutions with $\lambda(0) \neq 0$, see note 6 in § 3.3.

7. This equation can alternatively be obtained directly from relation (1.8.8a).

matter: By assuming that the metric does not change with time, we obtained a condition on the pressure gradient which guarantees that the matter will be at rest, in hydrostatic equilibrium.

In practice, a stellar model for a given equation of state, $P = P(\rho)$, can be constructed by integrating equation (10.6.10) numerically together with the definition (10.6.7) of $m(r)$. It is convenient to begin in the star's center by giving P_c and ρ_c; for small r the solution will always be regular, $m = \frac{4}{3}\pi\rho_c r^3$, $P = P_c - \text{const.}\ r^2$. The numerical integration will lead eventually to $P = \rho = 0$ for some particular radius $r \equiv R$. Just as in the nonrelativistic case, we can obtain the family of all possible stellar models for our given equation of state by varying the central pressure P_c from 0 to ∞. However, we cannot know beforehand the masses for which solutions will be obtained.

As is explained in detail in Part II, one must distinguish between the total density of mass energy, ρ, which includes all forms of internal energy as well as rest mass, and the quantity $\rho_0 = n/A$, where n is the number density of baryons and A is Avogadro's number. We shall call ρ_0 the rest-mass density.

The proper volume of a small fluid element and the volume of a spherical layer in the star are given by (cf. eq. [10.6.1])

$$dV = e^{\lambda/2}r^2 dr\ d\cos\theta\ d\phi\ ; \quad dV = 4\pi e^{\lambda/2}r^2 dr\ . \tag{10.6.12}$$

Consequently, the total number of baryons in the star is

$$N = \int n\,dV = 4\pi \int_0^R n(r)e^{\lambda/2}r^2 dr\ . \tag{10.6.13}$$

In place of N it is conventional to deal with a quantity M_0, which is proportional to N but has the dimensions of a mass:

$$M_0 \equiv N/A = \int \rho_0 dV = 4\pi \int_0^R \rho_0(r)e^{\lambda/2}r^2 dr\ . \tag{10.6.14}$$

The quantities P, ρ, and ρ_0 are linked by the first law of thermodynamics (when the entropy is held constant, e.g., for cold matter, $S = 0$):

$$d(\rho/\rho_0) = (1/c^2)P d(1/\rho_0)\ , \quad \rho/\rho_0 = 1 \text{ for } \rho_0 = 0\ ;$$
$$\rho = \rho_0(1 + E/c^2)\ , \quad E = \int (P/\rho_0^2)d\rho_0\ . \tag{10.6.15}$$

From this equation we see that, for any star, M_0 is the mass which the star's baryons would have if they were all dispersed throughout a volume so large that all types of interactions between them could be neglected, and if in the process of dispersal they were put into the form of the most tightly bound nucleus (^{56}Fe) together with its electrons. The quantity

$$\mathfrak{E} = (M - M_0)c^2 \tag{10.6.16}$$

is the star's total energy with its rest mass-energy subtracted off; clearly, $\mathfrak{E}$ must be negative for a stable star. The energy released when a star is constructed from matter of initially small density is equal to $\mathfrak{E}c^2$.

10.6 Relativistic Equations of Stellar Equilibrium

The quantity ν plays the role of a gravitational potential. The redshift of a photon emitted at a point r and received at ∞ is determined by $\nu(r)$. The derivatives ν' and ν'' appear in the Einstein field equations (10.6.4)–(10.6.6), but ν itself does not appear. Hence, we can freely add a constant to ν; this corresponds to renormalizing the time coordinate t. It is conventional to normalize ν in the same manner as we normalize the Newtonian gravitational potential: $\nu(\infty) = 0$. This corresponds to setting coordinate time equal to proper time at $r = \infty$.

In the general case when the entropy is not constant throughout the star, the equation of hydrostatic equilibrium (10.6.11) is equivalent to a variational principle in which one extremizes the total energy (mass) while holding baryon number and the distribution of entropy per baryon fixed. Mathematically, the integral (10.6.9) must be extremized with the constraint of the constancy of the integral (10.6.13); and in the extremization process the density must be expressed as $\rho = \rho(n, S)$, where S is given as a function of the number of baryons inside a given shell, F:

$$F \equiv \int n \, dV = 4\pi \int_0^r n e^{\lambda/2} r^2 dr \; ; \quad S = S(F) \; . \qquad (10.6.17)$$

The quantity $\lambda(r)$ is absent from the integral (10.6.9). But when the radial distribution of density is being varied, it is nevertheless necessary to take into account the variation of λ because λ appears in the constraint integral (10.6.13).

In performing the variation it is most convenient to take the number of baryons inside a given shell, F, as the integration variable. In effect, F is a "Lagrangian radius" of the shell of matter. We must then regard $r(F)$ as an unknown function, and write

$$n(F) = \frac{e^{-\lambda/2}}{4\pi r^2} (dr/dF)^{-1} \; , \qquad (10.6.18a)$$

$$m(F) = \int_0^F \rho e^{-\lambda/2} dV = \int_0^F \rho e^{-\lambda/2} (dF/n), \quad M = m(N) \; , \quad (10.6.18b)$$

$$e^{-\lambda} = 1 - \frac{2Gm(F)}{c^2 r(F)} \; . \qquad (10.6.18c)$$

Here, $\rho = \rho(n, S)$; $n(F)$ is given by expression (10.6.18a); and $S(F)$ is specified as an initial condition for the problem. From equation (10.6.18b) it is evident that, when $r(F)$ is varied,

$$r(F) \to r(F) + \delta r(F) \; ,$$

the variation of $\lambda(F)$ makes a contribution to the variation of the mass. By performing the extremization of M (eq. [10.6.18b]), we obtain as our Euler-Lagrange equation the equation of hydrostatic equilibrium (10.6.10).

Strictly speaking, to obtain an equilibrium state we need demand only that M be stationary as a functional of $r(F)$ (extremization means that the first functional derivative is zero). If, in addition, M is a minimum, i.e., its second functional derivative is positive, the equilibrium will be stable.[8]

In the important case of radially constant entropy per baryon, the solutions of the equation of hydrostatic equilibrium have physically obvious properties. In this case we can write $P = P(\rho_0)$ and $\rho = \rho(\rho_0)$, and we can introduce the chemical potential of the matter as a single-valued function of its rest-mass density:

$$\mu = c^2 \left(\frac{\partial \rho}{\partial \rho_0}\right)_s = c^2 \frac{\rho}{\rho_0} + \frac{P}{\rho_0} = c^2 + E + PV . \quad (10.6.19)$$

The chemical potential can be defined as the amount by which the energy of a system increases when one unit of rest mass is added to it at constant entropy per unit rest mass. The thermodynamic relation between the energy E per unit rest mass and the volume V per unit rest mass, $dE = -PdV$, is responsible for the expression for μ corresponding to classical thermodynamics. The additional term, c^2, in expression (10.6.19) is needed because μ includes the rest mass, while E is defined with the rest mass subtracted off.

It follows from the equation of hydrostatic equilibrium for a star with constant entropy that

$$\mu e^{\nu/2} = \text{const.} = c^2 e^{\nu(R)/2} = c^2 e^{-\lambda(R)/2} = c^2(1 - 2GM/Rc^2)^{1/2} . \quad (10.6.20)$$

The right-hand side is obtained by inserting the values at the surface of the star:

$$r = R , \quad \nu = \nu(R) = -\lambda(R) , \quad P = 0 ,$$
$$E = 0 , \quad \rho = \rho_0 , \quad \mu = c^2 . \quad (10.6.21)$$

Let us apply the variational principle to a system which consists of an equilibrium star (with mass M and rest mass M_0) plus an amount of matter δM_0, which is being added to the star. Since only a small amount of matter is added to the star, the star will be in near-equilibrium after the addition. The variational principle guarantees that the total mass $M' = M + \delta M$ after the addition will not change when its matter distribution shifts slightly; i.e., it must be precisely the same, independently of where the matter is added. When the matter is added at the surface,

$$\delta M_0 = \rho_0 \delta V \, e^{\lambda(R)/2} ,$$
$$\delta M = \rho \, \delta V = \rho_0 \delta V = \delta M_0 \, e^{-\lambda(R)/2} = \delta M_0 (1 - 2GM/Rc^2)^{1/2} . \quad (10.6.22)$$

If the matter is added not at the surface, but to a layer at radius r, then one must take account not only of the change in $\rho(r)$ but also of the change in $\lambda(r')$ for $r < r' < R$. By taking account of equations (10.6.18a, b, c) and equation (10.6.20) which follows from them, one can discover that the

---

8. For a mathematical treatment of this stability criterion, see Appendix B of Harrison, Thorne, Wakano, and Wheeler (1965).

change in mass δM is, indeed, independent of r; it is always given by expression (10.6.22).

Let us now examine the nonrelativistic limits of our relativistic formulae. The equation of hydrostatic equilibrium (10.6.10) takes on its nonrelativistic form (10.6.11) when we omit the term $r^3 P/c^2$ by comparison with $m(r)$, and the term Gm/c^2 by comparison with r; in this case $m(r)$ has the same meaning as in the nonrelativistic theory. As for the other equations: if we limit ourselves to the first terms in a series expansion of equations (10.6.4)–(10.6.6) for small ν and λ, we obtain

$$\frac{\nu'}{r} - \frac{\lambda}{r^2} = \frac{\kappa}{c^2} P , \qquad (10.6.23)$$

$$\tfrac{1}{2}\nu'' + \tfrac{1}{2}\frac{\nu'}{r} - \tfrac{1}{2}\frac{\lambda'}{r} = \frac{\kappa}{c^2} P , \qquad (10.6.24)$$

$$\frac{\lambda'}{r} + \frac{\lambda}{r} = -\frac{\kappa}{c^2}\epsilon . \qquad (10.6.25)$$

By adding equations (10.6.23), (10.6.25), and twice (10.6.24) we obtain

$$\nu'' + 2\frac{\nu'}{r} = \frac{\kappa}{c^2}(\epsilon + 3P) . \qquad (10.6.26)$$

The left-hand side is the Laplacian $\nabla^2\nu$ for $\nu = \nu(r)$, so, since $P \ll \rho c^2$,

$$\nabla^2\nu = \frac{8\pi G}{c^4}(\rho c^2 + 3P) = \frac{2}{c^2}4\pi G\rho . \qquad (10.6.27)$$

Hence, we see that the Newtonian gravitational potential and ν are connected (for $\nu \ll 1$) by the formula

$$\phi(r) = \frac{c^2}{2}\nu(r) . \qquad (10.6.28)$$

We show, finally, how the Newtonian expression for the star's energy can be derived from the formulae of GTR. We must calculate the quantity $\mathfrak{E}$ defined in equation (10.6.16):

$$\mathfrak{E} = (M - M_0)c^2 = c^2 \smallint (\rho e^{-\lambda/2} - \rho_0)dV , \qquad (10.6.29)$$

where $e^{-\lambda/2}dV$ is used in place of $4\pi r^2 dr$, and where we have made use of expression (10.6.9). By writing

$$\rho = \rho_0(1 + E/c^2) ; \qquad e^{-\lambda/2} = (1 - 2Gm/rc^2)^{1/2} \approx 1 - Gm(r)/rc^2$$

and by neglecting all second-order and higher quantities, we obtain

$$\mathfrak{E} = c^2 \smallint \left(\rho_0 + \rho_0\frac{E}{c^2} - \rho_0\frac{Gm}{rc^2} - \rho_0 \right) dV$$

$$= \smallint E dm - G\smallint \frac{m\,dm}{r} , \qquad (10.6.30)$$

where $dm = \rho_0 dV$.

The Equilibrium and Stability of Stars

The difference between GTR and Newtonian theory arises in the next order. The relativistic correction terms contain c^2 in their denominators; the calculation of them is more difficult than the above, and will be delayed until later.

In concluding this section, let us discuss the condition of thermodynamic equilibrium in GTR at nonzero temperatures.

It is most convenient to study the equilibrium conditions of thermodynamics in an external gravitational field. The equilibrium conditions can be formulated with the help of variational principles: the maximization of the entropy for a given energy and given number of conserved particles; or the minimization of the energy for a given entropy and given number of particles. In any situation, in order to verify that equilibrium is actually achieved, one can mentally transfer any (small) number of particles or small quantity of energy from one place to another, and calculate the corresponding energy or entropy change.

In studying the thermodynamic equilibrium in an external field, one need not take account of any change in the gravitational field when transferring the energy or particles. Moreover, thermodynamic-equilibrium considerations in an external field are equivalent to the same considerations in the matter's own field, because only first variations are considered. However, there is no such equivalence for the second variation—i.e., for the second-order changes in energy and entropy, which involve terms that are proportional to the square of the variations in thermodynamic variables. And it is the second variation which determines the stability of the equilibrium! Thus, in studying the stability one must take into account that in the case of a star the matter is in its own gravitational field, not in an external field; and one must include the changes in the gravitational field when studying perturbations in the mass and energy distribution.

However, we deal here only with the simple problem of equilibrium; and hence, we can idealize the matter as being in an external gravitational field.

The energy of a particle E_0, as measured by a distant observer, can be expressed in terms of its locally measured energy E by

$$E_0 = E(-g_{00})^{1/2} = E\, e^{\nu/2} . \qquad (10.6.31)$$

The quantity conserved during free-particle motion is just E_0; so we ask that the sum (integral) of E_0 over all particles (over the entire body) be an extremum. If the number of particles in any fixed region of space changes, then $\partial(En)/\partial n\,|_s = \mu$ is just the definition of the chemical potential there. Consequently, the extremization condition

$$\delta\,\mathfrak{E}_0 = \delta \int E_0 n\, dV = \int \frac{\partial(En)}{\partial n}\, e^{\nu/2} \delta n\, dV = \int \mu e^{\nu/2} \delta n\, dV$$

262

together with the additional constraint

$$N = \int n dV = \text{const.} \,, \quad \delta N = \int \delta n \, dV = 0 \,,$$

gives the result

$$\delta\mathfrak{E}_0 - \Lambda\delta N = \int (\mu e^{\nu/2} - \Lambda)\delta n \, dV = 0 \,,$$

$$\mu e^{\nu/2} = \Lambda = \text{const.} \,,$$

where Λ is a Lagrange multiplier. This relation was derived earlier in this section in a more complicated manner.

Let us next consider energy transport from one region to another without any change in the density of particles. For a local observer the formula $dE = TdS$ is always valid, where dS is the change in entropy per baryon and T is the temperature which the local observer measures. It is essential that in any energy exchange the total energy be conserved. Consequently, the equilibrium condition gives

$$\delta\int E_0 n dV = \delta\int E e^{\nu/2} n dV = 0 \,, \quad \delta\int Sn dV = 0 \,;$$

$$\int (e^{\nu/2}\delta E - \Lambda dS)n dV = \int \left(e^{\nu/2}\frac{\partial E}{\partial S} - \Lambda \right)\delta Sn dV$$

$$= \int (e^{\nu/2}T - \Lambda)\delta Sn dV = 0 \,.$$

It follows that the condition for thermal equilibrium is

$$e^{\nu/2}T = T(g_{00})^{1/2} = \text{const.} \tag{10.6.32}$$

Note the contrast with the nonrelativistic condition, $T = \text{const.}$ It is obvious that the two conditions agree in the limit where g_{00} is nearly unity everywhere. The relativistic expression for the heat flux due to thermal conductivity will similarly contain the term $\nabla(T\sqrt{g_{00}})$ instead of ∇T. For further discussion see, e.g., chapter 3 of Thorne (1967).

How can we understand simply the fact that the condition for thermodynamic equilibrium is not $T = \text{const.}$ in GTR? It can be understood by remembering that the condition for thermal equilibrium is that the heat flux due to thermal conductivity vanishes. The thermal conductivity is proportional to the mean free path of an atom (in a neutral gas), of an electron (in a plasma), or of a photon (in a high-temperature plasma). One must examine how the gravitational field affects the free motion of the particles which carry the heat. The gravitational redshift of the photon frequency corresponds to a temperature change proportional to $g_{00}^{-1/2}$. By examining the motion of collisionless particles with finite rest mass, one finds that the same result holds there also.

10.7 RELATIVISTIC EQUATIONS FOR ROTATING STARS[9]
(BY K. S. THORNE)

The last section dealt with the equilibrium of a precisely spherical, non-rotating, relativistic star. When rotation is taken into account, the theory becomes much more complicated. As we proceed through this book, we shall use a variety of approximation schemes to help us understand the physical effects of rotation (see, e.g., §§ 11.12 and 11.14). But approximation schemes are not always necessary. Much can be said about the exact properties of a fully relativistic, rapidly rotating star. The keys to such a discussion are energy considerations and a variational principle analogous to those of the last section. But before introducing them, we need a metric.

We shall assume that our star's rotation is stationary and axisymmetric—i.e., that the metric coefficients and the thermodynamic variables are indedependent of time t, and of angle ϕ about the rotation axis. If we reverse the direction of time flow, we also reverse the direction of the star's rotation; so the metric will change. But if we reverse the direction of time ($t \to -t$) and then reverse the direction of rotation ($\phi \to -\phi$), we return the star to its original state. Thus (with an appropriate choice of coordinates) the metric must be invariant under $t \to -t$, $\phi \to -\phi$; i.e., $g_{t1} = g_{t2} = g_{\phi 1} = g_{\phi 2} = 0$, so that

$$ds^2 = g_{tt}dt^2 + 2g_{t\phi}dtd\phi + g_{\phi\phi}d\phi^2 + \Sigma_{A,B}g_{AB}dx^A dx^B . \quad (10.7.1)$$

Here capital indices run over the two coordinates x^1 and x^2; and the metric coefficients are functions only of x^1 and x^2. With an appropriate choice of x^1 and x^2, our coordinates become ordinary, flat-space coordinates far from the star:

$$x^1 \to r , \quad x^2 \to \theta ,$$
$$ds^2 \to c^2 dt^2 - dr^2 - r^2(d\theta^2 + \sin^2\theta\, d\phi^2) . \quad (10.7.2)$$

Recall that the nonvanishing function $g_{t\phi}$ is a manifestation of the star's "dragging of inertial frames": It causes a gyroscope in or near the rotating star to precess relative to distant inertial frames—i.e., relative to the "distant stars" (§ 1.10; § 4.3). It also causes the gravitational Zeeman effect (§ 4.10). The other off-diagonal term in the metric, g_{12}, has no physical significance; in fact, it can be removed by an appropriate choice of x^1 and x^2.

One can easily determine the total mass-energy M and angular momentum K of the rotating star by observing how the star curves spacetime far away—i.e., by examining to higher precision than expression (10.7.2) the asymptotic form of the metric:

9. This section is based in part on the author's recent review of the theory of rotating, relativistic stellar models (Thorne 1970), in part on recent papers by Bardeen (1970) and Hartle (1970), and in part on remarks made to the author by Ya. B. Zel'dovich in Moscow, September 1969.

10.7 Relativistic Equations for Rotating Stars

$$ds^2 \rightarrow (1 - 2GM/rc^2)dt^2 + (4GK/c^3r^2)dt\,d\phi$$
$$- [1 + O(GM/rc^2)][dr^2 + r^2d\theta^2 + r^2 \sin^2 \theta \; d\phi^2] \,. \tag{10.7.3}$$

Here $O(GM)/rc^2)$ is a correction term which depends on the precise choice of coordinates. The key points are that (i) as for a nonrotating star, so also here, it is the Newtonian correction, $-2GM/rc^2$, to g_{tt} that reveals the star's total mass; and (ii) as in the weak-field limit (§ 1.10), so also here, it is the off-diagonal $g_{t\phi}$ term that reveals the total angular momentum.

The "fluid" inside the star is assumed to rotate in the ϕ-direction. Thus, its four-velocity must have the form

$$u^t = u^t(x^1, x^2) \; ; \quad u^\phi = u^\phi(x^1, x^2) \; ; \quad u^1 = u^2 = 0 \,. \tag{10.7.4}$$

If a distant observer with X-ray vision watches a particular fluid element at (x^1, x^2), he sees it rotate with an angular velocity of

$$\Omega(x^1, x^2) = d\phi/dt = u^\phi/u^t \,. \tag{10.7.5}$$

(Recall that the proper time measured by his clock is the same as coordinate time t.) By using the normalization condition $u^j u_j = 1$, we can re-express the fluid's four-velocity as

$$u^t = [g_{tt} + 2g_{t\phi}\Omega + g_{\phi\phi}\Omega^2]^{-1} \; ; \quad u^\phi = \Omega u^t \,. \tag{10.7.6}$$

Notice that rigid rotation corresponds to $\Omega(x^1, x^2) = $ constant. In that case the distant observer sees the star's fluid to rotate as a solid body, with the distance between neighboring particles remaining forever constant.

Let us now turn to the energy considerations and variational principle that we promised.

We begin by calculating the change in the star's total mass-energy M when a unit amount of rest mass is added to it. We expect an answer which reduces to equation (10.6.22)—i.e., $\delta M/\delta M_0 = (g_{tt})^{1/2}$—when the star is nonrotating and the mass is added at its surface.

We add the rest mass δM_0 by the following idealized process. An astrophysicist far from the star drops a bundle of rest mass δM_0 down an idealized pipe, which is inserted into the star, to a colleague who rides with the fluid ring at (x^1, x^2). The colleague catches the mass, inserts it into his fluid ring, and then throws any energy that remains back up the pipe to the distant astrophysicist. The astrophysicist then uses conservation of mass-energy to determine the change in the star's mass.

The energy balance for this injection process is this:

1. The astrophysicist is careful to drop the bundle of rest mass δM_0 so that it has zero initial angular momentum and zero initial kinetic energy. Thus, its initial four-momentum is

$$p_t^{(in)} = \delta M_0 c^2 \,, \quad p_\phi^{(in)} = p_1^{(in)} = p_2^{(in)} = 0 \,. \tag{10.7.7}$$

2. As the mass falls (with the pipe carefully shaped so the mass never hits its walls!) it moves along a geodesic of the star's metric (10.7.1). The geodesic equation guarantees that, because the metric coefficients are independent of t and ϕ, the momenta p_t and p_ϕ "conjugate to" these coordinates are conserved. Thus, when the falling mass reaches the colleague on the ring at (x^1, x^2), and he catches it, it has four-momentum components

$$p_t^{(c)} = p_t^{(in)} = \delta M_0 c^2 , \qquad p_\phi^{(c)} = p_\phi^{(in)} = 0 ,$$

$$p_1^{(c)} \neq p_1^{(in)} , \qquad p_2^{(c)} \neq p_2^{(in)} . \qquad (10.7.8)$$

3. The total energy which the colleague catches, as measured by him, including rest mass-energy $\delta M_0 c^2$ and kinetic energy of fall, is the projection of the bundle's four-momentum on the colleague's four-velocity:

$$\delta W^{(c)} = p_j^{(c)} u^j = p_t^{(c)} u^t + p_\phi^{(c)} u^\phi = p_t^{(in)} u^t$$

$$= u^t(x^1, x^2)\delta M_0 c^2 . \qquad (10.7.9)$$

4. Of this total energy caught, our colleague uses up a total amount (including rest mass)

$$\delta W^{(inject)} = \mu(x^1, x^2)\delta M_0 = (\partial \rho / \partial \rho_0)_s \, \delta M_0 c^2$$

$$= \delta M_0(c^2 + E + PV) \qquad (10.7.10)$$

when he injects it into the star. Here $\mu(x^1, x^2)$ is the chemical potential (eq. [10.6.19]). In using this formula we assume that the rest mass is in thermodynamic equilibrium with its surroundings when it is injected—i.e., that it has the same entropy per unit rest mass, s, as its surroundings. Note that $\delta M_0 c^2$ is the rest mass inserted, $\delta M_0 E$ is the internal energy inserted, and $\delta M_0 PV$ is the work that must be done against the pressure P of the surrounding fluid in order to open up a volume $\delta M_0 V$ for the rest mass.

5. After the injection our colleague has left a total mass-energy

$$\delta W^{(left)} = \delta W^{(c)} - \delta W^{(inject)} . \qquad (10.7.11)$$

He throws this mass-energy back up the pipe with zero angular momentum and with kinetic energy that he takes from $\delta W^{(left)}$ itself.

6. By analogy to equation (10.7.9), the energy which the distant astrophysicist finally receives, as measured by him, is

$$\delta W^{(final)} = \delta W^{(left)}/u^t(x^1, x^2) . \qquad (10.7.12)$$

(Note: u^t is a "redshift factor" analogous to $(1 - 2GM/rc^2)^{-1/2}$ for the nonrotating case, which is to be used whenever energy is transferred *with zero angular momentum* between (x^1, x^2) and infinity.)

7. The total mass-energy that has been added to the star is the amount

266

$\delta M_0 c^2$ initially dropped down the pipe, minus the excess amount $\delta W^{(\text{final})}$ received back:

$$\delta M c^2 = \delta M_0 c^2 - \delta W^{(\text{final})}$$

$$= \delta M_0 c^2 - [u^t \delta M_0 c^2 - \mu \delta M_0]/u^t$$

$$= (\mu/u^t)\delta M_0 .$$

Let us focus attention carefully on what has been held fixed during this injection process. We have held fixed the entropy per unit rest mass (i.e., per baryon) of the fluid at (x^1, x^2), and also its *total* angular momentum— not angular momentum per unit rest mass! To make this clearer, let us think of x^1 and x^2 inside the star as Lagrangian coordinates which are attached to rings of rotating fluid; and let us denote by

$$\Delta K = k(x^1, x^2)\Delta x^1 \Delta x^2$$

the total angular momentum in the ring of coordinate cross-section $\Delta x^1 \Delta x^2$. Then it is the entropy per unit rest mass, $s(x^1, x^2)$, and the angular momentum of the ring, $\Delta K(x^1, x^2)$, which are held fixed in the above injection process:

$$(\delta M c^2)_{s,\Delta K} = [\mu(x^1, x^2)/u^t(x^1, x^2)]\delta M_0 . \tag{10.7.13}$$

Notice that, as promised, when the star is nonrotating and the mass is added at its surface $[\mu(x^1, x^2) = c^2, u^t = (g_{tt})^{-1/2} = (1 - 2GM/c^2R)^{-1/2}]$, this formula reduces to equation (10.6.22).

Suppose that instead of adding rest mass to the fluid ring at (x^1, x^2) we add only entropy—i.e., only heat. If the amount of rest mass in the ring is ΔM_0 and the entropy per unit rest mass that is added is δs, then the total entropy added is

$$\delta S = \Delta M_0 \delta s . \tag{10.7.14}$$

(Recall: entropy, temperature, and other thermodynamic variables are always measured in the rest frame of the fluid! Also note our inconsistency: elsewhere S is entropy per particle; only in this section is it the total entropy in a ring.) The addition of this entropy entails the addition of an amount of heat given by

$$\delta W^{(\text{inject})} = T(x^1, x^2)\delta S , \tag{10.7.15}$$

as measured by our colleague riding with the fluid element. (Here T is temperature.) And if the angular momentum is held constant during the addition, this corresponds to a mass-energy change of

$$\delta M c^2 = \delta W^{(\text{inject})}/u^t(x^1, x^2) \tag{10.7.16}$$

as measured by the distant astrophysicist. Thus, we can write

$$(\delta M c^2)_{\Delta M_0, \Delta K} = [T(x^1, x^2)/u^t(x^1, x^2)]\delta S \tag{10.7.17}$$

for the change in total mass-energy when an entropy δS is added to the ring at (x^1, x^2), while holding its rest mass ΔM_0 and angular momentum ΔK fixed.

Next, suppose that we add angular momentum to our ring without changing its rest mass or entropy. We can do this by an injection process similar to that used above: Our astrophysicist friend throws a bundle of mass with initial energy and angular momentum

$$\delta W^{(\text{in})} = p_t^{(\text{in})} , \qquad \delta K = -p_\phi^{(\text{in})} \qquad (10.7.18)$$

down to his colleague on the ring at (x^1, x^2). The colleague catches the bundle, which, according to his measurements, has a total energy of

$$
\begin{aligned}
\delta W^{(\text{c})} &= p_j^{(\text{c})} u^j = p_t^{(\text{c})} u^t + p_\phi^{(\text{c})} u^\phi \\
&= p_t^{(\text{in})} u^t + p_\phi^{(\text{in})} u^\phi = \delta W^{(\text{in})} u^t - \delta K u^\phi .
\end{aligned}
\qquad (10.7.19)
$$

He does not wish to keep any of the energy $\delta W^{(\text{c})}$, as that would change the rest-mass and/or entropy of his ring. But he does wish to keep the angular momentum $\delta K = -p_\phi^{(\text{in})} = -p_\phi^{(\text{c})}$. Therefore, he throws the bundle back up to the astrophysicist, but with zero angular momentum so that the familiar redshift law holds:

$$\delta W^{(\text{final})} = \delta W^{(\text{c})}/u^t . \qquad (10.7.20)$$

The total mass-energy which went into the star, as measured by the distant astrophysicist, is

$$
\begin{aligned}
\delta M c^2 &= \delta W^{(\text{in})} - \delta W^{(\text{final})} \\
&= \delta W^{(\text{in})} - [\delta W^{(\text{in})} u^t - \delta K u^\phi]/u^t = \Omega \delta K ,
\end{aligned}
\qquad (10.7.21)
$$

i.e.,

$$(\delta M c^2)_{\Delta M_0, s} = \Omega(x^1, x^2)\delta K . \qquad (10.7.22)$$

This result is in precise agreement with the corresponding Newtonian result.

We can now combine equations (10.7.13), (10.7.17), and (10.7.22) into one grand equation which describes how the total mass-energy of the star changes when small changes are made in the rest mass ΔM_0, entropy per unit rest mass, s, and angular momentum, ΔK, of the ring at (x^1, x^2):

$$\delta M c^2 = \frac{\mu(x^1, x^2)}{u^t(x^1, x^2)} \, \delta M_0 + \frac{T(x^1, x^2)}{u^t(x^1, x^2)} \, (\Delta M_0 \delta s) + \Omega(x^1, x^2)\delta K . \qquad (10.7.23)$$

If very small amounts of rest mass, entropy, and angular momentum are added to many different rings distributed over the entire star, then the net change in total mass will be the sum (or integral) of (10.7.23) over all rings. (Obviously, our division of the star into many rings is an artifice designed to make the exposition clearer. One could equally well treat the star's matter as a continuum.)

Actually, equation (10.7.23) is accurate only to first order, $\delta M/M$, in the fractional mass added. This is because, after rest mass, entropy, or angular momentum has been added to a ring, the star will be slightly out of hydrostatic equilibrium; and it will start to oscillate with an amplitude $\delta R/R \sim \delta M/M$ and an energy $\delta W^{(\text{oscil})} \propto (\delta R/R)^2 \propto (\delta M/M)^2$. Eventually, the oscillations will damp and the star will settle down into a new equilibrium state, for which δM, δM_0, δs, and δK satisfy equation (10.7.23) only to first order.

Of course, it is the first-order changes that are crucial for equilibrium, and the second-order changes that are crucial for stability.

A mass-energy variational principle for the equilibrium of a rotating, relativistic star has been developed by Bardeen (1971). He envisions a star constructed from a large number of fluid rings, which are labeled by the Lagrangian coordinates x^1 and x^2. Each ring contains a fixed rest mass, $\Delta M_0(x^1, x^2)$, a fixed entropy per unit rest mass, $s(x^1, x^2)$, and a fixed angular momentum, $\Delta K(x^1, x^2)$. At an initial moment of time the space is given all conceivable metrics $g_{ij}(x^1, x^2)$ compatible with the initial-value equations of general relativity. (Note that by fixing the metric, rest mass, entropy, and angular momentum, one automatically determines all other physical quantities associated with the star—e.g., the density, pressure, angular velocity, etc., in each ring.) Of all such "initial-value" configurations, those and only those which extremize the total mass-energy M are in hydrostatic equilibrium. Put differently, the configurations of extremal M are the solutions to Einstein's time-independent, axisymmetric field equations. This variational principle, and its specialization to the case of constant entropy and uniform angular velocity (Hartle and Sharp 1967) may be useful in future numerical studies of rotating, relativistic stars.

In studying *hydrostatic equilibrium* one can make quite arbitrary choices of the rest mass, entropy, and angular momentum of each ring. However, there are other equilibrium conditions associated with ΔM_0, s, and ΔK, which must be satisfied if the star is to maintain its same structure forever.

Consider an arbitrary configuration in hydrostatic equilibrium. Let us redistribute its angular momentum by taking a small amount δK away from one ring and putting it into another, while holding the rest mass and entropy of each ring fixed. Equation (10.7.23) tells us that, if Ω is smaller at the point where δK is added than where it was removed, then we can extract an amount of energy

$$-\delta Mc^2 = (\Omega_{\text{removed}} - \Omega_{\text{added}})\delta K \qquad (10.7.24)$$

from the star in the process of the move. Alternatively, the star itself can make the angular-momentum transfer by means of viscous forces, converting the excess energy into heat. And, of course, the transfer *will* be made in practice, because there are always nonzero forces, however small, linking all fluid elements.

The direction of angular-momentum transfer is always that which releases energy; and according to equation (10.7.23) it also always equalizes the angular velocities. Thus, the transfer will continue until $\Omega(x^1, x^2)$ has been made constant throughout the star, at which point no further energy can be released. The final state

$$\Omega(x^1, x^2) = \text{constant} \qquad (10.7.25)$$

is in *equilibrium against angular-momentum transfer* with fixed ΔM_0 and s.

We can similarly consider a redistribution of entropy (i.e., heat) in which the angular momentum and rest mass of each ring are held fixed. In this case we can obtain energy release unless

$$T(x^1, x^2)/u^t(x^1, x^2) \equiv T\,[g_{tt} + 2g_{t\phi}\Omega + g_{\phi\phi}\Omega^2]^{1/2} = \text{const.} \qquad (10.7.26)$$

Thus, condition (10.7.26) is the criterion for *thermal equilibrium* with fixed ΔK and ΔM_0.

Finally, we can redistribute the rest mass with fixed entropy per unit rest mass and fixed angular momentum per ring. The condition for no energy release, i.e., for *convective equilibrium* with fixed ΔK and s, is

$$\mu(x^1, x^2)/u^t(x^1, x^2) \equiv \mu[g_{tt} + 2g_{t\phi}\Omega + g_{\phi\phi}\Omega^2]^{1/2} = \text{const.} \qquad (10.7.27)$$

Of course, any actual physical process inside a star leads to a transfer of all three quantities: angular momentum, entropy, and rest mass; so our division of the equilibrium conditions into three distinct cases is rather artificial. However, we can say with complete precision that full equilibrium requires: (i) extremal M with ΔM_0, s, and ΔK fixed in each shell (hydrostatic equilibrium); (ii) $\Omega = \text{const.}$ (angular-momentum equilibrium); (iii) $T/u^t = \text{const.}$ (thermal equilibrium); and (iv) $\mu/u^t = \text{const.}$ (convective equilibrium). If the star's matter is not chemically homogeneous, other constraints may be needed.

The above conclusions are valid in the Newtonian limit as well as in general-relativity theory. In Newtonian theory, when the above constraints are violated, there is great complexity in the resultant flow patterns and processes which transfer angular momentum, heat, and matter. For example, although $\mu/u^t = \text{const.}$ is the criterion for convective equilibrium in a uniformly rotating star, nevertheless rotation can strongly suppress convection when this criterion is violated.[10] A slight violation of this criterion in a uniformly rotating star activates far fewer modes of convective motion than in a nonrotating star; moreover, the few modes activated all break the star's axial symmetry. Hence, for a rapidly rotating star the time required to achieve $\mu/u^t = \text{const.}$ might be so long that this criterion for equilibrium would be of little physical significance. A satisfactory understanding of such nonequilibrium questions for rotating stars in Newtonian theory has thus far eluded theorists; and in general relativity there is no

---

10. See, e.g., Cowling (1951). To make the connection between Cowling's paper and the above discussion, it is useful to know that for uniformly rotating stars $\mu/u^t = \text{const.}$ is equivalent to isentropy ($s = \text{const.}$), which in turn is equivalent to the Schwarzschild criterion for convection; see, e.g., § 3.6.5 of Thorne (1967).

hope for an adequate understanding in the near future. We must content ourselves with the above criteria for equilibrium, but continuously warn ourselves that they might be of little value for rapidly rotating, relativistic stars.

We conclude this section with a derivation of a useful formula for the rotational energy of a slowly, rigidly rotating star. For such a star we define

$$\mathfrak{E}_{rot} \equiv \begin{pmatrix} \text{total mass-energy} \\ \text{of rotating star} \end{pmatrix} - \begin{pmatrix} \text{total mass-energy of non-} \\ \text{rotating star, with same rest} \\ \text{mass and entropy in each} \\ \text{ring} \end{pmatrix} . \quad (10.7.28)$$

Thus, $\mathfrak{E}_{rot}$ is the total energy that can be extracted by halting the star's rotation without redistributing or removing any heat or rest mass.

We can calculate $\mathfrak{E}_{rot}$ from equation (10.7.23). Suppose that we remove angular momentum, and hence also energy, from all the rings simultaneously at such a rate as to keep Ω uniform over the entire star. Then extraction of a total angular momentum

$$\delta_{tot}K = \Sigma_{\text{all rings}}\delta K$$

from the star releases a total energy of

$$\delta\mathfrak{E} = \Omega\delta_{tot}K . \quad (10.7.29)$$

For a slowly rotating star the angular momentum is proportional to the angular velocity:

$$K = J\Omega + O(\Omega^3) . \quad (10.7.30)$$

Here J is the moment of inertia of the (fully relativistic) nonrotating star and $O(\Omega^3)$ are neglected terms associated with centrifugal flattening. By integrating

$$\delta\mathfrak{E} = \Omega\delta_{tot}K = J\Omega\delta\Omega$$

over angular velocity, we find for the total energy extracted

$$\mathfrak{E}_{rot} = \tfrac{1}{2}J\Omega^2 + O(\Omega^4) . \quad (10.7.31)$$

Notice that the moments of inertia J which enter into the angular-momentum relation (10.7.30) and the energy relation (10.7.31) are identical. This is as true for fully relativistic stars as it is for Newtonian stars!

10.8 THEORY OF COLD WHITE DWARFS

A. Newtonian Theory

We will now consider a star at the very endpoint of its evolution, when nuclear reactions have been completed and the star has cooled completely. We will calculate the mass of the cold star as a function of its central density. We first rewrite the equilibrium formula (10.1.3) in the form

$$M = \frac{b\langle P\rangle^{3/2}}{G^{3/2}\langle\rho\rangle^2} , \quad (10.8.1)$$

where the numerical value of the dimensionless constant b depends on the method of averaging the density (and the pressure associated with it) over the star. It is more convenient to write equation (10.8.1) in terms of the central values of density and temperature:

$$M = \frac{b_1 P_c^{3/2}}{G^{3/2} \rho_c^2} . \qquad (10.8.2)$$

If the equation of state of the matter throughout the entire star is determined by the expression $P = A\rho^\gamma$ (where A = const., γ = const.), then b_1 can be computed by using the Emden function (see § 10.5). For $\gamma = \frac{5}{3}$, we have $b_1 = 3.0$, and for $\gamma = \frac{4}{3}$ we have $b_1 = 4.55$. When $\rho \ll 10^6$ g cm^{-3}, the pressure is produced predominantly by the nonrelativistic, degenerate electron gas (see § 6.2), and $\gamma = \frac{5}{3}$. Substituting this equation of state into equation (10.8.2), and using $b_1 = 3.0$, we find

$$M = 2.8 \times 10^{-3} \mu_e^{-5/2} \rho_c^{1/2} M_\odot .$$

Recall that μ_e is the number of nucleon masses per electron in the star.

With increasing ρ, the electron gas gradually becomes relativistic. For the time being, we will not consider the neutronization of matter (see chapter 6) or the effects of GTR. Then, when $\rho_c \gg 10^6$ g cm^{-3}, we must use instead of the equation of state of a nonrelativistic electron gas, the equation of state for matter whose pressure is produced by ultrarelativistic, degenerate electrons, and whose mass density is produced by atomic nuclei (see chapter 6):

$$P = 1.23 \times 10^{15} (\rho/\mu_e)^{4/3} . \qquad (10.8.3)$$

Since the atomic nuclei are still not degenerate and since $P \ll \rho c^2$ everywhere, the Newtonian theory is applicable. Substituting this equation of state into (10.8.1) and by using $b_1 = 4.55$, we obtain

$$M_{cr} = 5.75 \, M_\odot / \mu_e^2 .$$

Thus, for cold matter in Newtonian equilibrium, there exists an upper limit on the mass (called the Chandrasekhar limit), which equals $5.75 \, \mu_e^{-2} \, M_\odot$, and which is reached in the limit as $\rho \to \infty$ (then γ precisely equals $\frac{4}{3}$ over the entire star).

The overall picture from the viewpoint of the energy approach is shown in Figure 36. It is convenient to plot $\rho_c^{1/3}$ on the abscissa; then at large ρ_c the energy curves become straight lines. When $M > M_{cr}$, there exist no equilibrium solutions which satisfy the condition $\partial E/\partial \rho_c = 0$. When $M < M_{cr}$, there exists only one such solution. As $M \to M_{cr}$, the equilibrium solution has $\rho_c \to \infty$, since the curve $E(\rho_c)$ becomes horizontal at large ρ_c when $M = M_{cr}$. In this case the energy of the star approaches the value E_{cr}, which equals $-m_e c^2$ per electron (see the derivation in Appendix 1 of this section).[11] This corresponds to $(5 \times 10^{17}/\mu_e)$ ergs g^{-1}, where μ_c is the atomic weight per electron: $\mu_e = A/Z$.

Table 10 (from Schatzman 1958) gives the characteristics of stellar models constructed from an ideal, degenerate gas.

At large densities of the order of $\rho \approx 10^9$–10^{10} g cm^{-3} the above picture

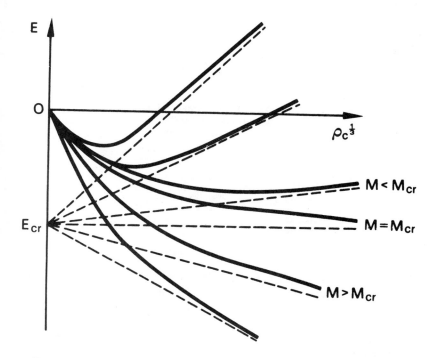

Fig. 36.—Energy curves for a cold star with pressure due to an ideal gas of degenerate electrons, with effects of GTR and of neutronization of matter neglected.

is changed by two phenomena: the process of neutronization of matter, and the onset of effects of GTR. We will begin with a discussion of the effect of GTR.

B. *Influence of Effects of the General Theory of Relativity*

A white dwarf at high density has a very small margin of stability $(\gamma - \frac{4}{3} \ll 1)$. Thus, as was noted in § 10.5, even small corrections for GTR are sufficient to disrupt the stability when the radius of the star is considerably larger than the Schwarzschild radius. Let us find that critical value of

11. The existence of a finite energy of the star, E_{cr}, when $\rho_c \to \infty$ was pointed out by Emin-Zade (1959) and by Savedoff (1963).

the central density at which the onset of instability takes place as a result of GTR. Introducing a correction for GTR requires a careful analysis of the concept "correction at a given density distribution," since one must consider both the non-Euclidean nature of space and the difference between the rest-mass density and the density of total mass-energy, which includes the internal energy divided by c^2.

Rather lengthy computations (see Appendix 2 to this section) lead us to the following form for the correction to the energy of the star as a result of the effects of GTR:

$$\Delta E_{\mathrm{GTR}} = -0.93(G^2/c^2)M^{7/3}\,\rho_c^{2/3}\ . \tag{10.8.4}$$

TABLE 10

PARAMETERS OF STARS COMPOSED OF AN
IDEAL, DEGENERATE GAS

ρ_c/μ_e (g cm^{-3})	$\mu_e R$ (cm)	$M\mu_e^2/M_\odot$
1.23×10^5	2.79×10^9	0.88
9.82×10^5	1.93×10^9	2.02
3.50×10^6	1.51×10^9	2.95
2.65×10^7	9.92×10^8	4.33
3.37×10^8	5.44×10^8	5.32
2.76×10^9	3.10×10^8	5.61
3.10×10^{10}	1.53×10^8	5.72
2.48×10^{11}	7.99×10^7	5.75

Notice that ΔE_{GTR} is of the order of $-GM^2/R$ multiplied by the ratio of the gravitational radius of the star to its actual radius:

$$\Delta E_{\mathrm{GTR}} \approx (-GM^2/R)(r_g/R)\ . \tag{10.8.5}$$

With the effects of this correction included, the energy curves have the shapes shown in Figure 37. The extremal points of E (i.e., the static configurations which satisfy the equilibrium equations) are marked by heavy vertical bars. On the curve that corresponds to M''_{cr}, the maximum and the minimum coincide and yield an inflection at $\rho_c = \rho''_c$ (*double bar*). For ρ''_c, we obtain the following expression:

$$\rho''_c = 3.75(m_p\mu_e/m_e)(\rho_0\mu_e)\ , \tag{10.8.6}$$

where m_p is the mass of the proton, $\mu_e = A/Z$ is the atomic weight per electron, and m_e is the electron mass, so that $m_p\mu_e/m_e$ is the ratio of the total mass of matter to the rest mass of the electrons. The quantity $\rho_0\mu_e$ is the density at which the electron Fermi momentum reaches m_ec, i.e., the density at which the transition from a nonrelativistic to a relativistic electron gas takes place; numerically, $\rho_0 = 0.985 \times 10^6$ g cm^{-3}. Thus, when $\mu_e = \frac{56}{26} = 2.2$ (for iron), $\rho''_c = 3.3 \times 10^{10}$, $E_F = 25\ m_ec^2 = 12.5$ MeV, and the corresponding radius of the star is $R \approx 1000$ km (whereas the gravitational radius of the star is 3.6 km); hence, the onset of instability takes place when $R \gg r_g$.

Figure 38 shows the geometrical location of the extrema (i.e., the energies of the equilibrium configurations) as a function of the density. On this diagram, different points of the curve correspond to stars with different masses and different numbers of nucleons. The minimum E_e (e for equilibrium) coincides with the horizontal inflection point of the curve $E(\rho_c, M = M''_{cr})$ of Figure 37.

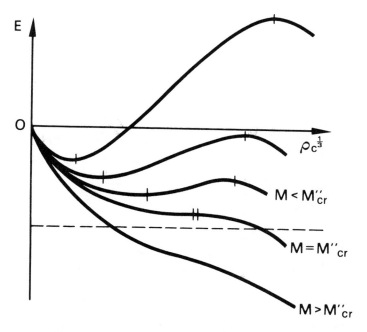

Fig. 37.—Modifications of the energy curves shown in Fig. 36 due to the effects of GTR or neutronization of matter. Broken line shows $E = E_{cr}$ (cf. Fig. 36).

In the upper part of Figure 38 is shown the curve corresponding to the mass of the star; the ordinate indicates the mass for which equilibrium is attained at a value of the central density which is plotted on the abscissa. Since the curve may have two extrema with the same value of M (see Fig. 37), the curve $M(\rho_c)$ passes through a maximum, so that the horizontal line $M = $ const. intersects it twice. The maximum $M(\rho_c)$ is attained at the same $\rho_c = \rho''_c$ at which the inflection $M(\rho''_c) = M''_{cr}$ takes place. The maximum separates the domains of stability and instability.

In his recent contributions, Chandrasekhar has attributed considerable theoretical significance to the existence of a maximum ρ''_c and a corresponding $R'' \sim 1000$ km for extreme white dwarfs; he considers this conclusion a unique confirmation of GTR beyond the domain of weak fields. However, it should be noted that a theory with Newtonian gravitation in flat space, but with the inclusion of the weight of the energy, would bring us to qualitatively the same result. Even more important, the process of neutronization takes place before the density ρ''_c is reached, i.e., before the composition of the nuclei is changed by the high-energy electrons which are at the

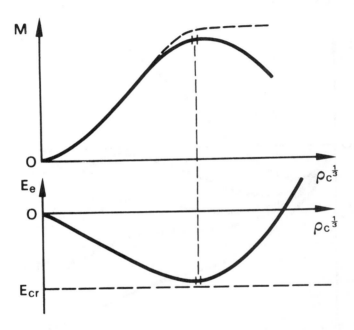

FIG. 38.—Variation of the energy of a cold equilibrium star, and of its mass, as functions of the central density. Dashed lines in upper part of diagram indicate the shape of the curve $M = M(\rho_c)$, considering the pressure of the degenerate electron gas but ignoring the effects of GTR and of neutronization of matter (Chandrasekhar curve). When $\rho_c \to \infty$ along this curve, the mass approaches the Chandrasekhar limit. Energy indicated by dashed line in bottom diagram corresponds to this limit.

boundary of the Fermi distribution in momentum space (see § 6.5). Therefore, there exists another factor which limits the increase of ρ_c and the decrease of R as the critical mass is approached, even without the inclusion of GTR effects.

C. *Influence of the Neutronization of Matter*

In addition to the effects of GTR, the maximum density of a cold white dwarf is limited by the inverse β-process between stable nuclei and electrons which are at the edge of the Fermi distribution; this process begins at a certain critical density (see § 6.5), and it has the form

$$(Z, A) + e^- = (Z - 1, A) + \hat{\nu}.$$

The neutrinos escape freely from the star. In isolation the nucleus $(Z - 1, A)$ is unstable. It undergoes β-decay

$$(Z - 1, A) \rightarrow (Z, A) + e^- + \nu.$$

However, as we discussed in detail in § 6.5, this decay cannot take place in a star of the density considered, since the nuclei are submerged in a degenerate electron gas, and all cells of phase space below and at the momentum of the newly produced electrons are already filled. Thus, no electron can be produced.

The inverse β-process brings about a decrease in the total number of electrons per gram of matter and an increase in the number of neutrons in the nuclei. Finally free neutrons appear. The possibility of the creation of neutron configurations was suggested by Baade and Zwicky (1934), Hund (1936), and Sterne (1933), and was computed by Landau (1938).

What influence will the phase transition associated with the neutronization have upon the structure of a star? We consider models of cold stars with a continuously increasing central density, ρ_c. After ρ_c reaches the critical value for the beginning of neutronization, ρ_{cr}, a small core of matter with nuclei (Z_2, A_2) forms at the star's center, while the rest of the star maintains its composition of (Z_1, A_1). The pressure in this core is smaller than it would have been if a phase transition had not taken place. Therefore, the effective γ for the star is decreased by the phase transition. Is this decrease sufficient to make the star unstable? It was demonstrated by Ramsey (1950) in the context of the theory of planets that when $\rho_c = \rho_{cr}$, there will immediately appear an inflection of the energy curves, but the stability of the star will be disrupted at ρ_{cr} only if the density discontinuity during the phase transition is sufficiently large:

$$\frac{\rho_2}{\rho_1} = \frac{A_2 Z_1}{A_1 Z_2} > 1.5.$$

If, however, the discontinuity is smaller (as it is, for example, in the Fe $\rightarrow$ Cr reaction), then stars with $\rho > \rho_{cr}$ are still stable. However, with increasing density, the size of the core with (Z_2, A_2) composition increases; and, moreover, there occurs a chain of further inverse β-processes between electrons and nuclei, which eventually produce an onset of instability.

Table 11 shows the critical density at the beginning of neutronization for a number of nuclei, and the critical density (eq. [10.8.6]) at which GTR produces instability for a star which is made completely of a given chemical element. As a rule, $\rho_{neutr} < \rho_{GTR}$. The beginning of neutronization at the center of the star, as we have seen, does not imply an onset of instability;

that occurs at somewhat higher densities. Nonetheless, it turns out that loss of stability due to neutronization occurs at densities that only slightly exceed the initial value given in Table 11. It is rather complicated to compute this with precision because it is necessary to consider simultaneously several reactions. Apparently, the stability limit for white dwarfs depends more on neutronization than on the effects of GTR. It is predominantly neutronization that causes the curves $E(\rho_c)$ to bend downward, and it is also neutronization that causes the transition from the diagram shown in Figure 36 to the diagram shown in Figure 37. For stars made of light nuclei, particularly ^{4}He, the story would be different. However, light nuclei will not exist in

TABLE 11

COMPARISON OF THE CRITICAL DENSITIES FOR NEUTRONIZATION AND
FOR GTR EFFECTS IN STARS COMPOSED OF
DIFFERENT ATOMIC NUCLEI

Element	Neutronization Potential (MeV)	Critical Density for Neutronization, $\rho/(10^9 \mathrm{g\ cm^{-3}})$	Maximum Central Density for Stability due to GTR $\rho/(10^9 \mathrm{g\ cm^{-3}})$
$^{56}_{26}\mathrm{Fe} \rightarrow ^{56}_{25}\mathrm{Mn}$	3.7	1.15	31.3
$^{32}_{16}\mathrm{S} \rightarrow ^{32}_{15}\mathrm{P}$	1.7	0.145	27
$^{28}_{14}\mathrm{Si} \rightarrow ^{28}_{13}\mathrm{Al}$	4.6	1.92	27
$^{24}_{12}\mathrm{Mg} \rightarrow ^{24}_{11}$	5.5	3.15	27
$^{20}_{10}\mathrm{Ne} \rightarrow ^{20}_{9}\mathrm{F}$	7.03	6.2	27
$^{16}_{8}\mathrm{O} \rightarrow ^{16}_{7}\mathrm{N}$	10.4	19	27
$^{12}_{6}\mathrm{C} \rightarrow ^{12}_{5}\mathrm{B}$	13.37	39	27
$^{4}_{2}\mathrm{He} \rightarrow ^{3}_{1}\mathrm{H} + ^{1}_{0}n$	20.6	132	27
$^{3}_{2}\mathrm{He} \rightarrow ^{3}_{1}\mathrm{H}$	0.018	2.8×10^{-5}	15.2

stars with high density. This is due to the fact that such densities are attained only at late stages of evolution when light nuclei have already been transformed into heavy nuclei by the thermonuclear reactions which take place at earlier stages of the evolution. According to Öpik's (1957) calculations (see also Hayashi 1966), white dwarfs with $M < 0.5\ M_\odot$ must consist predominantly of elements with intermediate atomic weight ($A \approx 24$).

The indeterminancy of the precise value of ρ_{cr}, as well as of the chemical composition, has very little influence upon the maximum mass, M_{max}, corresponding to ρ_{cr}. The curve $M = M(\rho_c)$ of the Newtonian theory, plotted without any corrections, is already near its asymptotic value, the Chandrasekhar limit, when $\rho \approx 10^9$–$10^{10}\ \mathrm{g\ cm^{-3}}$.

The maximum mass, according to Harrison, Wakano, and Wheeler (1958), Hamada and Salpeter (1961), and Saakyan and Vartanyan (1964), equals approximately $1.2\ M_\odot$ (Fig. 39). This result is obtained by numerical integration of the equation of hydrostatic equilibrium (10.1.2), and by taking into account the changes of the equation of state as one moves outward from the dense interior of the star to the less dense surface. Thus far we have not considered the effects of rotation on the star. Rigid-body rotation does not significantly change either the critical mass or the critical density of a white dwarf; see Krat (1950). However, differential rotation can make significant changes (see Ostriker and Bodenheimer 1968).

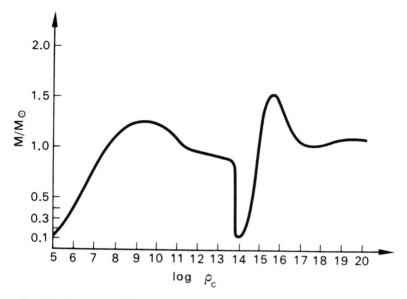

Fig. 39.—Mass of a cold star in equilibrium as a function of central density, with the interaction of baryons (Saakyan and Vartanyan 1963) taken into account.

On the descending branch of the curve $M = M(\rho_c)$ (Fig. 39), the equilibrium configurations correspond to maxima of the energy curves and are unstable.

10.9   NEUTRON STARS

Let us trace the further variation of M with increasing ρ_c.

It was demonstrated in chapter 6 that, with increasing density, free neutrons occur in matter. When $\rho > 10^{12}$ g cm^{-3}, the pressure (as well as the density) is due primarily to the degenerate neutron gas. If the neutrons did not interact with each other, then this gas would be ideal; and so long as it was nonrelativistic, its adiabatic index would be $\gamma = \frac{5}{3}$ (and γ would always be greater than $\frac{4}{3}$, even in the relativistic regime). However, it is known that there exist attractive forces between neutrons; and while these forces are not sufficient for the formation of nuclei consisting of neutrons, they nonetheless make a negative contribution to the energy. Thus, as before, γ is smaller than $\frac{4}{3}$. The equilibrium states are unstable, and the curve $M = M(\rho)$ continues to descend (see Fig. 39).

At small distances between baryons, the forces of attraction are replaced by forces of repulsion which make a positive contribution to the pressure. Consequently, when $\rho \sim 2 \times 10^{14}$ g cm^{-3}, the effective γ for the entire star again becomes larger than $\frac{4}{3}$.

Thus, the mass M reaches a minimum value when $\rho_c \approx 2 \times 10^{14}$ g cm^{-3}. This minimum mass, $M_{\min}$, can be estimated once we know the pressure at $\rho_c = 2 \times 10^{14}$ g cm^{-3}; using formula (6.6.1) for the pressure, and substituting into equation (10.8.1) with $b \approx 3$, we obtain

$$M_{\min} \approx 0.05 \, M_\odot \, .$$

We recall that this is only an order-of-magnitude estimate because in reality a star cannot consist entirely of neutrons. In the external regions, the pressure is insufficient for the existence of stable neutrons, so the external shell consists of nuclei and electrons. For more details, see Saakyan and Vartanyan (1964).

Let us comment on equilibrium configurations with positive energy. In Figure 37, they correspond to maxima in the domain $E > 0$. The existence of these $E > 0$ maxima is due to the fact that the neutronization of matter produced an extremum (maximum) where one did not exist previously; the energy, even without the neutronization correction, was positive. Such configurations obviously are unstable; a complete scattering of the matter of such a star is possible.

When densities are high (of the order of nuclear densities), there are powerful forces of repulsion between the baryons; these forces cause the curves $E(\rho_c, M)$ to bend upward (Fig. 40). Curves with M larger than $M_{\min} \approx 0.05 \, M_\odot$ show secondary minima which correspond to stable stellar configurations. These are the neutron stars. The corresponding curve $M = M(\rho_c)$ for equilibrium configurations is shown in Figure 39. The curves $E(\rho_c, M)$ that have secondary minima are bounded from above by a curve with a minimum mass $M_{\min} \approx 0.05 \, M_\odot$. For smaller masses there exist no equilibrium configurations at high density.

Computations with an approximate equation of state indicate that, for small masses, the secondary minima are in the domain of positive energy (see the contribution by Bisnovatyi-Kogan and Zel'dovich 1966). They correspond to stable states. In the preceding section, we pointed out that unstable equilibrium states (with positive energy) occur for stars of nonideal gas even without the effects of GTR, and that this is not really unusual or strange. Here we see that even if the influence of GTR is neglected, there can exist metastable states with positive energy which are stable with respect to small perturbations.

With a further increase of density, the effects of GTR become dominant. They cause the curves $E(\rho_c, M)$ to bend downward (Fig. 41), which—as was demonstrated in § 10.8—leads at lower densities to an inflection, and at higher densities to a maximum in the curves $E(\rho_c, M)$, and also to a second maximum in the equilibrium curve $M = M(\rho_c)$. Let us consider in some detail the reasons for the appearance of a secondary maximum in the curve

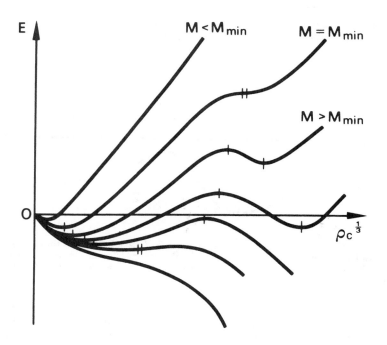

Fig. 40.—Energy of a cold star as a function of central density, with the elasticity of the nuclear fluid taken into account.

$M = M(\rho_c)$. The first reason was mentioned in passing at the end of paragraph B of § 10.8 and is expressed qualitatively by the phrase "energy has weight." For static configurations in GTR, the conservation laws are equations of hydrostatic equilibrium. The equation of hydrostatic equilibrium for a spherical field has the form

$$dP/dr = -\tfrac{1}{2}(\rho c^2 + P)(d\nu/dr) , \qquad (10.9.1)$$

where the metric is expressed in the form shown in § 3 of chapter 3, ρ is the density of mass-energy, and P is the pressure. An alternative form for the equation of hydrostatic equilibrium is equation (10.6.10). In the weak-field approximation $\tfrac{1}{2}\nu = \phi/c^2$, where ϕ is the Newtonian potential. Recalling that in this case $\rho c^2 \gg P$, we obtain the formula of hydrostatic equilibrium of Newtonian theory

$$dP/dr = -G\rho m(r)/r^2 .$$

It is clear from equation (10.9.1) that the equation of equilibrium includes effects of the density of mass-energy ρ and the pressure P.

At large densities, the principal contribution to the energy density (and consequently to the total density of mass-energy) is no longer the rest energy of the particles, but instead is the energy of their motion and of their interaction. We denote the baryon number density by n, and the adiabatic

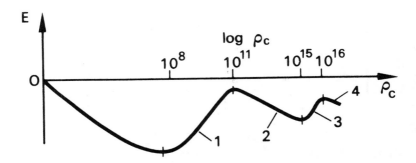

FIG. 41.—Typical energy curve for a sphere of fixed mass. Behavior of the curve shape at various densities ρ_c is determined by (1) the compressibility of degenerate electrons, (2) the neutronization of matter, (3) the compressibility of nuclear matter, and (4) the effects of GTR.

index by $\gamma_1 = d \ln P/d \ln n$. For a degenerate electron gas γ_1 is always larger than $\frac{4}{3}$; and, in principle, it may reach the value of $\gamma_1 = 2$ for mutually repulsive particles (Zel'dovich 1961), as was demonstrated in detail in § 6.12.

However, the equation of hydrostatic equilibrium involves not n, but rather the density of mass-energy ρ and the pressure P. Because the equation of state has the asymptotic form $P \sim \rho c^2$, the effective index of the second type, $\gamma_2 = d \ln P/d \ln \rho$, becomes smaller than $\frac{4}{3}$ and approaches the limit $\gamma_2 \to 1$.

Figure 42 shows γ_1 and γ_2 for an ideal neutron gas. The change from $\gamma_2 > \frac{4}{3}$ to $\gamma_2 < \frac{4}{3}$, as has been demonstrated in preceding sections, leads in Newtonian theory to a maximum in $M(\rho_c)$. Thus taking into account the "equivalent mass of the energy" produces a maximum in $M(\rho_c)$ even in the Newtonian, flat-space theory (see contributions by Saakyan 1962 and Guseynov 1965).

Another reason for the maximum is that in the Einsteinian theory, the law of gravitation changes for large ϕ, and space ceases to be Euclidean; this was discussed in detail in §§ 3.2 and 3.3.

10.9 Neutron Stars

Both phenomena are of the same order and act in the same direction. They were considered in detail as small corrections in the preceding section (see Appendix 2 to § 10.8).

The first numerical calculations of the structures of superdense stars were undertaken by Oppenheimer and Volkoff (1938), who used the equation of state of an ideal Fermi gas (degenerate neutron gas). The results of their computations are shown in Figure 43 (the curve M; regarding curves M_0 and M_1, see § 10.10); the maximum mass was found to be $M_{max} = 0.72\ M_\odot$.

However, the equation of state of an ideal gas at such densities is at best

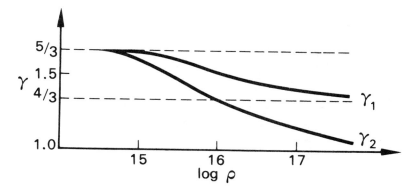

FIG. 42.—Dependence of $\gamma_1 = d \ln P/d \ln n$ and $\gamma_2 = d \ln P/d \ln \rho$ on the density ρ for a cold, ideal neutron gas.

only a rough approximation (see chapter 6). The investigations of the equation of state of a real gas by Cameron (1959a), Saakyan and Vartanyan (1964), and Tsurata and Cameron (1966) yield the value

$$M_{max} = 1.6\text{--}2\ M_\odot .$$

Figure 39 shows the curve $M = M(\rho)$ according to the computations of Saakyan and Vartanyan (1964). It covers the domain of densities corresponding to white dwarfs, and the domain of neutron stars (see also Inman 1965).

In these calculations, the authors took into account the variation of the equation of state in the transition from the dense interior of the star to its surface. Interestingly, in the domain of high densities beyond the maximum in the mass curve, where no real equilibrium cold stars can exist, the total mass of a star in hydrostatic equilibrium has a periodic dependence on ρ_c as $\rho_c \to \infty$ (see Fig. 43). This was first demonstrated by Dmitriev and Kholin (1963). Later (and independently), demonstrations were given by other investigators. We will not go into detail here, but the reader will find the details in Harrison, Thorne, Wakano, and Wheeler (1965).

We will refer to the maximum mass for cold stars at $\rho \approx 10^{15}$ g cm^{-3} as the OV maximum (Oppenheimer and Volkoff) to distinguish it from the Chandrasekhar maximum which occurs for ρ of the order of 10^9 g cm^{-3}. At the surface of a star with $M = M^{\text{OV}}_{\text{max}}$, e^ν takes the minimum possible value

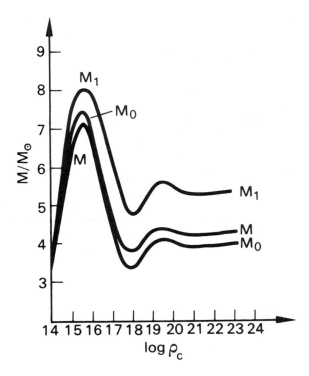

Fig. 43.—The quantities M, M_0, and M_1 as functions of ρ_c for a cold, ideal Fermi gas of neutrons.

for a stellar surface. This minimum value is $(e^{\nu/2})_{\text{min}} \approx 0.7$. Therefore, theoretically, the maximum gravitational redshift which can be observed in the spectrum of a star is

$$(\omega/\omega_0)_\gamma = (e^{-\nu/2})_{\text{min}} \approx 1.4 \ . \tag{10.9.2}$$

The neutrino sources are located in the center of the star, so the gravitational redshift for a neutrino is determined by the value of e^ν at the center. This value in the star considered is $(e^{\nu/2})_{\text{min},0} \approx 0.4$. Consequently, for a neutrino, the maximum gravitational redshift of frequency is

$$(\omega/\omega_0)_\nu = (e^{-\nu/2})_{\text{min},0} \approx 2.5 \ . \tag{10.9.3}$$

Figure 44 shows the run of $e^{\nu/2}$ and $e^{\lambda/2}$ for two stars, one with $\rho_c = 5.5 \times 10^{14}$ g cm^{-3} and $M = 0.64\ M_\odot$, and the other with $\rho_c = 3.6 \times 10^{15}$ g cm^{-3} and $M = M^{\mathrm{OV}}_{\mathrm{max}} = 1.55\ M_\odot$. The curve $e^{\lambda/2}$ characterizes the deviation of the space geometry from Euclidean near the star and inside it. The

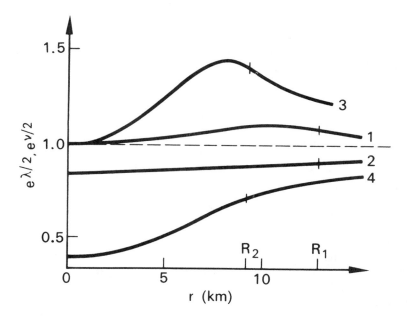

F$_{\mathrm{IG}}$. 44.—Dependence of $e^{\lambda/2}$ and $e^{\nu/2}$ on r for stars with $M_1 = 0.64\ M_\odot$ and $M_2 = 1.55\ M_\odot$. Curve 1 corresponds to $e^{\lambda/2}$ and curve 2 to $e^{\nu/2}$ for M_1; curve 3 to $e^{\lambda/2}$ and curve 4 to $e^{\nu/2}$ for M_2. Radii R_1 and R_2 correspond to the surfaces of these stars.

coordinate radii for these stars, $R = r_{\mathrm{surf}} = (S/4\pi)^{1/2}$ (where s is the surface area of the star) are 13 and 9.3 km, respectively. The distances of the surfaces from the center,

$$R = \int_0^{r_{\mathrm{surf}}} e^{\lambda/2} dr \,,$$

are 13.8 and 11.5 km, respectively (Saakyan and Vartanyan 1964).

The quantity $e^{\nu/2}$, as noted previously, is analogous to the Newtonian potential. It shows directly the retardation of the flow of time, as compared with the flow at infinity. By contrast with $e^{\lambda/2}$, the value of $e^{\nu/2}$ does not approach unity at the center of the star. This is due to our choice of normalization: we selected the time coordinate t so that at infinity it will always coincide with the time measured by an observer's clock; therefore, $(e^{\nu/2})_\infty = 1$. The value of $e^{\nu/2}$ at the center of the star is as many time smaller than $(e^{\nu/2})_\infty$ as the rate of time flow at the center of the star is smaller than at infinity.

285

10.10   THE MASS DEFECT

Let us write down the expression for the total energy of a star, E, in a case when the densities are small, and the Newtonian theory is applicable:

$$E = E_0 + W + U .$$

We have included in this expression $E_0 = Nmc^2$, the rest mass-energy of the nucleons which compose the star; W, the energy of motion and interaction of the nucleons; and U, the potential energy of self-gravitation, which is negative. We denote $E_0 + W = E_1$. In the relativistic domain, we have analogously

$$E = Mc^2 = 4\pi c^2 \int_0^R \rho r^2 dr , \tag{10.10.1}$$

$$E_0 = M_0 c^2 = c^2 \int_V mn dV = Nmc^2 , \tag{10.10.2}$$

$$E_1 = M_1 c^2 = c^2 \int_V \rho dV , \tag{10.10.3}$$

where the element of volume is $dV = 4\pi e^{\lambda/2} r^2 dr$.

Figure 43 shows the graphs of M, M_0, and M_1 as functions of ρ_c for a degenerate, ideal, neutron gas.

We recall that the density of mass-energy, measured locally, includes not only the rest mass but also the internal energy of the nucleon motion and the interaction energy (other than gravitational!) of the particles, per cubic centimeter. The total mass of the star, M, does not equal the sum of the masses of elements of its volume, M_1; in fact, since $e^{\lambda/2} \geq 1$, it follows that

$$M < M_1 .$$

The difference $\Delta_1 M = M_1 - M$ is called the total gravitational mass defect. The origin of $\Delta_1 M$ is obvious: when combining the mass elements $dm = \rho dV$ (which already have the assigned density ρ) into a star, we must take into account their energy of gravitational interaction. This binding energy (which is not included in eq. [10.10.3], but is in eq. [10.10.1]), and the mass that corresponds to it, are negative. Hence $\Delta_1 M > 0$. In the Newtonian approximation, $c^2 \Delta_1 M = -\Omega$. The ratio (in the general case, and not just in the Newtonian approximation)

$$a_1 = \Delta_1 M / M$$

is called the coefficient of gravitational packing. It defines the ratio between the gravitational energy and the total energy. Figure 45 shows the relationship between a_1 and ρ_c for stars consisting of a real gas, as calculated by Saakyan and Vartanyan (1964). For small ρ_c, the value of a_1 is small; it approaches zero as $\rho_c \to 0$. For denser configurations, $a_1 \approx 0.5$.

The difference $\Delta_2 M = M_0 - M = Nm - M$ is called the *partial* mass defect or just the mass defect. The energy that corresponds to $\Delta_2 M$ is the energy which is released during the formation of a dense star from initially rarefied matter. It is apparent from the physics of this process that for a completely stable static star originating from diffuse matter,[12] $\Delta_2 M > 0$.

In the Newtonian approximation, $c^2 \Delta_2 M = -(W + U)$. The ratio (in the general case, and not just in the Newtonian approximation)

$$a_2 = \Delta_2 M / M$$

represents the total fraction of the rest mass-energy emitted during the formation of the star. The curve of $a_2(\rho_c)$, computed for stars composed of a real gas by Saakyan and Vartanyan (1964), is shown in Figure 45. At large densities, a_2 becomes negative. This will be discussed further in § 10.11.

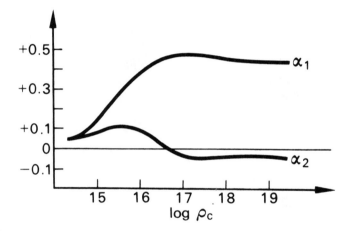

Fig. 45.—Dependence of $a_1 = \Delta_1 M / M$ and $a_2 = \Delta_2 M / M$ on the central density ρ_c of the star.

The gravitational mass defect is sometimes incorrectly referred to as gravitational screening. Such a term does not reflect the essence of the matter because the phenomenon considered is not at all similar to the effect of a screen. For example, when we combine two particles, we obtain a mass for the system which is smaller than the sum of the masses of the particles. However, (1) this weakening of gravitation has no directionality (which should have been the case if the second particle had really been a screen), and (2) any binding forces have the same property of reducing the total mass of the particles—in this respect, gravitation is not unique. The mass of a deuteron is smaller than the sum of the masses of a proton and a neutron, but because

12. We consider here the mass defect only for static configurations. If the configuration is nonstatic, then in principle the total mass M of an assigned number of nucleons can be arbitrarily small (see § 11.7). For example, M for a closed cosmological model is zero (see Vol. II).

of this we are not about to assert that the neutron has a gravitational screen-effect upon the proton.

When particles are combined into a bound system, an energy which equals the mass defect is emitted in the form of photons, neutrinos, gravitational waves, etc. The distant observer will detect the mass defect (the decrease of the mass) gravitationally, not at the moment of the combination of the

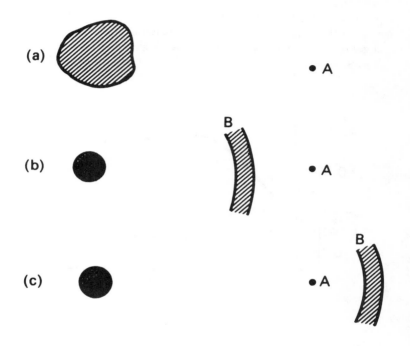

Fig. 46.—Variation of the mass of matter as measured gravitationally by an observer at A during the course of formation of a dense star. (*a*) Diffuse matter prior to contraction into a star. (*b*) Matter contracted; the emitted energy (region B) has not yet passed the observer, so he does not detect any decrease in the mass of the body. (*c*) The wave has passed the observer; he notices a decrease in the mass of the body by ΔM.

particles, but after the emitted energy passes him (Fig. 46). Up to this moment, any energy transformation will have no effect whatsoever on the mass of the star as measured by the observer (this is a direct consequence of the law of conservation of energy).

10.11 STABILITY OF NEUTRON STARS

Stable equilibrium implies a minimum energy of a star for a given entropy and number of particles. The simple-minded analysis of stability

(§ 10.5), in which the energy is viewed as a function of one parameter ρ_c, is asymptotically precise only when $\gamma \approx \frac{4}{3}$ and the corrections for GTR are small. The variational principle in GTR has been discussed in detail by Chandrasekhar (1964a, b, 1965), as well as by Harrison, Thorne, Wakano, and Wheeler (1965). For a review see chapter 4 of Thorne (1967). We present here another analysis (Zel'dovich 1963a), which is valid both in the relativistic domain and in the Newtonian domain, but which specializes to cold stars. (It can easily be generalized to hot, isentropic stars.)

First, we consider how the mass of an equilibrium configuration changes with the addition of one particle, brought in to a radius r from infinity, where its energy was mc^2. In other words, we derive dM/dN. The energy of such a particle, which has fallen freely in the gravitational field to a radius r, has the value[13]

$$E = mc^2 \exp\left[-\tfrac{1}{2}\nu(r)\right]. \tag{10.11.1}$$

The difference $E(r) - \mu(r)$, where $\mu(r)$ is the chemical potential of the particles of the cold star, is emitted, for example, as γ-rays. Because the γ-rays lose energy through the gravitational redshift (see § 4 of chapter 3), the energy

$$\Delta E = (E - \mu) \exp\left[\tfrac{1}{2}\nu(r)\right] \tag{10.11.2}$$

escapes to infinity.

On the other hand, it follows from the equation of hydrostatic equilibrium for cold stars that (cf. eq. 10.6.20)

$$\mu(r) \exp\left[\tfrac{1}{2}\nu(r)\right] = \text{const.} = mc^2 \exp\left[\tfrac{1}{2}\nu(R)\right]. \tag{10.11.3}$$

By combining equations (10.11.2) and (10.11.3), we find that

$$dM/dN = m \exp\left[\tfrac{1}{2}\nu(R)\right] = \text{const.}$$

The change in M does not depend on the place in the equilibrium configuration at which the particle is added. Notice from equations (10.11.2) and (10.11.3) that always

$$dM/dN < m. \tag{10.11.4}$$

The fact that dM/dN is independent of the place where the particle is added means that if we assign a perturbation to the particle distribution, $\delta n(r)$, without changing the total number of particles, i.e., so that $\delta N = 0$, then to first order we will also have $\delta M = 0$; i.e., $(\delta M/\delta n)_{N\text{=const.}} = 0$. This is indicative of the fact that the equilibrium state corresponds to an extremum of the mass, i.e., an extremum of the total energy of the system. If this extremum is a minimum, the state is stable. Consider the segment

13. The energy is measured by a local observer and does not include the potential energy of the particle in the gravitational field. The total energy, including potential energy, of the particle naturally does not change during its fall (the radiation of gravitational waves is not taken into account since it approaches zero when $m/M \to 0$, i.e., in the case of a test particle of small mass).

of the curve $M = M(\rho_c)$ near an extremum, M_{ex} (see Fig. 43). It follows from equation (10.11.4) that the extrema of $N(\rho_c)$ and $M(\rho_c)$ occur at the same central density, $\rho_c = \rho_{crit}$. (The extrema of $M(\rho_c)$ and $M_0(\rho_c) = mN(\rho_c)$ are shown in Fig. 43.)

It follows that to the left and the right of ρ_{crit} one can select two different stationary stellar models, with different densities ρ_{c_1} and ρ_{c_2}, but with the same number of baryons N. The density distribution for one of these models can be represented as a perturbation of the distribution for the other model

$$\rho_2(r) = \rho_1(r) + \delta\rho . \tag{10.11.5}$$

In the most general case, the dynamical behavior of small perturbations can be expressed as a sum of eigenfunctions of a linearized problem. In this series, the dependence of the ith harmonic upon time is given by the expression

$$\delta\rho_i = \phi_i(r)e^{\omega_i t} . \tag{10.11.6}$$

Our particular perturbation $\delta\rho$, which transforms the stationary solution ρ_1 into another stationary solution ρ_2, does not depend on time; consequently, $\omega_1 = 0.$[14]

Thus, at the extremum of the curve $M(\rho_c)$

$$\omega_1 = \omega_1^2 = 0 .$$

Obviously, the case with $\omega_1^2 = 0$ is at the boundary between $\omega_1^2 < 0$, where ω_1 is imaginary, and $\omega_1^2 > 0$, where ω_1 is real. If all $\omega_i^2 < 0$, then the solution is stable, and the emergence of a positive ω_1^2 indicates an onset of instability. This reasoning is applicable to a stellar model constructed with GTR as well as to a nonrelativistic model. Let us emphasize that so far we have considered only small perturbations.

Now let us analyze the stability of cold stars. Proceeding from the principle of maximum entropy for a stable star, Zel'dovich (1963a) demonstrated that a precise criterion for stability in the nonrelativistic domain is the non-intersection of the functions $r(m)$ which characterize the stellar density distribution for two stars of nearly equal mass (see appendix to § 10.1). It follows that on the section of the curve $M(\rho_c)$ where $dM/d\rho_c < 0$ all equilibrium models are certainly unstable. In the domain of "white dwarfs" ($\rho_c < 10^{10}$ g cm^{-3}), $dM/d\rho_c > 0$; and as is well known, these stars are stable (see § 10.8). Here, all $\omega_i^2 < 0$. A transition through the Chandrasekhar maximum indicates an onset of instability against perturbations which compress the entire star as a whole (see § 10.1); i.e., ω_1^2 for the fundamental mode becomes positive. All other ω_i^2 are negative, and the star has finite restoring forces with respect to variations in the shape of its density distribution. At the minimum of $M(\rho_c)$, where $\rho_c \approx 10^{14}$ g cm^{-3}, the stability of the star is restored; here

14. The method which we use here has been developed systematically by Barenblatt and Zel'dovich (1958).

$\omega_1{}^2$ again changes sign, and one can clearly see that all other $\omega_i{}^2$ are still less than zero. At the OV maximum, we find that $\omega_1{}^2$ again becomes positive. As was demonstrated by Dmitriev and Kholin (1963), and subsequently by Harrison, Thorne, Wakano, and Wheeler (1965), soon after the OV maximum, for increasingly large ρ_c, the density distribution functions for stars of nearly equal mass begin to intersect many times. As in the Newtonian theory, this must indicate instability. Not only $\omega_1{}^2$ but also higher harmonics become positive. As $\rho_c \to \infty$, one passes through an unlimited number of maxima and minima, at each of which the sign of one more $\omega_i{}^2$ becomes positive, producing one more instability. Thus, beyond the OV maximum all configurations are unstable. A more rigorous proof of these assertions can be found in the work of Harrison, Thorne, Wakano, and Wheeler (1965).

10.12   CONFIGURATIONS WITH POSITIVE ENERGY

In this section we return to the problem of positive energy of an equilibrium star. In the relativistic theory, where the mass and the energy of a star are only different expressions of the same property, positivity of the energy means that $\Delta_2 M < 0$.

For stable stars which have originated directly from diffuse matter, the gravitational mass defect is negative; $\Delta_2 M > 0$. However, in the general case, one cannot make a definitive a priori assertion as to the sign of

$$\Delta_2 M \,=\, 4\pi \int\limits_0^R (mne^{\lambda/2} - \rho)r^2 dr$$

in an equilibrium configuration because, on one hand, the energy of motion of the nucleons and their interaction cause $nm < \rho$, and, on the other hand, $e^{\lambda/2} \geq 1$. The problem of the sign of $\Delta_2 M$ must be resolved by a specific computation of the stellar model.

Models with negative $\Delta_2 M$ certainly cannot be formed directly by condensation of diffuse matter. In principle, this does not exclude the possibility of reaching such a state by a release of nuclear energy: energetically, a state can be formed whose energy is less than the energy of rarefied hydrogen, but greater than the energy of rarefied iron vapor which is dispersed at infinity. In such a state the body is obviously unstable in the sense that it can explode, dispersing its matter to infinity. However, in order to judge the possibility of the existence of such a body, one must investigate its stability with respect to *small* perturbations.

The formation of a body with positive energy in the course of stellar evolution is hardly probable, but it is within the realm of physical law. A body that is unstable with respect to small perturbations is incapable of existence in general.

Interest in states with positive energy, especially in the highly relativistic domain, at a density many times greater than nuclear has recently been linked (Saakyan 1965) to Ambartsumyan's (1960) idea that evolution proceeds from a superdense state to a diffuse state. States with positive energy can explode and pass from a dense state to a diffuse state. It is necessary to emphasize, however, that equilibrium states with positive energy exist only for a mass not greater than 2–3 $M_\odot$ and have no relationship to the grandiose explosions of the nuclei of galaxies and quasars. We dwell on this problem only because it is frequently discussed in the literature.

Consider the relationship of the mass M of a star to the number of nucleons, N, that it contains. First, it is obvious that a curve of M versus N passes through the origin $N = 0$, $M = 0$. Moreover, in § 11 of this chapter it was demonstrated that $dM/dN < m$. Hence, it appears at first glance that we always have $M < Nm$ and $\Delta_2 M > 0$. However, such is not the case.

The curves $M(\rho_c)$ and $N(\rho_c)$ pass through a maximum at the same density ρ_{crit}, and dM/dN is finite everywhere and has no singularities (see § 10.11). Thus, it follows that the plot of M versus N will have a cusp corresponding to the total maximum M and N. This behavior is shown in Figure 47, which is based on calculations by Saakyan and Vartanyan (1964) for superdense configurations.[15] Everywhere on the curve, $dM/dN < m$; but there exists a segment where $Nm < M$ and $\Delta_2 M < 0$. Naturally, the configurations on this segment are unstable; small perturbations will make them either contract or expand. If the mass of the star is scattered, the matter will have a kinetic energy at infinity which is nonzero.

The physical reason that $\Delta_2 M < 0$ is as follows. At very high density, the energy of motion and of repulsion of the baryons is substantially greater than their rest energy, $\rho > mn$. Therefore, despite the fact that inclusion of the negative energy of the gravitational field somewhat decreases this difference, nonetheless

$$\Delta_2 M = \int_V (mn - \rho e^{-\lambda/2}) dV < 0 .$$

The effect of the negative energy of the field is contained in the factor $e^{-\lambda/2}$. Figure 47 shows the existence of configurations with $\Delta_2 M < 0$ for an equation of state of a real gas. This is also apparent from the curve of a_2 in Figure 45. In Figure 43, the curves M_0 and M for an ideal gas intersect; in this case, too, $\Delta_2 M < 0$ when $\rho_c > 5 \times 10^{16}$.

It is clear that the equilibrium solutions with negative binding energy must be unstable. Moreover, the stability against small perturbations is lost at the point of greatest (positive) binding energy, so the configurations of negative binding energy must lie deep in the unstable region.

To what extent is this situation caused by the laws of GTR? In the nonrelativistic theory of the simplest case, the power-law equation of state

15. According to the comments in § 10.9, the curves $M(\rho_c)$ and $N(\rho_c)$ have an unlimited number of maxima as $\rho_c \to \infty$; however, the amplitude of the oscillations in M and N is attenuated. There is a corresponding infinite, damped sequence of cusps in the curve $M = M(N)$. Figure 47 also shows only two cusps $(4.3 \times 10^{16}$ and $2.3 \times 10^{17})$; additional cusps are not shown because of the smallness of the scale.

$P = A\rho^\gamma$ (Lane-Emden theory; cf. § 10.5), the binding energy is proportional to $\gamma - \frac{4}{3}$, the same quantity as determines stability against small perturbations. In this case, also, solutions of negative binding energy exist, and they are unstable.

Negative binding energy is a criterion for instability against a definite type of perturbation—dispersal of the star's matter to infinity. Does this always imply instability against *small* perturbations, as is suggested by the few examples given above (Oppenheimer-Volkoff and Lane-Emden cases)? No!—as one can see from the following counterexample of a nonrelativistic

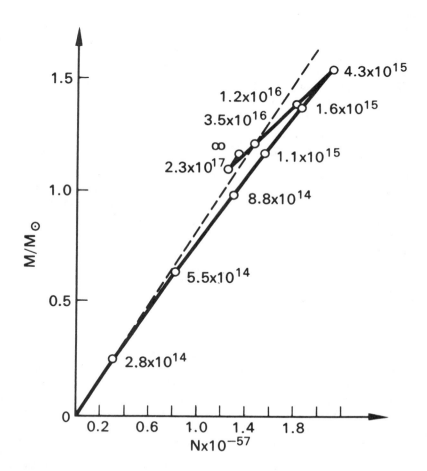

Fig. 47.—Relationship between the mass M of a cold star and the total number of baryons N contained in it. Central density of the star is indicated beside the circles on the line. Dashed line is $M = Nm_H$.

equilibrium solution with negative binding energy which is stable against small perturbations.

To obtain our counterexample we take a sophisticated equation of state:

$$P = \rho^\gamma , \quad \rho > \rho_0 ; \quad P = A\rho_0{}^\gamma (\rho/\rho_0)^{1+\zeta} , \quad \rho < \rho_0 ;$$

$$E = \frac{1}{\gamma - 1} A\rho^{\gamma-1} + A\rho_0{}^{\gamma-1} \left(\frac{1}{\zeta} - \frac{1}{\gamma - 1} \right) = \frac{1}{\gamma - 1} A\rho^{\gamma-1} + E_0 , \quad \rho > \rho_0 ;$$

$$E = \frac{1}{\zeta} A\rho_0{}^{\gamma-1} (\rho/\rho_0)^\zeta , \quad \rho < \rho_0 .$$

Let A and γ have "normal" values, with $\gamma > \frac{4}{3}$; and focus attention on the limit of this equation of state as the constants ζ and ρ_0 approach zero, but at such a rate that $E_0 = A\rho_0{}^{\gamma-1}[1/\zeta - 1/(\gamma - 1)]$ remains finite. The bulk of the star is then described by the Emden solution corresponding to a normal value of γ, but the energy contains the additive constant ME_0.

By using the formulae of § 10.3 of the appendix to § 10.5, we can calculate the following formula for the binding energy of a star made of matter with the above equation of state:

$$\mathfrak{E} = -\frac{3 - n}{5 - n} \frac{GM^2}{R} + ME_0 = -\frac{3\gamma - 4}{5\gamma - 6} \frac{GM^2}{R} + ME_0 .$$

Here R is the radius which the star would have if its equation of state had the simple form $P = A\rho^\gamma$ everywhere. In our case of very small ρ_0 and ζ, $\rho \approx \rho_0$ at $r = R$. Outside this radius, at $\rho < \rho_0$, there extends an atmosphere; the condition of the existence of a finite external radius R_1 for the atmosphere places a limit on E_0. Neglecting the mass of the atmosphere, we can use for the gravitational potential in it $\phi = -GM/r$ and thereby obtain

$$H + \phi = (1 + \zeta)E - GM/r = (1 + \zeta)E_0 - GM/R = -GM/R_1 < 0 ,$$

so that $E_0 < GM/r$, which means that

$$\mathfrak{E} < \frac{2(\gamma - 1)}{5\gamma - 6} \frac{GM^2}{R} .$$

Since $\gamma > \frac{4}{3}$, there is ample possibility for constructing solutions which are stable against small perturbations but which have negative binding energy.

Consider a *Gedankenexperiment* in which such a configuration is expanded homologously. Its energy rises at first, but then begins to sink when its mean density reaches the regime $\langle \rho \rangle < \rho_0$; it is this energy barrier which ensures the stability against small perturbations.[16] When the atmosphere does not have a finite outer radius R_1, then, rather than exploding, the star

16. *Editor's note* (KST).—A situation analogous to this actually occurs in certain low-density, relativistic neutron-star models: see, e.g., the low-mass H-W-W neutron-star models in Table 3 of Hartle and Thorne (1968), which have negative binding energy but are stable against small perturbations.

leaks matter outward by means of a stellar wind (Bisnovatyi-Kogan and Zel'dovich 1966). This is the simplest example of adiabatic, continuous mass loss.

10.13 THE METASTABILITY OF EVERY EQUILIBRIUM STATE

Let us digress somewhat from the collapse problem and make two remarks on problems of principle.

For stars composed of a cold Fermi gas and having a number of nucleons N less than the OV maximum, N_{max}, there always exist one or several static equilibrium configurations. Among these configurations is one with minimum total energy. It is stable with respect to small perturbations. Does that mean that the nucleons cannot be regrouped (without changing their number)! so that the resulting configuration (certainly nonstatic) would have a total energy (and consequently a mass M) smaller than the initial one? We will demonstrate below that such a conclusion would be incorrect; the minimum of energy corresponding to the stationary state is actually only a local minimum (Zel'dovich 1962b).

By compressing the mass with external pressure, in principle we can bring it so close to its gravitational radius[17] that the gravitational forces (which approach infinity at r_g) exceed the pressure forces (which can grow no faster than ρ) and force the mass to contract further, without the influence of the external forces; in other words, they force it to collapse. From this description, it would appear to follow that the collapse of a small mass, while possible, is still separated from the equilibrium state by a gigantic energy barrier. We will demonstrate that this is false; rather, the energy barrier can be arbitrarily small in terms of its absolute value.

We will begin by proving the last assertion. Naturally, the smaller the initial mass, the less the energy that must be expended to compress it to $R \sim r_g$, where collapse sets in. In particular, the density to which the matter must be compressed increases with decreasing mass:

$$\rho \approx 2 \times 10^{16} (M_\odot/M)^2 .$$

Work is done during the compression, and this increases the mass. When a small mass is compressed to high density, practically all the mass is created by the work, so the energy barrier is $M \sim r_g$.

Take a cold configuration in equilibrium. Compress its small central part, forcing it to collapse. Then the layer at the boundary of the collapsing core will lose its interior support and will begin to contract toward the center, also setting in motion the more external layers. Because of the nature of relativistic collapse, the inner layers will fall forever, as measured by the clock of an external observer, never finding support from beneath. Consequently, the external layers will not stop, either. Thus, the entire star will

17. Since we are doing work during the compression, the mass of the matter increases.

become involved in the contraction; i.e., it will collapse. The smaller the region of the initially contracted core, the less energy must be expended to force the entire star to contract from its stable state.

Thus we have proved that the energy barrier separating collapse from equilibrium is infinitely small;[18] however, the perturbations which trigger the collapse are not small at all, and the compression of the matter prior to the onset of collapse is greater, the smaller the required input. For example, a star with a mass equal that of the Sun can be forced to collapse by compressing in its center a core with a mass equal the mass of the Earth. But in order to force such a core to collapse, the compression must be to a density of

$$\rho \approx 2 \times 10^{16} (M_\odot/M_\delta)^2 \approx 2 \times 10^{27} \text{ g cm}^{-3} \text{ !}$$

Naturally, such "fluctuations" cannot just occur—either in a thermodynamic or in a quantum-mechanical way. It is also obvious that we could not have discovered such a possibility of transition to collapse in the linearized theory of small perturbations, which we examined earlier in this chapter. Notice that, although the density perturbations are large here, the perturbations in the star's total energy are small.

Now we will demonstrate that a given number of nucleons, N, can always be arranged together in such a manner that their total energy will be arbitrarily small—i.e., that the mass M measured by an external observer will be arbitrarily small. Assume, for the sake of brevity, that the cold star consists of an ideal Fermi gas. Let N be the number of baryons. We pack them together to sufficient density so that the expression for an ultrarelativistic gas is valid:

$$\rho = \tfrac{3}{4}\hbar(3\pi^2)^{1/3}n^{4/3}/c . \qquad (10.13.1)$$

For M and N we have the formulae for matter at rest (see § 3.3):

$$M = 4\pi \int_0^R \rho(r)r^2dr , \qquad (10.13.2)$$

$$N = 4\pi \int_0^R n(r)e^{\lambda/2}r^2dr . \qquad (10.13.3)$$

18. Einstein's gravitation theory is a nonquantum theory. Thus, proceeding from considerations of dimensionality, we can pinpoint the limits of its applicability (Wheeler 1960) (see chapter 2). From the constants $\hbar$, G, and c^3 one can construct a quantity with the dimensions of length $L^* = (\hbar G/c^3)^{1/2} = 1.6 \times 10^{-33}$ cm.

On a scale smaller than L^*, the quantum fluctuations of the metric must become important. Consequently, a mass with a gravitational radius $r_g = L^*$ is the smallest mass which one could compress to dimensions r_g without addressing himself to the quantum theory. This mass is $m = 10^{-5}$ g, which corresponds to an energy of 10^{16} ergs. This energy constitutes a lower limit on the barrier against collapse, if this barrier is increased by quantum effects. Harrison, Thorne, Wakano, and Wheeler in their book (1965) estimate a possible lower limit on the barrier by stipulating that r_g must be larger than the Compton wavelength (of an electron or of a nucleon). It seems to the authors that such a restriction is unnecessary.

We distribute the baryons so that

$$\rho = a/r^2, \quad r < R; \quad \text{and} \quad \rho = 0, \quad r > R, \quad (10.13.4)$$

where a is an arbitrary constant. Using formulae (10.13.1)–(10.13.4), we obtain $\lambda = \text{const.}$ when $r < R$, and

$$M = \text{const. } N^{2/3}a^{1/2}[1 - (8\pi G/c^2)a]^{1/3}. \quad (10.13.5)$$

The distribution (10.13.4) has the properties $\rho \to \infty$ as $r \to 0$, and ρ discontinuous at $r = R$. It is easy to verify that these singularities can always be smoothed out in such a way that the relationship (10.13.5) changes by an arbitrarily small amount. The resultant distribution has no singularities in either the metric or the density.

It follows from expression (10.13.5) that, for any assigned N, the mass $M \to 0$ as $a \to c^2/8\pi G$. This proves our assertion. Understandably, the resultant configuration is nonstatic; in fact, its mass is close to zero and is certainly smaller than the static configuration of minimum mass for the given N. The nucleons packed in this manner are at rest at the initial moment, but the acceleration is nonzero, and the configuration will collapse.

We see that in principle a machine could be designed which would construct configurations with mass defects arbitrarily close to M_0. The energy released from matter by this machine would almost equal $M_0 c^2$, which is considerably greater than the nuclear energy 0.01 $M_0 c^2$.

Of course, the design of such a machine to operate with masses much smaller than $M^{\text{OV}}_{\text{max}}$ is an impossible task, because matter would have to be compressed to fantastic densities.

For a mass close to the OV limit, the characteristic densities are not at all fantastic, and a transition to collapse is possible; e.g., a star of $M \approx 1.5$ $M_\odot$ can pass inertially through its stable neutron-star state and collapse onward, during the hydrodynamic contraction which is initiated near the Chandrasekhar maximum.

The collapsed mass at the star's center is, in essence, a catalyst for the collapse of the surrounding matter. It is a bottomless hole down which the matter falls. The rate of fall depends upon the motion of the surrounding matter, and on its pressure and density. The relevant formulae will be given in chapter 13.

10.14 EQUILIBRIUM OF A SUPERMASSIVE STAR

A. *Introductory Remarks*

Having considered the equilibrium of cold stars of small mass, we proceed to the completely opposite case, the equilibrium of hot, supermassive stars

The first attempt to explain quasars, undertaken by Hoyle and Fowler (1963*a*, *b*), involved supermassive stars with $M \sim 10^5$–10^9 $M_\odot$ and applied to such stars all the conventional concepts and approximations used in conventional stellar theory. The theory of such hypothetical objects is obviously of interest independently of quasars.

The first contributions did not take into account rotation or turbulence. The effects of macroscopic mass motion have been discussed more recently, and the resultant theory has been used to explain quasars (Fowler 1966; Roxburgh 1965; Ozernoy 1966a, b; Layzer 1965; Pacholczyck 1965; Anand 1965; Bardeen and Anand 1966; Bisnovatyi-Kogan, Zel'dovich, and Novikov 1967).

We recall that a star in its normal state is in hydrostatic equilibrium; the release of nuclear energy (if it is going on) is slow and is not included in the conditions of hydrostatic equilibrium. The energy leaks from the central regions to the surface and is radiated into the surrounding space. Initially, the star consists primarily of hydrogen. The star is on the main sequence of the Hertzsprung-Russell diagram. As its hydrogen burns out, the parameters of the star change slowly. A limiting case is a star which has almost completely exhausted its energy supplies and consists almost wholly of iron. However, we will see below that for stars with $M \geq 5 \times 10^5 \, M_\odot$, in the absence of rotation and turbulent motion, nuclear reactions generally turn out to be nonessential.

In a star in hydrostatic equilibrium with a mass of over 100 $M_\odot$, the entropy is so large that the pressure and internal energy are provided primarily by radiation; and the plasma pressure and energy[19] are relatively small. This was noted as early as 1926 by Eddington; see § 10.2. This characteristic accounts for the difference in structure and evolution between a supermassive star and a normal star (where the determining factor is the plasma energy and the contribution of radiation is relatively small).

For pure radiation, the adiabatic index is $\gamma = \frac{4}{3}$ (which is the critical value for stellar equilibrium; see § 10.1). Because of this, the adiabatic index in massive stars differs very little from $\frac{4}{3}$, and therefore, just as in the theory of white dwarfs (see § 10.8), one must analyze accurately the deviations of γ from $\frac{4}{3}$. The deviation of γ from $\frac{4}{3}$ is related to the fact that the plasma contributes to the pressure; at high temperatures an additional contribution is made by the creation of electron-positron pairs, and by the dissociation of iron, $^{56}Fe \rightarrow 13 \, \alpha + 4 \, n$. It is because of the influence of the plasma that $\gamma > \frac{4}{3}$, and the star can be in stable hydrostatic equilibrium.[20] Just as in the theory of white dwarfs with masses approaching the critical limit, small effects of GTR which can disrupt the stability of the star are important; in the theory of supermassive stars, these factors are important in spite of the fact that the gravitational potential is small, i.e., $\phi \ll c^2$. The

19. The plasma energy is the energy of nuclei and electrons. The plasma pressure is the pressure caused by these particles.

20. A special form of instability of large stars, related to isothermal perturbations, is discussed in § 11.11.

plasma makes a positive contribution to the stability of the star, whereas the e^+e^- pairs and GTR effects make a negative contribution.

The theory of large stars (without consideration of the e^+e^- pairs) was developed by Fowler (1964*a*, *b*) and subsequently and independently by Zel'dovich and Novikov (1965), who used another method and took into account the e^+e^- pairs. A comparison of the two methods will be given in an appendix to this section. The role of e^+e^- pairs was also analyzed by Sato (1966).

In our discussion of supermassive stars we will use the method of successive approximations. First, we will find the equilibrium of the star by using Newtonian theory and considering only the radiation component of the equation of state. Thereafter, we will consider consecutively the effects of plasma, of pair creation, and of GTR. It appears that processes of dissociation of iron and of neutronization of matter are not essential for the equilibrium state of a supermassive star. The effect of rotation upon equilibrium will be considered in §§ 11.12–11.16 and 11.18.

B. Stellar Equilibrium with $\gamma = \frac{4}{3}$

The total energy E of a star may be expressed in the form

$$E = \int_V E_1 \rho dV - G\int_V (m\rho/r)dV . \qquad (10.14.1)$$

The first integral is the internal energy; the second is the gravitational energy. E_1 is the internal energy per unit mass, and m is the mass inside a sphere with radius r, $m = 4\pi \int \rho r^2 dr$, where the integration is from 0 to r.

Consider as a zero-order approximation the energy of radiation (however, disregard the equivalent mass of the radiation as compared with the mass of the plasma). The specific internal energy E_1 per unit mass, specific entropy per unit mass, and pressure can be expressed in the form

$$E_1 = aT^4/\rho ; \quad S = \tfrac{4}{3}aT^3/\rho ; \quad P = aT^4/3 = b\rho^{4/3} ; \quad (10.14.2)$$

where

$$a = 7.7 \times 10^{-15} \text{ erg cm}^{-3} \text{ deg}^{-4} ; \quad b = (3/256a)^{1/3}S^{4/3} .$$

From this we express E_1 in terms of ρ and S:

$$E_1 = 3^{4/3}4^{-4/3}a^{-1/3}S^{4/3}\rho^{1/3} = 3b\rho^{1/3} . \qquad (10.14.3)$$

If the distribution of matter in the star, $\rho = \rho(r)$, is known, then upon substituting equation (10.14.3) into equation (10.14.1), and integrating, we obtain

$$E = k_1 bM\rho_c^{1/3} - k_2 GM^{5/3}\rho_c^{1/3} , \qquad (10.14.4)$$

where ρ_c is the central density, and the constants k_1 and k_2 depend on the distribution of matter in the star (see Appendix 1 to this section). In our case, the relationship between pressure and density at constant[21] S has a polytropic form, $P \sim \rho^{4/3}$, with a polytropic index $n = 1/(\gamma - 1) = 3$. Using the density distribution for a polytrope with $n = 3$ (see § 8 of this chapter), we obtain the numerical values of the constants: $k_1 = 1.75$ and $k_2 = 0.638$. The equilibrium of the star is determined by the extremization of E at constant mass (to be more precise, at constant number of nucleons) and at constant entropy S. The only quantity that varies in equation (10.14.4) is ρ_c. The condition of equilibrium (the condition of extremal E), $dE/d\rho_c = 0$, is satisfied only when

$$k_1 bM - k_2 GM^{5/3} = 0 . \tag{10.14.5}$$

The equilibrium state is neutral; it does not depend on ρ_c.

We emphasize that neutral equilibrium takes place only with respect to contraction and expansion of the star as a whole, i.e., with respect to a homologous motion of the star's matter. The star is stable with respect to any deformation of its density distribution.

For a star in equilibrium, we find from equation (10.14.5) that

$$b_{eq} = (k_2/k_1)GM^{2/3} = 0.364 \, GM^{2/3} , \tag{10.14.6}$$

and from this value of b_{eq} we obtain for the value of the entropy (measured in ergs per degree Kelvin), which is unique for the given mass,

$$S_{eq} = 7.85 \times 10^7 (M/M_\odot)^{1/2} . \tag{10.14.7}$$

In this approximation, the total energy of the star is identically zero, and the density and temperature at any point are connected by the relation

$$T = 1.97 \times 10^7 \, °\text{K} \, (M/M_\odot)^{1/6} \rho^{1/3} . \tag{10.14.8}$$

If the entropy is not equal to the equilibrium entropy, then from the general expression for the energy (10.14.4), for $|S - S_{eq}| < S_{eq}$ we obtain

$$E = 2.3 \times 10^{40} (M/M_\odot)^{7/6} (S - S_{eq}) \rho_c^{1/3} .$$

When $S > S_{eq}$, the energy grows monotonically with ρ_c; and when $S < S_{eq}$, the energy drops monotonically (cf. Fig. 48).

C. *The Influence of the Plasma*

Next we consider the effects on the equation of state produced by the energy and pressure of the plasma. At a given temperature, the effect of the plasma is to increase the internal energy. Instead of expression (10.14.2) for E_1, we can write

$$E_1 = aT^4/\rho + \tfrac{3}{2}\Re T/\mu ,$$

21. Generally speaking, S may vary over the star, but it cannot decrease toward the surface; otherwise a convection takes place and equalizes S again. Qualitatively, the variability of S over the stellar matter does not affect the result; however, to be definite, assume $S = $ const. throughout the star. In large stars, convection probably makes S constant.

where the second term on the right is the plasma energy, and μ is the mean molecular weight. However, at a given entropy the effect of the plasma is to decrease the energy. Qualitatively, this is obvious from a general principle: The state of thermodynamic equilibrium corresponds to the maximum entropy at a given energy, or, expressed differently, to the minimum energy at a given entropy. The zero-order approximation, in which the

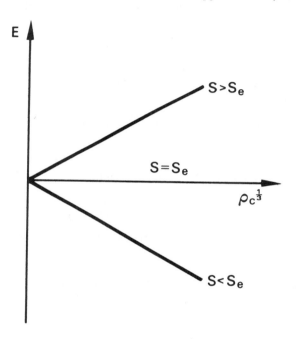

Fig. 48.—Energy E of a star with fixed entropy, as a function of $\rho_c^{1/3}$, where ρ_c is the central density. The pressure and thermal energy are due to the radiation field.

energy of the plasma is not taken into account, corresponds to a state consisting of radiation and cold plasma with zero total energy. A transition to a fully equilibrium state at the given entropy can only decrease the energy. Numerically, the plasma correction to the internal energy of the matter can be computed from the condition of thermodynamic equilibrium. At constant entropy it is expressed by

$$\Delta E_{1,\,\text{plasma}} = -3.85 \times 10^{12}\, \frac{S^{1/3}}{A}\, \rho^{1/3} \left(\ln \frac{aS^{1/2}}{\rho^{1/2}} + Z \ln \frac{bS^{1/2}}{\rho^{1/2}} \right), \quad (10.14.9)$$

where $a = 8.6 \times 10^3\, g\, A^{5/2}$, $b = 2.18 \times 10^{-1} A/Z$, A is the atomic weight, Z is the charge of the nucleus, and g is the statistical weight of the nucleus.

This expression is valid in the domain where plasma corrections to the energy and entropy are small, and where the plasma is a nondegenerate ideal gas. In an equilibrium star with a mass of 10^4–$10^8 \, M/M_\odot$, these restrictions are satisfied to adequate accuracy.[22]

Now we can compute the energy of the entire star, taking into account the influence of the plasma. As was noted in § 10.8, the correction to the equation of state (of the order of $a = \Delta E_1/E_1 \ll 1$) not only changes the energy in a given mass element but also induces a change in the distribution of the matter in the star which is also proportional to a; this variation in density must be included in the energy computation. However, in view of the extremal properties of the density distribution, this variation causes a a variation of the order of a^2 in the total energy of the star. Therefore, to compute the correction for the total energy of the star to order a, one may integrate ΔE_1 over the distribution for the zeroth approximation (over the unperturbed Emden function); this will give the first term (of the order of a) in the expansion of the energy in powers of a.

Integrating equation (10.14.9) over the density distribution of the star, we obtain:

$$\Delta E_{\text{plasma}} = -\rho_c^{1/3}\{a_1 - b_1[1.176 \ln \rho_c - 1.615]\} \,,$$
$$a_1 = 4.5 \times 10^{45} S^{1/3}(M/M_\odot)A^{-1}\{\ln (8.6 \times 10^3 g A^{5/2}S^{1/2})$$
$$+ Z \ln (2.2 \times 10^{-1}S^{1/2} \, A/Z)\} \,, \qquad (10.14.10)$$
$$b_1 = 1.925 \times 10^{45} S^{1/3}(M/M_\odot)A^{-1}(1 + Z) \,.$$

This expression is valid only under the restrictions mentioned above.

Instead of Figure 48, we now obtain the series of curves $E(\rho_c)$ shown in Figure 49 for various values of the entropy. All the curves correspond to the same value of the mass; the entropy is the distinguishing parameter. The curves now have minima. These minima correspond to equilibrium states of the star and are marked by vertical bars. The broken line is the locus of the minima $E_{\text{eq}}(\rho_c)$. In the coordinates of Figure 49, the curves $E(\rho_c)$ may be obtained from one another by a similarity transformation; the broken line is a straight line; and (according to the virial theorem)

$$E_{\text{eq}} = -\tfrac{3}{2} \left(\Re\langle T\rangle/\langle\mu\rangle\right) M \,;$$

that is, the energy of the star equals minus the thermal energy of the plasma. Since $\langle T\rangle/T_c \approx 0.6$, we obtain

$$E_{\text{eq}} = -5 \times 10^{41}(M/M_\odot)\langle T\rangle/\langle\mu\rangle = -3 \times 10^{41}(M/M_\odot)T_c/\langle\mu\rangle \,.$$

Substituting the expression for T_c from equation (10.14.8), we obtain

$$E_{\text{eq}} = -5.9 \times 10^{48}(M/M_\odot)^{7/6}\rho_c^{1/3}/\langle\mu\rangle \,.$$

22. For reference purposes, without derivations, the first-order correction for the entropy of a hydrogen plasma is $S_{\text{plasma}} = 8.3 \times 10^7 \, (15.5 + 2 \ln T^{3/2} - 2 \ln \rho)$, where the temperature is in degrees Kelvin and the density is in grams per cubic centimeter. For an equilibrium hydrogen star, this gives $S_{\text{plasma}}/S \approx 60(M/M_\odot)^{-1/2}$; for the pressure we have $\Delta P_{\text{plasma}}/P \approx 8.6(M/M_\odot)^{-1/2}$. Since the additive constant in the entropy is not essential, it is possible to develop a more complex approximation with an error of the order of $(\Delta P/P)^2$. We will not pursue this further here.

D. The Creation of Electron-Positron Pairs

At temperatures of the order of 5×10^8 ° K ($kT/m_e c^2 \approx 0.1$), there is a significant number of electron-positron pairs in equilibrium. At a given temperature, the creation of pairs increases the energy of the matter; however, as noted previously, it follows from thermodynamics that at a given entropy the creation of pairs decreases the energy.

It will be demonstrated below that the creation of pairs causes an instability of a massive star at a temperature which (in energy units) is 10–15

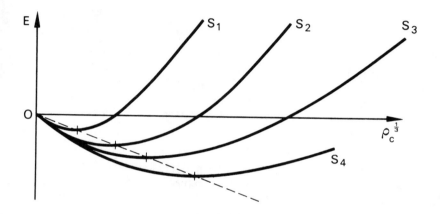

FIG. 49.—Energy E of a star, including the contribution of the plasma (electrons and nuclei) to the pressure. $S_1 > S_2 > S_3 > S_4$. Minima of the curves correspond to the equilibrium positions of a star of a given entropy. *Dashed line*, locus of equilibrium positions.

times smaller than the energy of one pair, $2m_e c^2$. (For reference: $2m_e c^2 = 1.02$ MeV, $\theta = 2m_e c^2/k = 11.9 \times 10^9$ ° K). When $T \sim 10^9$, the number of positrons in the entire star is no greater than several percent of the number of electrons, and even at the center of the star $n_+ < 0.25\, n_-$. Therefore, we use asymptotic formulae (see § 8.3, and formula [8.3.9]) for a nonrelativistic, nondegenerate gas

$$n_+ n_- = \frac{4(2\pi m_e kT)^3}{(2\pi \hbar)^6} \exp\left(-2m_e c^2/kT\right) . \qquad (10.14.11)$$

Here we have assumed that $n_+ \ll n_-$, $n_- = n_{-0} = \rho/\mu_e m_p$, $n_+/n_- = n_+ n_-/n_{-0}^2$ where n_{-0} is the number of electrons in matter of a given density when pair production is not taken into account. Substituting numerical values, we may write

$$n_+/n_- = 1.1 \times 10^{14}\, \mu_e^2 (1/\rho^2)(T/\theta)^3 \exp\left(-\theta/T\right) .$$

The quantity θ is defined above.

Using the zeroth approximation (see formula [10.14.8]) and expressing the density of the star in terms of its mass and temperature, we obtain

$$n_+/n_- = \tfrac{1}{430} (M/M_\odot)\mu_e^2 (T/\theta)^{-3} \exp(-\theta/T) .$$

We can substitute the local temperature into this formula and obtain the local value of n_+/n_-. In particular, the formula is valid at the center, where $T = T_\bullet$.

For averaging any quantity x (which changes rapidly with varying temperature or density) over a star, there exists a convenient formula:

$$\langle x \rangle = x_c \times 3.2 (d \ln x/d \ln T)_c^{-3/2} .$$

The derivation of this formula is based on the fact that near the center the temperature (as well as the density) has a maximum of the form

$$T = T_c(1 - ar^2) .$$

The domain in which x changes by a factor e is the domain where T changes by $T/(d \ln x/d \ln T)$. The mass in this domain is proportional to r^3. From this the form of the formula follows. In determining the coefficients, we have used the Emden distribution. Using this formula, we find that, for the case with which we are concerned,

$$\langle n_+/n_- \rangle = (n_+/n_-)_c \times 3.2(\theta/T_c - 3)^{-3/2} \approx (n_+/n_-)_c \times 3.2 (T_c/\theta)^{3/2} .$$

Consider now the thermodynamic aspects of this situation. The additional energy which is created at a given temperature as a result of the creation of pairs is $2m_ec^2n_+$ per unit volume, or $\Delta E_1|_T = 2m_ec^2n_+/\rho$ per unit mass. We neglect terms of order T/θ (in this case, the kinetic energy of positrons compared with their rest mass); this automatically requires that we also neglect the pressure of the pairs.

Changes in the energy at fixed entropy are related to changes in the energy at fixed temperature by

$$\Delta E_1|_S = -T \frac{d\Delta E_1|_T}{dT} = -(T/\theta)\Delta E_1|_T = -kTn_+ /\rho .$$

Let us calculate the energy of the entire star for a given entropy. Instead of $E(S, \rho_c)$, it is convenient to calculate $E(S, T_c)$. Expressing ρ in terms of T by means of formula (10.14.8) in the zeroth approximation, we first obtain for the energy in terms of mean temperature $\langle T \rangle$:

$$E = A(S - S_{eq})\langle T \rangle + \tfrac{3}{2} (R\langle T \rangle/\mu)M(B + \ln \langle T \rangle) - k\langle T \rangle N_+ , \quad (10.14.12)$$

where N_+ is the total number of positrons in the star. It is convenient to re-express this formula as

$$E = A'(S - S_{eq})T_c + DT_c[\ln T_c - \tfrac{2}{3}(\mu/\mu_e)(n_+/n_-) + B']$$

$$= A'(S - S_{eq})T_c + DT_c\Big[\ln T_c - \tfrac{2}{3}\mu\mu_e \frac{M/M_\odot}{430} \qquad (10.14.13)$$

$$\times (T/\theta)^{-3/2} \exp(-\theta/T) \times 3.2(T/\theta)^{3/2} + B'\Big] .$$

For iron, $\mu \approx \mu_e \approx 2$, so

$$E = A'(S - S_{eq})T_c$$

$$+ DT_c[\ln T_c - \tfrac{1}{50} M/M_\odot (T_c/\theta)^{-3/2} \exp(-\theta/T_c) + B'].$$

The contribution of pairs to the energy (which is negative in sign) drastically increases in absolute value with increasing temperature. When we take the pairs into account, the curves of Figure 49 are changed to the form shown in Figure 50.

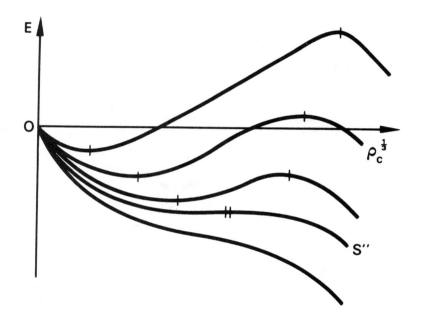

Fig. 50.—Change in the energy curves of Fig. 49 due to the creation of e^+e^- pairs (for $M < 10^4\, M_\odot$) or to the effects of GTR (for $M > 10^4\, M_\odot$).

In addition to the minima, maxima also appear now on the isentropic curves; the minima and maxima are marked by a vertical bar. On a curve that corresponds to some particular entropy—the "critical entropy" S_{cr}—the maxima and minima coalesce, giving a point of horizontal inflection (marked by two bars). When $S < S_{cr}$ (for a fixed mass), there is no extremum of $E(\rho_c)$, i.e., there is no equilibrium. The equilibrium corresponding to the maximum of any curve is unstable.

The locus of the extrema of $E(\rho_c, S)$, i.e., the curve of the equilibrium energy $E_e(\rho_c)$, is shown in Figure 51. The minimum of this curve corresponds to the horizontal inflection point of Figure 50; the descending branch of E_e corresponds to stable equilibrium; and the ascending branch corresponds to unstable equilibrium. On the ascending branch there is a domain where $E_c > 0$. We recall that the corresponding positive-energy states, which lie at the maxima of isentropic curves (Fig. 50), arose because of the negative correction to the energy. This correction caused the appearance of extrema which did not exist previously without the correction (Fig. 49); the energy itself was positive without the correction, of course.

The critical state, as noted previously, is reached when

$$\partial E/\partial \rho_c = \partial^2 E/\partial \rho_c{}^2 = 0 , \quad \text{i.e., } \partial E/\partial T_c = \partial^2 E/\partial T_c{}^2 = 0 .$$

We examine primarily the condition $\partial^2 E/\partial T_c{}^2$, inasmuch as the condition on the first derivative can always be satisfied by choosing an appropriate S. All

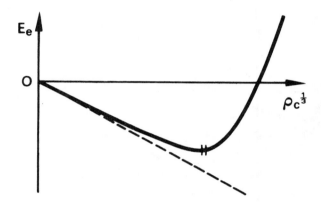

FIG. 51.—Energy E_e of an equilibrium star, as a function of central density. Dashed line indicates the energy of an equilibrium star, including radiation and plasma but neglecting the creation of pairs and the effects of GTR.

the constants A'_1, S_e, D, B' drop out of equation (10.14.13) for E when condition $\partial^2 E/\partial T_c{}^2 = 0$ is imposed, leaving the condition

$$1 - \tfrac{1}{50} M/M_\odot \, (T_c/\theta)^{-7/2} \exp\,(-\theta/T_c) = 0 . \qquad (10.14.14)$$

Table 12 shows the parameters of the critical states for three typical stars made of iron.

The numbers cited in Table 12 confirm the validity of the assumptions $T_c/\theta \ll 1$, and $(n_+/n_-)_c < 1$.

For a hydrogen star with $\mu = \tfrac{1}{2}$, $\mu_e = 1$, the data are shown in Table 13.

The largest masses shown in Tables 12 and 13 (6000 $M_\odot$ for iron; 24000 $M_\odot$ for hydrogen; on the average $\sim 10^4 M_\odot$) are the limits where the effects of GTR become strongly pronounced; we will consider these effects below.

The corrections in the equation of state due to nuclear dissociation (for example, $^{56}\text{Fe} \rightarrow 13 \, \alpha + 4n$) generally require a much higher temperature than corrections due to the creation of pairs; in the theory of the equilibrium of supermassive stars, these processes (and also the processes of neutronization of matter) are not essential.

The correction to the energy of a hot star with large mass due to GTR is given by the same formula as in the case of white dwarfs $\Delta E_{\text{GTR}} = -0.93\ (G^2/c^2)M^{7/3}\ \rho_c{}^{2/3}$. The sign of this correction, and the nature of its influence upon the overall picture is about the same as for pairs e^+e^-. Thus, there are two reasons for the transition from Figure 49 to Figure 50. For

TABLE 12

PARAMETERS OF THE CRITICAL STATE CAUSED BY e^+e^-
PAIRS IN A STAR MADE OF IRON

PARAMETER	$M/M_\odot$		
	300	3000	6000
$T_c\ (10^9\ {}^\circ\text{K})$........	1.2	0.92	0.87
$\rho_c\ (\text{g cm}^{-3})$.........	10000	1600	800
θ/T_c..............	10	13	14
$e^{-\theta/T_c}$..............	5×10^{-5}	2.4×10^{-6}	8×10^{-7}
$\langle n_+/n_-\rangle$...........	0.022	0.013	0.011
$(n_+/n_-)_c$...........	0.22	0.18	0.18

TABLE 13

PARAMETERS OF THE CRITICAL STATE CAUSED BY
e^+e^- PAIRS IN A STAR MADE OF HYDROGEN

PARAMETER	$M/M_\odot$	
	2400	24000
$T_c\ (10^9\ {}^\circ\text{K})$..............	1.2	0.92
$\rho_c\ (\text{g cm}^{-3})$..............	3600	550

NOTE.—At a given temperature the quantities shown in Table 12 but not here are independent of composition.

practical purposes, only one of the reasons is significant for a given mass. When the mass is less than $10^4\ M_\odot$, the pair effect dominates; when the mass is larger than $10^5\ M_\odot$, the GTR effect dominates. Such a pronounced separation is a consequence of the strong dependence of the equilibrium number of pairs upon temperature. With increasing density, in equilibrium stars of mass greater than $10^4\ M_\odot$, the effects of GTR change the shape of the isentropic curve long before a temperature sufficient for intensive pair creation is reached.

Thus, for stars with $M > 10^4\ M_\odot$, the transition from Figure 49 to Figure 50 is due to GTR. Consider again the curve E_e of Figure 51. For stars with $M > 10^4\ M_\odot$, the appearance of a minimum and of an ascending branch in this curve is related to GTR effects. The presence here of a domain where $E_e > 0$ is also caused by GTR. However, we emphasize again (see § 10.8) that positive energies in an equilibrium state are not necessarily related to GTR; they may arise for other reasons as well.

The critical state, corresponding to the minimum E_e for a given mass, and marked by two bars on Figure 51, is extremely important. It corresponds to the last equilibrium state in the sequence with decreasing entropy. In this state, the star has the minimum possible equilibrium energy for a given mass, and the maximum possible temperature and density.

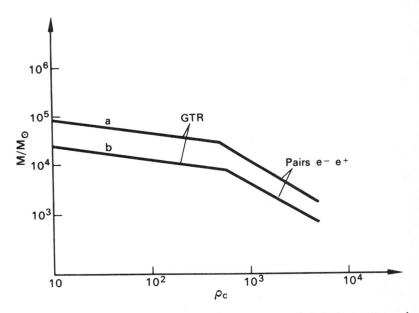

FIG. 52.—The critical states of massive stars; (a) corresponds to hydrogen stars and (b) to iron stars.

For $M/M_\odot > 10^5$, the critical state is determined by the effects of GTR. In this case, the following formulae (see Fowler 1964a) are valid:

$$\rho''_c = 2.43 \times 10^{17}(1/\mu^3)(M/M_\odot)^{-7/2} \text{ g cm}^{-3} , \qquad (10.14.15)$$

$$T''_c = 1.23 \times 10^{13}(1/\mu)(M/M_\odot)^{-1} \,^\circ\text{K} , \qquad (10.14.16)$$

$$E'' = -0.93 \times 10^{54}(1/\mu^2) \text{ ergs} . \qquad (10.14.17)$$

The energy E'' of the critical state does not depend on the mass. Comparing these expressions with the critical values for pairs, we find that the influence of both effects becomes equal when $M \approx 8 \times 10^3$ for iron, and when $M \approx 3 \times 10^4$ for hydrogen. The locus of critical states in the M-ρ_c diagram is shown in Figure 52.

The critical states of supermassive stars, as well as the entire evolution of such stars, depend in an essential manner on the rotation of the star. We will examine the effects of rotation in the following chapter.

10.15   CRITICAL STATES OF STARS WITH INTERMEDIATE MASS

We turn our attention now to stars with a mass that is intermediate between $1.2\ M_\odot$ and $10^3\ M_\odot$. Let us concentrate attention on the critical states for such stars; i.e., let us calculate the curve bb' of Figure 34. Some considerations regarding the factors that lead to instability have been discussed by Hoyle and Fowler (1960, 1965), Fowler and Hoyle (1964), and Zel'dovich (1963a). Specific numerical values with certain simplifying assumptions as to the chemical composition have been contributed by Bisnovatyi-Kogan (1966) and by Bisnovatyi-Kogan and Kazhdan (1966). Below we will cite the conclusions of this work.

A star with a mass greater than the mass limit of white dwarfs may be in equilibrium only at a nonzero temperature;[23] and it becomes unstable as $T \to 0$.

We begin with an investigation of stars whose mass only slightly exceeds the limit on the mass of a cold white dwarf. For such stars, the temperatures near the critical state are still so small that the principal contribution to the energy and pressure is made by degenerate electrons; this degeneracy, because of the high density, is a relativistic degeneracy. We will regard as small corrections the contribution to the energy due to nonzero temperature of the electrons, due to the difference between this degeneracy and purely relativistic degeneracy, due to nondegenerate nuclei, and due to the effects of GTR.

The thermal conductivity of degenerate electrons is extremely high, and therefore we will consider the star to be isothermal.[24] The pressure is provided primarily by degenerate electrons, and depends only to a small extent upon T, $\gamma \approx \frac{4}{3}$; consequently, we consider the distribution of matter in the star to be represented by an Emden polytrope with $n = 3$. As emphasized previously, because of the extremum properties of the structure functions the result is only slightly sensitive to this choice of density distribution. We use now the general result of the energy approach derived in the appendix to § 10.14.

The total energy of the star is

$$E = M \int_0^1 E_1(\rho, T)dz - 0.639\ GM^{5/3}\rho_c^{1/3} - 0.93(G^2M^{7/3}/c^2)\rho_c^{2/3}, \quad (10.15.1)$$

23. To find the critical state for stars with $M > M_{\text{Chandra}}$, naturally we consider stars with ρ_c of the same order as in white dwarfs, and in particular we require that the density be less than nuclear density. As we have seen (§ 10.9), at nuclear density in equilibrium, there exist cold stars with $M < \sim 2\ M_\odot$.

24. Naturally, we exclude the narrow surface layer, which is not essential to the energy balance.

where $z = m/M$, the first term is the energy of the plasma, the second term is the Newtonian gravitational energy for the polytrope of index $n = 3$, and the third term is the correction for GTR. At present we will not include changes in the stellar energy due to neutronization.

By using general thermodynamic expressions for the energy and entropy (see Part II), one can derive the following expressions for nearly degenerate, ultrarelativistic electrons:

$$E_1 = \frac{(3\pi^2)^{1/3}}{4} \frac{\hbar c \rho^{1/3}}{(\mu_e m_p)^{4/3}} + \frac{1}{4}\left(\frac{3}{\pi}\right)^{2/3} \frac{m_e^2 c^3}{\hbar(\mu_e m_p)^{2/3}\rho^{1/3}}$$
$$+ \frac{\pi}{2}\left(\frac{\pi}{3}\right)^{1/3} \frac{(kT)^2}{\hbar c(\mu_e m_p)^{2/3}\rho^{1/3}} + \frac{3}{2}\frac{kT}{A m_p}, \tag{10.15.2}$$

$$S = \pi\left(\frac{\pi}{3}\right)^{1/3} \frac{kT}{\hbar c(\mu_e m_p)^{2/3}\rho^{1/3}}$$
$$+ \frac{k}{A m_p}\left\{\frac{5}{2} + \ln\left[\left(\frac{kT A m_p}{2\pi\hbar^2}\right)^{3/2}\frac{g A m_p}{\rho}\right]\right\}. \tag{10.15.3}$$

Here ρ is the mass density of baryons, A is the atomic number of the nucleus, and g is the statistical weight of the nucleus. For the time being, we will neglect the electron mass. The difference between the total mass and the rest mass is taken into account in equation (10.15.1) in the correction for GTR (see appendix to § 10.8) and is not included in equation (10.15.2) or (10.15.3).

These expressions are valid at temperatures below the temperature of electron degeneracy and at densities greater than the density of relativistic degeneracy (see Part II). For an iron star, these conditions yield

$$T < 5 \times 10^7 \rho^{1/3}\,{}^\circ\text{K}, \tag{10.15.4}$$

$$\rho > 2.2 \times 10^6 \text{ g cm}^{-3}. \tag{10.15.5}$$

The conditions for the equilibrium of the star are obtained by equating to zero the derivative of the total energy (eq. [10.15.1]) with respect to $\rho_c^{1/3}$, while holding the entropy of each element of matter fixed:

$$\frac{\partial E}{\partial \rho_c^{1/3}} = 3\rho_c^{-4/3}\int_0^1 P(\rho, T)\frac{dx}{\phi(x)} - 0.639\, GM^{5/3}$$
$$- 1.86(G^2 M^{7/3}/c^2)\rho_c^{1/3}. \tag{10.15.6}$$

Here $(\partial E_1/\partial\rho)_S = P$; and $\phi(x)$ is the Emden function for $\gamma = \frac{4}{3}$. Using expressions (10.15.2) and (10.15.3) for the thermodynamic functions, we obtain a series of equilibrium states for a star of given mass, in which ρ_c is a parameter (Fig. 53).

In the sequence of equilibrium states for fixed mass there exists a critical point which separates stable states from unstable states. To find this critical state, we add the condition that the second derivative of E with respect to $\rho_c^{1/3}$ be zero, to condition (10.15.6) and obtain

$$\frac{\partial^2 E}{(\partial \rho_c^{1/3})^2} = 9\rho_c^{-5/3} \int_0^1 (\gamma - \tfrac{4}{3}) P(\rho, T) \frac{dx}{\phi(x)}$$

$$- 1.86(G^2 M^{7/3}/c^2) = 0 . \qquad (10.15.7)$$

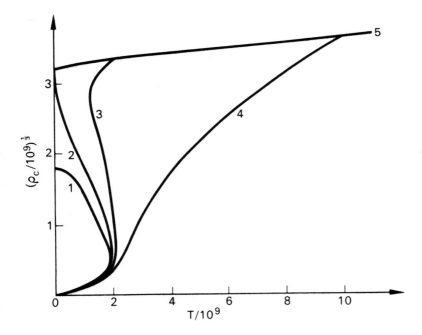

Fig. 53.—Sequences of stable equilibrium states for $M = $ const., and the curve of critical states for stars of intermediate mass. Curve 1, $M = 1.19\ M_\odot < M_{\mathrm{Chandra}}$; curve 2, $M = 1.2\ M_\odot = M_{\mathrm{Chandra}}$; curve 3, $M = 1.23\ M_\odot > M_{\mathrm{Chandra}}$; curve 4, $M = 1.36\ M_\odot > M_{\mathrm{Chandra}}$; curve 5, critical states at which instability sets in.

We emphasize again that at present we are ignoring neutronization, so that in the integrand of equation (10.15.7), it is always true that $\gamma > \tfrac{4}{3}$, and the loss of stability is due to the small effects of GTR (the last term in eq. [10.15.7]).

The curve of critical states is shown on the ρ_c-T_c diagram of Figure 53. Above this curve, there are no stable states. Figure 54 shows the curves

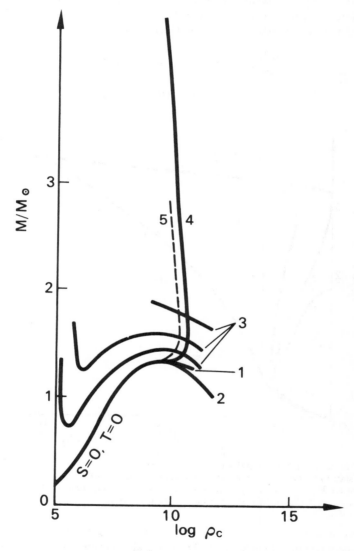

Fig. 54.—Isothermal contours and the curve of critical states on the (ρ_c, M)-diagram. (1) $S = 0$, $T = 0$ without inclusion of neutronization, but with the effects of GTR included; (2) $S = 0$, $T = 0$, but with both neutronization and GTR included; (3) isothermal contours (schematic); (4) the curve of critical states without taking neutronization into account but including GTR; (5) the curve of critical states including both neutronization and GTR.

$M = [M(\rho_c)]_{T=\text{const.}}$ and a curve indicating the onset of instability. (Such curves, without inclusion of GTR, were calculated by Baglin 1965; regarding their stability, see Baglin 1966.) The lines $M = \text{const.}$ are the lines of evolution of a star without mass loss. When $M < M_{\text{Chandra}}$, the line of evolution ends at $T = 0$ and does not reach the curve of instability. When $M > M_{\text{Chandra}}$, the hydrostatic evolution ends with a loss of stability.

It should be emphasized that the maxima of the constant-temperature curves are not really the critical points. We have emphasized frequently that the criterion for stability involves $(\partial M/\partial \rho_c)_{S=\text{const.}}$, rather than $(\partial M/\partial \rho_c)_{T=\text{const.}}$. The loss of stability takes place to the right of the maxima of the curves $T = \text{const.}$

At low temperatures ($T < 10^9 \, ^\circ$K), the principal contribution to the correction of the energy is due to the nondegenerate nuclei. For nuclei, the adiabatic index is $\gamma = \frac{5}{3}$, so they cause a stabilizing action even though they have very little influence upon the critical mass. As a result, the critical density increases and the curve of stability loss (Fig. 54) moves to the right, almost horizontally. When $T > 10^9 \, ^\circ$K, the fundamental role in the energy corrections to T is played by electrons; the critical mass increases rapidly; and the curve of instability moves upward and to the left.

As yet we have not considered neutronization. For an iron star,[25] when a small increase of mass above the critical point occurs, it is neutronization that must cause instability because (as was demonstrated in § 10.8) when $T = 0$ and $S = 0$, the neutronization of iron begins at a density 30 times smaller than the critical density due to GTR. Calculations of the onset of instability in white dwarfs due to neutronization have not yet been made. They require a consideration of nuclear processes which take place at a nonzero temperature.

Qualitatively, the curve that denotes the loss of stability must retain its shape, but it will shift leftward, approximately as indicated by the broken line of Figure 54.

Let us now consider the interval of masses $5 \, M_\odot < M < 10^3 \, M_\odot$. The central temperatures for such stars close to the critical state are so large that temperature effects can no longer be regarded as small corrections to the energy. Over the entire span of evolution the electron gas does not become degenerate. However, new factors enter the picture—dissociation of iron into α-particles, protons, and neutrons (see chapter 8), and (for large masses, $M \approx 10^3 \, M_\odot$) the creation of e^+e^- pairs. The effects of GTR are still considered small.

Bisnovatyi-Kogan and Kazhdan (1966) have solved equations (10.15.6)

---

25. At $T \approx 3 \times 10^9 \, ^\circ$K and $\rho \approx 10^9$ g cm^{-3} nuclear reactions instantaneously result in combustion of the entire supply of nuclear fuel.

The Equilibrium and Stability of Stars

and (10.15.7) by numerical integration, using the equation of state of Imshennik and Nadezhin (see chapter 7, and § 8.4). In these calculations, the stars are considered isentropic since the electrons are nondegenerate, and their terminal conductivity is low.

The results of the computation are shown in Tables 14 and 15.

Table 15 gives the total energies of stars, E_{cr}, in the critical state, and their energies per unit mass, E_{cr}/M.

TABLE 14

VALUES OF ρ_c, T_c, AND S AT THE ONSET OF INSTABILITY FOR STARS OF VARIOUS MASSES

$M/M_\odot$	$\rho_c/(10^7$ g cm$^{-3})$	$T_c/(10^9 \,^\circ$ K$)$	$S/(10^9$ ergs [g $^\circ$ K]$^{-1})$
5	10.2	6.72	0.205
10	4.2	6.40	0.316
50	1.0	5.96	0.682
100	0.94	6.38	0.99
500	0.31	5.99	2.13
1000	0.00063	1.12	2.76

TABLE 15

TOTAL ENERGIES AND ENERGIES PER UNIT MASS FOR STARS AT THE CRITICAL STATE

	$M/M_\odot$			
	1.2	5	10	10^3
$\dfrac{E_{cr}}{10^{50}\text{ ergs}}$	-4.74	-13.4	-18.8	-37
$\dfrac{E_{cr}/M}{10^{17}\text{ ergs g}^{-1}}$	-2.0	-1.3	-0.94	-0.018

Figure 55 shows the relationship between the central density of an iron star in its critical state and its mass, for the interval $1.2\,M_\odot < M < 10^5\,M_\odot$.

For masses close to $1.2\,M_\odot$, the onset of instability is caused by the neutronization of matter; GTR also plays a role. When $5\,M_\odot < M < 500\,M_\odot$, the critical state is determined by the dissociation of iron into ^{4}He, p, and n, and by the effects of GTR. With increasing mass, the role of GTR decreases. When the mass is a little smaller than $10^3\,M_\odot$, the loss of stability occurs because of the formation of e^+e^- pairs, since the pressure in such a star is primarily due to radiation. In smaller masses, the contribution of radiation to the pressure is small; therefore, the creation of pairs does not cause a critical state. When the creation of e^+e^- pairs becomes important for the critical state, ρ_{crit} rapidly decreases. Finally, when $M > 10^4\ M_\odot$, the critical state is again determined by the effects of GTR (see § 14).

The calculations indicate that for large masses, stable equilibrium states at high densities, $\rho > \rho_{\text{crit}}$, are impossible, as was also the case for a cold star in the neutron state. The effects of GTR cause an instability in spite of the fact that $\gamma > \frac{4}{3}$ is possible in a hot plasma when T_c is large.

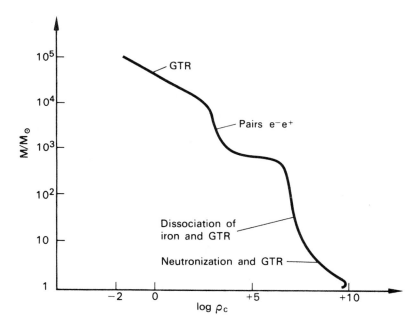

Fig. 55.—Instability curve for iron stars; causes of instability are indicated in individual areas.

It follows that for large hot masses the curve $M(\rho_c)|_{S=\text{const.}}$ has one maximum rather than two as for cold stars.[26] The vanishing of the second maximum with increasing entropy occurs by means of its combining with the minimum. At some entropy S_0, the second maximum and the minimum vanish; this is demonstrated schematically in Figure 56. The point of merger

26. The oscillations of the curve $M(\rho_c)|_{s=0}$ beyond the second maximum are not important (see § 10.11).

of the second maximum with the minimum has been computed by Bis-novatyi-Kogan (1968a). This point is at $M = 70\,M_\odot$, $\rho_c \approx 2 \times 10^{10}\,\mathrm{g\ cm^{-3}}$.

This completes the task posed in § 10.5: We have determined the boundaries of possible stable equilibrium states of stars in the mass-density diagram.[27]

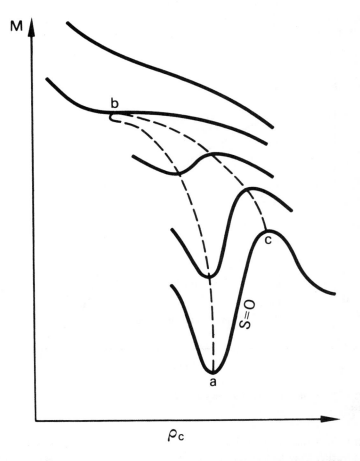

Fig. 56.—Domain of stability of neutron stars. *Solid lines*, isentropes; *ab*, locus of the minima; *bc*, locus of the maxima.

27. Of course, this is only a rough picture. Other types of instabilities, which yield additional restrictions, will be discussed in subsequent sections.

APPENDIX TO § 10.1

We will demonstrate that in Newtonian theory the condition that the total energy of a star be an extremum (with constant chemical composition and conservation of entropy in each mass element) is equivalent to the condition of hydrostatic equilibrium. The total energy of the star, in the absence of macroscopic motions, may be expressed in the form

$$E = \int_0^M E_1(S, \rho)dm - \int_0^M (Gm/r)dm \,, \qquad (10.1A.1)$$

where $m = \int 4\pi r^2 \rho dr$ (the integration is from 0 to r) is the mass inside a sphere with radius r.

Using the thermodynamic identity

$$P = - \left[\frac{\partial E_1}{\partial(1/\rho)}\right]_{S=\text{const.}} \,, \qquad (10.1A.2)$$

where P is the pressure, we can calculate the first variation of the total energy:

$$\delta E = + \int_0^M (P/\rho^2)\delta\rho dm + G \int_0^M (m/r^2)dm\delta r \,. \qquad (10.1A.3)$$

We transform the first integral in equation (10.1A.3):

$$\int_0^M \frac{P}{\rho^2} \delta\rho dm = - \int_0^M P\delta\left(\frac{1}{\rho}\right)dm = - \int_0^M P\delta\left(4\pi r^2 \frac{dr}{dm}\right)dm$$

$$= - \int_0^M 8\pi P \frac{dr}{dm} r\delta r dm - \int_0^M 4\pi P r^2 \delta\left(\frac{dr}{dm}\right)dm$$

$$= - \int_0^M 8\pi P \frac{dr}{dm} r\delta r dm + \int_0^M 4\pi \frac{d}{dm}(Pr^2)\delta r dm$$

$$= \int_0^M 4\pi r^2 \frac{dP}{dm} \delta r dm = \int_0^M \frac{dP}{\rho dr} \delta r dm \,.$$

Substituting the expression obtained into equation (10.1A.3) and setting $\delta E = 0$, we find

$$\frac{dP}{dr} + \frac{\rho Gm}{r^2} = 0 \,,$$

which is the equation of hydrostatic equilibrium. Thus, the condition of an extremum of the energy is actually the simple condition of hydrostatic equilibrium.

We now write the expression for the total energy of the star, not assuming that the velocity of motion of the stellar matter is zero:

$$E = \int_0^M \left[E_1(S, \rho) - \frac{Gm}{r} + \tfrac{1}{2}u^2\right]dm \,,$$

where u is the velocity of a mass element. Obviously, the state established above, in which there exists an extremum of the energy, will be stable if the extremum is a minimum. Actually, for stability, any state must be more energetic than the equilibrium state whether it has $u = 0$ or $u^2 > 0$.

Consequently, the investigation of the stability is reduced to finding the conditions for which $\delta^2 E > 0$. We will limit ourselves to perturbed states which are momentarily at rest, $u = 0$.

In the expression for the second variation, the coefficient of $(\delta r')^2$, where $r' = dr(m)/dm$, is proportional to

$$\frac{\partial}{\partial \rho}\left(\rho^2 \frac{\partial E_1}{\partial \rho}\right) = \frac{\partial P}{\partial \rho}\,.$$

Therefore, a necessary condition for stability is $\partial P/\partial \rho > 0$. Otherwise, by taking a small but rapidly oscillating δr so that $(\delta r)^2$ is small and $(\delta r')^2$ is large, we could obtain $\delta^2 E < 0$.

The physical significance of this condition is obvious: matter with $\partial P/\partial \rho < 0$ is unstable at given pressure, irrespective of the effects of gravity.

In texts on the calculus of variations (e.g., Gel'fand and Fomin 1961) it is demonstrated that when $\partial P/\partial \rho > 0$, a necessary and sufficient condition for the second variation $\delta^2 E$ to be positive (or negative) definite is the non-intersection of the functions $r(m)$ corresponding to adjacent extrema; i.e., the nonintersection of the solution of the equations obtained from the condition $\delta E = 0$.

Translating this theorem into the language of the problem at hand, we obtain the following condition for the stability of the star. Let $r_0(m)$ be the equilibrium solution which corresponds to the total mass M_0. (Of course, at the edge of the star, i.e., when $m = M_0$, the condition $P = 0$ must be satisfied.) Let $r_1(m)$ be the solution corresponding to another mass M_1, close to M_0. Then the solution is stable if and only if for all m

$$\frac{r_1(m) - r_0(m)}{M_1 - M_0} < 0\,. \tag{10.1A.4}$$

Consequently, for stability it is necessary that with increasing mass ($M_1 - M_0 = \Delta > 0$), i.e., with the addition of a mass Δ from outside, each internal mass element of the star must move toward the center ($\Delta r = r_1 - r_0 < 0$).

This condition is quite natural and it can be regarded as a generalization of the condition $\partial P/\partial \rho > 0$. The volume of the star and of each part of the star must decrease with the application of external pressure. It is important to note that this condition is not obtained intuitively, but is a precise mathematical result, a formal proof of which is given in the text cited above (Gel'fand and Fomin 1961).

When integrating the equation of hydrostatic equilibrium, it is convenient to choose the density at the star's center. Then, as a result of the integration, we obtain the relationship $M(\rho_c)$. Since for small m,

$$r(m) = (3m/4\pi\rho_c)^{1/3}, \qquad (10.1A.5)$$

it is easy to see that condition (10.1A.4) will not be satisfied unless $dM/d\rho_c > 0$.

This presents a rigorous proof that the solutions located on the descending branch of the curve $M(\rho_c)$, where $dM/d\rho_c < 0$, are unstable with respect to small perturbations. This result was derived in the main portion of this section in a rather approximate manner, and its precise proof is an argument in favor of the qualitative correctness of our rough approximation. At the same time, it should be noted that the condition $dM/d\rho_c > 0$ is necessary but, generally speaking, not sufficient for stability.

In parts of the star, the matter may have $\gamma < \frac{4}{3}$, and the star may be stable nevertheless, since stability demands only $\gamma > 0$. How may we obtain an effective average γ, which would permit us to estimate stability? The plotting of a pair of curves $r_0(m)$ and $r_1(m)$, for neighboring central densities ρ_{c0} and ρ_{c1} and correspondingly close masses M_0 and M_1, always permits us to check uniquely stability by means of condition (10.1A.4), and thus yields a precise, exhaustive, and convenient solution to the problem. Another proof that the maximum of the curve $M(\rho_c)$ plays the role of the boundary of stability is given at the end of § 10.11.

For a review of other exact methods of determining stability, valid in GTR as well as in Newtonian theory, see Thorne (1967).

APPENDIX TO § 10.5

A Derivation of the Virial Theorem and of the Expression for the Gravitational Energy Using the Variational Principle

By using the variational principle, we will derive two useful relationships: (1) between the total energy of a star and its gravitational energy (virial theorem), and (2) between the radius of the star and its gravitational energy.

These relationships apply to a star consisting of matter with a polytropic equation of state $P = A\rho^{1+1/n}$; from this equation of state, it follows that the energy of a unit mass is $E_1 = nA\rho^{1/n} = nP/\rho$ (the energy of matter cooled by adiabatic expansion to zero density is set to zero). For the first relationship, it is inessential whether or not $A = A(S)$ is a constant throughout the star; for the second relationship, the constancy of A is necessary. Both relationships are well known in the classical theory of stellar equilibrium (see formulae [10.1.11]–[10.1.13]), where they are derived from the differential equation of hydrostatic equilibrium. A derivation of these relationships from the variational principle is useful as an exercise in applying the variational principle, and also because the meaning of the relationships appears in a new light.

We express the total energy of the star in the form

$$E = \int E_1(m)dm - G\int \frac{m\,dm}{r} = W + U,$$

where m is the mass inside a given layer, and integration is performed from $m = 0$ (center of star) to $m = M$ (external surface; M is the total mass of the star). Here W is the total internal energy of the matter, U is the gravitational energy, and

$$dm = \rho\, 4\pi r^2\, dr.$$

The density distribution is fully determined once the function $r(m)$ is given (i.e., once the radius r of the shell containing a mass of matter m is assigned). In hydrodynamic terminology, r is an Eulerian coordinate and m is a Lagrangian coordinate of the particles of matter. Knowing $r(m)$, we find

$$\rho = \frac{1}{4\pi r^2}\left(\frac{dr}{dm}\right)^{-1}.$$

The specific energy of the matter depends on the density.

According to the variational principle, E has a minimum[28] in the state of equilibrium, for a given mass M. Consequently, for any variation of $r(m)$, the first-order change in E must be zero.

We will consider a homologous perturbation, i.e., an expansion $(a > 1)$ or contraction $(a < 1)$ of the star, which preserves the shape of $r(m)$: $r(m) = ar_0(m)$. For such a perturbation we find:

$$dE/da|_{a=1} = \int(dE_1/d\rho)(-3\rho)dm + G\int(m/r_0)dm = 0.$$

When $E_1 = A_1\rho^{1/n}$, we obtain

$$dE/da|_{a=1} = -(3/n)W - U = 0; \quad W = -(n/3)U;$$
$$E = W + U = [(3 - n)/3]U.$$

For an arbitrary equation of state

$$\frac{dE_1}{d\rho} = -\frac{1}{\rho^2}\frac{dE_1}{d(1/\rho)} = \frac{P}{\rho^2},$$

it is true that

$$\int(dE_1/d\rho)(-3\rho)dm = -3\int(P/\rho)\,dm = -3\int P\,dV,$$

so we obtain

$$U = -3\int P\,dV.$$

28. The minimum corresponds to a stable state. For what follows it is sufficient that E be an extremum.

Next we will consider stars for which $E_1 = nA\rho^{1/n}$ and A is constant (see § 10.3). It is apparent from dimensional considerations that the distributions of density in such stars with different masses are identical (the distribution may depend only on the dimensionless parameter n; from A and G one cannot construct a dimensionless quantity). Therefore,

$$E = ManA\rho_c^{1/n} - Gb\frac{M^2}{(M/\rho_c)^{1/3}} = aM\rho_c^{1/n} - \beta M^{5/3}\rho_c^{1/3} .$$

From the conditions of hydrostatic equilibrium we find

$$dE/d\rho_c = (1/n)\, aM\rho_c^{(1/n)-1} - \tfrac{1}{3}\beta M^{5/3}\,\rho_c^{1/3-1} = 0 ,$$

$$\rho_c = \left(\frac{n}{3}\frac{\beta}{a}M^{2/3}\right)^{3n/(3-n)} = \gamma M^{2n/(3-n)} .$$

When $n > 3$ (i.e., for a stable equilibrium), the central density ρ_c always increases with increasing M. For the stellar radius we have $R \sim (M/\rho_c)^{1/3} \sim M^{(1-n)/(3-n)}$. Consequently, when $1 < n < 3$, the radius decreases with increasing mass; for $n = 1$, $p = A\rho^2$, the radius does not depend on the mass; and only for $n < 1$ does the radius increase.

Substituting the expression for ρ_c into the formula for E, we get

$$E = \delta M^{(5-n)/(3-n)} .$$

From the preceding virial theorem we know that

$$W = -[n/(3-n)]E , \quad U = [3/(3-n)]E ;$$

and it is obvious that $E < 0$ and $\delta < 0$.

Consider now two methods of increasing the mass of the star: (1) From an equilibrium configuration with mass M, we pass to another equilibrium configuration with mass $M + dM$. Obviously,

$$E(M + dM) - E(M) = dE = \frac{dE}{dM}dM = \frac{5-n}{3-n}\frac{E}{M}dM .$$

(2) We add the mass dM to an equilibrium configuration with mass M, placing the mass dM at the surface of the star where the pressure is zero. The internal energy of the added mass is zero (since $P = 0$), and the gravitational energy is obviously $-(GM/R)dM$.

Consequently,

$$dE = -(GM/R)dM .$$

Now, on the basis of the variational principle, we assert that the two expressions for dE must be equal: from the second method, we obtained a density distribution differing from an equilibrium distribution for the mass $M + dM$, since the mass was added on the surface. However, in view of the fact that the equilibrium distribution is extremal, a deviation from the equilibrium distribution will cause a variation ol E which is of second order (i.e., proportional to $(dM)^2$ when the added mass dM is small).

Thus, by using the expression for U, we obtain

$$\frac{5-n}{3-n}\frac{E}{M} = -\frac{GM}{R}, \qquad U = \frac{3}{3-n}E = -\cdot\frac{3}{5-n}\frac{GM^2}{R}.$$

The last relationship remains valid also when $n = 3$, whereas in the preceding relation an indeterminacy arises. From the expression cited, it immediately follows that the solution of the equation of equilibrium when $n = 5$ is degenerate, and $R \to \infty$. This is of no practical significance, because for $n > 3$ the formal solutions of the equation of equilibrium, which satisfy all our conditions, are unstable.

Finally, let us note that for the case of an isentropic configuration with arbitrary equation of state—even when Newtonian theory is replaced by GTR—the relationship between the derivative of the energy with respect to the number of particles N and the gravitational potential on the surface of the star remains valid (see § 10.12). However, in view of the fact that there are no simple analytical expressions for $E(M)$ or $E(N)$ in the general case, such simple elegant formulae as the above are not obtained from this relationship.

APPENDIX 1 TO § 10.8

We here derive the value of the energy of a star, whose pressure is produced by a relativistically degenerate electron gas, in the limit $\rho \to \infty$.

We write out formula (6.2.11) for the energy per electron

$$E_e = \langle E \rangle - E_0 = m_e c^2 [\tfrac{3}{4}(\rho/\rho_0)^{1/3} - 1 + \tfrac{3}{4}(\rho_0/\rho)^{1/3}], \qquad \rho_0 = \mu_e\, 10^6.$$

Passing to the energy per gram of matter, we obtain the expression

$$E_1 = \frac{E_e}{m_p\mu_e} = \frac{5 \times 10^{17}}{\mu_e}[\tfrac{3}{4}(\rho/\rho_0)^{1/3} - 1 + \tfrac{3}{4}(\rho_0/\rho)^{1/3}].$$

Upon substituting this expression into the equation of equilibrium (see appendix to § 10.14), and upon passing to the limit $\rho \to \infty$, $\gamma = \tfrac{4}{3}$, we find that the last term in brackets approaches zero; the divergent part of the first term is canceled by the (negative) gravitational energy of the star; and the remainder gives for the total energy of the star per unit mass

$$E_2 = -(5 \times 10^{17}/\mu_e) \text{ ergs g}^{-1}.$$

APPENDIX 2 TO § 10.8

Corrections for the General Theory of Relativity

First we give the corrections to the energy for a given arbitrary configuration of matter which, in general, is not in equilibrium. We assume, however, that the matter is momentarily at rest—i.e., that its instantaneous velocity is zero. However, the instantaneous acceleration, generally speaking, is not zero because of the absence of equilibrium.

We must take into account the dependence of the density of total mass-energy on the energy. We denote the density of rest mass by ρ_0, so that the density of total mass-energy, ρ, is $\rho = \rho_0(1 + E_1/c^2)$, where E_1 is the specific energy (in addition to the rest mass) per unit rest mass.

We must also take into account the non-Euclidean nature of space (see § 3.3):

$$dV = e^{\lambda/2}4\pi r^2 dr , \qquad V = 4\pi \int_0^r e^{\lambda/2}r^2 dr > \tfrac{4}{3}\pi r^3 ,$$

where r is the "coordinate" radius, such that the circumference about the star's center is $2\pi r$, and the area of a sphere is $4\pi r^2$. An invariant characterization of a configuration which contains a given total number of baryons is the function $\rho_0(V)$, where V is the instantaneous volume. The equilibrium state corresponds to an extremum of the observed mass of the star,

$$M = 4\pi \int_0^R \rho r^2 dr ,$$

for a given rest mass (i.e., for the given number of baryons)

$$M_0 = \int_0^R \rho_0(V)dV ,$$

and for a given entropy, which enters into the relationship

$$E_1 = E_1(S, \rho_0) .$$

Calculating the energy from the rest mass of the star, we get

$$E = c^2(M - M_0) = c^2 \int_0^R (\rho e^{-\lambda/2} - \rho_0)dV .$$

This expression should be compared with the Newtonian expression,

$$E_N = \int E_1\rho_0 dV - G\int (m'dm'/r') ,$$

where m' is the instantaneous "Newtonian" mass (computed without corrections for the relationship between mass and energy), r' is the "Newtonian" or Euclidean radius, and

$$dm' = \rho_0 dV ; \qquad m' = \int_0^V \rho_0 dV ; \qquad r' = (3V/4\pi)^{1/3} .$$

The function $\rho_0(V)$ is the same in Newtonian theory and in relativity theory.

The difference $\Delta E = E - E_N$ will be regarded as the correction for GTR, and we will compute the first nonvanishing term in expansion of ΔE powers of G. Obviously, the dimensionless-expression parameter should be

$$\frac{r_g}{R} \sim \frac{GM}{Rc^2} \sim GM^{2/3}\rho_c^{1/3}c^{-2} .$$

The ratio $E_1/c^2 \sim P/\rho c^2$ is of the same order as r_g/R; so terms of first order in G are already included in the Newtonian approximation.

We use the only relationship which does not require equilibrium:

$$e^{-\lambda} = 1 - \frac{2Gm}{rc^2} , \quad e^{-\lambda/2} = \left(1 - \frac{2Gm}{rc^2}\right)^{1/2} = 1 - \frac{Gm}{rc^2} - \tfrac{1}{2}\left(\frac{Gm}{rc^2}\right)^2 .$$

We obtain with the necessary accuracy

$$\Delta E = \int dV\left[-E_1\rho_0 \frac{Gm}{c^2 r} - \tfrac{1}{2}\rho_0 \frac{G^2 m^2}{c^2 r^2} + \rho_0 G\left(\frac{m'}{r'} - \frac{m}{r}\right)\right], \quad (10.8\text{A}.1)$$

$$\frac{m'}{r'} - \frac{m}{r} = \frac{m' - m}{r'} - \frac{m(r' - r)}{rr'} , \quad (10.8\text{A}.2)$$

$$m' - m = -\frac{1}{c^2}\int E_1\rho_0 dV + \frac{G}{c^2}\int \frac{\rho_0 m}{r} dV , \quad (10.8\text{A}.3)$$

$$r' - r = \frac{G}{r^2 c^2}\int mr dr . \quad (10.8\text{A}.4)$$

By using these relationships, we obtain the final correction in the form of a sum of five integrals in which we can everywhere identify the density, the volume, and the radius with the corresponding Newtonian-Euclidean quantities; the error involved is of the order of the lowest neglected power of the expansion parameter:

$$\Delta E = I_1 + I_2 + I_3 + I_4 + I_5 ,$$

$$I_1 = -\frac{G}{c^2}\int E_1 \frac{m dm}{r} , \qquad I_2 = -\tfrac{1}{2}\frac{G^2}{c^2}\int \frac{m^2 dm}{r^2} ,$$

$$I_3 = -\frac{G}{c^2}\int(\int E_1 dm)\frac{1}{r} dm , \quad I_4 = +\frac{G^2}{c^2}\int\left(\int \frac{m dm}{r}\right)\frac{dm}{r} ,$$

$$I_5 = -\frac{G^2}{c^2}\int(\int mr dr)\frac{m dm}{r^4} .$$

The integrals are taken over the entire mass of the star; the inner integrals in I_3, I_4, and I_5 are taken from the center to the instantaneous m, or r. The integrals are given in the order which follows naturally from formulae (10.8A.1)–(10.8A.4). This expression for ΔE simplifies substantially when it is applied to an equilibrium distribution of gas with an adiabatic index $\gamma = \tfrac{4}{3}$, i.e., when we assume that

$$E_1 = 3P/\rho , \quad dP/dr = -Gm\rho/r^2 .$$

In this case, after several integrations by parts, we obtain $I_3 + I_4 = -\frac{2}{3}I_1 + 2I_2$, $I_5 = \frac{1}{3}I_1$, and finally $\Delta E = \frac{2}{3}I_1 + 3I_2$. This expression agrees precisely with that derived by Fowler (1964a, b) in connection with the theory of supermassive stars. (Note that Fowler has considered the equilibrium configuration.)

By using the Emden function with $n = 3$ to calculate the integrals, we finally get

$$\Delta E_{\text{GTR}} = -0.93(G^2/c^2)M^{7/3}\rho_c{}^{2/3} .$$

APPENDIX 1 TO § 10.14

The General Features of the Energy Approach and a Comparison with the Method of Fowler (1964a)

The Newtonian theory.—We write the Emden solution with $n = 3$ as a dimensionless function of the ratio of the "partial mass" m to the total mass M:

$$\rho = \rho_c \psi(m/M) = \rho_c \psi(z) , \qquad m = 4\pi \int_0^r \rho r^2 dr ;$$

$$M = 4\pi \int_0^R \rho r^2 dr , \qquad z = m/M .$$

We consider first the formulae for fixed entropy (e.g., for $S = 0$). In this section we write $E_1(\rho)$ instead of $E_1(\rho, S)$, and $(d/d\rho)$ instead of $(\partial/\partial\rho)_S$. Giving $E_1(\rho)$ determines the pressure

$$P = -dE_1/dV = \rho^2 \, dE_1/d\rho .$$

We write[29]

$$E_1 = A\rho^{1/3} , \qquad P = \tfrac{1}{3}A\rho^{4/3} . \tag{10.14A.1}$$

We substitute this E_1 into the general expression for the energy of the star

$$E = \int E_1 dm - G\int (mdm/r) , \tag{10.14A.2}$$

and use the Emden distribution, thereby obtaining

$$E = \int A(\rho_c\psi)^{1/3}dm - G\int \frac{mdm}{[(M/\rho_c)\xi]^{1/3}} \tag{10.14A.3}$$

$$\xi = (3/4\pi)\int_0^z \psi^{-1}(z)dz . \tag{10.14A.4}$$

Using the dimensionless integration variable z, we obtain

$$E = aA\rho_c{}^{1/3}M - \beta GM^{5/3}\rho_c{}^{1/3} , \tag{10.14A.5}$$

---

29. The additive constant in the energy is not essential for the equilibrium problem, since we are concerned with the derivative of the energy. It must be included if we are to be concerned with the actual magnitude of E.

where a and β are dimensionless numbers:

$$a = \int_0^1 \psi^{1/3}(z)dz , \qquad \beta = \int_0^1 \xi^{-1/3}(z)zdz . \qquad (10.14A.6)$$

We find the equilibrium mass M_{e0} (the subscript e denotes equilibrium, the subscript 0 denotes the absence of corrections) from

$$\frac{dE}{d\rho_c^{1/3}} = aAM - \beta GM^{5/3} = 0 , \qquad M_{e0} = (aA/\beta G)^{3/2} . \qquad (10.14A.7)$$

For a mass that is close to equilibrium, $M = M_e + \mu$, we find upon performing a series expansion

$$E = -\tfrac{2}{3}aA\mu\rho_c^{1/3} = -\tfrac{2}{3}\beta GM_e^{2/3}\mu\rho_c^{1/3} = -k\mu\rho_c^{1/3} , \qquad (10.14A.8)$$

where $\tfrac{2}{3}\beta GM_e^{2/3}$ is denoted by k. Suppose that an arbitrary small correction is introduced into the equation of state

$$E_1 = A\rho^{1/3} + f(\rho) , \qquad P = \tfrac{1}{3}A\rho^{4/3} + \rho^2 f'(\rho) ,$$
$$\Delta E_1 = f(\rho) , \qquad \Delta P = \rho^2 f'(\rho) . \qquad (10.14A.9)$$

The corresponding correction to the energy of the star is

$$\Delta E = \int_0^M f(\rho)dm = \int_0^M f(\rho_c\psi(z))dm . \qquad (10.14A.10)$$

We will consider both this correction and $\mu = M - M_e$ to be small. The correction is computed with $M = M_{e0}$ and added to the change in the energy due to the deviation of M from M_e. We obtain (the constant k is determined by formula [10.14A.8])

$$E = -k\mu\rho_c^{1/3} + \int_0^{M_{e0}} f[\rho_c\psi(m/M_e)]dm . \qquad (10.14A.11)$$

We find the condition of equilibrium to be

$$\frac{dE}{d\rho_c} = 0 = -\tfrac{1}{3}k\mu_{eq}\rho_c^{-2/3} + \int_0^{M_{e0}} f'[\rho_c\psi(m/M_e)]\psi(m/M_e)dm . \qquad (10.14A.12)$$

The integral in equation (10.14A.12) can be expressed in the following form (see definitions at the beginning of this appendix):

$$\int f'\psi dm = \frac{1}{\rho_c}\int f'\rho_c\psi dm = \frac{1}{\rho_c}\int f'\rho dm = \frac{1}{\rho_c}\int \frac{\Delta P}{\rho} dm . \qquad (10.14A.13)$$

From formulae (10.14A.12) and (10.14A.13), we obtain μ_e:

$$\mu_e = \frac{3}{k}\rho_c^{-1/3}\int \frac{\Delta P}{\rho} dm , \qquad M_e = M_{e0} + \mu_e , \qquad (10.14A.14)$$

and the expression for the equilibrium energy—which, by substituting equation (10.14A.14) into equation (10.14A.11) and denoting $1/\rho = V$, we put into the form

$$E_e = -3 \int (\Delta P/\rho)dm + \int f dm = \int_0^{M_{e0}} (\Delta E_1 - 3V\Delta P)dm . \quad (10.14A.15)$$

The energy and the pressure in the zero-order approximation satisfy the relationship

$$E_{1,0} - 3VP_0 = 0 . \quad (10.14A.16)$$

Consequently, we can add equation (10.14A.16) to the integrand, and obtain

$$E_e = \int_0^{M_{e0}} (E_1 - 3VP)dm . \quad (10.14A.17)$$

Finally, following Fowler, we introduce a dimensionless quantity

$$\bar{\epsilon} = \frac{E_1}{3PV} = \frac{E_1\rho}{3P} , \quad E_1 - 3VP = \frac{\bar{\epsilon}-1}{\bar{\epsilon}} E_1 = (\bar{\epsilon}-1)E_1 \quad (10.14A.18)$$

(the last equality follows from the assumption that $\bar{\epsilon} - 1 \ll 1$, and the entire theory is constructed to first order in $\bar{\epsilon} - 1$), and we find

$$E_e = \int (\bar{\epsilon}-1)E_1 dm = (\langle\bar{\epsilon}\rangle - 1)\int E_1 dm$$
$$= (\langle\bar{\epsilon}\rangle - 1)\int E_{1,0} dm = (\langle\bar{\epsilon}\rangle - 1)\int (Gm/r)dm . \quad (10.14A.19)$$

This expression coincides with Fowler's expression. We cite his conclusion: Take

$$E = \int E_1 dm - \int (Gm/r)dm \quad (10.14A.20)$$

and transform the first integral by parts:

$$\int E_1 dm = \int E_1\rho \, dV = -\int Vd(E_1 \rho) = -3\int Vd(\bar{\epsilon}P)$$
$$= 4\pi\int r^3\bar{\epsilon}dP - 4\pi\int r^3P \, d\bar{\epsilon} . \quad (10.14A.21)$$

Fowler substitutes the expression for dP from the equation of hydrostatic equilibrium into the first integral

$$dP = -\rho \frac{Gm}{r^2} dr = -\frac{Gmdm}{4\pi r^4} , \quad (10.14A.22)$$

while he neglects the second integral; finally, he obtains the following expression:

$$E = \int \bar{\epsilon} \frac{Gmdm}{r} - \int \frac{Gmdm}{r} = \int (\bar{\epsilon}-1) \frac{Gmdm}{r}$$
$$= (\langle\bar{\epsilon}\rangle - 1)\int \frac{Gmdm}{r} . \quad (10.14A.23)$$

When $\bar{\epsilon} = $ const., Fowler's result (10.14A.23) agrees with our result (10.14A.19). However, when $\bar{\epsilon} = \bar{\epsilon}(m)$, the neglect of $4\pi \int r^3 P d\bar{\epsilon}$ results in an error of the same order as that of the phenomenon considered; as a result, in Fowler's solution, the averaging of $\bar{\epsilon}$ follows an incorrect law with a weight of mdm/r, instead of the weight $E_1 dm \sim PdV$. More important, however, is the fact that Fowler limits himself to a consideration of $E_e(\rho_c)$, whereas a complete understanding of equilibrium, of its stability and other properties, requires a knowledge of the entire function of two or three variables [$E(M, \rho_c)$ when $S = 0$, or $E(S, M, \rho_c)$ in general] near equilibrium.

APPENDIX 2 TO § 10.14

General Relationships between E_e and E

It is frequently the case that in the zero approximation the energy is proportional to $\rho_c^{1/3}$, so that the coefficient of $\rho_c^{1/3}$ vanishes for the state of neutral equilibrium. Then the problem is as follows. For a cold star, when $S = 0$, the energy is expressed in the form

$$E = -k\mu\rho_c^{1/3} + \phi(\rho_c) , \qquad (10.14A.24)$$

where $\mu = M - M_e$ and where ϕ takes into account the deviation from the zero-order approximation. For a hot star with $M = $ const. and $s = S - S_e$,

$$E = bs\rho_c^{1/3} + \phi(\rho_c) . \qquad (10.14A.25)$$

It is required to find the equilibrium, with the correction function ϕ taken into account. When only corrections for the equation of state are taken into account, the function ϕ is expressed by

$$\phi(\rho_c) = \int (E_1 - 3PV)dm = \int (\bar{\epsilon} - 1)E_1 dm . \qquad (10.14A.26)$$

Integration is performed along the Emden curve $n = 3$ of the zero-order approximation for $M = M_e$ or $S = S_e$; the only free parameter on which the total energy depends is ρ_c. Taking into account the contribution to ϕ from GTR, we have

$$\phi(\rho_c) = \int (E_1 - 3PV)dm = -0.93 \ (G^2 M^{7/3}/c^2)\rho_c^{2/3}$$
$$= \phi_0(\rho_c) - \text{const.} \ \rho_c^{2/3} . \qquad (10.14A.27)$$

When we take GTR into account, only a change of the form of the function ϕ results; the nature of equation (10.14A.25) does not change. In formula (10.14A.27) ϕ_0 is a quantity computed from equation (10.14A.26) without taking into account GTR; the meaning of ϕ is clear from formula (10.14A.27) itself. The statement of the problem is as follows: Given the equation $E(S, \rho_c)$ which enters into the determination of equation (10.14A.25), find $E_c(\rho_c)$ from the condition of equilibrium, and find the boundary of the existence of equilibrium configurations. It is convenient to introduce $x = \rho_c^{1/3}$ as a variable. We obtain[30]

$$E = bsx + \phi(x) , \quad \partial E/\partial x = 0 = bs + \phi'(x) ,$$

$$s_e = -\phi'/b , \qquad\qquad (10.14A.28)$$

$$E_e(x) = -x\phi'(x) + \phi(x) = -x^2(d/dx)(\phi/x) .$$

The horizontal point is the boundary on the existence of solutions:

$$\partial E/\partial \rho = (\partial^2 E/\partial \rho^2) = 0 \rightarrow \partial E/\partial x = (\partial^2 E/\partial x^2) = 0 , \quad (10.14A.29)$$

$$bs_{cr} + \phi'(x_{cr}) = 0 , \qquad\qquad (10.14A.30)$$

$$\phi''(x_{cr}) = 0 . \qquad\qquad (10.14A.31)$$

Equation (10.14A.31) permits us to find x_{cr} (i.e., the critical density), and then equation (10.14A.30) gives us the critical value of the entropy (or the critical value of the mass μ in an analogous problem with fixed entropy and variable mass). We have

$$\frac{dE_e}{dx} = -x\phi''(x) - \phi'(x) + \phi'(x) = -x\phi''(x) = -x\frac{\partial^2 E}{\partial x^2}\bigg|_s .$$

Consequently, the point of horizontal inflection on the curve $E(x, s)$, when $S = $ const., coincides with the minimum of the curve of equilibrium states,

$$E_e(x) = E[x, s_e(x)] .$$

The stability of each equilibrium state obviously depends on the sign of $\partial^2 E/\partial x^2$ at the point where $\partial E/\partial x = 0$. It follows from the formula that on the descending branch of the curve $E_e(x)$, the solutions are stable ($\partial E_e/\partial x < 0$, $\partial^2 E/\partial x^2 > 0$). Thus the assertion made in the text of this section is formally proved.

---

30. The correction for GTR is proportional to $\rho_c^{2/3} \sim x^2$ and is negative in the given configuration: $\Delta E = -nx^2$. Hence, $\Delta E_e = -x\phi' + \phi = +nx^2$; i.e., the correction for GTR to the equilibrium energy, ΔE_e, is positive: it is equal in magnitude and opposite in sign to the correction to the energy of a given configuration, ΔE:

$$\Delta E = -\Delta E_e .$$

11 STELLAR EVOLUTION

11.1 EVOLUTION OF A STAR UP TO THE LOSS OF STABILITY OR THE WHITE-DWARF STAGE

We can now attempt to analyze the evolution of stars. We shall not discuss all stages of the evolution in detail; for that we refer the reader to monographs cited in the introduction to Part III of this book. Instead, we shall review the general situation and make certain comments which are necessary for further exposition.

According to the contemporary viewpoint, stars are formed from an initially rarefied medium by gravitational condensation of diffuse matter, which consists primarily of hydrogen. After condensing out, a star evolves by contracting and radiating energy. One can say that in the course of the contraction the star's luminosity is a direct result of the release of gravitational potential energy. This energy source was pointed out long ago by Kelvin. The temperatures are still low, and the release of nuclear energy is negligibly small. The star is in hydrostatic equilibrium (see § 10.1) without internal sources of energy. The duration of this phase is relatively short; it amounts to

$$\tau \approx 5 \times 10^7 (M_\odot/M)^2 \text{ years}.$$

As the star contracts and radiates, the temperature in its interior increases in accordance with its negative heat capacity, and finally rises so high that nuclear reactions involving the transformation of hydrogen into helium begin to take place. The star, at this stage in its development, has arrived at the main sequence of the Hertzsprung-Russell diagram (temperature-luminosity diagram). Ambartsumyan (1960, p. 179) and his collaborators have proposed an alternative possible evolution of stars prior to the beginning of nuclear reactions (i.e., before they reach the main sequence). They suggest that protostars arise not from diffuse matter, but from superdense D-bodies, whose nature as yet is unknown. This does not change the subsequent evolutionary development, which is determined by the star's mass and by its initial chemical composition, as well as by its angular momentum and magnetic field. When nuclear reactions begin, the star is in a state of hydrodynamic and thermal equilibrium. The subsequent "main sequence" stage is the longest period in the active life of the star. Its duration is determined by the hydrogen reserve in the core (where the temperature is sufficiently high for nuclear reactions), and by the rate of conversion of hydrogen to helium. Obviously, this period, τ, is proportional to M/L, where L is the luminosity of the star. Calculations indicate that the mass of the stellar core in which hydrogen can be exhausted by nuclear burning is of the order of 0.1 M, whence it follows that

$$\tau \approx 10^{10}(L_\odot/L)(M/M_\odot) \text{ years} . \tag{11.1.1}$$

The inhomogeneity of the chemical composition, caused by the burning of hydrogen at the center, leads to a restructuring of the star; its external envelope expands and its core contracts.

In sufficiently massive stars ($M > M_\odot$ and $M \sim M_\odot$) the temperature in the core increases to such an extent that reactions involving triple collisions of α-particles to form ^{12}C take place:

$$3 \, ^4\text{He} \rightarrow \, ^{12}\text{C} + \gamma . \tag{11.1.2}$$

A detailed computation of the further evolution is extremely difficult. On one hand, the inner structure of the star becomes very complex: there arise energy sources in the spherical shells surrounding the core, and the structures of the zones of radiative and convective energy transport become very complicated. On the other hand (and this is especially important), various instabilities occur. These instabilities and their significance will be discussed in detail in § 11.3.

It is possible that at this stage of its development the star can lose mass as a result of slow, stationary flow from the surface (§ 10.3 and § 13.7), and also as a result of explosions (§ 10.4). An explosion may lead to the destruction of the entire star as well. We are not going to consider these processes now, but will return to the problem later in this chapter.

Let the mass of the star, after all the above stages of evolution, be M. After the nuclear fuel is exhausted, the further emission of energy leads to a contraction of the star (because of its negative heat capacity) and to an increase in its temperature.

The equilibrium states of the star (minima of the energy curves) move in the M-ρ diagram (Fig. 57) from left to right along a horizontal line of constant M. We first consider the evolution of stars with $M < 1.2 \, M_\odot$ (e.g., M_1 on Fig. 57). In this case the contraction continues until electron degeneracy begins to occur over most of the star. After that, the contraction is drastically retarded, since there is only a weak dependence of pressure upon temperature; during the subsequent evolution, the pressure P decreases by approximately a factor of 2, and the star settles down at the point D_1.

During the contraction of the star before the onset of degeneracy in the electron gas, the temperature increases since the specific heat is negative. After the onset of electron degeneracy, the continued (but slowed) contraction and the radiative loss of energy cool the star off. The star's temperature, after passing through a maximum, decreases. The specific heat of the star is now positive. The sequence of equilibrium states of a star with $M = 1.19\ M_\odot$ is shown on a ρ_c-T_c diagram in Figure 53 (curve 1). In accordance with the positive specific heat of the star, the central temperature T_c decreases, in the late stages, with increasing ρ_c. In order of magnitude, the temperature maximum corresponds to the degeneracy energy of electrons at the final state D_1. For stars with mass $\approx M_\odot$, the maximum T_c is $\approx 10^9$ ° K.

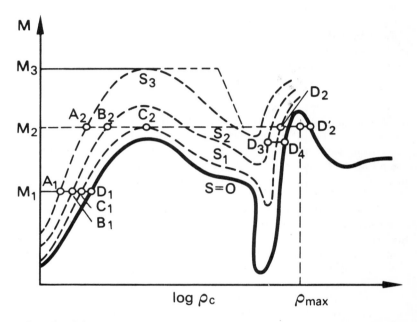

Fig. 57.—Schematic diagram of the evolution of stars with masses M_1 smaller than the OV limit and M_2 larger than the limit. M_2 loses mass during the collapse. Isentropes are shown by dashed curves.

During further cooling, all thermonuclear reactions that involve a net release of energy cease, and the electron degeneracy becomes stronger. This last stage of the life of a star is called the white-dwarf stage. White dwarfs cool slowly, with their radiation feeding principally off the thermal energy of the atomic nuclei which are still in a nondegenerate state.

The time required for a white dwarf to cool depends on its particular temperature and chemical composition; for $T \sim 5 \times 10^6$ and for $A = 20$, it is about 10^9 years (see, e.g., Schwarzschild 1958; Schatzman 1958; Mestel 1965). Recently several authors (Mestel and Ruderman 1967; Van Horn 1968; Ostriker and Aksel 1970; Greenstein 1969) have considered the possibility of crystallization of the matter in white dwarfs at moderately low temperatures. For a temperature $T \ll T_D \approx 4 \times 10^6 \, (\rho/10^6)^{1/3}$, Debye degeneracy of the crystalline lattice occurs (Mestel and Ruderman 1967), the heat capacity decreases sharply, and cooling takes place more quickly than according to the classical formulae. However, Bisnovatyi-Kogan and Seidov (1970) have pointed out that, for white-dwarf masses near the critical one in a range $\Delta M/M \sim 2 \times 10^{-4}$, an essential role is played by gradual heat release from an equilibrium β-reaction in the center. As a result, the decrease in heat capacity due to Debye degeneracy becomes unimportant, and the classical estimates are valid.

Very recently Greenstein has pointed out that an essential role in the cooling of white dwarfs is played by convection in their surface layers. The cooling time will be very sensitive to this convection.

As we see, during the entire evolution the star moves slowly from left to right on the M-ρ_c diagram, approaching the curve that corresponds to $T = 0$ $(S = 0)$.

The final chemical composition of a white dwarf depends on which nuclear reactions take place during the stages of contraction and heating; and the possibility for a particular reaction to take place depends, in turn, on the temperature. Certainly in all stars with $M \gtrsim 0.3 \, M_\odot$ the temperature attained during evolution is considerably higher than $T = 10^7$ ° K; at such temperatures, the nuclear reactions which transform H to ^{4}He take place. Let us compute the maximum temperature which is attained in a star of mass M. We have already stated that this temperature, in order of magnitude, is equal to the degeneracy temperature of the electron gas $T_{\max} \sim T_{\mathrm{degen}}$, which is proportional to $\rho^{2/3}$. By using the equation of state and the radially averaged equation of hydrostatic equilibrium of the star, it is easy to derive that $T_{\max} \propto (\mu^2 M)^{4/3}$. Numerical calculations of Öpik (1957) give for $T_{\max}$ at the center of the star

$$\log_{10} T_c = 8.9 + \tfrac{8}{3} \log_{10} \mu + \tfrac{4}{3} \log_{10} (M/M_\odot) . \qquad (11.1.3)$$

According to Öpik's estimate, the temperature remains near maximum for $\sim 10^{14}$ seconds. Hence, we can estimate how nuclear reactions of helium with ^{12}C and other elements will change the chemical composition of the star. If $T > 3 \times 10^8$ ° K, then the process of triple collisions of α-particles (e.g., reaction [11.1.2]) leads to the formation of ^{12}C. Simultaneously with this reaction, the following reactions take place:

$$
\begin{aligned}
^{12}\mathrm{C} + {}^4\mathrm{He} &\to {}^{16}\mathrm{O} + \gamma , & {}^{20}\mathrm{Ne} + {}^4\mathrm{He} &\to {}^{24}\mathrm{Mg} + \gamma , \\
^{16}\mathrm{O} + {}^4\mathrm{He} &\to {}^{20}\mathrm{Ne} + \gamma , & {}^{24}\mathrm{Mg} + {}^4\mathrm{He} &\to {}^{28}\mathrm{Si} + \gamma .
\end{aligned}
\qquad (11.1.4)
$$

According to Öpik's calculations, white dwarfs with $M > 0.5 \, M_\odot$ must consist primarily of ^{24}Mg; heavier nuclei are not formed because all the helium is consumed. For a star with a mass of about 0.4–0.45 $M_\odot$, in the central regions a considerable amount of helium is exposed to such transformation;

however, when $M < 0.4\,M_\odot$, a white dwarf must consist primarily of helium[1] (for the most recent calculations, see Takarada, Sato, and Hayashi 1966).

Let us now trace the final stages of the evolution of a star with mass $M \gtrsim 1.2\,M_\odot$. As its entropy decreases, such a star also moves slowly from left (point A_2) to right (point C_2) along the quasi-equilibrium line $M_2 =$ const. or $M_3 =$ const. of Figure 57. It was noted in § 10.15 that when $M > 1.2\,M_\odot$, the onset of instability occurs to the right of the maximum of the curves $M(\rho_c)|_{T=\text{const.}}$ (Fig. 54). Thus, in the course of the star's final, near-degeneracy evolution, it reaches a minimum temperature, at which it again acquires a negative heat capacity; its temperature during the subsequent contraction increases again, as the star approaches its onset of instability. The sequence of equilibrium states of such a star is shown on the ρ_c-T_c diagram of Figure 53 (curves 3 and 4). Upon the star's reaching the critical point C_2 (Fig. 57), the instability sets in, producing a catastrophic contraction at a rate of the order of the velocity of free fall

$$dR/dt \sim [(GM/R^2)(R_c - R)]^{1/2}\,.$$

Let a star be at the "collapse" point (maximum of a curve of constant entropy in the M-ρ_c diagram of Fig. 58). Perturbations which shift the star to the right or to the left (points A and B on Figure 58) bring it out of equilibrium. In both cases, the equilibrium configuration which corresponds to the perturbed density is below the perturbed configuration in Figure 58; i.e., it has a smaller mass (points A' and B'). This means that the gravitational forces at A and B exceed the pressure forces and cause the star to contract, thus increasing its density. But, while contraction returns the star to equilibrium C_2 from state A, from state B it leads the star further and further from equilibrium, producing collapse. The rate at which the collapse begins in a star at the "collapse point" is determined by the rate of the slow evolution of the star, i.e., the velocity with which it approaches the critical point C_2 and, continuing its motion, proceeds rightward from

1. The time of evolution for a star with $M < 0.3\,M_\odot$, according to equation (11.1.1), is longer than the age of the Metagalaxy. The smaller M is, the smaller is the maximum possible temperature. Therefore, in stars with sufficiently small M, the temperature in the future (when the Metagalaxy is old enough that their evolution has progressed sufficiently far) can ensure the occurrence of nuclear reactions which stop at ^{3}He. It would appear that at the end of their evolution (when $s = 0$) these stars will consist of ^{3}He. But, as noted in § 10.8, the electron energy required for neutronization of ^{3}He is very small; it amounts only to 18 keV. Therefore, in the course of the evolution of such stars, ^{3}He will be transformed into tritium; and the latter, by a conventional thermonuclear reaction, will be transformed into ^{4}He. Thus, these stars at the end of their evolution will consist of ^{4}He. On the other hand, for a mass of $M < 0.1\,M_\odot$ and with an initial chemical composition of H and He, electron degeneracy will occur before the temperature can rise sufficiently for any nuclear reactions to take place. Such stars shine at the expense of gravitational energy; at the end of their revolution they will have the same chemical composition as they had in the beginning, i.e., primarily hydrogen. Only pycnonuclear reactions can change this chemical composition (see § 6.5).

this point.[2] However, after a somewhat noticeable departure from the equilibrium state, the forces of gravitation exceed the forces of pressure by a finite value, and the acceleration of the contraction is a significant fraction of the free-fall acceleration. Thus, very soon after the onset of collapse, the star is contracting with nearly the acceleration of free fall. This, of course, is a very rough first approximation. It should be emphasized especially that when the state C_2 is reached, the rate of further contraction does not depend at all on the rates of those processes which, in the course of slow evolution, brought the star to its critical state.

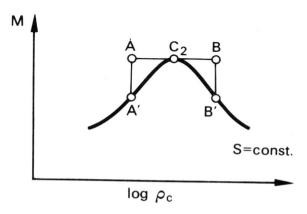

FIG. 58.—A star at the "collapse" point C_2. A deviation from equilibrium to the left (point A) produces a force which returns the star to C_2. A deviation to the right (point B) causes collapse, i.e., hydrodynamic contraction of the star.

Here one must note the following important circumstance. When we analyze the onset of the instability in a star, and its subsequent hydrodynamic contraction, we assume that the processes that cause the instability (i.e., that make $\gamma < \frac{4}{3}$) take place in a time which is considerably smaller than the hydrodynamic time t_H (time of free fall; see § 10.1). With respect to such processes, the matter must be in a state of equilibrium. These processes take place almost adiabatically, and consequently at constant entropy. An example of such a process is the creation of electron-positron pairs at high temperature.

Very recently Chechetkin and Imshennik (1969) (building upon the work of Imshennik, Nadezhin, and Pinaev) have investigated in detail the equilibrium and stability properties of hot matter.

To be exact, no true thermodynamic equilibrium is possible, because neu-

2. As noted in § 10.2, as the star approaches the point C_2, one must, strictly speaking, regard the evolution as dynamical, since the rate of loss of entropy is not small compared with the hydrodynamic velocity.

trinos and antineutrinos are constantly pouring out. What Chechetkin and Imshennik found is the stationary state in which the creation of particles and nuclei of every species is precisely balanced by corresponding annihilation and decay. No inverse reactions induced by ν or $\bar{\nu}$ were taken into account, because no ν's and $\bar{\nu}$'s are present; they leave the system freely. For example, $p + e^- \rightarrow n + \nu$ is compensated by $n + e^+ \rightarrow p + \bar{\nu}$, and also by a complicated cycle involving n-capture, β-decay, and the nuclear photoeffect; but not by the inverse reaction $\nu + n = p + e^-$. For a sequence of stationary states all with the same entropy, the relation $P(\rho)$ determines the adiabatic index γ. The result is not much different from previous equilibrium calculations.

The authors calculated also the index γ_r (r for "rapid") for changes of density which are so fast that no weak interactions occur; in these changes the ratio $p:n$ (including p's and n's bound in nuclei) is constant.

This γ_r is greater than γ. The critical value $\gamma = \frac{4}{3}$ is attained when $\gamma_r > \frac{4}{3}$. In this situation the evolution proceeds with a speed determined by the weak-interaction rate. Therefore, an important intermediate phase of the evolution arises. In the earlier, quiet phase, the evolution rate is determined by the energy losses. The early part of this quiet evolution is accompanied by exothermic nuclear reactions which compensate the heat losses so that the evolution cannot proceed to a new stage until the nuclear fuel has been exhausted. (This early part comprises the main sequence.) In the second part of the quiet phase the energy losses lead to a shrinkage of the star, and a consumption of gravitational energy. However, the matter is still in an equilibrium or stationary state.

After the limit $\gamma = \frac{4}{3}$ (but with $\gamma_r > \frac{4}{3}$) is reached, the intermediate phase mentioned above begins. As it enters this phase, the star begins to depart more and more from its previous, stationary state.[3] With an increasing departure, Δ, from the stationary state the speed of the weak interaction processes increases rapidly ($\sim \Delta^4$ or Δ^5). The end of the intermediate period occurs when the weak interaction is no longer a brake against hydrodynamic collapse.

The neutronization and the emission of ν and $\bar{\nu}$ during hydrodynamic collapse are considered in §§ 11.8 and 11.11. The intermediate period is badly in need of accurate calculations that take exact simultaneous account of nuclear interactions, weak interactions, and hydrodynamic processes.

The foregoing discussion of stability and evolution was carried out in terms of the star's (mean) density, (mean) pressure, and (mean) composition. No detailed consideration was given to the effects of radial variations in these quantities, nor to the stellar rotation and magnetic fields.

Naturally, in this approximation it is not possible to take into account

3. Near the beginning the influence of the slow, weak interactions can be described by an effective "second" viscosity of the matter.

the occurrence of shock waves during rapid contraction. The shock waves cause increasing entropy at the "hydrodynamic" rate, i.e., in a time of the order of t_H. Nor is it possible to take into account thermonuclear detonation,[4] the absorption of neutrinos in an outer envelope, and an ejection of part of the mass. The destruction of the entire star can be produced by a nuclear explosion. Consideration of all these phenomena requires a specific computation of nonstationary processes in the star. We will dwell on this in more detail and will cite the literature in §§ 11.3 and 11.4. But in the meantime we will assume that if a nuclear explosion takes place during the hydrodynamic collapse, then it does not destroy the whole star but only ejects part of the envelope.

After the onset of collapse at point C_2, the density grows (horizontal or falling dashed lines in Fig. 57) at the hydrodynamic rate (i.e., with a characteristic time t_H), until the star reaches a new stable state at point D_2, if its mass is less than 1.6 $M_\odot$ at this moment. Since the onset of collapse takes place at finite entropy and since, moreover, the entropy increases in the course of contraction, it follows that point D_2 is not on the limiting curve $S = 0$, but to the left of it. The contraction will continue past point D_2 because of the huge inertia of infall. If the kinetic energy of contraction is sufficiently high, the star can "penetrate" the energy barrier and come out at point D_2', after which unlimited contraction will continue. If the kinetic energy of contraction is not sufficient for crossing the barrier, then, after reaching a certain maximum density ρ_{max} ($\rho_2 < \rho_{max} < \rho_3$), the star will perform damped oscillations about D_2.

The damping is caused by the aforenamed process of entropy increase, and by the radiation of energy (in the form of photons from the surface, neutrinos from the interior, and gravitational waves from the entire star). Also, the star may eject part of its mass. As a result, the star passes from the isentrope at which collapse began to a higher isentrope (from S_1 to S_2). If the mass losses are appreciable, then, in addition, the position of the star in the M-ρ_c diagram drops, and the star arrives at point D_3 in a state of equilibrium. With further loss of energy by radiation (see § 11.2) the star gradually approaches $S = 0$ (point D_4) through a series of equilibrium states.

The state of the star in the interval D_3–D_4 is a *neutron* or *baryon* state. Here a question arises. The evolution of the star begins from hydrogen, i.e., from protons; as a result of thermonuclear reactions, the protons combine into complex nuclei and release tremendous energy (of the order of 0.01 Mc^2). At the endpoint of the evolution, at the stage of the baryon star, the matter again consists of individual baryons, with the rest mass of a neutron being even greater than the mass of H. Where did the energy, lost

---

4. Although fuel may be exhausted at the center of the star, it may have been preserved in the envelope.

by radiation during the evolution, come from? The answer is obvious. Gravitation created a large density; this brought about a neutronization of matter and forced complex nuclei to break up into individual baryons; consequently, the gravitational energy in the final state compensates for the energy which was radiated by the star into the surrounding space.

The idea of Kelvin and Helmholtz, that the luminosity of a star is due to its gravitational energy, is not correct during the lengthy stage in which nuclear reactions occur. The star shines because of these reactions. However, in the final phase of the evolution the gravitational energy breaks the nuclear bonds, and finally gravitation becomes responsible for the entire radiated energy. In this sense, the idea of those two great physicists of the past century is correct.

The evolution of a star which, after loss of stability, enters the superdense regime with mass $M > 1.6\ M_\odot$ is qualitatively different from the above; it will be considered in § 11.3 below.

11.2 INSTABILITY OF MASSIVE STARS WITH NUCLEAR SOURCES OF ENERGY

In this section, we consider the specific form of the stellar instability which leads to the existence of an upper limit on mass for normal stars.

As far back as 1941, Ledoux demonstrated that sufficiently massive stars with nuclear sources of energy will be unstable with respect to the growth of oscillations ("pulsational instability"). Pulsational instabilities have been studied by Schwarzschild and Härm (1959), who have used new models of the inner structures of massive stars, obtained by numerical techniques. For the most recent calculations see Stothers and Simon (1968). The mechanism of secular instability is as follows. The star, being in stable equilibrium, has a natural vibration frequency $\omega^2 \approx GM/R^3$. Assume that such oscillations are excited. There exist mechanisms of amplifying the oscillations, as well as mechanisms of attenuating them. The chief amplifying mechanism is the variation of nuclear energy release at the center of the star. Because of the strong dependence of the energy release upon temperature, most of the energy release takes place at maximum contraction. In effect, the star receives an impulse which increases the amplitude of its vibrations. At maximum expansion, the rate of nuclear reactions is reduced, and this enhances the "acceleration" of the oscillations during the contraction phase.

Damping of the oscillations takes place because of changes in the heat flow through the star. It can be demonstrated that these changes create a force which in this case retards the oscillations.

The rate of variation of the oscillation energy, L_1, can be expressed in the form

$$L_1 = L_2 - L_3\,,$$

where L_2 is the rate of increase of oscillation energy due to nuclear reactions, and L_3 is the rate of loss of oscillation energy. If $L_1 > 0$, the oscillations increase, and the star is unstable. If $L_1 < 0$, then the star is stable with respect to pulsations.

Schwarzschild and Härm performed calculations for stars consisting of 75 percent H and 22 percent He in the range of masses between $M = 218\ M_\odot$ and $M = 28\ M_\odot$. The results of their computations for main-sequence stars are shown on Table 16.

For main-sequence stars of small mass, the oscillation amplitude at the center, as compared with the amplitude at the surface, is small. Accordingly, the driving rate L_2 is small compared with the damping rate L_3, and L_1 is

TABLE 16

SECULAR INSTABILITY FOR MASSIVE STARS

	$M/M_\odot$			
	218.3	121.1	62.7	28.2
$\log_{10}(L/L_\odot)$	6.64	6.24	5.76	5.04
$(\delta R/R)_c$	0.632	0.583	0.312	0.392
t_{years}	930	1800	44000	$-$ 1400

NOTE.—The headings give the mass of the star; the first row, the total luminosity; the second row, the amplitude of the oscillations at the center if $\delta R/R = 1$ at the surface; and the third row, the e-folding time for increases in the amplitude of oscillation. The minus sign indicates that the oscillations are damped.

negative; thus the star is stable. For large masses $(\delta R/R)_c$ is twice as large as for small masses, and L_1 is positive; i.e., the star is unstable.

The critical value of the mass above which stars are unstable is about 60 $M_\odot$. This value depends on the chemical composition: $M_{cr} \propto \mu^{-2}$. The numbers in Table 17 are approximate, of course; and the value of $M_{cr} = 60\ M_\odot$ is not exact. However, one can apparently assert that stars with $M > 100\ M_\odot$ and with nuclear sources of energy are pulsationally unstable. If such objects are formed from rarefied gas (§ 11.1), then they can exist only during the Kelvin contraction time, i.e., for

$$t = 5 \times 10^7\ (M_\odot/M)^2\ \text{years} .$$

At the onset of hydrogen burning, they should begin to pulsate and to eject matter intensely.[5] Interestingly enough, the upper limit of the masses of known stars is close to $M_{cr} \approx 60\ M_\odot$.

5. *Editors' note.*—Recent hydrodynamic analyses indicate that such objects can exist with finite amplitude oscillations for at least about 10^3 times longer than the linear theory would suggest, and perhaps for their whole nuclear burning time.

11.3 STABILITY OF STELLAR EVOLUTION

Astronomical observations reveal a great qualitative variety in the stellar world. This variety applies, for example, to the chemical and the isotopic composition of stars: Stars have been observed in which rare-earth elements are a thousand times more abundant than in average stars; other stars have a ratio of $^{13}C:^{12}C \approx 1$ (instead of 0.01 on Earth); and finally, one star is known with $^3He/^4He = 4$ (instead of the conventional ratio of 10^{-7}). Other anomalies exist, also. Some stars have anomalously large magnetic fields. There are stars whose brightness varies periodically (Cepheids), stars which flare up regularly, and finally stars which experience catastrophic explosions (supernovae). A well known example of the flare-up of a supernova is the explosion that produced the Crab Nebula.

Roughly speaking, however, all recently formed young stars are similar to each other; they consist of approximately 70 percent H, 29 percent He, and 1 percent heavier elements. All properties of such stars are fully determined by the mass; these stars constitute a single-parameter family. On the spectrum-luminosity diagram they form the Hertzsprung-Russell "main sequence." The age scale for a star is determined by its rate of fuel consumption; a star with $M \sim M_\odot$ reaches its middle age after 5×10^9 years, while a star with a mass of 30 $M_\odot$ exhausts its hydrogen and ages in 6×10^6 years. A star which has almost exhausted its hydrogen still does not differ substantially from a very young star, in terms of its external properties; after some time, however, it begins to change rather rapidly.

It is during the period of evolution after hydrogen is exhausted that the great variety of observed properties and of stellar behavior occurs. Problems relating to this stage of evolution are not fully clarified and apparently are not necessarily related to relativistic effects. Therefore, we will discuss these stages only very briefly.

What are the parameters that create this variety of stellar properties if, as we are assuming, all stars are initially condensed from gas of approximately the same composition? In fact, the differences in the initial gas clouds are unimportant; all stars on the main sequence have passed through a lethic[6] stage: they have "forgotten" the asymmetry, the turbulence, and the temperature of the initial gas clouds from which they condensed.

What, then, does the star "retain" that may influence its further development? We mentioned above that the fundamental characteristic of a star is its mass. During the period of hydrogen burning on the main sequence, the mass loss is negligible; the mass remains constant. A second quantity which is conserved during the formation of a star is the initial chemical composition. A third conserved quantity is the angular momentum. The

6. This newly coined word (Bird 1964) is derived from the name of the mythical river Lethe, separating the Kingdom of the Living from the Kingdom of the Dead. Lethe is the river of oblivion.

magnetic configuration might also be regarded as a conserved property of the star, i.e., as an invariant. Incidentally, this is not at all obvious; as yet it has not been clarified to what extent the magnetic field of the star is a result of amplification of the magnetic field of the interstellar medium during the condensation of the star. Another possibility is the appearance of magnetic fields as a result of internal convective motions (the so-called dynamo effect; see Braginsky 1964; cf. the Bachelor theorem on the magnetic field in a turbulent conductive fluid). Finally, the fate of a star may be substantially influenced by the presence of a nearby star with which it forms a close binary system (Martynov 1965; Kippenhahn and Weigert 1967; Paczynski 1966). Theoreticians frequently forget that binary stars are quite numerous among certain classes of stars; in fact, the hypothesis exists that all "novae" are binaries. In this context note the considerations about the ages of the two components of Sirius (see § 13.7) and the restrictions imposed upon the parameters of double stars by the radiation of gravitational waves (Braginsky 1965).

Let us proceed now from the enumeration of parameters which characterize a star and determine its evolution, to the exploration of those processes which may produce some form of instability. Apparently, of all the enumerated factors the rotation, the magnetic field, and the presence of a binary component are small perturbations compared with the gravitational force which depends on the total mass. Therefore, these factors substantially influence the stability of the star only during periods when the margin of stability is small.

It is asserted frequently (but not precisely) that a rapid absorption of heat may produce a catastrophic contraction[7] and that the emission of neutrinos is similar to other energy-consuming processes such as the dissociation of iron, $^{56}_{26}Fe \rightarrow 13\ ^4He + 4n$, and the creation of electron-position pairs. In fact, neutrino emission is different; it causes a change in entropy. The rate of neutrino emission determines $dS/dt = -(1/\rho T)(du_\nu/dt)$. For stable configurations, which depend on S as a parameter, the rate of entropy change determines the rate of evolution. The speed of contraction is always less than the velocity of light. So long as gravitational self-closure has not occurred, the neutrinos escape from the star (if the star's matter is transparent to neutrinos). Thus, neutrino emission is fundamentally a nonequilibrium process, and as such it differs substantially from the creation of e^+e^- pairs, and from the dissociation of iron.

In hot matter, the time for establishing equilibrium between the radiation field and e^+e^- pairs is insignificant, whatever scale is applied; for example, when $T_9 = 6$, this time is of the order of 10^{-18} sec. Consequently, at each moment of time and at each point in the star the pairs are in com-

---

7. This contraction is known as *implosion* (as opposed to explosion, the ejection of matter).

plete equilibrium; their number is not determined by the rates of pair formation and annihilation. The heat consumed by pair formation does not disappear; when the matter expands and the temperature falls, the number of pairs decreases and the consumed heat is discharged back. Equilibrium creation of pairs is not a factor that changes the entropy. The creation of pairs changes only the form of $P = P(\rho, S)$; i.e., it changes the relationship between the pressure and the density at given entropy. The same applies to the dissociation of iron and of helium.

As a result of pair formation or the dissociation of iron, $\gamma = (\partial \ln p / \partial \ln \rho)_s$ becomes less than $\frac{4}{3}$ in a certain temperature and density domain, and stars become unstable. The source of this behavior is, of course, the consumption of energy to form the rest masses of the e^+e^- pairs, or the consumption of energy to break nuclear bonds. The ratio between the additional pressure of the new particles and the additional energy density turns out to be small—in fact, less than $\frac{1}{3}$. However, the description of this behavior by introducing special quantities such as dQ/dt (where Q is the density of thermal energy) is a distressing consequence of underestimating the power of thermodynamic methods, and of underestimating the clarity and the simplicity which result from using the concept of entropy.

In the (ρ, T)-plane one can draw a curve where $\gamma = \frac{4}{3}$ (see Fig. 33, a). This curve separates the region of stability from the region of instability. In the roughest approximation to the stellar structure, one can draw curves $P(\rho, T) = aM^{2/3}\rho^{4/3} = \text{const.} \ \rho^{4/3}$ corresponding to stellar evolution at mean hydrostatic equilibrium. The intersection of such a curve with $\gamma = \frac{4}{3}$ signals the onset of instability in a star of given mass.

For example, according to computations of Imshennik and Nadezhin (1965), for a star with $M = 20 \ M_\odot$, $\gamma = \frac{4}{3}$ is reached when $\rho \approx 6 \times 10^6$ g cm^{-3}, $T_9 \approx 4.8$. However, in the course of further adiabatic contraction after $\gamma < \frac{4}{3}$, the curve $S = \text{const.}$ intersects the second line $\gamma = \frac{4}{3}$ and again reaches a region of stability. The reason for this is the formation of a great number of nonrelativistic particles during the dissociation of iron, Fe $\rightarrow \alpha + n \rightarrow p + n$.

Thus, after rapid nonstationary contraction, the existence of a new region of stable equilibrium could conceivably halt the contraction if the larger part of the mass turned out to be in the $\gamma > \frac{4}{3}$ region, and if the destabilizing effects of GTR were not too strong. During the halt of contraction, there would appear shock waves which, in propagating toward the surface of the star's atmosphere, would transport energy into a mass that becomes smaller, and would strip off the outermost layer. Such is the schematic mechanism of the flare of a supernova, as analyzed by Nadezhin and Frank-Kamenetsky (1962, 1964a, b, 1965), and by Imshennik and Nadezhin (1964, 1965). These authors investigated in detail the hydrodynamics of the process of establishing new equilibrium and the hydrodynamics of the ejection of the shell by a shock wave, without going into the forces that caused the initial onset of instability. The external properties of the calculated process agree generally with the observational data.

As the shock wave emerges through the star's surface, the ejection of the outer shell of matter or even the disintegration of the entire star may be caused by a second very important factor which changes the course of the evolution, and which has been emphasized by Hoyle and Fowler (1960, 1965) for many years (see also Fowler and Hoyle 1964). This factor is related to the nonuniform chemical composition of the star. In the absence of convective mixing, when full thermodynamic equilibrium has at last been reached at the center of the star, matter near the center has been completely transformed into iron, while the next layer outward contains oxygen and carbon, the next contains helium, and in the outermost layer there remains unburned hydrogen.

The transformation of hydrogen into helium requires weak interactions (since half of the protons must be transformed into neutrons) and can thus never proceed rapidly; therefore, we will not take into account energy release from hydrogen burning. However, even without hydrogen, the transformation energy of heavier nuclei is greater than the negative energy of the star as a whole. This means that the supply of nuclear energy, e.g., of the reaction $2\ ^{16}O \rightarrow ^{32}S$, along with the thermal energy of the star, is sufficient to overcome gravitation and scatter the entire star to infinity. The processes $3\ ^4He \rightarrow ^{12}C$, $2\ ^{12}C \rightarrow ^{24}Mg$, $2\ ^{16}O \rightarrow ^{32}S$ do not require the proton-neutron transformation; they take place solely by the strong interaction (of nuclear forces).

At sufficiently high temperatures, which weaken the effect of the Coulomb repulsion of nuclei, these processes can take place in a time that is shorter than the free-fall time; i.e., they can proceed explosively. Fowler and Hoyle (1964) have developed the schematic outline of an explosion caused by implosion: The shock wave, passing through appropriate layers, produces nuclear reactions in these layers, which are accompanied by a release of heat. In other words, the shock wave becomes a detonation wave. All layers above the detonation point are ejected with gigantic velocities. But one must not think that the iron core can survive when a sufficient amount of nuclear energy has been released. Even though the core will initially contract because of the increased pressure when the nuclear reaction is ignited in the shell, later the core may also expand and be disrupted because of the removal of the external pressure of the shell that is ejected.

One must keep in mind that the entropy of the matter in the star's core is that needed for equilibrium at the given ρ_c only when the core is under the pressure of the overlying layers of the star. However, this entropy is considerably greater than the equilibrium S for the smaller mass that remains after the shell is ejected. Of course, if the collapse has progressed far enough when the external shell is ejected, then the core will continue to collapse (Colgate and White 1966). Thus, at each moment of its evolution, almost until the complete exhaustion of all fuel, a star is "sitting on a powder keg"; it contains a suicidal amount of fuel.

Recently, detailed computer analyses of the explosion of a supernova have been published (Colgate and White 1966; Arnett 1966, 1967; Imshennik and Nadezhin 1967). Apparently, a major role in the flare-up of a supernova is played by neutrino radiation and its absorption in the shell (for more details on this see § 11.4).

Is implosion the only mechanism capable of blowing up a star? How stable with respect to thermal explosion is a configuration which is totally stable in a hydrodynamical sense?

In our overview of the total course of evolution, we linked thermal stability to the negative heat capacity of a star as a whole. There exist two causes which, under certain conditions, create thermal instability.

1. Negative heat capacity is typical of a nondegenerate plasma in its own gravitational field. At high density and not very high temperature, when substantial electron degeneracy occurs, the heat capacity of the star becomes positive (this was mentioned in § 10.15, and in § 11.1; see also Fig. 53). In a star with $M = M_\odot$, as the entropy decreases, the temperature initially increases, and then decreases. At a low temperature (compared with the temperature of electron degeneracy for a given density), the result is a white dwarf. A decrease in temperature with decreasing entropy indicates positive heat capacity. This circumstance brings about a rapid decrease of the nuclear reactions and a congealing of the composition of the white dwarf. As the temperature drops, the nuclear reaction rates decrease; reactions lag behind the heat losses, creating a condition of falling entropy; for a positive heat capacity this leads to a further decrease in the temperature and consequently does not produce an explosion.

2. Consider another case. Negative heat capacity is a concept that refers to the star as a whole: it is a result of the restructuring of the entire star due to a change of entropy. Each individual small layer of matter in the star has positive heat capacity, equal to C_p: each layer is under constant pressure produced by the surrounding matter. Thus, theoretically, a thermal explosion of an individual layer is possible. This process is inhibited because the given layer is in thermal contact with matter above and below. On the other hand, if we consider a layer that is sufficiently thick, then an increase in entropy in this layer will be accompanied by a significant decrease in pressure; the heat capacity increases with increasing thickness, and in the limit it passes through $C = \pm \infty$, becoming negative.

For reactions with sufficiently strong temperature dependence, taking place in a thin layer between the burned-out core and the envelope, an instability toward the growth of thermal perturbations may develop. This type of instability was considered by Gurevich and Lebedinsky (1955); it was independently discovered and analyzed by Schwarzschild and Härm (1965) in the course of a numerical computation of the evolution of a star with $M = 1.0 \, M_\odot$. The growth of the entropy in the layer is inhibited by the development of convection, and results in convective mixing. It is possible that thermal explosions, whose development is stopped short by intensified convection, play a role in multiple flares of some stars.

The anomalous compositions of some stellar atmospheres indicate the occurrence of mixing of matter that has never been exposed to stellar burning (hydrogen and probably also some helium) with matter that has been in the stellar interiors and contains heavy nuclei (Hoyle and Fowler 1965). Some of the elements observed can be produced only by combining neutrons with nuclei of elements in the middle of the periodic table; i.e., they require a temperature at which hydrogen cannot survive. Finally, there are certain properties of the composition, (e.g., ^{3}He/^{4}He > 1) which, in the judgment of several authors (Sargent and Jugaku 1961; Fowler, Burbidge, Burbidge, and Hoyle 1965; Reeves 1965; Wallerstein 1962; Novikov and Syunyaev 1967) indicate extremely strong irradiation of matter by particles with energies of many MeV, i.e., particles such as cosmic rays which are not in thermal equilibrium.

This exposition does not claim to be an exhaustive description of nonstationary phenomena. However, we hope that even such a brief review will convey to the reader a conception of the possible nature of future theoretical developments which at the present time are being pursued extensively at various scientific centers.

Next we turn our attention to modern theories of the catastrophic outbursts of supernovae.

11.4 SUPERNOVA OUTBURSTS (BY V. S. IMSHENNIK AND D. K. NADEZHIN)

In the last few years a number of papers have appeared which represent a great advance in the theory of the late stages of evolution of massive stars ($M > M_\odot$) and their gravitational collapse. It is of interest to establish the relationship between these theoretical investigations and the observed outbursts of supernovae. The classical theory of stellar evolution, which takes into account all of the details of nuclear burning, convection, and radiation diffusion, had encountered by 1965 deep difficulties at the stage of helium burning (Hoffmeister, Kippenhahn, and Weigert 1964a, b; Iben 1964). For this reason, the later stages of the evolution––those involving the burning of carbon, oxygen, etc., up to the formation of iron—had to be considered in an approximate manner, i.e., by using polytropic models. Fowler and Hoyle (1964) made a significant attack on this problem. One of the most important results of their study was the construction of an approximate model of a pre-supernova star, i.e., a star at the moment immediately preceding the onset of instability and gravitational collapse. In a gaseous sphere with a polytropic index of $n = 3$, all of the gas particles of the star move along the same evolutionary track in the density-temperature plane. This circumstance allowed Fowler and Hoyle to restrict themselves to a consideration of the local evolution at the center of the star, taking into account the most important physical processes that occur at high temperatures and densities—e.g., the effects of relativistic electron-positron pairs, of radiation, of neutrino emission, of various nuclear reactions, and of β-reactions. The quantitative conclusions derived by Fowler and Hoyle apply only to stars of very large mass $(M > 10\ M_\odot)$, where electron degeneracy can be neglected. The pre-supernova stellar model of 30 $M_\odot$, which was taken by Fowler and Hoyle, consists of a central iron core (3 $M_\odot$), an oxygen mantle (17 $M_\odot$) (the mantle and the core have a polytropic structure), and a rarefied hydrogen-helium envelope (10 $M_\odot$). Fowler and Hoyle demonstrated that the immediate cause of the onset of instability and of the subsequent implosion is the photodissociation of iron nuclei into helium nuclei plus nucleons. For a star with a total core-plus-mantle mass of 20 $M_\odot$, the implosion begins when the central temperature reaches $T \approx 5 \times 10^9\ {}^\circ$ K and the central density reaches $\rho_c \approx 10^7$ g cm^{-3}. Prior to this stage, the central regions of the star have burned to iron as the star has evolved quasistatically through successive states of hydrostatic equilibrium. Although the evolution has been quasistatic, in the period when $T_c \gtrsim 0.5 \times 10^9\ {}^\circ$ K the principal mechanism of energy loss–neutrino radiation–has produced a much faster evolution than would result solely from photon emission off the star's surface.

In this manner, the star has approached the critical state described above. The most important question to be asked about the subsequent evolution is this: What is the origin of the explosion in which the massive envelope is ejected? Attainment of the critical state manifests itself by the beginning of a rapid contraction of the star, or at least of a rapid contraction of the central region. It seems clear that the external layers, deprived of support, must collapse inward initially. Fowler and Hoyle concluded that the observed supernova explosion might be caused by the implosion detonating the unburned oxygen in the star's mantle. It is important that the existence of an oxygen mantle outside the iron core depends on the assumption of the absence of mixing of stellar matter in the course of the previous evolution.

Thus, in explaining the explosion we have to consider the nuclear reaction

$$^{16}O + {}^{16}O \rightarrow {}^{32}S + 16.54\ \text{MeV}. \tag{11.4.1}$$

The characteristic energy of the outburst of a Type II supernova (supernovae of Type II are related to massive stars) is of the order of 10^{52} ergs. Such an energy can be obtained by the combustion of 10 $M_\odot$ of oxygen. Thus, Fowler and Hoyle come up with a specific solution to the problem of how the implosion could trigger an outburst of the right magnitude and scatter the outer part of the star into space. The dynamics of the implosion per se could not then be analyzed because all estimates were made by using the approximate assumptions of free fall and of adiabatic processes.

Colgate and White (1966) were the first to perform computations of the implosion dynamics for various stellar masses (10 $M_\odot$, 2 $M_\odot$, 1.5 $M_\odot$, and 2 $M_\odot$, plus the shell of a red giant with $\sim$ 8 $M_\odot$). The computations took into account the effects of relativistic degeneracy of the electrons and of nonrelativistic degeneracy of the nucleons. The nature of the neutrino radiation that was considered (a modified URCA process was implied, even though it was not taken into account properly) was such that the central region of the star was cooled to low temperatures during the implosion. The collapse of this central region was halted at near-nuclear density ($\rho_c \sim 10^{15}$ g cm^{-3}) by the mounting pressure of nonrelativistically degenerate neutrons. However, the halt of the collapse of the core of the star by itself could not cause a substantial ejection of mass from the external envelope (though subsequent processes could; see below). Moreover, the halt took place in the region where GTR effects could become substantial, but the implosion dynamics were calculated in the framework of conventional nonrelativistic hydrodynamics with gravitation in the Newtonian approximation.

The important innovation of Colgate and White was to recognize that the star should be opaque to electron neutrinos during the late stages of its implosion. The cross-section for neutrino absorption and scattering by matter varies with the neutrino energy as $\sigma \propto E^2$; for the neutrino mean free path to be much smaller than the star, the neutrino energy must be high. Colgate and White assumed that high-energy neutrinos ($\sim$ 50 MeV, $\sim 5 \times 10^{11}$ ° K) are created in the powerful shock front that separates the star's cold, static neutron core from the matter falling freely onto it from outside. These neutrinos are sufficiently energetic to be absorbed in the star's envelope; consequently, they carry the energy released at the core boundary out to the star's outer layers. Such a form of energy transport was called deposition by Colgate and White. Deposition leads to a heating of the envelope, and thence to the creation of an outward-moving shock wave in it. This, then, is a solution to the main problem of modern supernova theory—the transition from contraction to explosion. It should be stressed that the numerical calculations of deposition made by Colgate and White are only illustrative of the essential role of deposition; they are not firm proofs of the nature of a supernova outburst because the mathematical model used is based on intuitive considerations rather than on completely rigorous investigations. Thus, the quantitative conclusions as to the mass ejected by the star (which amounts to a considerable portion of the entire mass of the model), the nature of the outgoing shock wave, the fraction of the mass converted into cosmic rays, the electromagnetic radiation emitted in the optical range—all of these factors may be considerably modified when the physical details of the model are examined in greater detail.

Colgate and White have also estimated the effects of oxygen detonation against a background of stellar implosion; they conclude that the detonation plays no important role in the energy balance of the supernova model. However, subsequent investigations have shown that the hypothesis, due to Fowler and Hoyle, that oxygen detonation does play an important role may still be correct in certain circumstances (see below).

According to Colgate and White, the implosion of stars with small masses ($2 M_\odot$ and $1.5 M_\odot$) proceeds in the same way as the implosion of stars of large mass ($10 M_\odot$), even though the immediate cause of the implosion is not the decomposition of iron, but the absorption of degenerate electrons by nuclei (at densities $\rho \geq 10^{11}$ g cm^{-3}, the Fermi energy of electrons becomes ≥ 20 MeV, which is quite adequate for the neutronization of any nuclei, including even ^{4}He).

Shortly after the publication of the Colgate-White work, two interesting articles dealing with neutrino processes were published by Arnett (1966, 1967). Arnett found, in disagreement with Colgate and White, that the star's core does not form cold; rather it is heated during the explosion so that its temperature reaches the order of $3 \times 10^{11}\,^\circ$K ($kT \approx 24$ MeV). The shock wave separating the collapsed core from the infalling envelope is not the only site for the creation of energetic neutrinos according to Arnett. The neutrinos are created and absorbed throughout the entire core. Arnett criticized the neutrino-deposition approximation, and he used another method of treating energy transport by neutrinos. After the onset of neutrino nontransparency, Arnett used the approximation of neutrino-equilibrium diffusion, constructed in analogy to radiative-equilibrium diffusion of photons, to describe the process of energy transport. It should be stressed that, while this work represents a further development of the Colgate-White concept, it also appears to be more consistent from a physical viewpoint. Unfortunately, our insufficient knowledge of neutrino physics, together with the immense complexity of neutrino interactions, does not yet permit us to take full advantage of the diffusion approximation. In the determination of the Rosseland mean neutrino opacity, Arnett considered only neutrino-electron scattering, assuming that this leads to estimates of minimum values of the opacity. These estimates are very approximate, including the assertion that they are minimal. The method of determining the electron number density is especially questionable. In addition, it is demonstrated in the same paper that the end result, the discharge of the shell, is rather sensitive to the magnitude of the neutrino energy transport. Arnett (1966) computed the sources of neutrino radiation in more detail than was done previously and discovered that the neutron core of the star should be hot after the implosion has stopped, rather than cold as predicted by Colgate and White. In agreement with Colgate and White, Arnett demonstrated that, without taking into account the neutrino opacity, an ejection of the envelope essentially does not occur. In his first paper Arnett concluded that his results agreed with those obtained by Colgate and White. To check the structural sensitivity of the results an additional computation of the implosion was performed by using an entirely different model of a pre-supernova (derived previously by Chiu 1966*b*). The Chiu model, obtained by an evolutionary sequence without the use of the polytropic approximation but with inclusion of neutrino losses, has a very dense (almost isothermal) core with a density at the center of $\rho \sim 10^9$ g cm^{-3}. In this case the implosion is caused by electron capture, as in the Colgate-White models of small mass. The final results, in spite of the considerable difference in the structure of the pre-supernova star (a condensed model versus a polytropic model with $n = 3$) do look alike generally. Arnett (1966) points out the possibility of a large role played by muon neutrinos, to which the star is transparent regardless of its density, and which are produced by spontaneous decay of μ-mesons, which in turn are born in pairs in the hot neutron core of the star:[8]

$$\mu^+ \rightarrow e^+ + \nu_e + \bar{\nu}_\mu \; ; \quad \mu^- \rightarrow e^- + \bar{\nu}_e + \nu_\mu \,. \qquad (11.4.2)$$

In a later paper Arnett (1967) refined the equation of state of the matter, which was rather approximate in the earlier work: he included the radiation of muon neutrinos, he took into account the modified URCA process, and he also included some other physical refinements of the overall problem. Upon computing a series of masses (32 $M_\odot$, 8 $M_\odot$, 4 $M_\odot$, 2 $M_\odot$), he arrived at different conclusions than previously. For the two largest masses, the implosion of the central region of the star did not stop, and no mass ejection took place. Apparently the result here was substantially affected by the muon neutrinos, which carried away a tremendous amount of energy, $\sim 10^{54}$ ergs (while electron neutrinos carried away only 10^{52} ergs), and which did not halt the implosion. The muon losses of energy had a smaller influence upon the models of 4 $M_\odot$ and 2 $M_\odot$; there the results were similar to the ones obtained previously. For a model of 2 $M_\odot$, an ejection of mass occurred, and a hot neutron nucleus with a mass of 0.57 $M_\odot$ remained; after cooling, this hot neutron nucleus can become a stable neutron star. Thus, according to Arnett (1967), stars with a mass somewhat greater than 4 $M_\odot$ finally experience a relativistic collapse and do not produce any supernova outburst. The implosion may be accompanied by a supernova outburst only for stars of sufficiently small mass. In evaluating Arnett's final conclusions, we should perhaps recall the extreme sensitivity of the computed results to the assigned magnitude of the neutrino opacity. For this reason, the indicated critical mass (4 $M_\odot$) might change considerably.

Here important new results due to Domogazky (1969) should be noted. Muon neutrinos created in reaction (11.4.2) have energies of $\sim$35 MeV.

---

8. This process becomes important only at very high temperatures. For comparison, it should be pointed out that $m_\mu c^2/k = 1.2 \times 10^{12}$ ° K. It is easy to demonstrate that the effective exponential temperature dependence of the energy losses by muon neutrinos is $-1.2 \times 10^{12}/T$, and accordingly the temperature must be very high in the region where this process becomes important.

Such neutrinos cannot interact with nucleons because any such interaction requires a threshold energy of ~ 100 MeV corresponding to the rest mass of a μ-meson. However, it was shown by Domogazky that the concentration of μ-mesons at temperatures $T_9 > 200$ (according to Arnett's calculations the temperature can reach $T_9 \sim 300$) is so high that the neutrinos created by reactions (11.4.2) will be absorbed by the μ-mesons:

$$\bar{\nu}_\mu + \mu^- \rightarrow e^- + \bar{\nu}_e,$$
$$\nu_\mu + \mu^+ \rightarrow e^+ + \nu_e. \tag{11.4.3}$$

These reactions, by contrast with reactions (11.4.2) are without threshold.

The mean free path of the muon neutrinos at a temperature $T_9 = 300$ is $\sim 10^5$ cm, which is much smaller than the size of the star's collapsed core. Consequently, the core will be opaque to the muon neutrinos!

Moreover, for densities $\gtrsim 10^{11}$ g cm^{-13} (such densities are reached quickly during the implosion) the main neutrino-producing reactions are not (11.4.2), but rather

$$\mu^- + p \rightarrow n + \nu_\mu,$$
$$\mu^+ + n \rightarrow p + \bar{\nu}_\mu. \tag{11.4.4}$$

Neutrinos produced by these reactions have energies near the muon rest mass; consequently, if the motion of the nucleons is taken into account, the mean free path of these neutrinos for the reactions inverse to reactions (11.4.4) is less than 10^5 cm, even at $T_9 \geq 100$.

In view of the above results of Domogazky, Arnett's conclusion about the peculiar effects of muon neutrinos upon the dynamics of collapse must be reexamined.

An analysis of the implosion dynamics of a massive star ($10\ M_\odot$, $30\ M_\odot$) similar to but independent of Arnett's analysis, was undertaken by Ivanova, Imshennik, and Nadezhin (1967). They found, in agreement with Arnett, that the core of a massive star is heated during the implosion. The density-temperature relation in the contracting core is strongly dependent upon the rate of neutrino emission. In the collapse of a massive star (10–$30\ M_\odot$) the neutrino emission is due primarily to the URCA process on free nucleons. Ivanova, Imshennik, and Nadezhin showed that the characteristic time for the capture of electrons and positrons onto free nucleons in the imploding core is less than one-tenth the hydrodynamic time scale. Consequently, kinetic equilibrium between the direct and indirect β-processes is maintained during the implosion. A large fraction of the electrons vanishes due to neutronization, and the electron gas becomes only partly degenerate ($\mu_e/kT = 1$–2).

The "freezing" of β-processes, which was assumed by Colgate and White, does not occur under these conditions; consequently, their formula for neutrino emission and their result that the core of the contracting star is kept cool are incorrect. The density-temperature curve for the center of an imploding star, as calculated by Ivanova, Imshennik, and Nadezhin (1967), is shown in Figure 59. The cooling curve (curve 3) in that figure was produced by using Colgate and White's incorrect formula for the neutrino losses; by contrast, in computations with the correct formula for neutrino losses, the central regions of the star are heated (curve 2). The rate of energy loss due to the URCA process, in the case of total transparency to neutrinos and antineutrinos, can be written in the form (Ivanova, Imshennik, and Nadezhin 1967):

$$\epsilon_{\nu\beta} = 1.3 \times 10^9 \, T_9{}^6 a(T_9)\Phi(\rho/T^3) \text{ ergs (g sec)}^{-1} . \qquad (11.4.5)$$

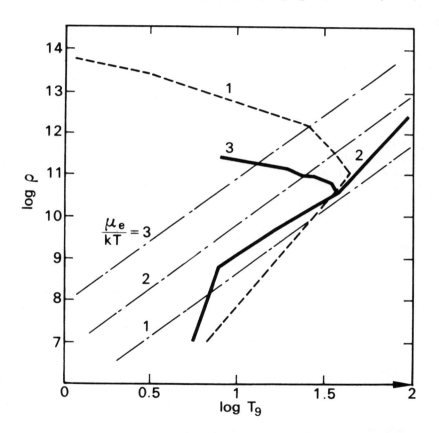

Fig. 59.—Influence of neutrino radiation on the physical conditions in the imploding core of a star with mass 10 $M_\odot$. Curve 1, calculations by Colgate and White (1966); curve 2, calculations with the rigorous formula for neutrino radiation; curve 3, control calculations with Colgate and White's formula.

Here a is a quantity which takes account of the temperature-dependent chemical composition (at low temperatures the composition is iron; at high temperatures the iron has disintegrated into free nucleons). For Fe, $a \sim 1$; for free neutrons $a = 56 \, (ft)_{Fe}/(ft)_n = 612$. Here we have made use of Fowler and Hoyle's (1964) maximal estimate $(ft)_{Fe} = 1.3 \times 10^4$ sec. The quantity Φ is a slowly varying function which takes account of degeneracy. In the absence of degeneracy $\Phi = 1$, and with increasing degeneracy Φ decreases.

Ivanova, Imshennik, and Nadezhin (1967), in contrast with Arnett (1966, 1967), took account of the absorption of neutrinos and antineutrinos in the central regions of the star by introducing an effective coefficient. At each time-step in the integration this coefficient was calculated on the basis of a known solution to the transport equation and the star's known distribution of density and temperature at that time-step. The computation assumed, on the basis of prior analysis, that the antineutrino absorption is dominated by interactions with protons, and the neutrino absorption by interactions with neutrons:

$$\bar{\nu} + p \rightarrow e^+ + n \,, \quad \nu + n \rightarrow p + e^- \,. \tag{11.4.6}$$

(Neutrino and antineutrino scattering on electrons is not so important as was assumed in the work of Colgate and White and of Arnett.) Ivanova, Imshennik, and Nadezhin used the approximation that the value of the effective absorption coefficient calculated for the center of the star is correct everywhere else in the star as well; of course, this approximation amounts to ignoring the effects of neutrino deposition. Another difference between their work and Arnett's is that they did not take account of muon neutrinos.

Ivanova, Imshennik, and Nadezhin were the first to consider oxygen detonation (reaction [11.4.1]) on the implosion background. The oxygen-burning equation can be written in the form

$$dX_{16}/dt = -\rho X_{16}^2 \, (T_9/5.3)^{26} \,, \tag{11.4.7}$$

where X_{16} is the concentration by weight of oxygen. For the temperature range $3 < T_9 < 5$, where intensive burning of oxygen occurs, the right-hand side of equation (11.4.7) approximates the data of Fowler and Hoyle with a fractional precision of 1.5 or better. Oxygen detonation is energetically capable of causing a disintegration of the entire star: if we take into account the nuclear energy, the total stellar energy is positive at the moment of the loss of stability. In this respect there is a fundamental theoretical difference between neutrino energy transport and detonation processes in supernovae: when the total energy is negative, only a part of the mass can be ejected during the neutrino-transport phase, and the remnant either passes into a new equilibrium state or else it collapses. The energy release in oxygen combustion is

$$\mathfrak{E}_a = (\Delta q/2Am_p)|dX_{16}/dt| \,, \quad \Delta q = 16.54 \text{ MeV} \,, \quad A = 16 \,. \tag{11.4.8}$$

To justify that this is the dominant mode of oxygen destruction, the time of the competing process, i.e., oxygen photodissociation, was estimated. The estimate indicates that at the characteristic temperatures and densities of oxygen combustion, the required time for the photodissociation of oxygen is considerably longer than the time required for combustion. According to the computation, oxygen combustion acquires features of detonation and leads to a substantial release of energy.

Equation (11.4.7) was solved numerically together with the hydrodynamical equations describing the stellar implosion. The computation assumed an initial stellar model of 10 $M_\odot$, of which the outermost 3.6 $M_\odot$ was pure oxygen. In this model almost all of the oxygen (80–90 percent) was burned, and the energy released was $\sim 3 \times 10^{51}$ ergs.

Let us now describe the sequence of events which leads to the outburst of a massive ($M \geq 10\,M_\odot$) supernova, as elucidated by the work cited above.

The initial stellar model becomes unstable as a result of the photodisintegration of iron which decreases the adiabatic index $\gamma = (\partial \ln P/\partial \ln \rho)_S$. The onset of instability occurs when the temperature and density in the star's center have reached 5.5×10^9 ° K and 10^7 g cm^{-3}, respectively, and the star's radius is $\sim 4 \times 10^{-2}\,R_\odot$.[9] (The boundary of the stability regime for stars of various masses was found by Bisnovatyi-Kogan and Kazhdan [1966], who used the approximate energetic method.)

The initial stage of the contraction is only weakly sensitive to the form of the equation of state, because the gravitational forces in the contracting matter are several times larger than the pressure forces. The various different approximations used for the equation of state in the works cited above lead to approximately the same density and temperature histories inside the star in these early stages of contraction.

By the time the temperature reaches $\sim 20 \times 10^9$ ° K (at which point the central density is $\sim 10^{10}$ g cm^{-3}), a large number of free nucleons have been created and the rate of neutrino emission is beginning to increase sharply. The neutrino emission is due largely to the URCA process on free nucleons. Despite the fact the adiabatic index γ now increases since photodisintegration is finished, and becomes greater than the critical value $\frac{4}{3}$, the star continues to contract rapidly. The rate of energy loss due to neutrino emission is so large that the star's core must collapse with hydrodynamical velocity in order to replenish these energy losses by the release of gravitational energy. If the rate of neutrino emission and the equation of state both have power-law forms, then the contraction, as affected by neutrino emission, can be described by a self-similar solution (Nadezhin 1968). This means not only that one can use the usual methods for investigating self-

9. The numerical data on which our overview of the supernova model is based were taken from Ivanova, Imshennik, and Nadezhin (1967) for the case of a star with $M = 10\,M_\odot$. The differences between those data and the results of Colgate and White (1966) and of Arnett (1966, 1967) will be discussed later.

similar solutions but also that the implosion stage as affected by neutrino emission is only weakly sensitive to the initial state of the star's core. (The initial conditions will be "forgotten.")

At a temperature of about 40×10^9 ° K (a density of 3×10^{11} g cm^{-3}) the "optical" thickness of the star's core to antineutrinos becomes of order 1. At that moment the optical thickness to neutrinos is already several times larger than 1 because the concentration of neutrons, which are the chief absorbers of neutrinos, is much greater than the concentration of protons, the chief absorbers of antineutrinos. The spectrum of the neutrinos and antineutrinos emitted by the star just before absorption becomes important is shown in Figure 60. The maximum emission is at an energy of ~ 8 MeV. As the neutrino and antineutrino opacity increases, the energy emission by neutrinos and antineutrinos decreases. The star's total neutrino-antineutrino luminosity reaches a maximum of 3×10^{53} ergs sec^{-1} and then quickly declines (see Fig. 61). The neutrino light curve has the form of a sharp peak with a typical width of 0.03 sec and a total energy output of 8×10^{51} ergs. Arnett derived an energy output from electron neutrinos that was 3 times larger than this. More recent Russian work agrees with Arnett for a star of mass ~ 2 $M_\odot$; but it gives an energy output 100 times greater than either Arnett's or the original Russian work for a mass of 10 $M_\odot$. Arnett's calculations included losses from muon neutrinos, which are only weakly sensitive to the concrete assumptions of the model. (The same muon-neutrino losses are obtained when one considers free infall to nuclear densities without account of electron neutrinos; cf. Guseynov 1968.)

As the neutrino losses are cut back by rising opacity, the contraction of the star's core slows down sharply, because the adiabatic index approaches $\gamma \approx \frac{5}{3}$ (hot, nondegenerate neutron gas).[10] That part of the star's core which is opaque to neutrinos at the moment of maximum neutrino output has a mass of ~ 0.1 $M_\odot$ (cf. Fig. 61); this is only 1 percent of the star's total mass. The star's central density at this moment is $\sim 5 \times 10^{12}$ g cm^{-3}. The radius of the neutrino "photosphere," as determined by the condition $\tau_\nu = 1$, is 3×10^6 cm; and the temperature of this photosphere is 80×10^9 ° K. The Fermi distribution in the photosphere produces a mean energy of the emitted neutrinos and antineutrinos of roughly 20 MeV.

The contraction of the core halts when the central density and temperature reach 3×10^{13} g cm^{-3} and 140×10^9 ° K, respectively. If muon neutrinos are taken into account in the approximation of complete transparency (Arnett 1967), the core's collapse stops at a higher temperature ($\sim 300 \times 10^9$ ° K) and density. However, even in this case the effects of GTR are of no consequence. They become important only at a later stage, when the massive collapsed core cools down and enters a stage of general-relativistic collapse.

10. The equilibrium energy density of the neutrino gas, and that of the photon gas, are two orders of magnitude less than that of the neutron gas at this point.

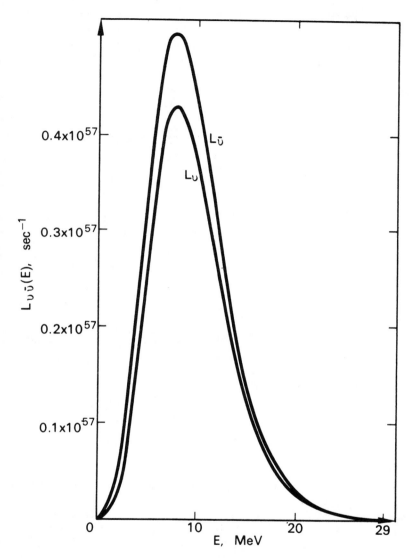

Fig. 60.—Spectrum of the neutrino and antineutrino radiation, $L_\nu(E)$ and $L_{\bar\nu}(E)$, at a moment just before the beginning of neutrino absorption inside the star ($\tau_\nu = 0.6$; $\tau_{\bar\nu} = 0.08$; $T_9 = 21.1$; $\rho_c = 9.2 \times 10^9$ g cm^{-3}).

An outgoing shock wave begins to form in the star's envelope at the moment the core halts, as a consequence of neutrino deposition or of oxygen detonation. A shock wave can also be formed in the absence of deposition and detonation, as a result of a simple hydrodynamical reflection at the boundary of the halted core, and the build-up of the outgoing hydrodynamical waves in the atmosphere's region of decreasing density. However, in this case the force of the shock wave is so small that its effects can hardly correspond to the observed supernova outbursts.

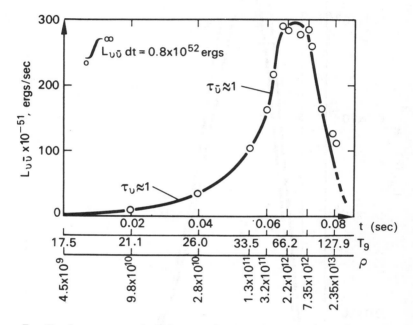

Fig. 61.—Supernova neutrino light curve. Density and temperature scales shown are for the center of the imploding star.

The principal parameter for comparison of supernova theory with the observations is the energy carried off as light and as kinetic motion of the ejected envelope. According to Shklovsky (1966) the energy output of a Type II supernova is about 10^{51}–10^{52} ergs; and for a Type I supernova, 10^{50} ergs. According to Colgate and White (1966) and Arnett (1967) neutrino deposition in a massive supernova will give the necessary explosive energy of $\sim 10^{52}$ ergs. In the calculations of Ivanova, Imshennik, and Nadezhin (1967) without deposition but with oxygen detonation, an energy of 3×10^{50} ergs was obtained. (The pure hydrodynamical effect without deposition or detonation gives only $\sim 10^{49}$ ergs.) Recall that the total energy released by combustion of all the oxygen is 3×10^{51} ergs. Only a small part of this energy (~10 percent) gets transformed into kinetic energy of the ejected envelope; most of the energy must be used to lift the matter out of the star's gravitational field. In this sense the result agrees with Colgate and White's conclusion that detonation is ineffective. However, the energy of the ejected envelope in this case of detonation is, in fact, near the lower limit of the observed energy release for supernova outbursts. Recently, Poveda and Woltjer (1968) have concluded from analyses of the observational data that the average energy of a Type II supernova outburst is ~4 × 10^{49} ergs. This energy, which is low compared to Shklovsky's estimates, can also be calculated from Gordon's (1967) results.

We should emphasize that it is quite possible to construct a stellar model in which the oxygen detonation occurs in a region with sufficiently small gravitational potential to yield the larger energy outputs suggested by Shklovsky. However, for this to happen the detonation cannot be caused by photodissociation-induced contraction of the star, because such a contraction will necessarily pull the oxygen into a region of large gravitational potential before igniting it. As an example, Fraley (1968) has shown that, for a star of very large mass ($\geq 40\ M_\odot$), oxygen burning in a central convective core during the normal stellar evolution would be characterized by an explosion; this explosion releases sufficient energy to eject a large fraction of the star.

The stellar rotation, which can have a strong influence upon the dynamics of the implosion, has not been taken into account adequately in the past computations. According to rough estimates (Colgate and White 1966; Ivanova, Imshennik, and Nadezhin 1969) rotation can halt the collapse at the moment when the neutrino opacity is beginning to be significant; and in the case of differential rotation with an angular velocity which increases inward, it can halt the collapse in the earlier stage of absolute transparency to neutrinos. In this case the only mechanism for explosion is the detonation of nuclear fuel. The first attempt to calculate the contraction of a rotating star by using two-dimensional hydrodynamical equations was made recently by Djachenko, Zel'dovich, Imshennik, and Paleychik (1968).

The above considerations show that the Fowler-Hoyle hypothesis of the role of nuclear-fuel detonation as the energy source of supernovae is perfectly reasonable, and is not inferior to the Colgate-White hypothesis of neutrino deposition.

The theory of the explosion of small-mass stars (1.5–2 $M_\odot$), which are probably connected with Type I supernova outbursts, is less fully developed than the theory for massive stars. The investigation of Type I supernovae is of particular interest now because they are probably the origin of the stable neutron stars which are identified with pulsars.

There are many unclear issues connected with the equation of state, with neutrino emission, and with the relation between the neutronization time scale and the hydrodynamical time scale. Chechetkin (1969) has studied the neutronization of iron-group elements which are overloaded with neutrons in a hot stellar core. Imshennik and Chechetkin (1969) have investigated the thermodynamics of matter under conditions in which neutrinos and antineutrinos escape freely from the system; that investigation was based on the kinetic equilibrium of β-processes for temperatures $5 \leq T_9 \leq 20$ and densities 10^8 g cm^{-3} $\leq \rho \leq 10^{12}$ g cm^{-3}. Because of the extremely slow character of β-processes, the neutronization of matter cannot lead to a rapid development of hydrodynamic stellar contraction. Neutrino emission due to hot neutronization of nuclei overloaded with neutrons has been discussed by Tsuruta and Cameron (1965) and by Nadezhin and Chechetkin (1969). The role of nucleon-nucleon collisions concomitant with β-processes has been considered by Kopyshev and Kuz'min (1968).

In contrast to Colgate and White and to Arnett (1966, 1967) (where the explosion of a small-mass star was calculated on the basis of equilibrium neutronization and neutrino deposition), Hansen and Wheeler (1969) (see also Arnett 1969) investigated the dynamics of an explosion caused by carbon detonation. They took as their initial model an isothermal white dwarf with mass 1.42 $M_\odot$, temperature 10^{8} ° K, central density 7×10^9 g cm^{-3}, and chemical composition 90 percent ^{12}C and 10 percent ^{24}Mg. They suggested that all the ^{24}Mg becomes gradually transformed into ^{24}Na by electron capture in the star's core of mass 0.5 $M_\odot$. This electron capture produces a gradual increase in the molecular weight of the core, and as a result induces hydrodynamical collapse. The increase of the temperature during the collapse initiates a powerful "carbon ignition and outburst." The star is totally disrupted, with an average expulsion velocity of 7000 km sec^{-1} and with an energy output of 8×10^{50} ergs. Neutrino emission by the URCA process was not taken into account in this model. Other details of the connection between the physical conditions inside white dwarfs and supernova outbursts have been considered in a number of papers (Hoyle and Fowler 1960; Schatzman 1963; Bisnovatyi-Kogan and Seidov 1969; Arnett 1969).

The role of magnetic fields in supernova outbursts is of considerable interest. On the basis of indirect considerations about the presence of magnetic fields in pulsars (Gold 1968) and on the basis of measurements of the polarization of the radio waves from the supernova remnant Cas A (Mayer 1968), one concludes that very strong magnetic fields ($\sim 10^9$–10^{12} gauss) originate in the implosion. Bisnovatyi-Kogan (1970) has suggested that such magnetic fields can transport angular momentum from the central regions of a collapsing, rotating star to the outside. Thus, the magnetic fields might play the same role in supernova dynamics as neutrino deposition: they might be an effective means of transporting the gravitational energy released by the core's contraction out to the intermediate layers of the star. Asymmetries in the ejection of matter can also be explained by the presence of a magnetic field in a rotating collapsing star (Ivanova, Imshennik, and Nadezhin 1969).

It is clear that the modern theory of supernovae has many degrees of freedom; however, the principal feature of the theory is the loss of stability by a star in the late stages of its evolution, after most of its nuclear fuel has been exhausted. The late stages of evolution must necessarily be accompanied by a large amounts of neutrino radiation. The detection of this radiation (Domogazky and Zatsepin 1969a, b) would mean a great stride forward in supernova theory and in the overall theory of stellar evolution.

A very important question is that of the light curve of a supernova. A comparison of the light curves obtained from theoretical computations with observational data must yield valuable information on the parameters which describe the star immediately before the outburst, i.e., its mass, its radius, and the rate of decrease of density near its surface. The propagation of a strong shock wave in a medium with decreasing density has the property that the values of the gas-dynamic quantities (pressure, velocity, etc.) have a very weak dependence on the formation and early history of the shock wave. The principal characteristics determining the process are the total energy contained in the moving gas and the rate of decrease of density before the shock hits. For a power-law form of density decrease, which is approximately fulfilled in the external layers of the star, the propagation of a strong shock wave is self-similar (Gandel'man and Frank-Kamenetsky 1956) and the result does not depend at all upon the specific form of the initial conditions. Because of this, in a theoretical analysis of the explosion of a supernova we can consider the propagation of the shock wave separately from the mechanism of shock-wave generation near the center of the star. The connecting parameter is the total energy of the gas moving behind the shock front. Shell ejection by a shock wave has been examined by Nadezhin and Frank-Kamenetsky (1962, 1964b). Imshennik and Nadezhin (1964) have attempted to explain the light curve of a Type II supernova by computing the escape of a shock wave, taking into account both the radiative transport of energy and hydrogen recombination in the expanding shell. The main idea behind this work is that the light curve of a supernova is the result of the release of the thermal energy of the external layers of the star after they have been heated by a strong shock wave. Polytropes with index $n = 3$, and with masses 15 and 50 $M_\odot$ and radii 9 and 20 $R_\odot$, were taken as presupernovae. The escape of a shock wave to the surface is accompanied by an abrupt peak in the luminosity. Although the luminosity at the maximum in these models of a Type II supernova agrees with the observed magnitude ($M_{bol} \sim -21$ mag), the width of the luminosity peak is insignificantly small (approximately 20 minutes instead of approximately 20 days). The luminosity drops due to rapid expansion and adiabatic cooling of the external layers (since the density is still quite high, the radiative diffusion can be neglected). As a result of further decrease in temperature, hydrogen recombination sets in; energy transport by radiation becomes the principal cooling mechanism, and a cooling wave originates. (Cooling waves were first described by Zel'dovich, Kompaneets, and Rayzer 1958.) The calculated light curve has a plateau with a value $M_{bol} = -13$ to -15 mag. The luminosity is constant for approximately 50–100 days. The radiation is maintained by means of the energy being released during the recombination of hydrogen. This plateau in the theoretical light curve is in good agreement (in magnitude as well as in duration) with the "hump" after the well-known drop in the observed light curves of Type II supernovae.[11] The main part of the energy is radiated by a supernova for approximately 20 days near the light-curve maximum $[(M_{bol})_{max} = -21$ mag], but a subsequent plateau-type behavior is nonetheless clearly noticeable on the observed light curves. Thus, if one could in some way expand in time the 20-minute luminosity peak on the theoretical curve to approximately 20 days, then the observed and theoretical curves would coincide in their principal parameters. Apparently the only way is to increase the radius of the initial model of the star, which would cause an increase in the time of adiabatic cooling of the very external layers of the star, and thus would also increase the duration of the luminosity near the maximum. To this end, the computation for an initial model (yellow supergiant) with $M = 30\ M_{\odot}$, $R = 500\ R_{\odot}$, $L = 2 \times 10^5\ L_{\odot}$ was performed.

An increase of the radius by approximately 50 times in comparison with the initial models of Imshennik and Nadezhin (1964) brought about an increase of the luminosity peak to approximately 1 day, whereas the light maximum remained about the same (~ -21 mag). In this computation, the energy of the moving gas behind the shock wave, which is the free parameter in the problem, was chosen so that the velocity of the disintegrating shell coincided with the velocity that is observed in Type II supernovae (~ 5000 km sec^{-1}). It follows from the computation that the maximum luminosity of a supernova is attained when the radius has increased by only a factor of approximately 1.5 compared with the initial radius. Thus, the ascending branch of the light curve of a supernova is caused mainly by a heating of the surface layers by a shock wave, rather than by an increase of the area of the emitting surface. Thus, by using the observed luminosity (~ -21 mag) and temperature ($\sim 4 \times 10^{4\,\circ}$ K) of a supernova near maximum light, we can estimate the radius of a pre-supernova; this radius turns out to be $\sim 10^4\ R_{\odot}$. Such large radii can occur in red supergiants of the M2 spectral class, as well as in the recently discovered infrared stars. The computation of the passage of a shock wave through the atmosphere of such a star is complicated by the fact that there are as yet no satisfactory models of red supergiants. However, this deficiency in the theory of stellar structure is not so serious an obstacle as to make one refuse to investigate the spread of powerful shock waves in an extensive atmosphere. When the shock wave

11. Imshennik and Nadezhin (1964) used obsolete data on Type II supernovae, as a result of which this plateau in the light curve was erroneously classified as the maximum light of a supernova. This error was pointed out by Shklovsky (1966).

has a velocity of $\sim(5\text{--}10) \times 10^3$ km sec^{-1} and the gas is heated to a temperature of the order of $(50\text{--}100) \times 10^3$ ° K, it is not important whether the atmosphere before the shock hit was in equilibrium or whether it had a modest velocity of $\sim10\text{--}100$ km sec^{-1}; and it is not important what the temperature of the atmosphere was. It should be stressed, in connection with this, that an extensive atmosphere can be formed by the outflow of matter from the star in the pre-supernova evolutionary stage (Bisnovatyi-Kogan and Zel'dovich 1968). In the work of Poveda and Woltjer (1968) there were some indications (unfortunately based on indirect arguments) that the pre-supernova star has an extensive atmosphere.

The propagation of a shock wave in an extremely rarefied atmosphere is substantially different from the propagation in a dense atmosphere. A heat wave, which has considerably more extended dimensions than the thickness of the shock front, moves ahead of the shock wave. This problem was analyzed in detail in the monograph by Zel'dovich and Rayzer (1966). The emergence of a heat wave in the computation for a yellow supergiant with a radius of $R = 500 \, R_\odot$ also contributed to a strong increase in the duration of the luminosity near maximum—from 20 minutes to 1 day. Moreover, the yellow supergiant, after the light maximum, also had a "hump" of the light curve with approximately the same parameters as before. Thus, unlike the region near maximum, the plateau of the light curve resulting from the cooling wave is only weakly sensitive to the radius of the pre-supernova. As demonstrated by Imshennik and Nadezhin (1964), the primary determining parameter of the cooling wave is the ionization potential of the matter.

Recently, the spreading of a shock wave in an extended atmosphere with radius 5000–10000 $R_\odot$ has been calculated using hydrodynamical equations together with equations for radiative energy transport (Grasberg and Nadezhin 1969a, b). The resultant theoretical light curve and the characteristics of the supernova photosphere are given in Figure 62. The initial increase in luminosity (segment A–B) results from the heating of the gas by the thermal wave which spreads out in front of the viscous discontinuity. Segment B–C is due largely to the spreading of the ionization wave. At the moment corresponding to point C the hydrogen has been completely ionized. The subsequent growth of the luminosity to its maximum (point D) is due to the intensive heating of the envelope by radiative energy, transported outward from the viscous discontinuity to the star's surface. The maximum luminosity occurs at the moment (D) when the losses of energy by radiation from the envelope begin to exceed the competitive process of heating. From then until point E a cooling wave develops in the expanding envelope.

These theoretical characteristics of supernovae agree to a certain extent with observations.

A different set of light curves, derived by Imshennik and Nadezhin (1964) for the case of the explosion of a compact stellar model, agree well with the observations of the irregular supernovae NGC 5457, NGC 6946 (1948), and NGC 5236 (Minkowski 1964). It seems that the pre-supernova stars in these cases for some reason had no extensive atmosphere. It is possible that the frequency of such "abnormal" supernovae might be quite large, since observational selection effects may militate against noticing such outbursts (these supernovae are 5–6 magnitudes less bright than typical Type II supernovae).

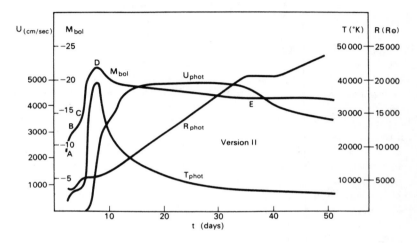

Fig. 62.—Photon light curve for a Type II supernova. Also shown are the changes with time of the photosphere's radius and temperature, and of the velocity of the matter passing through the photosphere. (The photosphere is defined to be $\tau = \frac{2}{3}$.)

As a result of the energy losses from a shock wave spreading in an extended atmosphere and the transformation of the energy loss into a non-stationary heat wave which moves in front of the viscous discontinuity, the matter in the atmosphere is compacted together into a thin, dense layer and is ejected by the shock wave (Grasberg and Nadezhin 1969b). In this case, of course, a Raleigh-Taylor instability may develop; its consequences need investigation. By contrast, when the shock wave moves outward through an ordinary, compact atmosphere, a density minimum does not develop between the ejected envelope and the rest of the atmosphere (Nadezhin and Frank-Kamenetsky 1964b; Imshennik and Nadezhin 1964). (The results described above were based on the assumption of local thermodynamic equilibrium between the radiation and the matter.)

A number of papers have been published[12] which attempt to explain

12. Borst (1950); Baade, Christy, Burbidge, Fowler, and Hoyle (1956); Anders (1959).

the exponential decay of the light curves of supernovae (primarily Type I) by the decomposition of unstable isotopes of californium, or of elements of the iron group. Such isotopes might occur in the external layers of the star due to nuclear reactions which take place at the high temperature behind a powerful shock wave. After the emergence of the shock wave, the unstable isotopes decay and serve as a source of radiated energy.

The problem of the maximum of the light curve has also been touched upon by Colgate and White (1966). Using some rough estimates, they concluded that the emission of thermal energy from the external layers, heated by the shock wave, is not sufficient to explain the observed light maxima. (It should be recalled that their pre-explosion stellar model had $R \sim 0.01$–$0.1\ R_\odot$. Consequently, their results are essentially the same as those of Imshennik and Nadezhin for compact models.) In their opinion the energy source of the luminosity of supernovae may be the β-decay of unstable nuclei ejected together with the shell from the interior of the star.

A most recent contribution to the nonthermal theory of supernova light curves is a paper by Colgate and McKee (1969). They point out that the main features of supernova light curves can be explained by the ejection of an envelope enriched with 0.25 $M_\odot$ of the radioactive nucleus ^{56}Ni. A total radioactive energy of 10^{49} ergs is then released by the two-stage decay ^{56}Ni $\rightarrow$ ^{56}Co $\rightarrow$ ^{56}Fe with mean lifetimes of 6.1 days and 77 days.

Very briefly we mention the possibilities of observation of electron neutrinos from supernovae. According to Arnett (1966, 1967) and Ivanova, Imshennik, and Nadezhin (1967), neutrinos are radiated with a total energy of 10^{52} ergs for 3×10^{-2} sec with a mean particle energy of $\sim$10–20 MeV. Then, following the reasoning of Arnett (1966), one can show that in order to produce more than about ten counts on the Davis (1968) measuring device, a supernova must be at a distance of less than about 5 kpc, i.e., within the boundaries of our Galaxy.

Finally, a few words on the origin of cosmic rays: Colgate and White (1966) concluded on the basis of their computations that the outermost 10^{-4} of the mass of the star is ejected with relativistic energies exceeding $2m_pc^2$. To provide the energy density of cosmic rays, 5×10^{-14} ergs cm^{-3} in the galactic volume of 5×10^{68} cm^3 for a lifetime of 2×10^8 years, one outburst of a supernova of 10 $M_\odot$ is required every 1.5×10^4 years.

The acceleration of particles to relativistic energies by the shock wave as it moves through the stellar atmosphere has also been studied by Nadezhin and Frank-Kamenetsky (1964a). They found that the yield of relativistic particles depends strongly on the value of the star's escape velocity. If the pre-supernova has the size of a main-sequence star, then the yield of relativistic particles due to direct gas-dynamical acceleration is negligible. Thus, the role of supernovae in the origin of cosmic rays depends upon a detailed elucidation of the pre-supernova structure.

11.5 THE PHYSICS OF NEUTRON STARS

A neutron star at point D_3 of Figure 57 is in a quasi-equilibrium state. Considerable recent research has been devoted to the theory of such stars. The state of the theory before the discovery of neutron stars as pulsars was reviewed by Wheeler (1966b); the review also includes a detailed bibliography. For a discussion of pulsars see § 13.10.

As noted previously, the neutron core of the star is surrounded by a shell consisting of nuclei and degenerate electrons; the outermost surface layer consists of an ordinary plasma. (Actually, the picture can be more complicated than this; see, e.g., Bayme, Pethick, Pines, and Ruderman 1969.) Immediately upon formation (after its collapse), a neutron star will be extremely hot. Regardless of the processes which might have brought the star to point D_3, the temperature of its interior, as pointed out by Chiu (1964), cannot long remain greater than several billion degrees, since at such temperatures intensive creation of neutrino-antineutrino pairs (which instantaneously leave the star) rapidly cool it. In all inner portions of the star, where the electron gas is degenerate, the thermal conductivity is extremely high. Therefore, the core of a neutron star is isothermal [or, more precisely, according to relativistic thermodynamics $T \sim (g_{00})^{-1/2}$]; and only in the outermost shell is there a temperature gradient. The structure of this outermost shell can be computed if the radius and the mass of the neutron configuration are known and if a definite temperature of the stellar interior has been specified. As a result, one can determine the surface temperature of the star and its luminosity.

In the first calculations the effective surface temperature turned out to be approximately 10^{-2} of the central temperature (see, for example, Bahcall and Wolf 1965a, b; Tsuruta and Cameron 1966). If $T_c \approx 10^8 °$ K, then $T_e \approx 10^6 °$ K (see Table 17).

The cooling time of a neutron star after its formation depends on its reserve of thermal energy and on the processes by which the thermal energy is lost. According to Bahcall and Wolf (1965a, b), the supply of thermal energy is given by the expression

$$E_T = 5 \times 10^{47} \left(\frac{T_c}{10^9}\right)^2 \left(\frac{\rho}{3.7 \times 10^{14}}\right)^{-2/3} \left(\frac{M}{M_\odot}\right) \text{erg} . \qquad (11.5.1)$$

This value was computed without taking into account the possible superfluidity of the neutron-rich nuclear matter of which the star is composed. The possibility of superfluidity was pointed out by Ginzburg and Kirzhnits (1964). If a substantial part of the mass of the stellar matter really turns out to be superfluid, then E_T will be considerably less than the above estimate, and the cooling time will be somewhat less than that cited below. For more on the influence of superfluidity and superconductivity in stars see § 13.11.

The cooling of a neutron star takes place by the radiation of neutrinos from the central region, where the temperature is high, and by photon emission from the surface (see Tsuruta and Cameron 1966). At very high interior temperatures, the most important process that leads to the creation of neutrino-antineutrino pairs is, as noted above, the annihilation of electron-positron pairs:

$$e^+ + e^- \rightarrow \nu_e + \bar{\nu}_e .$$

However, under the conditions found in a neutron star this process is suppressed by the high Fermi energy of the electrons, which is of the order of

TABLE 17

COMPARISON OF THE SUN'S RADIATION AND THE RADIATION FROM
A NEUTRON STAR WITH $R = 10$ km AND $T_e = 10^6$ ° K

Parameter	I Neutron Star	II Sun	Ratio I/II
Radius...............	10^6 cm	7×10^{10} cm	1.4×10^{-5}
Surface area..........	1.3×10^{13} cm²	6.4×10^{22} cm²	2×10^{-10}
Surface temperature...	10^6 ° K	5790° K	1.7×10^2
Wavelength of the maximum of the spectrum (on the frequency scale).............	40 Å	4600 Å	8.7×10^{-3}
Luminosity...........	7×10^{32} ergs sec⁻¹	4×10^{33} ergs sec⁻¹	0.175
Absolute bolometric magnitude (mag)....	6.5	4.6	$\Delta m = 1.9$
Fraction of the radiated energy in the interval 3000–10000 Å.......	5.4×10^{-6}	0.68	8×10^{-6}
Luminosity in the interval 3000–10000 Å....	3.8×10^{27} ergs sec⁻¹	2.7×10^{33} ergs sec⁻¹	1.4×10^{-6}

SOURCE.—Adapted from Wheeler (1966*b*).

20 MeV. At more moderate temperatures, neutrino production takes place primarily by the reactions

$$n + n \rightarrow n + p + e^- + \bar{\nu}_e , \quad n + p + e^- \rightarrow n + n + \nu_e .$$

The variation of the stellar temperature T_e with time, caused by the emission of neutrinos and photons, is shown in Figure 63.[13] At high temperatures of the order of $T_e > 2 \times 10^6$ ° K the star cools essentially by neutrino radiation; beginning at $T_e \approx 2 \times 10^6$ ° K, photon emission from the surface dominates. Figure 63 shows that a surface temperature of $T_e \approx 2 \times 10^6$ ° K is retained by the star for a period of almost 10^4 years. The period of existence of temperatures larger than 2×10^6 ° K is considerably shorter.

Bahcall and Wolf (1965*a*, *b*) have pointed out the possibility of cooling by neutrino production due to the interaction of pions with neutrons (π-n process). If such a process were to take place in a neutron star, it would cool

13. The diagram shows the effective cooling time (for a substantial decrease of temperature) as a function of temperature.

within 24 hours. However, it is apparently true that even in stable neutron stars the density is still not so high that the matter in equilibrium would include an appreciable number of π^- mesons, so this cooling process does not take place.

Of considerable interest is the investigation of pulsations of neutron stars about the point D_3 of Figure 57 (Hoyle, Narlikar, and Wheeler 1964; Cameron 1965*b;* Finzi 1965; Meltzer and Thorne 1966; Hansen and Tsuruta 1967; Finzi and Wolf 1968; Langer and Cameron 1969; Thorne 1969*a*). The angular frequency of oscillation of a uniform star, as one sees from dimensional considerations, is $\omega \approx (GM/R^3)^{1/2}$. For neutron stars this equation gives about 10^4 sec^{-1} for ω. Hence we find the oscillation energy to be

$$E_{\text{oscil}} = \tfrac{1}{2}(R\omega)^2 M(\delta R/R)^2 \approx 10^{53}(M/M_\odot)(\delta R/R)^2 \text{ erg} , \qquad (11.5.2)$$

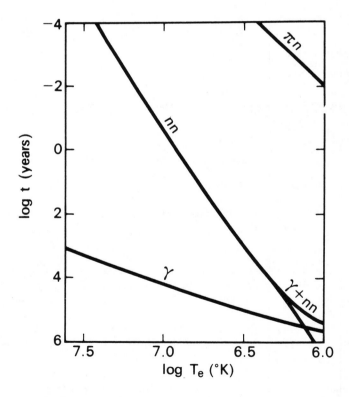

Fig. 63.—The cooling of a neutron star according to Bahcall and Wolf (1965). Plotted vertically is the time for a substantial decrease of the effective surface temperature (at a given temperature) due to various processes.

where $\delta R/R$ is the relative amplitude of oscillation. Dissipation of this energy is produced by the following processes: (1) the effective viscosity, caused by weak-interaction nuclear processes, which take place at the same rate as, or slowly compared to, the rate of the stellar pulsation; (2) the transport of energy by shock waves escaping through the surface of the star; (3) radiation reaction due to the emission of gravitational waves by the pulsating star.

Let us consider the first process: For matter in equilibrium near nuclear density, the higher the density is, the greater is the number of protons and electrons required for equilibrium with the neutrons. During contraction, the reaction

$$n + n \rightarrow n + p + e^- + \bar{\nu}_e \qquad (11.5.3a)$$

takes place, and antineutrinos escape from the star. During expansion,

$$n + p + e^- \rightarrow n + n + \nu_e \qquad (11.5.3b)$$

occurs, and neutrinos escape. The lag of equilibration of the nuclear composition, caused by the slowness of the weak interactions, leads to an irreversible loss of energy.[14]

This loss has been calculated by Finzi (1965), by Meltzer and Thorne (1966), and by Hansen and Tsuruta (1967). Finzi, with an error corrected by Meltzer and Thorne, has suggested the following estimate of the energy loss:

$$dE_{\text{oscil}} \approx 10^{55}(\rho_c/3.7 \times 10^{14})^5(M/M_\odot)(\delta R/R)^8 \text{ ergs sec}^{-1} . \quad (11.5.4)$$

Thus, for an amplitude of $\delta R/R \approx 0.01$ and a central density of 3.7×10^{14} g cm^{-3}, the oscillation energy is $E_{\text{oscil}} \approx 10^{49}$ ergs; and the loss rate is $dE_{\text{oscil}}/dt \approx 10^{39}$ ergs sec^{-1}. If no other damping forces were to exist, then the time required to attenuate the oscillations to this level would be $t_{\text{attenuation}} \approx 300$ years.

Damping by reactions (11.5.3) feeds roughly half the pulsation energy into outgoing neutrinos, and roughly half into heat. It is clear from equations (11.5.1) and (11.5.2) that a neutron star can store much greater energy in its vibrations (until they are damped) than in heat. Consequently, so long as the star is vibrating, its temperature will be kept high by the constant conversion of pulsation energy to heat. Numerical computations of this effect and of the rate of pulsation damping by reactions (11.5.3) have been performed by Hansen and Tsuruta (1967), and by Finzi and Wolf (1968). For a particularly simple neutron-star model Finzi and Wolf find that the continual conversion of pulsation energy to heat maintains a constant ratio of thermal energy to oscillation energy, $E_T/E_{\text{oscil}} = \text{const.} \approx \frac{1}{7}$.

Langer and Cameron (1969) have shown that in high-density neutron stars the process

$$n + n \rightarrow p + \Sigma^-, \quad p + \Sigma^- \rightarrow n + n \qquad (11.5.5)$$

14. The remainder of this section was written by K. S. Thorne.

is much more effective than the URCA process (11.5.3) at damping pulsations. This process can proceed only at densities high enough (ρ greater than about 1×10^{15} g cm^{-3}) for the Fermi energy of the neutrons to stabilize Σ^- particles. At lower densities no Σ^- are present, so the URCA reactions (11.5.4) dominate. The Σ^--process (11.5.5) damps pulsations in the same manner as the URCA process—by maintaining compositions which lag the equilibrium composition as the density rises and falls. When the density is rising, the Σ^- abundance is too low; and the reaction $n + n \rightarrow p + \Sigma^-$ proceeds, producing protons and Σ^- particles with energies somewhat above the Fermi sea, which then go into heat. When the density is falling, the Σ^- abundance is too high; and the reaction $p + \Sigma^- \rightarrow n + n$ produces "hot" neutrons.

Langer and Cameron point out that this process is a much more effective damping mechanism than the URCA process, because it proceeds much more rapidly. Both are weak-interaction processes; but the URCA process involves six particles, while the Σ^--process involves only four; and the URCA process involves two interactions (strong and weak) with a rate that is the product of the reaction probabilities, while the Σ^--process involves only a weak interaction. Numerical computations by Langer and Cameron for a neutron-star model of $M \approx 1.8\ M_\odot$, which has Σ^- particles in the innermost one-eighth of its volume, reveal this: A time of only 1 second is required to damp the pulsations to $\delta R/R \approx 10^{-3}$, $E_{\mathrm{oscil}} \approx 10^{48}$ ergs; and after ~ 1000 years the pulsations have damped down to $\delta R/R < 10^{-8}$, $E_{\mathrm{oscil}} < 10^{38}$ ergs. Thus, all neutron stars with $\rho_c > 10^{15}$ g cm^{-3}, except those in the process of formation, must be "very dead" pulsationally.

The second process of dissipation, the escape of shock waves through the surface, which can lead to the ejection of mass and heating of the surface layers, has not been investigated yet. It is not known how rapidly the shock waves will attenuate the vibrations by converting their energy into radiation and into kinetic and thermal energy of ejected matter. The radiation of electromagnetic waves due to the changing magnetic field of the star during the pulsation is described in § 13.6.

The third dissipation process, the emission of gravitational waves, is thought to be the dominant damping mechanism for nonradial pulsations. Thorne (1969*a*) (see § 1.14 for discussion) has calculated the radiation damping for the quadrupole modes of a variety of neutron-star models. He finds that the amplitude dies exponentially with an *e*-folding time that ranges from 10 seconds for $M \approx 0.4\ M_\odot$ to 0.2 seconds for $M \approx 2\ M_\odot$.

For a rotating neutron star, rotational flattening makes the "radial" modes of pulsation slightly nonradial and thereby enables them to emit gravitational radiation. The resultant radiation-reaction forces damp the pulsations with an *e*-folding time of (see Wheeler 1966*b*)

$$\tau_{\mathrm{radial}} \sim (\Omega^2 R^3/GM)^2 \tau_{\mathrm{quadrupole}} \sim (\Omega^2 R^3/GM)^2 \text{ sec} . \qquad (11.5.6)$$

Here Ω is the star's angular velocity, R is its radius, and M is its mass. In high-mass stars, unless $\Omega^2 R^3/GM \approx 1$, this effect will be dominated by damping due to the Σ^--process (11.5.5). But in low-mass stars ($M \lesssim 1\,M_\odot$), which contain no Σ^- particles, it may be competitive with the URCA-process damping of equation (11.5.3).

In chapter 13 we shall discuss those properties of neutron stars which led to their discovery, as well as accretion of matter and other processes in the vicinities of neutron stars.

11.6 Evolution of a Star with a Mass Greater than the Oppenheimer-Volkoff Limit

Consider now the last stages of evolution of a star with mass larger than the Oppenheimer-Volkoff limit for superdense configurations ($M \gtrsim 1.6\,M_\odot$). The qualitative difference between this case and preceding cases is that for such great masses there exists no equilibrium configuration with $S = 0$ (and $T = 0$). This means that a cooled massive star cannot reach an equilibrium state without losing a substantial portion of its mass. On the other hand, we do not know whether evolutionary processes exist which necessarily produce sufficient mass loss to bring a massive star to a final equilibrium state with $S = 0$ and $M < 1.6\,M_\odot$. The possibility of such processes was discussed in the preceding section. If a massive star does not lose a substantial portion of its mass, then the concluding stage of its evolution will necessarily be nonstationary.

Let us trace the last stages of such evolution. The star slowly approaches the limit of stability along a sequence of successive quasi-equilibrium states. At the limit, the star becomes unstable and contracts with a hydrodynamic velocity. In the future, if the star does not lose the major portion of its mass as the result of an explosion (see §§ 11.3 and 11.4), then the contraction can never end in an equilibrium state. Such states do not exist. During a time of the order of $t_H = 1/(6\pi G\rho)^{1/2}$ the star will contract to the point that the gravitational potential at its surface becomes of the order of c^2 and effects of GTR begin to become pronounced. As of this moment, the star enters the phase of relativistic contraction, i.e., collapse.

One can foresee the possibility of a two-step collapse. As shown above, the Oppenheimer-Volkoff maximum for cold neutron stars ($M \sim 2\,M_\odot$) is situated in the region of stable hot neutron stars.

Therefore, the loss of stability and resultant collapse of a hot star with $M > 2\,M_\odot$ and initial density $\sim 10^7$ g cm^{-3} (cf. the discussion of γ and γ_r in § 11.1) could terminate in a hot neutron-star state, and only later, after sufficient energy loss, might it proceed further to relativistic collapse into the "frozen-star" state.

It is important to analyze what the observational properties of such a two-step collapse could be. Of course, such an analysis would require detailed hydrodynamical calculations which take account of both general relativity and particle physics.

It must be remembered that the reasoning which led us to relativistic collapse as the fate of a star above the OV limit was based on an over-simplified picture of homologous compression, with uniform composition, and without mass loss. Perhaps this picture would be valid for a star all mixed by convection, or made of iron from the beginning. In reality, however, there is plenty of unburned nuclear fuel in any star's outer layers, even after the completion of nuclear reactions at its center.

Cameron (1969), in a review entitled "How Neutron Stars Are Born," suggests that massive stars ultimately "die" not by relativistic collapse, but because of disruption by nuclear explosions. He argues that perhaps only stars with masses slightly exceeding the Chandrasekhar white-dwarf limit ($\sim 1.2 \ M_\odot$) have time enough to get rid of their nuclear energy quietly, and to proceed onward to the neutron-star state. In this hypothesis there are no frozen stars to be observed—though the possibility of neutron-star formation is not denied; after all, pulsars exist!

This controversy is an example of how difficult life is for theorists in an era like ours when things are so unsettled!

11.7 RELATIVISTIC COLLAPSE

The collapse proceeds, for practical purposes, with free-fall velocity, since the gravitational forces exceed the pressure forces by a finite (not small) amount. Near the Schwarzschild sphere, as demonstrated in § 3.2, the gravitational force approaches infinity, whereas the pressure remains finite. Thus, in analyzing the approach of the stellar surface to the Schwarzschild sphere, in first approximation the pressure can be neglected. These considerations are supported by a specific numerical computation of the relativistic contraction of a star by Podurets (1964a).

Consider the surface of a collapsing star. During the contraction, the mass M does not change; consequently, at the surface (if $dP/dr = 0$) particles simply fall in the vacuum gravitational field of a mass M. Thus, to delineate the nature of the collapse, it is sufficient to consider the free fall of a test particle in the field of a mass M. As demonstrated in § 3.4, an external observer sees the falling particle approach the gravitational radius at a rate described by the equation

$$r = r_g + (r_1 - r_g) \exp\left[-c(t_* - t_*^1)/2r_g\right], \qquad (11.7.1)$$

and he sees the brightness of light emitted radially by the falling particle decay in accordance with the equation

$$I = \text{const.} \exp\left[-2c(t_* - t_*^1)/r_g\right]. \qquad (11.7.2)$$

It follows from equation (11.7.1) that the surface of the collapsing star approaches r_g only asymptotically during an infinitely long time, as seen by the external observer; hence our use of the term *frozen* rather than *collapsed*.

The formula (11.7.2) for the decay of brightness is directly applicable only to the central point of the visible disk of a contracting star. It is considerably more involved to draw conclusions for the entire disk, since one must then consider rays moving at a large angle to the radius, and the paths of such rays near the star are rather complex. An analysis of this problem (Podurets 1964b; Ames and Thorne 1967) shows that for the luminosity of the entire star, L, there exists a formula which is analogous to formula (11.7.2), but has a somewhat different decay time

$$L = \text{const. exp} \left[-\frac{2c(t_* - t_*^1)}{3(\sqrt{3})r_g} \right], \qquad (11.7.3)$$

where r_g is the gravitational radius of the star. A particular photon can leave the star and escape to infinity only if it crosses the surface of the star before the star reaches the Schwarzschild sphere (gravitational self-closure).

For each point with a Lagrangian coordinate r_* there exists a last moment of proper time, $\tau(r_*)$, at which the point can emit radiation that will escape from the star before self-closure. The curve $\tau(r_*)$ may be called the line of the "last ray." Obviously, it is a null radial geodesic that intersects the surface of the star at the moment when the star crosses the Schwarzschild sphere.

Consider the simple case of the contraction, with parabolic velocity, of a uniform sphere of dust with mass M. This case is unique in that uniform matter remains uniform during contraction. The density depends on proper time according to the law $\rho = 1/[6\pi G(\tau_0 - \tau)^2]$, where τ_0 is the moment at which the entire sphere contracts into a point. The expression $\rho(\tau)$ does not depend on the initial radius of the sphere. In this case, the equation for the line of the "last ray" is

$$\tau^{1/3} - \tau_1^{1/3}(r_g/r_1)^{1/2} = (r_1/3\tau_1^{2/3})(R - 1),$$

where r_1 is the radius at the moment τ_1 and R is a dimensionless Lagrangian coordinate chosen such that the circumference is proportional to R^2. For the surface of the sphere $R = 1$ and throughout this equation we have set c equal to 1.

Above, in the analysis of the brightness decay of the contracting star, we considered sources at the star's surface. It is obvious that neutrino sources will occur at the center of the contracting star. The rate at which neutrino energy is emitted can be computed easily from a knowledge of the law of contraction; thereupon one can determine the law of decay of the neutrino luminosity measured by an external observer (it turns out to be analogous to eq. [11.7.3]), and the total mass carried off during the contraction by the neutrinos (Zel'dovich 1963a; Zel'dovich and Podurets 1964; Fowler 1964a). In these computations the concept of the line of the "last ray" is important. These problems are analyzed in detail in §§ 11.8 and 11.9. Let it be just noted at this point that the mass carried off by the neutrinos is small compared with the total mass of the star.

Self-closure ensures that the mass of a collapsing star cannot decrease substantially due to radiated energy; and it also ensures that the major portion of the gravitational energy is not radiated in the form of light or neutrinos, but is transformed into kinetic energy of the contracting body. Hence, we can draw the following conclusions. The distant observer sees that a catastrophically collapsing star, when its dimensions are still much larger than r_g, contracts with hydrodynamic velocity, i.e., very rapidly. When $(R - r_g) \sim r_g$, the star continues to contract impetuously; in finite proper time it reaches r_g and continues inward. However, as seen by the external observer, the contraction becomes drastically retarded because of the effects described above, and the radius of the star approaches r_g only asymptotically (eq. [11.6.1]). The mean density of the star approaches

$$\rho_{max} = 2 \times 10^{16} (M_\odot/M)^2 \text{ g cm}^{-3} . \qquad (11.7.4)$$

The luminosity of the star abruptly falls off, despite the fact that photon creation in the star continues at almost the same rate (in fact, even at an increasing rate). Because of the gravitational redshift and other phenomena mentioned in § 3.5, the luminosity decays according to the law (11.7.1). The characteristic decay time is of the order of r_g/c, which is 2×10^{-5} sec for $M = 2 M_\odot$. For the external observer, the star becomes "frozen." As demonstrated in § 4.5, the collapse of a rotating sphere as seen by a distant observer is qualitatively similar to the collapse of a nonrotating sphere. Taking into account the pressure of matter does not change the conclusions qualitatively. Then, too, one encounters gravitational self-closure and the asymptotic picture of a "frozen" star, as described above. We stress that in the limit when $t \to \infty$, the observer sees the frozen star as nonrotating; however, in the external gravitational field, the terms which are due to the angular momentum K remain finite.

Thus, despite the fact that the gravitational field of a rotating star differs from the Schwarzschild field, its collapse takes place in a manner qualitatively similar to the collapse of a nonrotating star. The star asymptotically approaches the "frozen" state and undergoes only a finite number of revolutions before "freezing." The external observer never finds out what happens to the star when its radius becomes smaller than the gravitational radius.

Even though no radiation escapes from the star after gravitational self-closure, the star does not "disappear" from our world without leaving a trace. Its mass M and its static gravitational field are constant throughout the collapse. Such a "frozen" star interacts with surrounding bodies by its gravitational field (which is extremely strong near its gravitational radius).

We have found the final state of a star with mass greater than critical, $M > M^{ov}_{max}$. This state, while catastrophically nonstatic for the star itself, is "static" for an external observer, in the sense described above. This resolves the "paradox of large masses," which arose from the work of Oppenheimer and his collaborators (Oppenheimer and Volkoff 1938; Oppenheimer and Snyder 1939), and which has been discussed extensively in the literature (see the work of Wheeler and his colleagues, described in Wheeler 1960, 1966b; also the review by Chiu 1964). At first glance, this paradox appears rather unpleasant. A cooling star with mass $M > M^{ov}_{max}$ contracts infinitely after its loss of stability (there is no limit whatsoever to the contraction!). And then what? Wheeler views these difficulties as so substantial that he suggests (Wheeler 1960) that the "extra" nucleons (above the critical limit for a neutron star) annihilate during the contraction, and are transformed into radiation that escapes from the star, leaving its mass always less than critical. This assumption violates a fundamental law of physics—the conservation of baryon charge—and the critical density at which the violation must take place is rather moderate for large masses. For example, when $M = 10^8 \, M_\odot$, we have, according to equation (11.6.4), $\rho_{crit} = 2$ g cm^{-3}. The temperatures attained during contraction to the critical stage are not very large, either. Under such rather unremarkable conditions, certainly nothing utterly fantastic can take place—in short, nothing that cannot be observed in our terrestrial laboratories. While the gravitational potential and the gravitational field are unusually high, according to the principle of equivalence the gravitational field per se does not locally change the laws that govern physical processes.

In our opinion, there is no paradox whatsoever for the external observer. For such an observer, the collapse "stops" at $R \rightarrow r_g$, and there is no need to invent such fantastic violations of reliably established physical laws. Of course, there is the problem of what will happen to the matter that falls inside r_g—not from the viewpoint of the external observer, but from "its own" viewpoint (i.e., from the viewpoint of the comoving observer). We discussed this problem in detail in § 4.6.

11.8   NEUTRINO EMISSION IN THE COLLAPSE OF A COOL STAR

Let us return to our discussion of the catastrophic contraction which follows loss of stability. A star with a mass that only slightly exceeds the Chandrasekhar limit becomes unstable and begins to collapse when it has cooled nearly completely. During the hydrodynamic contraction of such a star, degenerate electrons with Fermi energy above a certain threshold are captured by the atomic nuclei (inverse β-decay). "Neutronization" of matter takes place.

If the matter is compressed slowly enough, then for each type of nucleus there exists a critical density at which neutronization occurs. This density corresponds to an electron Fermi energy which equals the reaction threshold of neutronization. We recall (see § 5.5) that proton neutronization,

$$p + e^- \rightarrow n + \nu \,,$$

takes place at a density of 1.6×10^7 g cm^{-3} (reaction threshold, $E - m_e c^2 = 0.78$ MeV); the neutronization of iron,

$$^{56}_{26}\text{Fe} + e^- \rightarrow {}^{56}_{25}\text{Mn} + \nu \,,$$

takes place at 6×10^8 g cm^{-3} (reaction threshold of 3.7 MeV); see Cameron (1959c) and Salpeter (1961).

For slow contraction, the closeness of the electron Fermi energy to the reaction threshold guarantees that the resulting neutrinos will carry away little energy.

The neutronization process is dominated by weak interactions, and is relatively slow. For fast contraction, neutronization lags behind its equilibrium at a given density. This means that the process will take place at a density considerably higher than the threshold density, and at a greater electron Fermi energy. The excess electron energy will be carried away by neutrinos. The neutronization is a mechanism for forming high-energy neutrinos.

Zel'dovich and Guseynov (1965a, b) have given approximate estimates of the neutrino energy. They have assumed that density varies according to the same law as for the free fall of uniform matter,

$$\rho = [6\pi G(t_0 - t)^2]^{-1} \,, \quad d\rho/dt = [3\pi G(t_0 - t)^3]^{-1}$$
$$= 2(6\pi G)^{1/2}\rho^{3/2} = \rho^{3/2}/450 \,. \tag{11.8.1}$$

The Fermi momentum and energy of the electrons are expressed in terms of the density ρ and the number of nucleons per electron, μ_e, by

$$P_\text{F} = m_e c(\rho/\mu_e 10^6)^{1/3} \,, \quad E_\text{F} = m_e c^2 [1 + (P_\text{F}/m_e c)^2]^{1/2}$$
$$\approx m_e c^2 (\rho/\mu_e \, 10^6)^{1/3} \,. \tag{11.8.2}$$

Denote by x the fraction of the original nuclei remaining unaffected at a given moment.

The probability of neutronization depends upon the properties of the initial and of the final nuclei Z_1 and Z_2 (with matrix element M_{12}). Under conventional laboratory conditions (without degenerate electrons), Z_1 is stable and Z_2 is β-radioactive. The decay rate of Z_2 permits us to find M_{21}; and, according to quantum mechanics, $|M_{12}| = |M_{21}|$. It is convenient to express the probability of neutronization of Z_1 by degenerate electrons in terms of τ, the half-life of Z_2, and in terms of the familiar function f of the decay energy Q.

When $E_F \gg Q$, one obtains

$$\frac{dx}{dt} = -\tfrac{1}{5}x \frac{(E_F/m_e c^2)^5 \ln 2}{f\tau} . \tag{11.8.3}$$

For allowed transitions, e.g., $n \to p + e^- + \bar{\nu}$, $f\tau = 800$ sec. Enough equations have been presented here for a complete solution of the problem. The simplest example of neutronization—cold hydrogen[15]—leads to the conclusion that $x = 0.5$ will be attained when $E_F \approx 7$ or 8 MeV, which considerably exceeds the threshold (1.25 MeV, including the rest energy). Hence, in this process the neutrinos carry away an energy of 5–7 MeV. The transformation of a proton into a neutron in a medium that consists of protons will start a chain of nuclear reactions ending with the formation of ^{4}He:

$$n + p \to D + \gamma , \quad D + p \to {}^3\text{He} + \gamma , \quad n + {}^3\text{He} \to T + p ,$$
$$^3\text{He} + e^- \to T + \nu , \quad n + {}^3\text{He} \to {}^4\text{He} + \gamma , \quad p + T \to {}^4\text{He} + \gamma .$$

The formation of one ^{4}He nucleus from four protons and two electrons is accompanied by the emission of 26 MeV; however, almost half of this energy is carried away by two high-energy neutrinos. Hydrogen neutronization during free fall takes place primarily at a density of 5×10^9 g cm^{-3}, even though the threshold density is only 1.6×10^7 g cm^{-3}.

Helium neutronization during catastrophic contraction (free fall) presents a more difficult problem. Helium has a rather high reaction threshold: $e^- + {}^4\text{He} \to T + n + \nu - Q$ ($Q = 21$ MeV). In addition, since a bound state of ^{4}H (i.e., $p + 3n$) does not exist (see the review by Baz', Gol'danski, and Zel'dovich 1965), there are three particles on the right-hand side of the reaction. Also, the probability of the reaction depends on the energy carried away by the neutron. Understandably, there exist no experimental data on the inverse process $n + T = {}^4\text{He} + e^- + \bar{\nu}$, since the probability of a weak interaction (with a free neutron) in flight is negligible. Faced with these difficulties, Zel'dovich and Guseynov (1965b) were forced to estimate the matrix element by using experimental results for the capture of a negative muon μ^-: $\mu^- + {}^4\text{He} \to T + n + \nu_\mu$. The rate of capture for μ^- in the $1s$ state around a ^{4}He nucleus is 370 sec^{-1}.

Assuming that the matrix element does not depend on the neutron energy, Zel'dovich and Guseynov found that

$$dx/dt = -660x \, y^2 (y - 1)^{3/2} ,$$

where

$$y = \frac{E_F}{Q} = \tfrac{1}{45}\left(\frac{\rho}{\mu_e 10^6}\right)^{1/3} . \tag{11.8.4}$$

Integration of this equation for x together with the law of free fall leads to the conclusion that the electron-capture reaction on ^{4}He takes place at $E_F \sim 45$ MeV, and at a density of $\sim 10^{12}$ g cm^{-3}.

15. Of course, the collapsing star cannot consist of hydrogen (it has all burned out). This computation is of a methodological nature; it is intended only to show the general nature of the process.

The slow neutronization of helium is followed by a much faster reaction with a smaller threshold ($\sim$10 MeV):

$$e^- + T \rightarrow 3n + \nu . \tag{11.8.5}$$

By these neutronization processes neutrinos with energies of 30–40 MeV are formed during collapse. Rough estimates show that the average cosmic flux of such neutrinos may reach 0.01 of the flux of the energetic solar neutrinos from the decay $^8Be \rightarrow {}^8B + e^- + \nu$, which have a maximum energy of 14 MeV. Since the probability of neutrino detection increases with increasing neutrino energy, we cannot rule out the possibility of an experimental discovery of energetic cosmic neutrinos originating from the collapse and neutronization of matter. In this framework, of special interest are those designs of experiments in which the energy and the direction of the neutrino will be measurable (see Reines and Wood 1965).

There are just two more comments to be made on this subject. The neutronization computations were performed for densities of freely contracting matter. A pressure gradient will slow the contraction of the central core. On the other hand, during the infall the density of matter in the outer layers will increase initially more slowly, but then faster, than described by the formula for free fall. (Here we compare the derivative $d\rho/dt$ at a given ρ. A comparison at a fixed moment of time would make no sense.) Thus, the law of density variation on which the computations are based cannot be considered an upper limit; deviations in both directions are possible.

Could the gravitational self-closure of a star affect the possibility of neutrino detection? We saw that self-closure takes place at an average density of $2 \times 10^{16}(M/M_\odot)^{-2}$. However, the density relevant to neutrino emission is that along the line of the "last ray" (see § 11.6). The maximum density at the center of the star on this line is somewhat smaller than the value given above. In the simple example of the contraction of a uniform star without pressure, $\rho_c = [2 \times 10^{16}(M/M_\odot)^{-2}]/2.55$. Helium neutronization takes place at $\rho \approx 10^{12}$ g cm^{-3}. Hence, for the overwhelming majority of stars, the neutrinos will emerge experiencing only a small redshift. Moreover, for stars with large masses we deal with hot plasmas. Neutrino emission during collapse and in supernova outbursts was discussed in § 11.4, above. Neutrino emission from hot plasmas during the collapse of supermassive stars will be discussed in § 11.11, below.

11.9 THE EVOLUTION OF A SUPERMASSIVE STAR: GENERAL REMARKS

A stable star in equilibrium, of a given chemical composition and a given entropy, is at the minimum of the energy curve $E(\rho, S)$ (see § 10.14). The equilibrium energy is negative. If the reserves of nuclear energy of the star have not been exhausted, then the nuclear energy is incomparably greater than the absolute value of the equilibrium energy E. In reality, the reserve of nuclear energy per gram of matter is of the order of $q \approx 10^{19}$ ergs g^{-1}, and for the entire star[16]

$$E_{\text{nuclear}} \approx 10^{19} M = 10^{52}(M/M_\odot) \text{ ergs },$$

which for the masses considered, 10^4–10^9 $M_\odot$, is much greater than the characteristic equilibrium energy at the critical point of stability loss (§ 10.14), $E'' = -3.56 \times 10^{54}$ ergs.

If the nuclear energy were released instantaneously, the star would be disrupted. However, the star has a self-adjustment mechanism for its energy sources. When formed from rarefied matter, it contracts until the release of nuclear energy near its center equalizes the radiation from the surface. This determines its central temperature, T_c, and consequently also its initial position on the curve E_e of Figure 51. The star slowly consumes its nuclear fuel in this state.

If the temperature is too low for nuclear reactions, then the evolution of the star (without mass loss) consists of radiation of light, which gradually decreases the entropy and energy, and gradually moves the star down the curve E_e of Figure 51. The minimum of this curve corresponds to a transition to states in which the equations of hydrostatic equilibrium have no solution. A catastrophic contraction takes place at a rate which is determined by equations of hydrodynamics. Rotation, as will be demonstrated in §§ 11.12–11.18, has a substantial influence upon the equilibrium states. The evolution with rotation taken into account is discussed in § 11.18.

Let us next calculate the rate and duration of the equilibrium evolution without rotation and without nuclear sources of energy; then we will look at the changes resulting from a consideration of the nuclear sources of energy; and at the conclusion of this section we will analyze the stage of catastrophic contraction.

11.10   RADIATIVE EQUILIBRIUM

The nature of radiative equilibrium in stars was analyzed in a classic book by Eddington (1926). In a supermassive star, the gravitational force is balanced by radiation pressure. The conditions of equilibrium for a radiation-dominated star can be considered in a straightforward manner by inserting the expression $P = P_m + P_r = nkT + \frac{1}{3}aT^4 \approx \frac{1}{3}aT^4$ for the pressure into the equation of hydrostatic equilibrium. Clearly, the gradient of the pressure will be proportional to the temperature gradient, $\partial P/\partial r = \frac{4}{3}aT^3 \partial T/\partial r$. However, the temperature gradient governs the diffusion of heat. Consequently, in a radiation-dominated star there is an intimate

---

16. To estimate E_{nuclear}, we consider the mass of the entire star and not only of the nucleus where the temperature is high. This is due to the fact that convection probably exists in massive stars and causes rather complete mixing of the matter.

relation between the state of hydrostatic equilibrium and the luminosity. This relation has a particularly simple and elegant form whenever the law of scattering and the opacity are simple. Such is the case for Compton scattering.

Consider the forces in the surface of the star. In a low density, strongly ionized plasma, Compton scattering on electrons is the fundamental source of opacity. Let us compute the force of the light pressure acting upon one free electron. This force is obviously

$$F_e = -\frac{1}{n_e}\frac{dP}{dr} = -\frac{1}{3n_e}\frac{d\epsilon}{dr}, \qquad (11.10.1)$$

where n_e is the number density of electrons and ϵ is the radiation energy density. In a medium whose optical thickness is greater than unity, the radiation flux q is given by

$$q = -D d\epsilon/dr, \qquad (11.10.2)$$

and the diffusion coefficient is

$$D = \tfrac{1}{3}c/(n_e\sigma_e), \qquad (11.10.3)$$

where σ_e is the scattering cross-section: $\sigma_e = 6.7 \times 10^{-25}$ cm^2. The cross-section does not depend on the frequency of the photon if

$$\hbar\omega \ll m_e c^2.$$

Substituting equations (11.10.3) and (11.10.2) into equation (11.10.1), we find

$$F_e = \sigma_e q/c. \qquad (11.10.4)$$

Notice that expression (11.10.4) does not depend on the assumption that the optical thickness is large. In reality, the time-averaged force of the radiation flux q on each electron equals expression (11.10.4), independent of the angular distribution of the radiation photons. The same force acts upon an isolated electron in the radiation field of a point source. Because of the charge neutrality of the plasma, the mass per electron is μ_e/a, where $a = 6 \times 10^{23}$ and μ_e is the mean molecular weight. In equilibrium, the force of the radiation pressure per electron, F_e, equals the gravitational force per electron:

$$GM\mu_e/ar^2 = \sigma_e q/c. \qquad (11.10.5)$$

Hence, expressing q in terms of the luminosity L, $q = L/2\pi r^2$, and substituting numerical constants, we obtain, for hydrogen plasma,

$$L = 1.3 \times 10^{38}(M/M_\odot) \text{ ergs sec}^{-1}, \quad L/L_\odot = 3 \times 10^4(M/M_\odot),$$
$$L/M = 3 \times 10^4(L_\odot/M_\odot). \qquad (11.10.6)$$

It should be emphasized that formula (11.10.6) for L gives an upper limit on the luminosity of any static star (not necessarily supermassive); a larger flux of radiation would expel the surface layers.

In massive stars, where the pressure is produced by radiation, expression (11.10.6) is not only an upper limit; it is also the actual luminosity. Digressing somewhat, we note that in such stars the condition of equilibrium (11.10.5) must be satisfied not only at the surface but also throughout the entire star. Therefore, the equality

$$L_r = 1.3 \times 10^{38}(M_r/M_\odot) \text{ ergs sec}^{-1} \tag{11.10.7}$$

must be valid. Here M_r is the mass inside a sphere of radius r, and L_r is the total flux of radiation through the sphere.

Obviously, nuclear sources of energy (if they exist) are located near the center, so the energy flux for the entire star cannot grow proportionally to mass as is suggested by equation (11.10.7). At first glance it may appear that a star with radiation the dominant source of pressure and with a central energy source cannot exist at all. But the point is that a central energy source will cause convection in a star, because without such a source the star is already on the edge of convective instability. Convective energy transport provides the required heat flux in a supermassive star. Whatever might cause the energy transport inside the star, at the surface itself the transport must be accomplished by radiant thermal conductivity, because the energy leaves the surface and escapes into surrounding space in the form of light rays. Therefore the relationship (11.10.6) remains valid for the total luminosity.

11.11   The Evolution of a Supermassive Star without Turbulence or Rotation

A. Cooling Time to the Critical State

Let us return to a star without nuclear sources of energy. As we have seen from formula (10.14.16), a star with $M > 10^5 \ M_\odot$ does not have a sufficiently high temperature at the center, even in the critical state, to produce neutrino radiation. Therefore, the cooling of the star is produced solely by the photon luminosity of equations (11.10.6). This luminosity determines the rate of evolution, i.e., the rate of the star's motion along the curve E_e of Figure 51.

When the stellar matter was in a dispersed state, its energy was zero. To reach the critical state, the star must emit an energy $-E'' = 3.56 \times 10^{54}$ ergs (see end of § 10.14). Thus, the time of evolution is given by the formula

$$t = -E''/L = 2.4 \times 10^8 \mu^{-2}(M/M_\odot)^{-1} \text{ years} . \tag{11.11.1}$$

As we have emphasized previously (see § 10.14), $|E''|$ is much smaller than the thermal energy of the star. Therefore, $t_{\text{evolution}}$ is much smaller than the time of thermal relaxation, $t_{\text{cooling}} = Q/L$.

Obviously, a star can be considered to be in quasistatic equilibrium only when the time evolution up to E''_{cr} substantially exceeds the characteristic time for hydrodynamic processes (see § 10.1). This time, in order of magnitude is

$$t_{\text{hydrodynamic}} = (6\pi G \langle \rho \rangle)^{-1/2} = 10^3 \langle \rho \rangle^{-1/2} \text{ sec.} \quad (11.11.2)$$

Substituting $\langle \rho \rangle = \frac{1}{54}\rho_c$ (an expression which is valid for a polytrope with $n = 3$), and replacing ρ_c by expression (10.14.15) for ρ''_c, we find

$$t_{\text{hydrodynamic}} = 5 \times 10^{-5} \mu^{3/2}(M/M_\odot)^{7/4}. \quad (11.11.3)$$

Comparing equations (11.11.3) and (11.11.1), we see that the times $t_{\text{evolution}}$ and $t_{\text{hydrodynamic}}$ are equal when $M \approx 10^8 M_\odot$. Thus, equilibrium nonrotating stars of such a large mass certainly cannot exist.

B. Nuclear Sources of Energy

We have noted previously that a star will contract in a quasi-equilibrium way until the temperature near the center becomes sufficiently high that the

TABLE 18

EFFECTIVE POWER GENERATION PER GRAM OF STELLAR MATTER,
AT THE CRITICAL STATE AT WHICH INSTABILITY SETS IN

	$M/M_\odot$			
	6×10^4	10^5	5×10^5	10^6
$\rho_c''(\text{g cm}^{-3})$	20	4	10^{-2}	10^{-3}
$T_c''/10^9 \,^\circ$ K	4×10^{-1}	2×10^{-1}	4×10^{-2}	2×10^{-2}
L/M (ergs/g sec)	6×10^4	6×10^4	6×10^4	6×10^4
$\langle A_{pp} \rangle$	4×10^3	4×10^1	8×10^{-2}	10^{-3}
$\langle A_{CN} \rangle$	5×10^{15}	3×10^{13}	3×10^3	5×10^{-3}
$\langle A_{3a} \rangle$	6×10^4	2×10^1	10^{-37}	10^{-84}

release of nuclear energy can compensate for the radiation from the surface. Table 18 shows the effective energy release per second per gram of stellar matter in the critical state

$$\langle A_{\text{nuclear}} \rangle = (\textstyle\int A_{\text{nuclear}} dM)/M \approx 0.1(A_{\text{nuclear}})_c.$$

The data for the carbon-nitrogen cycle are given under the conventional assumption that the concentration of carbon and nitrogen is about 0.5 percent; and the data for the $3a$ reaction and the p-p reaction are given for the composition H = 70 percent, He = 30 percent. We see from Table 18 that under this assumption the proton-proton reaction can be neglected entirely in the domain where the generation of nuclear energy $\langle A_{\text{nuclear}} \rangle$ is comparable to the rate, L/M, at which the star is radiating energy. (For L, formula [11.10.6] was used.) In the case of stellar matter without heavy elements (i.e., primordial matter from the "big bang" Universe, which has not undergone stellar evolution), only the $3a$ reaction is important.

Table 18 reveals that $\langle A_{\text{nuclear}} \rangle$ and L/M are comparable in the critical state when $M \approx 5 \times 10^5 \, M_\odot$ if there are heavy elements, and when $M \approx 6 \times 10^4 \, M_\odot$ if only H and He are present. Let us denote this critical mass by M_A. Then, when $M > M_A$, the temperature, even in the critical state, is not sufficiently high that the release of nuclear energy can compensate for the radiation losses; nuclear energy cannot keep the star in an equilibrium state for a long period of time. Thus, for stars with $M > M_A$, generation of nuclear energy is unimportant for the entire period of equilibrium evolution. The evolution of such stars after their formation is governed by the process of cooling described in the previous subsection.

For stars with $M < M_A$, the release of nuclear energy becomes significant before the critical state is reached. If there were no causes of instability, such a star could exist in equilibrium until its total store of nuclear energy

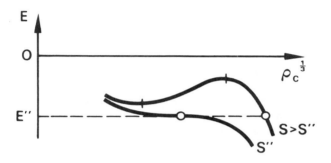

Fig. 64.—Shock waves, generated during collapse, increase the entropy of the star and shift it to the descending branch of the isentrope $S > S''$; i.e., they cannot halt the collapse.

was exhausted. This store of energy amounts to approximately $E_{\mathrm{nuclear}} \approx 10^{52} \, M/M_\odot$ ergs (see the beginning of this section). The time for this evolution would be of the order of $E_{\mathrm{nuclear}}/L \approx 10^6$ years.

C. Catastrophic Contraction

After the onset of catastrophic contraction, the star's energy may be regarded as practically constant if no nuclear reactions occur. Energy dissipation does not have time to reduce the entropy appreciably (the energy dissipation by neutrinos and the neutrino spectrum have been discussed in § 11.9).

The growth of entropy, at a given energy, due to shock waves, viscosity, etc., cannot stop the contraction after the critical state is reached. Actually, as is shown in Figure 64, a state with critical energy E'' and increased entropy S must be on that branch of the curve $E(\rho_c, S)$ on which unrestrained contraction takes place. Similarly, the ejection of part of the mass cannot stop contraction of the main part of the star when the star has passed the critical state. Because the ejected mass must have positive energy (otherwise it could not overcome the gravitational field of the star and escape to infinity), the ejection of the mass only decreases the (negative) energy of the remainder of the star; therefore, it is impossible to pass from a critical state to a state of stable equilibrium by an ejection of mass.

Contraction can be stopped only by the following factors.

First, it can be stopped as a result of fast, exothermic nuclear reactions at high temperatures. Under such conditions a release of heat can take place which will force the stellar matter to disintegrate completely. If the star has a mass greater than a critical value, M_A, then nuclear reactions cannot take place during the equilibrium stage of evolution, so the star will have sufficient nuclear fuel left for an explosion during the collapse. Calculations of the explosion during collapse for such a star consisting of 70 percent H and 30 percent ^{4}He (the abundances predicted for primordial matter in the "big bang" Universe) have been made by Bisnovatyi-Kogan (1968). He found that the nuclear explosion blows the star completely apart if its mass is only a little more than $M_A = 6 \times 10^4 \, M_\odot$—more particularly, if M is in the range $6 \times 10^4 \, M_\odot < M < 1.5 \times 10^5 \, M_\odot$. The fraction of the star's dispersing matter that is converted to heavy elements by the explosion is $Z \sim 0.04$ for $\mu \sim 1.5 \times 10^5 \, M_\odot$. Analogous calculations of heavy-element formation during explosion have been made by Wagoner, Fowler, and Hoyle (1967). A star with mass greater than $1.5 \times 10^5 \, M_\odot$ continues to collapse despite the nuclear explosion. Stars with mass less than M_A probably burn all their nuclear fuel during the stages of equilibrium evolution. However, if nuclear reactions have transformed the entire matter of the star into iron during the stage of equilibrium evolution, then nuclear outbursts during the collapse are certainly excluded because all possible nuclear energy has been released.

Second, contraction can be stopped when γ becomes larger than $\frac{4}{3}$ at high densities.

However, as we argued in § 10.15, stable states at high densities (the region *abc* of Fig. 56) are possible only for $M < 70 \, M_\odot$.

D. *Neutrino Emission by a Hot Plasma in the Collapse of a Supermassive Star*

During the contraction of a supermassive star, temperatures of the order of $T \approx 10^9 \, ^\circ$ K and higher are reached. At such temperatures, electron-positron pairs exist in equilibrium. According to present theory, the process

$$e^+ + e^- \rightarrow \nu + \bar{\nu} \tag{11.11.4}$$

must occur with a certain finite probability.

This reaction may be a powerful source of neutrino radiation during the collapse of a hot star.

How much energy in the form of neutrinos can be radiated by a collapsing star? Pertinent computations have been performed by Michel (1964),

Zel'dovich (1963d), Fowler (1964a), Zel'dovich and Podurets (1964), Chiu (1964), and Zel'dovich and Novikov (1965). Detailed computations from a work by Zel'dovich and Novikov (1965) are included in an appendix to this section. The calculations reveal that the amount of energy carried away by neutrinos certainly cannot be large. The key point is that for masses $M > 10^4 \, M_\odot$, the gravitational self-closure takes place before the temperature rises high enough to enable neutrinos to carry away a substantial portion of the stellar mass. The approximate formula for the mass carried off by neutrinos is

$$\Delta M/M \approx 5 \times 10^2 (M/M_\odot)^{-3/2} . \tag{11.11.5}$$

This formula applies to masses

$$10^4 < \langle M/M_\odot \rangle < 10^6 .$$

The temperature at the moment of self-closure is: $T_9 \approx 4 \times 10^3 (M/M_\odot)^{-1/2}$. This expression takes into account the limitation of the radiation by the

TABLE 19

APPROXIMATE ESTIMATES OF THE MASS LOST BY STARS DUE TO
NEUTRINO EMISSION DURING COLLAPSE

	$M/M_\odot$						
	10	100	10^3	10^4	10^5	10^6	10^8
$\Delta M/M$	0.05	0.1	10^{-3}	4×10^{-4}	10^{-5}	3×10^{-7}	10^{-14}

line of the "last ray." For large masses, the neutrino loss is even smaller than that given by equation (11.11.5). For $M \leq 2 \times 10^2 \, M_\odot$, before self-closure occurs the neutrino emission lowers the entropy of the star sufficiently to prevent further increase of the neutrino luminosity during the contraction. The radiated energy is not large in this case, either. The applicable approximate formula is

$$\Delta M/M \approx 8 \times 10^{-17} S_0^{6/7} ,$$

where S_0 is the specific entropy of the stellar matter at the onset of collapse, in ergs per 10^9 ° K per gram.

The last formula is particularly unreliable, since for small masses one must take into account the plasma energy and the electron degeneracy. Moreover, neutrino formation by other reactions not included in equation (11.11.4), the creation of μ-neutrinos, and the opacity of stellar matter to neutrinos also become important (see § 11.4).

Table 19 summarizes the mass loss due to neutrino emission during collapse. In reality, of course, Table 19 gives only a very rough approximation to the true picture. Let us summarize our conclusions about neutrino emission by collapsing ordinary stars and by collapsing supermassive stars:

1. The mass lost due to neutrino emission during collapse is always a small fraction of the mass of the star; hence, it cannot cause any abrupt weakening of the central gravitational field, nor can it cause the ejection of the star's external envelope by pressure that was previously balanced by gravitation (the Michel phenomenon [1964]); the impossibility of this mechanism was demonstrated by Zel'dovich (1963*d*). Even for the core of the star, the mass lost is less than 20 percent. (For a discussion of neutrino absorption by the envelope, see § 11.4.)

2. In the collapse of stars with $M \leq 3 \ M_{\odot}$, neutrinos with energies of 30–40 MeV are formed by the process of neutronization; the mass lost is up to about 1 percent.

3. In the collapse of stars with $100 \ M_{\odot} \geq M \geq 3 \ M_{\odot}$, neutrinos and antineutrinos are formed with a broad energy spectrum with a mean energy of the order of 30–50 MeV; the mass lost is up to $\sim$5 percent.

4. The general metagalactic density of neutrinos and antineutrinos released by collapse is never greater than approximately 5 percent of the mean density of stars collapsed in the Metagalaxy.

The detection of energetic neutrinos may well turn out to be a method for detecting spherically symmetric, "silent" stellar collapse (see also § 11.4).

11.12 ROTATION AND MASS SHEDDING: GENERAL RELATIONSHIPS

Theories of rotating massive stars have been given considerable attention lately: James (1964); Roxburgh (1965); Durney and Roxburgh (1965, 1967); Monaghan and Roxburgh (1965); Fowler (1966); Ozernoy (1966*a*); Bisnovatyi-Kogan, Zel'dovich, and Novikov (1967).[17]

The obvious reasons for this interest are as follows. In massive ($M/M_{\odot} > 10^4$) stars, the pressure of light exceeds the plasma pressure, the adiabatic index is close to the critical value of $\gamma = \frac{4}{3}$, and small corrections for GTR are sufficient to disrupt stability.

The rotation of a star as a whole is a nonrelativistic phenomenon and in this sense has an effect similar to motion of electrons and nuclei of the plasma; in fact, the contribution of rotation to the stability of a star is similar to that of a gas with an adiabatic index $\gamma = \frac{5}{3}$. However, in addition to the thermal motion of the plasma particles, there is also radiation pressure which grows faster with increasing temperature than does the pressure of the plasma. Most macroscopic motions of the stellar fluid (i.e., turbulence), which might work together with the plasma to produce stability at high temperatures, attenuate rapidly under stellar conditions. However, rigid rotation does not; it is characterized by a conserved quantity, the angular momentum, which prevents it from being transformed into other forms of energy.

17. The pioneering work is due to Chandrasekhar (1933).

Until recently, all computations of nonrotating stars were performed in approximations of hydrostatic equilibrium. Processes of rapid contraction and expansion, where particle accelerations are of the order of the acceleration of the force of gravity, have been considered quantitatively only in recent years, with the aid of high-speed computers.

In the case of rotating stars, it was natural to start with a consideration of equilibrium configurations. The additional assumption was made that the angular velocities become uniform rapidly, i.e., that the viscosity due to the turbulent and magnetic mechanisms of momentum transport is great.

It is known that equilibrium configurations rotating rigidly exist only for a narrow range of parameters, and that the average influence of the rotation upon the overall characteristics of a star (mean density, pressure, energy, etc.) in this range of parameters is not great.

To convince ourselves of this proposition, consider the configuration of a nonrotating star and find the angular velocity at which the centrifugal force at the equator equalizes the gravitational force. For this critical angular velocity, the kinetic energy of rotation of a particle at the equator equals half of its gravitational energy:

$$\omega^2 R = GM/R^2, \quad \tfrac{1}{2}v^2 = \tfrac{1}{2}\omega^2 R^2 = \tfrac{1}{2}GM/R. \tag{11.12.1}$$

But, when we consider quantities that are averaged over the whole star, we must also take into account the distribution of the matter within the star. As is well known, in the absence of rotation and with an adiabatic index $\gamma = \tfrac{4}{3}$, this distribution is rather nonuniform. The density at the center is larger than the average density by a factor of 54, $\rho_c/\langle\rho\rangle = 54$.

Another measure of the nonuniformity of the matter distribution is the magnitude of the gravitational energy U. Obviously,

$$U = -\tfrac{1}{2}G \int_M \int_M \frac{dm_1 dm_2}{r_{12}} = -\tfrac{1}{2}GM^2 \langle 1/r_{12} \rangle.$$

However, for an equilibrium polytropic configuration,

$$U = -\frac{3}{5-n}\frac{GM^2}{R}; \quad U_{n=3} = -\tfrac{3}{2}\frac{GM^2}{R}.$$

Consequently, when $n = 3$, $\langle 1/r_{12} \rangle = 3/R$; i.e., the mean distance between all pairs of particles is smaller than the radius by a factor of 3. For a uniform distribution of density, i.e., for an incompressible fluid, $\langle 1/r_{12} \rangle = 1.2/R$.

Yet another characteristic of the matter distribution is the mean square of the distance from the center, which enters into the expression for the moment of inertia:

$$I = \tfrac{2}{3}M \langle r^2 \rangle,$$

$$\langle r^2 \rangle = (1/M) \int_M r^2 dm = (1/M) \int_V \rho r^2 dV = (4\pi/M) \int_0^R \rho r^4 dr.$$

For the configurations of interest to us $\langle r^2 \rangle = 0.11\ R^2$. The smallness of the coefficient 0.11 indicates precisely that the major portion of the mass is concentrated close to the center; for a uniform distribution, the ratio is $\langle r^2 \rangle / R^2 = \frac{3}{5} = 0.6$.

Using these quantities, we find the ratio of the rotational kinetic energy to the gravitational energy of the entire star (here v is the velocity at the equator):

$$E_{\text{kin}} = \tfrac{1}{2} I \omega^2 = \tfrac{2}{3}\tfrac{1}{2} M v^2 (\langle r^2 \rangle / R^2) = \tfrac{1}{3} 0.11 G M^2 / R\ ,$$

$$|U| = \tfrac{3}{2} \frac{GM^2}{R}\ , \qquad \frac{E_{\text{kin}}}{|U|} = \tfrac{1}{3} \times \frac{0.11}{3/2} = 0.025 = \tfrac{1}{40}\ .$$

This small ratio is attained when, for particles at the equator, $E_{\text{kin}}/|U| = 0.5$ and the condition for mass shedding (equality of centrifugal force and gravitational force) is satisfied. In reality, a precise solution of the problem indicates that a static solution becomes impossible even earlier, i.e., for an even smaller ratio of the mean kinetic energy to the mean gravitational energy (when $E_{\text{kin}}/|U| = 0.007$). This is easily understood in qualitative terms by the fact that, in addition to the centrifugal force and the gravitational force, the particles are also acted upon by the pressure gradient.

Obviously, even before condition (11.12.1) is reached, the external surface of the star becomes deformed; it flattens at the equator. As a result, condition (11.12.1) must be written for a new radius R_1, where $R_1 > R$.

The shape of the star and the limit on the parameters of static solutions can be calculated easily in the Roche model. This model assumes that the rotation does not change the distribution of the main part of the mass. Consider the small part of the mass which is far from the center, i.e., the atmosphere of the star. The atmosphere is in the gravitational field of the central mass, $\phi_1 = -GM/r$. For fixed angular velocity ω (rotation about the z-axis), we can introduce the centrifugal potential

$$\phi_2 = -\tfrac{1}{2}\omega^2(x^2 + y^2)\ .$$

The equation of hydrostatic equilibrium has the integral form (the usual differential form is obtained by taking the gradient of this)

$$H + \phi_1 + \phi_2 = k\ . \tag{11.12.2}$$

Here H is the specific enthalpy of the matter, $dH = \rho^{-1}dP$; for the adiabatic index $\gamma = \tfrac{4}{3}$, this enthalpy is $H = 4P/\rho$. We assume that the constant k in equation (11.12.2) does not differ from its value in the absence of rotation.

Figure 65 shows the shape of the potential $\phi_1 + \phi_2$ along the z-axis (*dashed line*), and along an equatorial, radial direction—for example, along the x-axis (*solid line*).

In the equatorial plane, the sum $\phi_1 + \phi_2$ has a potential barrier and a maximum at $r_c = (GM/\omega^2)^{1/3}$; at this point $\phi(r_c) = \phi_c = \phi_1 + \phi_2 = -\frac{3}{2} GM/r_c$. Obviously, equation (11.12.2) has a physically meaningful solution only when $k < \phi_c$. However, in a nonrotating star, on the boundary $H(R) = 0$, $\phi(R) = -GM/R$; hence, $k = -GM/R$. From this we find $r_c = \frac{3}{2} R$. Thus, the maximum expansion of the star along the equator is only a factor of $\frac{3}{2}$. The corresponding angular velocity is smaller than the angular velocity found previously (eq. [11.12.1]) by a factor of 1.82. From the condition $r_c = \frac{3}{2} R = (GM/\omega^2)^{1/3}$, we find $\omega = (2/3)^{3/2} \sqrt{(GM/R^3)}$.

The ratio of the kinetic energy of rotation to the gravitational energy, derived previously, decreases by a factor of $1.5^3 = 3.4$, to 0.007.[18]

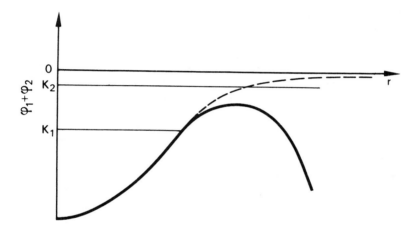

Fig. 65.—Shape of the potential $\phi_1 + \phi_2$ along a radial direction located in the equatorial plane (*solid line*), and in the polar (*z*) axis (*dashed line*). Horizontal lines K_1 and K_2 correspond to different values of the constant k in equation (11.12.2).

Because the kinetic energy of rotation for the entire star is small, the above assumptions that the rotation and resulting deformation of the star do not appreciably change either the gravitational field or the constant k, which characterizes the entire star, are justified.

18. The problem of the structure of a rotating body made from incompressible fluid has been investigated in considerable detail during the last century. There are a number of exact solutions in which the fluid takes the form of an ellipsoid of rotation. In such solutions, as the angular momentum K increases, the shape of the body changes from a sphere ($K = 0$) to a disk ($K \to \infty$); and the ratio of rotational kinetic energy to gravitational energy grows monotonically from 0 (for a sphere) to 0.5 (for a disk). The angular velocity and kinetic energy pass through maximum values along the sequence. It turns out, however, that the ellipsoid of rotation is unstable if the ratio of its half-axes (polar to equatorial) is less than $c/a = 0.58$.

For a uniformly rotating star, the condition of equatorial outflow puts an upper limit on the angular velocity and prevents entrance into the unstable regime.

It is useful to approach the problem of rotation from another, more or less global viewpoint. The equilibrium state of a spinning mass corresponds to a minimum of the total energy of the matter for a given equation of state $[P(\rho), H(\rho)]$, and for a given, fixed total angular momentum K. The condition of constancy of the angular momentum during the variation of the total energy obviously follows from the fact that the angular momentum is a conserved quantity. In the following section, proceeding from this principle, we will derive the density and shape which characterize a rotating star.

We must, however, stipulate immediately that the minimum of the energy which we are seeking is not an absolute minimum for a fixed mass and fixed angular momentum. Compare a star rotating as a solid body with a configuration that has the same mass M and angular momentum K but that consists of a nonrotating star of mass $M - m_1$ and a small mass m_1 which is in the equatorial plane at a large distance r_1 from the center and is revolving about the star. Assume that the small mass, with a small velocity of revolution, has the same angular momentum K as the rotating star, $K = m_1 r_1 v_1$. The kinetic energy of this mass is $K^2/2m_1 r_1^2$. Taking a sufficiently large distance r_1, one can make the kinetic energy as small as desired for any fixed small m_1 and constant K. For a small m_1, the potential energy of the mass m_1 is also small. Thus, for the given angular momentum we can always find a configuration whose energy is arbitrarily close to the energy of a nonrotating star with the same mass. For this, it is necessary only to remove a small part of the mass to a large distance and to put all of the angular momentum into it. If an equilibrium configuration A exists for a rotating star, then obviously its energy is greater than the energy of a nonrotating star B which has the same mass. On the other hand, an equilibrium configuration A which is stable with respect to any small perturbation is obviously a configuration of minimum energy; but the minimum is only a local one, since the energy of the two-piece $(M - m; m)$ state with the same angular momentum, K, can be made arbitrarily close to the energy of the nonrotating state B; and the energy of B is obviously lower than A. In this situation there exists an energy barrier[19] between states A and B. Its existence is guaranteed by the condition that A is a local minimum, not an absolute minimum, since B is below A.

It becomes obvious now that for certain values of the parameters (specifically, for large angular momentum and small entropy) the energy barrier may vanish. But in such cases there is no A-type minimum either. This

19. If we consider only one transition parameter between A and B, then the energy barrier is a maximum. However, since the state of a star is defined by many variables (an infinite number!) and since A is a minimum with respect to any perturbations, then between A and B there exists a path in configuration space which passes through a saddlepoint of minimum height.

means that the problem of the equilibrium configuration for a rotating star with a given mass, entropy, and angular momentum does not have a solution for all values of these parameters. The physical limit on the existence of a solution is related specifically to the onset (e.g., as the entropy gradually decreases below a critical value) of mass shedding.

These general considerations show that a star's rotational energy can be transformed into other forms of energy (relativistic particles, electromagnetic radiation, etc.) if a mechanism can be found to produce the right type of mass outflow with angular momentum.

As we have mentioned previously, equilibrium configurations are considered in the next section. The process of mass shedding, under conditions where no equilibrium configuration exists, will be considered in § 11.18.

11.13   EQUILIBRIUM AND THE SHAPE OF A ROTATING STAR: THE NEWTONIAN THEORY

Following Gurevich and Lebedinsky, we assume that the density of matter is constant on similar ellipsoids of revolution. In a nonrotating star the surfaces of constant density are spheres. We assume now that in a nonrotating star these spheres become transformed into ellipsoids which all have the same shape, and which retain the volume of each corresponding sphere. Consequently, the total volume of the star, like the volume between adjacent spheres, is constant. (This assumption is valid only for the case of an incompressible fluid. Although in realistic cases breakdowns in the assumption are large for the outer layers of a star, the assumption yields a correct general picture of the deformations of the central region, which contains most of the mass; it also yields adequate numerical estimates.) For a more precise analysis of the structure of rotating polytropes, see Hurley and Roberts (1965). Under our assumption, if a spherical star is compared with a deformed star that has the same central density, it will turn out that the density distribution as a function of mass contained inside a given layer is the same. Consequently, for this type of deformation, the total internal energy of the gas is precisely the same in the two configurations. Therefore, for $\gamma = \frac{4}{3}$ in a rotating star, as in a spherical star, $E_{\text{internal}} = 1.75\, Mb\rho_c{}^{1/3}$, where b is the coefficient appearing in the expressions $P = b\rho^{4/3}$, $E = 3b\rho^{1/3}$ (cf. § 10.14).

During deformation, the gravitational energy obviously decreases in absolute magnitude; contraction along the z-axis and flattening in the (x, y)-plane with preservation of volume brings about an increased mean distance between the particles.

A remarkable property of the gravitational potential of an ellipsoid is that inside a shell, the boundaries of which are two identical ellipsoids filled with matter of constant density (the mass per unit surface area is not constant!), the gravitational potential has a constant value.[20]

It follows in turn from this property that the law of variation of the gravitational energy in the deformation of a star is the same as in an ellipsoidal deformation of a sphere of incompressible fluid. To prove this assertion, we construct a star from identical layers of constant density, beginning with the external layer. In this process, each new layer is placed into a cavity where the potential is constant. The gravitational energy is smaller than the energy of an undeformed star by as many times as the potential inside the ellipsoidal layer is smaller than the potential inside a spherical layer with the same mass and the same cavity volume. Exactly the same considerations can be applied to a liquid with constant density (above, for a star, the density was assumed to be constant in each layer, but varying from one layer to another). The gravitational energy will vary by a similar ratio, which will depend on the degree of oblateness of the ellipsoid. We will define the oblateness by a parameter λ which is the ratio of the diameter along the axis of rotation to the diameter of a sphere of the same volume, $\lambda = c^*/(a^*b^*c^*)^{1/3}$, where $a^* = b^* > c^*$, so that $\lambda < 1$ and $\lambda = (c^*/a^*)^{2/3}$.

Thus, the gravitational energy is

$$U = -0.64GM^{5/3}\rho_c^{1/3}g(\lambda) .$$

The function $g(\lambda)$ appears on page 323 of Landau and Lifschitz (1962). Transforming that function into our notation, we put it in the form

$$g(\lambda) = \lambda^{1/2}(1 - \lambda^3)^{-1/2} \cos^{-1}(\lambda^{3/2}) .$$

Obviously $g(1) = 1$; for a value of λ which differs little from unity, the approximation

$$g(\lambda) = 1 - \tfrac{1}{5}(1 - \lambda)^2$$

is applicable. We shall use this below.

Finally, we find the moment of inertia and the kinetic energy of rotation. The moment of inertia is proportional to the square of the distance from the axis of rotation, measured perpendicular to that axis:

$$I = \text{const.}\ Ma^{*2} .$$

Comparing this with the moment of inertia I_m of a sphere of the same volume, we find

$$I/I_m = a^{*2}/(a^*b^*c^*)^{2/3} = (a^*/c^*)^{2/3} = \lambda^{-1} .$$

20. This is easy to demonstrate by noticing that the potential inside a complete ellipsoid of constant density has the form

$$\phi = \phi(0) + \alpha x^2 + \beta y^2 + \gamma z^2 , \quad \alpha + \beta + \gamma = \tfrac{1}{2}\Delta\phi = 2\pi G\rho .$$

The constants α, β, γ depend only on the ratio of the ellipsoid axes, but not on their absolute magnitude; only $\phi(0)$ depends on the absolute magnitude. Considering the potential of the layer as a difference between the potential of the external ϕ_1 and internal ϕ_2 of the ellipsoid, we find that $\phi = \phi_1(0) - \phi_2(0)$; the terms that are dependent on the coordinates cancel.

11.13 The Newtonian Theory of a Rotating Star

Accordingly, for a given angular momentum K, the kinetic energy of rotation of an oblate star, E_{kin}, is less than the kinetic energy of the sphere $(E_{kin})_{sphere}$:

$$E_{kin} = (E_{kin})_{sphere}\, \lambda < (E_{kin})_{sphere}\,,$$

since $\lambda < 1$. The flattening decreases the kinetic energy for a given angular momentum. This also suggests that the kinetic energy of rotation plays the role of a potential for the centrifugal force which acts to flatten the star. The moment of inertia and, accordingly, the kinetic energy of a spherical star are obtained by numerical integration of the Emden density distribution for the polytrope $\gamma = \frac{4}{3}(n = 3)$. The relevant dimensionless factors were given in the preceding section. Using them and the above effects of deformation, we find

$$E_{kin} = 1.25\,\lambda K^2 \rho_c{}^{2/3} M^{-5/3}\,.$$

Collecting all terms together, we obtain an expression for the energy of the star:

$$E = -k_2 GM^{5/3}\rho_c{}^{1/3} g(\lambda) + k_1 Mb\rho_c{}^{1/3} + k_3\lambda K^2 M^{-5/3}\rho_c{}^{2/3}\,,$$

$$k_1 = 1.75\,, \quad k_2 = 0.64\,, \quad k_3 = 1.25\,.$$

Let us denote by b_0 the value of b for which, without rotation, there is neutral equilibrium

$$k_2 GM^{5/3}\rho_c{}^{1/3} = k_1 b_0 M\rho_c{}^{1/3}$$

(cf. § 11.2), so that $b_0 = (k_2/k_1)GM^{2/3} = 0.364\,GM^{2/3}$; and let us introduce the dimensionless quantities $b/b_0 = h$ and $r^* = (\rho_c/\rho_0)^{1/3}$, where ρ_0 is the characteristic density constructed from GM and K:

$$\rho_0 = G^3 M^{10}/K^6\,.$$

With these notational changes, the expression for the energy can be rewritten in the form

$$E = (G^2 M^5/K^2)\,\{k_2 r^*[h - g(\lambda)] + k_3\lambda r^{*2}\}\,.$$

Denote the quantity in the braces by A. The factor in front of the braces is made up entirely of conserved quantities. To find the extremum of E, we vary λ and r^*. The quantity h is determined by the entropy and is thus held fixed in the variation; it changes slowly in the course of the evolution. The conditions of equilibrium (of extremal E) have the form $\partial A/\partial\lambda = 0$, $\partial A/\partial r^* = 0$. Using these two conditions, we can express λ and r^* in terms of h. A solution exists only when $h < 1$, which is quite natural. If $h > 1$, then the entropy is so large that the mass expands without limit, even without rotation; obviously, rotation cannot change this result. When $h < 1$, without rotation the gas contracts without limit. In this case, rotation stops the contraction. A decrease of h is accompanied by increasing density and increasing oblateness, described by the quantity $1 - \lambda$. For any $h < 1$, there exists a formal solution of the problem in which the surfaces of constant density are defined to be identical ellipsoids; when $h \to 0$, we have $\lambda \to 0$, $\rho \to \infty$, i.e., a solution corresponding to a flat disk. This solution is known to be unstable: the disk breaks up into clumps with dimensions of the order of the thickness of the disk.

As h decreases from unity it should eventually reach a critical value at which the star becomes unstable against a transformation of its ellipsoid of rotation into a triaxial ellipsoid, as is the case for an incompressible fluid. In our expressions, the dependence upon shape of the gravitational energy and rotational energy is characterized by a single parameter λ; therefore, for any arbitrary adiabatic index, instability will occur at the same λ as it does for an incompressible fluid. However, as we will see shortly, the solution loses its significance long before the triaxial instability sets in; it becomes unphysical when the oblateness is still small, when h is still close to unity, and accordingly when λ is also close to unity.

It is not difficult to find a complete set of solutions for the extremization of E by using λ as a parameter: from the condition $\partial A/\partial \lambda = 0$, we obtain $r^* = (k_2/k_3) [-g'(\lambda)]$; substituting into the condition $\partial A/\partial r^* = 0$, we find

$$h = g(\lambda) - (2k_3/k_2) \lambda r^* = g(\lambda) + 2\lambda g'(\lambda) .$$

By plotting appropriate graphs, it is easy to find λ and r^* for any given h. However, because of mass shedding, a solution in which matter is "forcibly" distributed with constant density on the surfaces of identical ellipsoids makes physical sense only for small oblateness, i.e., when λ differs little from unity and the corresponding h differs little from unity. Substituting the approximation for $g(\lambda)$ when $1 - \lambda \ll 1$, we obtain

$$h = 1 - \frac{(1 - \lambda)^2}{5} - \tfrac{4}{5}\lambda(1 - \lambda) = 1 - \tfrac{4}{5}(1 - \lambda) ,$$

$$(1 - \lambda) = \tfrac{5}{4}(1 - h) ; \quad r^* \approx \tfrac{2}{5}\frac{k_2}{k_3} (1 - \lambda) = \tfrac{1}{2}\frac{k_2}{k_3} (1 - h) = 0.25(1 - h) .$$

$$\rho_c = \frac{G^3 M^{10}}{(K^2)^3} 0.016(1 - h)^3 , \quad \langle \rho \rangle = \tfrac{1}{54}\rho_c .$$

Now let us determine the condition imposed upon the admissible values of λ and h by the shedding of gas at the equator. In the preceding section, it was established that, for an Emden sphere with corrections for the deformation of the external surface of the star, mass shedding corresponds to $z^* = E_{kin}/|U| = 0.007$, by contrast with $z^* = (E_{kin})_{sphere}/|U| = 0.025$ for a spherical star. However, when λ is close to unity, $z^* = (k_3 r^{*2})/(k_2 r^*) = (k_3/k_2)r^*$ for equilibrium. From the preceding formulae, we see that

$$\tfrac{1}{2}\frac{k_2}{k_3} (1 - h) = r^* = \frac{k_2}{k_3} z^* , \quad 1 - h = 2z^* , \quad 1 - \lambda = 2.5z^* .$$

Thus, the range of entropy h, in which uniform rotation can support stable equilibrium, is rather narrow: from $h = 1$ to $h = 0.95$ in the spherical approximation; or from 1 to 0.986 in the Roche model, which is more realistic. Mass shedding begins in the course of evolution when the entropy drops below a limit, which is of the order of 0.95 or 0.98 of the value for neutral equilibrium of the nonrotating star. The theory of mass loss will be considered in § 11.18.

Remaining within the framework of static solutions and Newtonian gravitation theory, we can state that the absolute value of the equilibrium density is not limited by anything. In the expression for the equilibrium density

$$\rho_c = (G^3 M^{10}/K^6) \, 0.016(1 - h)^3$$

the factor $(1 - h)^3$ is limited and small, but when K is small, ρ_c can be arbitrarily large—in principle! The same applies to the energy. It is easy to demonstrate that, in accordance with the virial theorem, $E = -E_{\text{kin}}$ for an equilibrium state. The condition of no mass shedding restricts $E_{\text{kin}}/|U|$ to a small value. But if ρ_c is not restricted when K is small, then the value of $|U|$ is not restricted either.

Here we are again confronted with a situation in which even small GTR corrections to the gravitational laws are essential for a qualitatively correct answer. As was demonstrated above, rotation which stabilizes the stellar equilibrium, being restricted by mass shedding, yields only small corrections to all expressions. Therefore, even small corrections for GTR may change the situation qualitatively. We will discuss this problem in the following section.

11.14   CORRECTIONS FOR GTR IN THE THEORY OF A ROTATING STAR

The effects of GTR will be analyzed by the energy method. However, we cannot limit ourselves to the first correction, which has the form $\Delta_{\text{GTR,1}} = -0.93 \, (G^2 M^{7/3}/c^2) \, \rho_c^{2/3}$ because of the more or less fortuitous circumstance that its dependence on density is the same as that of the rotational energy. (This did not occur in the comparison between GTR corrections and plasma corrections.) Therefore, a degeneracy appears, which does not exist in reality. One must take into account higher-order GTR terms in the expression for the energy. Of these higher-order terms, we retain only the largest:

$$\Delta E_{\text{GTR,2}} = -k_5(G^3/c^4)M^3\rho_c \, .$$

The computation of k_5 is rather difficult; this is already evident from the number of terms which were included in the calculation of the preceding correction, $\Delta E_{\text{GTR,1}}$, which was considerably simpler. From general considerations (in particular, from the fact that static solutions are impossible near the gravitational radius and for the characteristic density that corresponds to that size), it follows that $\Delta E_{\text{GTR,2}}$ is negative; i.e., k_5 is positive and of order unity. It was established in the preceding section that the change in the gravitational energy due to deformation is rather small:

$$1 > g(\lambda) > 1 - \tfrac{1}{5}(0.032)^2 = 0.9998 .$$

Moreover, we can neglect the influence of rotation upon the relativistic corrections. Thus, setting $\lambda = 1$, we investigate the expression

$$E = -0.64GM^{5/3}\rho_c^{1/3}(1 - h) + 1.25K^2M^{5/3}\rho_c^{2/3} - 0.93\,(G^2M^{7/3}/c^2)\rho_c^{2/3}$$
$$- k_5(G^3/c^4)M^3\rho_c ,$$

where h has the same significance as in § 11.13 For each fixed choice of the parameters M, h, K, the minima of the curve $E(\rho_c)$ are states of stable equilibrium. The evolution at the stage that interests us (prior to the beginning of mass shedding) consists of a decrease of h with constant M and K, and it is terminated by the vanishing of the minimum of $E(\rho_c)$ (the coalescence of the minimum with the maximum). At this stage, a point of horizontal inflection appears on the curve $E(\rho_c)$, and triggers a transition from slow evolution to collapse. The entire qualitative aspect of this matter has been discussed previously; see § 11.11.

Using the energy approach (based on the expression for the stellar energy obtained above), we will consider in detail below the problem of the equilibrium, stability and evolution of a rotating star.

In considering these problems, we will encounter two types of phenomena: the shedding of gas from the equator, which depends on the parameter z^* (ratio of rotational energy to gravitational energy),[21] and the hydrodynamic instability which leads to collapse. In principle, the loss of stability and the beginning of collapse for a rotating star may be analyzed by the same method as for a nonrotating star. However, the presence of two conserved quantities —i.e., two parameters (entropy and angular momentum) which characterize a star of given mass—has a significant influence on the onset of collapse.

For the case of a nonrotating star of a given mass the energy is $E = E(\rho_c, S)$. If all the states, including the nonequilibrium states, are considered, the curves $E(\rho_c, S)$ fill the (ρ_c, E)-plane. Through each point in this plane passes one curve $E(\rho_c, S)$; to each point there corresponds a certain entropy S. The conditions of equilibrium select a line in the plane. The evolution goes along that line and ends at the point where two conditions are satisfied: $\partial E/\partial\rho_c = 0$, $\partial^2 E/\partial\rho_c^2 = 0$. At this point lies the last possible equilibrium state; it is the point where stability is lost (cf. Fig. 66).

Let us compare this situation with the case of a rotating star, where

$$E = E(\rho_c, S, K) .$$

21. In this ratio we agreed to use the absolute value of the (negative) gravitational energy; and for simplicity we use the ratio to the Newtonian gravitational energy without including the GTR corrections.

If we do not restrict ourselves to equilibrium states, then each point in the (ρ_c, E)-plane can be attained in innumerable ways: The conditions $E = E_0$, $\rho_c = \rho_{c,0}$, together with $E = E(\rho_c, S, K)$, give only one equation linking the two parameters S and K. Adding the equilibrium condition $\partial E/\partial \rho_c = 0$, we obtain a system of equations which, in principle, has a solution for *any* pair E, ρ_c. Thus the totality of equilibrium solutions is the entire (ρ_c, E)-plane (or part of that plane), and not a line as it was in the case of a nonrotating star.

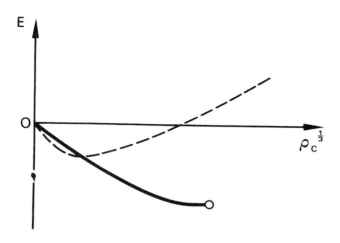

FIG. 66.—Energy E of a nonrotating star as a function of density ρ_c. Minimum of the dashed curve corresponds to the equilibrium state for a star of given entropy. Solid line traces out the equilibrium states for varying entropy. Small circle at end of curve is the termination of stable-equilibrium states.

The states which correspond to the onset of mass shedding, or the onset of collapse, form lines in the (ρ_c, E)-plane.

Which of these critical lines the star will approach in the course of its evolution is determined by the specific nature of its structure.

The general situation is shown in Figure 67, where the energy E is plotted on the ordinate. The entire region of interest to us is below the abscissa, where $E < 0$. On the abscissa we have plotted $x^* = \rho_c^{1/3}$. The solid line $OACB$ corresponds to the onset of mass shedding from the star's equator for the case of uniform rotation. As has been emphasized, only those states are considered which correspond to complete hydrostatic equilibrium.[22]

The condition of mass shedding corresponds to a particular constant ratio z^*_1 of the kinetic energy of rotation to the gravitational energy. If we set $z^* = z^*_1{}' > z^*_1$, then in the (ρ_c, E)-plane we obtain an analogous curve

22. An analytical computation of the curves shown in Figure 67 is included in the next section.

$OA'C'B'$, such that the region $OACB$ is completely enclosed in $OA'C'B'$. Beyond the region $OACB$ there is mass shedding, and exact hydrostatic solutions do not exist. Each point beyond [below] $OACB$ corresponds to an approximate solution in which a static core of the star feeds a stationary flow of gas, which is being shed from the equatorial zone. A constant value of $z^* = z^*_1{}'$ (line $OA'C'B'$) corresponds to a constant ratio of the rate of mass loss to the rate for establishing hydrostatic equilibrium. (This ratio of rates is small if $z^*_1{}' - z^*_1 \ll 1$; see below.)

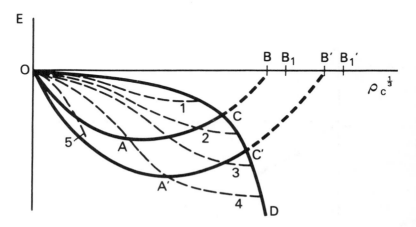

Fig. 67.—Diagram of E versus $x^* = \rho_c^{1/3}$ for the equilibrium states of a star of given mass and angular momentum. (For discussion of the significance of the various curves, see text.)

The line $OCC'D$ corresponds to the onset of collapse. Stable equilibrium states exist only to the left of and below $OCC'D$. There are no stable states between the abscissa and the region $OCC'D$; therefore, the curves CB and $C'B'$ are given as dotted lines. The dashed lines correspond to equilibrium states with a given constant angular momentum. These lines end at points which are on the boundary of collapse $OCC'D$. Each such line with a particular value of K is analogous to a line of equilibrium states in the problem of a nonrotating star; it intersects $OCC'D$ at a point where the line with constant K becomes horizontal.

As has been shown, the condition $\partial E_{eq}/\partial \rho_c = 0$ corresponds to $\partial E/\partial \rho_c = 0$ and $\partial^2 E/\partial \rho_c^2 = 0$ for the nonequilibrium curves $E(\rho_c, S, K)$ which are not shown in the diagram.

The greater the value of K, the lower the corresponding dashed line of Figure 67; these lines are denoted by numerals, and $K_1 < K_2 < K_3 < K_4$. During the evolution of a star its angular momentum is conserved until the onset of mass shedding. During this early period the evolution is caused by the loss of entropy as the star radiates; and the point depicting the star in Figure 67 moves to the right and downward along the dashed line (E decreases: ρ_c grows).

If the angular momentum is small (K_1), then the onset of collapse is reached before mass loss begins. The collapse takes place rapidly, with a velocity of the order of the velocity of free fall. During the collapse, exchanges of angular momentum between adjacent layers of matter cannot take place; the collapse proceeds too rapidly. Each particle retains the angular momentum which it had at the beginning of the collapse. Therefore, while the main mass of gas (the core) continues to collapse, the matter in the outer part of the star, near its equatorial plane, stops near an equilibrium orbit corresponding to its particular angular momentum. A detailed analysis of this process has not yet been given.

What happens when the star has a greater angular momentum, e.g., K_2, K_3, K_4, or K_5? It appears from Figure 67 that the early evolution, which conserves the angular momentum, brings the system to a state in which mass loss begins (to a point of intersection with line $OACB$). During mass loss, the angular momentum, the energy, and the mass decrease; they are removed by the escaping gas. The decrease of energy due to radiation emission also continues. The relationship between the changes in the various parameters depends on the specific assumptions. Strictly speaking, since the mass also changes, the evolution cannot be described by the motion shown in Figure 67, which is based on constant mass.

However, the variation of the mass is not large, as a rule, and therefore we can consider the trajectory in the (ρ_c, E)-plane in an approximate sense. For large angular momentum (K_5), upon the system's reaching $OACB$, the motion can take place approximately along the mass-shedding boundary $OACB$ with a transition from one dashed line to the next until the minimum point $OACB$ is reached.

However, if the angular momentum is equal to or smaller than K_4 (corresponding to the dashed line "4" which passes through the minimum A of curve $OACB$), then the evolution line must move away from $OACB$, since the evolution cannot take place with an increasing energy. In the course of evolution, the rate of mass loss increases; contraction takes place; and subsequently a state is reached in which stability is lost and collapse begins.

The precise situation at the onset of collapse must be determined by numerical integration of differential equations (see § 11.18); however, there are two conditions which narrow the domain of possible states prior to collapse. These conditions are: (1) the decrease of energy, and (2) the decrease of angular momentum in the course of the evolution.

We now proceed to compute the curves shown in Figure 67.

11.15 Approximate Theory of Equilibrium

We turn now to a semiquantitative analysis of the equilibrium of a rotating star, i.e., to calculating the curves shown in Figure 67. Specifically, we shall calculate the values of the energy, density, and temperature at several characteristic points, i.e., at the minimum of the mass loss curve and at the intersection of this curve with the collapse curve. For this we shall use the energy formula expressed in the form

$$E = -a^*qx^* + b^*x^{*2} - cx^{*2} - f^*x^{*3}, \qquad (11.15.1)$$

where the following notation is introduced: $x^* = \rho_c^{1/3}$; $-a^*x^* = -k_2 GM^{5/3}\rho_c^{1/3}$ is the gravitational energy; and $a^*(1-q)x^* = k_1 Mb\rho_c^{1/3}$ is the internal energy of the stellar matter. Consequently, $q = (1 - b/b_{\text{cr}}) = 1 - h$ is a dimensionless quantity which depends on the entropy S, and which equals zero at an entropy S_0 corresponding to neutral equilibrium for a nonrotating star. The quantity q increases with decreasing S. The contribution of the plasma to the energy is not included because we are interested in states whose stability is ensured mainly by rotation; the plasma alone cannot make the star stable. Taking the plasma into account in this situation does not change the estimates considerably. The expression $b^*x^{*2} = k_3 K^2 M^{-5/3}\rho_c^{2/3}$ is the rotational energy. In accordance with the results of § 11.13 the effect of the stellar deformation on the moment of inertia is also not taken into account. The expression $-c^*x^{*2} = -k_4 G^2 M^{7/3}c^{-2}\rho_c^{2/3}$ is the first correction for GTR (see § 10.8); $-f^*x^{*3} = -k_5 G^3 M^3 c^{-4}\rho_c$ is the second correction for GTR, i.e., the following term of the expansion in powers of $r_g/R \sim \rho_c^{1/3} \sim x^*$.

In investigations by Fowler (1966) and Roxburgh (1965), this term was not taken into account However, in the case of a star which is stabilized by rotation, the first correction for GTR, cx^{*2}, and the energy of rotation, bx^{*2}, depend on x^* in identical manners, so that inclusion of f^*x^{*3} is necessary in order to obtain correctly the curve at which stability is lost and collapse begins.

Let us find this curve. Its equation is obtained from two conditions:

$$\partial E/\partial x^* = -a^*q + 2(b^* - c^*)x^* - 3f^*x^{*2} = 0 ;$$
$$\partial^2 E/\partial x^{*2} = 2(b^* - c^*) - 6f^*x^* = 0 .$$

We use these conditions to express b^* and q in terms of x^*. The quantities a^*, c^*, f^* are determined by the mass of the star and are assumed to be known. Knowing b^* and q, we substitute in equation (11.15.1) to find the relation $E(x^*)$, which describes the onset of collapse. The result is

$$b^* = c^* + 3f^*x^* , \quad q = 3f^*x^{*2}/a^* ; \quad E = -f^*x^{*3} .$$

Only because of the second correction did we obtain $E < 0$ on the line of collapse. Let us turn to the line of the onset of mass shedding. We obtain its equation from the conditions of mass loss and equilibrium

$$(b^*x^{*2})/(a^*x^*) = z^* , \quad \partial E/\partial x^* = 0 .$$

11.15 Approximate Theory of Equilibrium

Here z^* is a known parameter which, as we have remarked, is small compared to unity (see § 11.12: the index 1 on z^*_1 is omitted here). We obtain

$$b^* = a^*z^*/x^*, \quad q = 2z^* - (2c^*x^*/a^*) - (3f^*x^{*2}/a^*),$$
$$E = -a^*z^*x^* + c^*x^{*2} + 2f^*x^{*3}.$$

This equation of line $OACB$ in the (E, x^*)-plane was obtained without any sophisticated methods; it is simply a cubic parabola. It is not difficult to determine the characteristic points of the curve, i.e., the point of minimum E (point A of Fig. 67), the point of intersection with the collapse curve (point C), and the values of x when $E = 0$ (point B).

We will not write out the tedious precise expressions with square roots; rather, we shall use the condition $z^* \ll 1$ to put them into simpler forms. For point A we obtain

$$x^*_A = (a^*/2c^*)z^*, \quad E_A = -(a^{*2}/4c^*)z^{*2}.$$

(Since $x^* \sim z^*$, the term $f^*x^{*3} \sim z^{*3}$ has been ignored here.) For points B and C we obtain

$$x^*_c \cong x_B \cong x^*_B{}' = a^*z^*/c^*,$$

where $x^*_B{}'$ was computed without taking into account the second correction for GTR. Since $x^* \sim z^*$, the term of highest order in x^* is simultaneously the term of highest order in z^*. However, the highest-order term must be taken into account wherever it is the only nonvanishing one; e.g., at point C, where

$$E_c = -fx^{*3} = -f^*x^*_B{}'^3 = -\frac{f^*a^{*3}}{c^{*3}}z^{*3}.$$

Replacing a^*, c^*, and f^* by their expressions in terms of the stellar mass M, we find

$$E_A = -\frac{(0.64GM^{5/3})^2z^{*2}}{4[0.93(G/c^2)M^{7/3}]} = -0.11Mc^2z^{*2},$$
$$E_c \approx 0.15Mc^2z^{*3}. \tag{11.15.2}$$

In computing E_c we have set to unity the unknown numerical coefficient f in the expression for the second correction for GTR. Finally, let us find the expressions for the evolution lines with constant angular momentum (the dashed lines in Fig. 67), i.e., the lines of constant b^*, which will be used as a label for them. The evolution along these lines is caused by the decrease of the entropy, i.e., by the growth of q. From the equilibrium condition $\partial E/\partial x = 0$, we obtain $a^*q = 2(b^* - c^*)x^* - 3f^*x^{*2}$; and substituting this result into the expressions for energy, we find

$$E = -(b^* - c^*)x^{*2} + 2f^*x^{*3}.$$

It is easy to demonstrate that x^* increases with decreasing entropy (i.e., as q increases) up to the point where the evolution line intersects the collapse curve. Thereafter x^* increases with increasing entropy. Such a coincidence is not accidental. Upon reaching the collapse curve, nonstationary contraction is necessary, for there exist no equilibrium solutions with an angular momentum and entropy smaller than those on the collapse curve.

11.16 ROTATING MASSIVE STARS AND QUASISTELLAR OBJECTS

The theory of rotating stars with very large masses is of interest in itself, since it is important to investigate theoretically the entire range of possible values for stellar parameters. However, supermassive stars are also interesting as a possible explanation for quasars (see chapter 14). In this context, the most important quantity is the energy that can be released by a rotating massive star.

The problem of the magnitude of the energy released is in turn subdivided into two separate problems: (1) the release of gravitational energy before the collapse, without taking account of nuclear reactions; and (2) the release of nuclear energy in stars with structure constrained by the conditions of mass loss and stability.

It is easy to solve these problems on the basis of the computations of the preceding section, if one limits oneself to configurations which precede the onset of mass shedding, i.e., for which $z^* \leq 0.007$. In that case, the minimum of the energy corresponds to

$$E_A = -\frac{a^{*2}}{4c^*}z^{*2} = -\frac{0.64^2}{4 \times 0.93}(0.007)^2 Mc^2$$

$$= -5.5 \times 10^{-6}Mc^2 = -5 \times 10^{15}M .$$

This energy is rather small compared with the nuclear energy of hydrogen burning, $\sim 6 \times 10^{-3}Mc^2$. Recalling the expression for the stellar luminosity when radiation pressure prevails,

$$L(\text{ergs sec}^{-1}) = 6 \times 10^4 M ,$$

we find that the time during which a star could emit light due to its stored energy E_A is

$$\tau = |E_A|/L = 10^{11} \text{ sec} = 3 \times 10^3 \text{ years} .$$

Consider now the conditions for the release of nuclear energy. The maximum density which corresponds to the given $z^* = 0.007$ is reached at point C, which is the intersection with the collapse curve (see Fig. 67). The abscissa of this point differs little from B'; i.e , it is approximately

$$x^* = \rho_c^{1/3} = \frac{a^*}{2c^*}z^* = \frac{0.64GM^{5/3}c^2}{2 \times 0.93 \times G^2 M^{7/3}}z^* = 0.34\frac{c^2}{GM^{2/3}}z^* ,$$

$$\rho_c = \frac{0.04z^{*3}c^6}{G^3 M^2} .$$

It is convenient to compare this expression with the density ρ_g which characterizes gravitational self-closure for the mass M:

$$\rho_g = \frac{3M}{4\pi R_g{}^3} = \frac{3}{32\pi}\frac{c^6}{G^3 M^2} = 0.03\frac{c^6}{G^3 M^2} = 1.8 \times 10^{16}(M/M_\odot)^{-2} .$$

We find

$$\rho_c = 1.3 z^{*3}\rho_g .$$

When $z^* = 0.007$ and $(M/M_\odot) = 10^8$,

$$\rho_c = 10^{10}(M/M_\odot)^{-2} = 10^{-6}\ \mathrm{g\ cm^{-3}} .$$

The corresponding temperature is obtained from the same formulae as one uses for nonrotating massive stars:

$$T_c = 2 \times 10^7 \rho_c{}^{1/3}(M/M_\odot)^{1/6} = 5.5 \times 10^{12}z^*(M/M_\odot)^{-1/2} .$$

Again substituting $z^* = 0.007$ and $M/M_\odot = 10^8$, we obtain $T_c = 4 \times 10^6$.

Obviously, such a temperature is entirely inadequate for the occurrence of nuclear reactions. According to Fowler's calculations, to provide the required luminosity by means of the CN cycle, a temperature of no less than $6 \times 10^{7\,\circ}$ K is required (Fowler 1966).

Thus, the rotating mass of gas may be like a quasar only when z^* exceeds by many times the threshold for mass shedding in solid-body rotation.

In this situation, one should expect that mass loss will take place. Suggestions have been made that z^* may be larger than 0.007 as a result of differential rotation, if the angular velocity of the internal layers is greater than that of the external layers (i.e., of the atmosphere of the star). In such cases, however, the viscosity, the turbulent friction, and possibly also the magnetic forces will equalize the angular velocities; and mass shedding will again take place.

But, one must not panic at the thought of mass loss! Most likely, quasars are really in the process of gradual mass loss. In the following section we will sketch a calculation of mass shedding from a rotating star on the assumption that solid-body rotation is maintained during the mass loss. Unfortunately, the problem of transfer of angular momentum in this process is not sufficiently well understood.

Before passing to more specific computations, we will give a semiquantitative discussion of the effects that can be expected.

Let us return to Figure 67. Before the star reaches line $OACB$, it evolves along the dashed lines without any loss of mass or angular momentum, but with gradually decreasing entropy. What happens when it reaches $OACB$? Obviously, mass shedding causes its mass, energy, and angular momentum to decrease. At the same time emission of radiation causes its entropy to decrease, unless its luminosity is compensated by nuclear-energy generation.

Suppose that energy losses dominate over angular-momentum losses as far as evolutionary effects are concerned. In this case the effects of mass shedding can be neglected, and any value of z^* can be produced by the evolution. Therefore, the star can achieve any energy emission

$$E = -0.15Mc^2z^{*3} ,$$

and any density and temperature

$$\rho_c = 2.4 \times 10^{16}z^{*3}(M/M_\odot)^2 , \quad T_c = 5.5 \times 10^{12}z^*(M/M_\odot)^{-1/2} ,$$

subject only to limits resulting from the onset of collapse. For example, when $z^* = 0.1$ and when $(M/M_\odot) = 10^8$, we find $T_c = 6 \times 10^7$, which is sufficient for the release of nuclear energy. To release as much energy from gravitation [23] as from the nuclear reactions would require $z^* = 0.34$, independent of the mass of the star.

In this case of $M/M_\odot = 10^8$ there is enough nuclear energy to provide the power requirements of a typical quasar for 3×10^6 years (with complete burn-out) or for 3×10^5 years (with 10 percent burn-out), which is quite a sufficient time to explain the existence of a quasar.

However, the absolute luminosity of the quasar 3C 273 is 10^{47} ergs sec^{-1}, which gives a mass closer to $10^9 M_\odot$. For such a mass the z^* required for nuclear reactions (~ 0.3) approaches the value $z^* = 0.34$, for which the release of gravitational energy is of the order of the release of nuclear energy. However, it is clear that a consideration of such large z^* (50 times larger than critical) is inconsistent with our neglect of the loss of angular momentum due to mass shedding. For such strong deviations from the conditions at the onset of mass shedding, the effects of angular-momentum loss will be significant compared with the effects of energy loss.

Consider now the opposite limiting case: Let the dominant process be the loss of angular momentum, which, however, can be accompanied by a loss of energy carried away in the escaping matter. If the angular momentum is large and the intersection with line $OACB$ takes place to the left of point A, then in the course of the loss of angular momentum, the star slides down to point A. At A it enters the domain of mass shedding (below line $OACB$) and moves somewhere between the line $K = $ const. (line $A - z^*_2$) and $E = $ const. (line $A - z^*_1$); see Figure 68. When it reaches the line of collapse, $OCz^*_1z^*_2$, the star has a vlaue of z^* in the range $z^*_1 < z^* < z^*_2$.[24]

Setting $z^*_H \equiv 0.007$, we find z^*_1 from the condition

$$0.15Mc^2z^{*3}_1 = 0.11Mc^2z^{*2}_A ; \quad z^*_1 = 0.032 .$$

To find z^*_2, we must first find the angular momentum, and the coefficient b^* which depends on it, at the minimum point A.

23. When we discuss the release of gravitational energy, we have in mind, of course, the difference between the gravitational energy of the star and the internal energy of its gas in equilibrium, i.e., $|E|$, rather than $|U|$.

24. The fact that the same notation is used for the dimensionless parameter z^* and for the points on Figure 68 should not lead to misunderstanding.

Using formulae of the preceding section, we find $b^* = 2c^* = 1.86(G^2/c^2)M^{7/3}$, and on the line of collapse

$$x^*_2 = (b^* - c^*)/3f^* = (c^*/3f^*), \quad E_2 = -c^{*3}/27f^{*2} = -0.03Mc^2.$$

Thus, in the framework of such a theory, using very rough estimates, we find that the amount of energy which can be released during the period of mass loss is sufficient for lengthy support of a quasar's luminosity (for more than 10^6 years!). However, rough estimates are not sufficient; it is necessary to calculate the evolution more exactly. Results of such calculations are included in § 18 of this chapter.

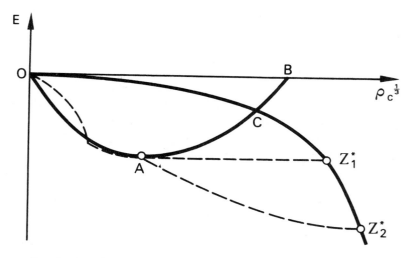

FIG. 68.—Probable courses of evolution of a star with angular-momentum loss

A key characteristic of our analysis is the shedding of mass at the star's equator. As is known from quasar observations (see chapter 14), especially of 3C 273, there is often a jet of relativistic, magnetized plasma escaping from the compact source of radiation. Apparently, the complete theory of quasars is more complex than that sketched above. This sketch is of interest primarily in connection with the energy balance of a quasar.

Let us turn our attention now to another model for quasars—one in which, instead of rotation, a different form of macroscopic motion is considered, namely, turbulence.

11.17   TURBULENCE

The possible role of turbulence in the theory of quasars (see chapter 14), has been analyzed by Layzer (1965); Ozernoy (1966a); Anand (1965); and

Pacholczyck (1965). Detailed calculations have been performed by Ozernoy (1966a).

Assume that the mean velocity of turbulent motion amounts to a certain fraction n^* of the sound velocity, $u = n_* a_{sound}$. Then, $\frac{1}{2}u^2 = (\frac{1}{2}n_*^2)(a_{sound})^2$. But with the adiabatic index of $\frac{4}{3}$ of a supermassive star, we have $(a_{sound})^2 = \frac{4}{3}(P/\rho) = \frac{4}{9}E_1$, where E_1 is the specific energy. Consequently, the kinetic energy of turbulent motion of the entire star is

$$\frac{1}{2}\int u^2 \rho dV = \frac{2}{9}n_*^2 \int E_1 dM = \frac{2}{9}n_*^2 h|U_g| \ ,$$

so that (since $h \cong 1$)

$$z^* = \frac{2}{9}n_*^2 \ ,$$

and finally (see formula [11.15.2]):

$$E_A \cong -5 \times 10^{-3}n_*^4 Mc^2 \ , \quad E_c = -3 \times 10^{-3}n_*^6 Mc^2 \ .$$

The result, as we see, is extremely sensitive to the value of n_*.

The central element in this model is the problem of the required time for the decay of turbulence ("turbulent decay time"). This time depends upon the value of n_*. The greater the turbulent decay time, all other conditions being equal, the greater the scale of the turbulence. In order of magnitude, the turbulent viscosity is characterized by a transport coefficient (analogue of the molecular coefficient of diffusion and kinematic viscosity), which is $\nu_t \cong u'l$, where u' is the turbulent velocity and l is the turbulent length scale. This formula is analogous to $\nu = D = \frac{1}{3}u\lambda$, where u is the velocity of the molecules and λ is their mean free path.

If we use the expression $\nu_t \approx u'l$ for ν_t, it is easy to show that the rate of dissipation of energy per unit volume is $\omega = \rho u'^3 l$, and the turbulent dissipation time is $t \sim l^2/\nu_t \sim l/u'$. It is essential that in contrast to problems of molecular (laminar) viscosity or thermal conductivity, we cannot take $t = R^2/\nu$, where R is the size of the system. In the case of turbulent motion, the scale of the nonuniformity L, defined by $\partial u/\partial x \sim u/L$, coincides in order of magnitude with the effective "mean free path" (turbulent scale length), $l \sim L$.

The decay of turbulence occurs most slowly when the largest-scale motions are involved. The energy of large-scale motions is not converted directly into heat, but first passes into motions of smaller scale, etc. As is well known, this produces a turbulence spectrum, a phenomenon first discovered by A. N. Kolmogorov.

This gradual process of energy degradation by going to smaller scales, however, does not change the general decay law $t \sim l^2/\nu_t \sim l/u'$. By setting $n_* = 1$, so that $u' = a_{sound}$, and by selecting a maximum scale which equals the radius of the star R, we obtain an order-of-magnitude estimate of the turbulent decay time $t \sim \tau \equiv R/a_{sound}$ Thus, t is of the order of the characteristic free-fall time for the star. When $R = 10^{16}$ cm and $M = 10^8 M_\odot$, we obtain $\tau = \sqrt{(R^3/GM)} = 10^7$ sec $\cong 0.3$ year.

The expression for the dissipation rate may contain a multiplicative factor of the order of 0.1 for normal turbulence, and of the order of 0.01 for magnetoturbulent motions. However, even $100\tau = 30$ years is an inadmissibly short time span to explain the existence of a quasar. The decay time increases as n_*^{-1}, so if we take $u \ll a_{\text{sound}}$, $n_* \ll 1$, we increase t somewhat. However, in this case the effectiveness of the turbulence as a stabilizing influence will significantly decrease.

On the other hand, if $n_* \gtrsim 1$, shock waves will be generated by supersonic gas motion. Gas masses moving in such a way with respect to each other do not flow around each other, but collide; and the energy dissipation strongly increases. It is certain that $n_* \geq 1$ cannot be attained.

The general rotation of a star can be interpreted as a turbulent motion on the maximum possible scale; in this sense, the theory of a rotating star connects directly to the theory of a turbulent star, though of course rotation is not turbulence.

On the other hand, the shedding of gas during the rotation of a star is accompanied by a redistribution of the angular momentum between the stellar particles, which is connected with the excitation of turbulence. Apparently, the future theory of quasars (if they really are supermassive stars with large-scale motions) will take into account both of these phenomena as well as the substantial role of magnetic fields in the transfer of angular momentum and in the stabilization of the turbulence.

A common property of any mechanism of stellar stabilization is macroscopic motion of matter at large velocities; according to the virial theorem $E = -T^*$, where T^* is the kinetic energy; the corrections for GTR violate the virial theorem and bring us to the conclusion that $T^* > |E|$. Hence, when $E = -0.006\ Mc^2$ (equal to the nuclear energy store),

$$T^* = \tfrac{1}{2}Mu^2 > 0.006Mc^2 ; \qquad \langle u^2 \rangle^{1/2} > 0.1\ c .$$

Observations of spectral lines formed at the surfaces of quasars indicate considerably smaller velocities, $u \leq 0.01\ c$. It would be very useful to devise methods for determining the velocities of the motion of the main mass of the quasar, which is hidden from us by its surface layers (atmosphere) .

11.18   The Evolution of a Rotating Star: Velocity of Mass Ejection

It was demonstrated in the preceding chapter that the shedding of matter due to centrifugal force occurs at a very small rotational kinetic energy. This is a consequence of the presence of an extended atmosphere, whose density is considerably lower than the density near the center of the star.

It is precisely in this situation that one can construct an effective approximate method for calculating the shedding of matter (Bisnovatyi-Kogan, Zel'dovich, and Novikov 1967). The essence of the method is that at each moment only a small part of the matter, that located near the equator and near the surface, can escape. For the escaping matter itself the acceleration is not small, since the matter is not in hydrostatic equilibrium: we are considering a situation for which no precise static solution exists.

However, the escaping matter has a small density; therefore, the rate of mass loss is small, and consequently, for the main mass of matter, conditions change rather slowly. As a consequence, the mass shedding produces a slow evolution of the star. The main mass at each moment is in a state that does not differ much from equilibrium.

Near the equator, a quasistationary flux develops. In this flux, each particle is exposed to a large acceleration, comparable to the centrifugal acceleration, to the acceleration of gravity, and to the acceleration produced by the pressure gradient. The state of the particle changes rather rapidly. However, the flux changes only insofar as the boundary conditions change in the region which feeds the flux with matter; in the main mass, conditions change only slowly. It is this slow change of the entire flux (in Eulerian coordinates) that permits us to interpret it as a stationary process. Such an analysis is accurate if the particle transit time through the critical region is small compared with the time required for a substantial change of the flux as a whole.

The analysis consists of joining two solutions: the equilibrium solution in the inner region, where the acceleration in the flow can be neglected by comparison with the acceleration of gravity; and the stationary solution in the outer region, where the flow dominates. It can be demonstrated that the choice of the join point has no effect whatsoever upon the final results of the analysis.

Before writing any equations, we state the assumptions which make up the framework of our computations.

1. Assume that up to a certain distance from the center of the star the particles retain a constant angular velocity. Obviously, a mechanism of angular-momentum transfer is required for this, since each particle's angular momentum must increase as it moves away from the center.

In principle, turbulent friction can be such a mechanism. However, turbulence cannot be sufficiently effective when the velocity of outflow approaches the velocity of sound or exceeds it. On the other hand, magnetic fields frozen into the escaping plasma (as suggested by Schatzman 1959, 1962; and by Hoyle and Burbidge may provide the transfer of angular momentum over large distances. We will not consider other manifestations of the magnetic field in the mass flux. The validity of this assumption requires further investigation.

2. The computation gives not just the rate of decrease of the mass of the star as a result of mass shedding but also the rates of decrease of its angular

momentum and energy. The angular momentum and the energy carried off by the escaping matter (in contrast to the quantity of matter per se) depend strongly on the point at which the exchange of angular momentum between the escaping matter and the star stops. In the absence of a specific theory of this phenomenon, we performed computations for several different decoupling points.

3. The third assumption may be considered harmless: In a system of coordinates rotating with the star, it is assumed that the motion is directed strictly radially. The absence of a v_ϕ component is equivalent to the above assumption of constant angular velocity. In addition, we assume that $v_\theta = 0$; this is likely if for no other reason than that the escape takes place in a narrow band of θ near the equator, $\mu = \cos\theta \ll 1$.

4. If the energy loss by mass ejection is small compared with the energy loss by radiation, heat flow in the escaping matter may be neglected.

In the framework of these assumptions, the equations for the stationary flow have a simple form in a rotating system of coordinates with the velocity u in the radial direction. They are:

1. *The equation of continuity*

$$\rho u r^2 = A ,$$

where A is a constant that is not specified beforehand.

2. *The energy equation* (Bernoulli equation)

$$\tfrac{1}{2}u^2 + H + \phi_1 + \phi_2 = k ,$$

where

$$\phi_1 = -GM/r , \quad \phi_2 = -\tfrac{1}{2}\omega^2 r^2 \mu^2 , \quad \mu = \sin\theta .$$

3. *The equation of entropy conservation* (in accordance with the fourth assumption), $S = $ const. When $\gamma = \tfrac{4}{3}$, it follows that $H = 4b\rho^{1/3}$.

According to the stipulations of § 12 of this chapter, the constant k in the energy equation equals $-GM/R$, where R is the radius of the Emden polytropic solution with the central density obtained in the preceding section. This central density, as noted previously, depends on the mass of the star, on the entropy of its matter, and on the angular momentum. The general nature of the solution is determined by the relationship between k and the maximum value of $\phi_1 + \phi_2$.

If $k < (\phi_1 + \phi_2)_{max}$, so that there exists an r such that $k = \phi_1 + \phi_2$ and consequently $u = 0$, $H = 0$, $\rho = 0$, then there is no mass shedding; and the relations $u = H = \rho = 0$ are conditions for the external surface of the star.

However, if $k > (\phi_1 + \phi_2)_{max}$, then there is mass shedding. Its theory can be developed easily along the same lines as the theory of accretion. Let us outline the basic features of the solution without getting involved in a formal proof.

A critical radius for the behavior of the flux is that point, $r = r_m$, at which the sum $\phi_1 + \phi_2$ reaches a maximum: at this radius a transition from subsonic flow (when $r < r_m$) to supersonic flow (when $r > r_m$) takes place. At this radius we find, noting that $H = 4P/\rho = 3(a_{\text{sound}})^2 = 3u^2$, all the flow characteristics:

$$\tfrac{1}{2}u^2 + H = 3.5u^2 = k - (\phi_1 + \phi_2)_{\text{max}} .$$

The density may be expressed in the simple form $H = \text{const.} \ \rho^{1/3}$ for a given entropy. From this and the above relation, we can find the flux per unit solid angle, $\rho u r^2$, which is a radially conserved quantity.

Since the centrifugal potential depends on angle, the condition $k > (\phi_1 + \phi_2)_{\text{max}}$ is achieved only within a certain range of angles near the equator.

Denoting by an index e the quantities that relate to the equator, we finally obtain (taking into account the range of angles in which shedding occurs)

$$dM/dt = \text{const.} \ [k - (\phi_1 + \phi_2)_{\text{max},e}]^4 .$$

The quasistationary theory, as well as the theory of corrections for GTR developed for the static (equilibrium) case, remains entirely applicable. The high power of the exponent on the bracket insures that even when z^* considerably exceeds the critical value at which shedding begins, the rate of mass loss remains small. Thus, when $z^* = 0.025$ or 0.05 (we recall that $z^*_{\text{crit}} = 0.007$), the time required to shed a significant fraction of the star's mass is 2×10^3 or 3×10^2 times larger than the characteristic hydrodynamic time,

$$\tau = (6\pi G\rho)^{-1/2} .$$

The transition from subsonic to supersonic flow has a great effect on the properties of the solution. Those phenomena taking place when $r > r_m$ cannot influence the flux; in particular, to determine $\partial M/\partial t$ one need not know how far outside r_m the constancy of angular velocity is retained. However, the angular momentum and the energy removed by the escaping gas depend very sensitively on this. When $r < r_m$ (subsonic flow), it follows from the condition of flux conservation, $\rho u r^2 = \text{const.}$, that u rapidly decreases with decreasing r. This means that (except for a small area adjacent to the critical surface) one can neglect $\tfrac{1}{2}u^2$ as compared to H. However, the condition $H + \phi_1 + \phi_2 = k$ obtained when $\tfrac{1}{2}u^2$ is neglected is nothing else but the condition of hydrostatic equilibrium. From thermodynamics, we know that when

$$S = \text{const.} ,$$

we obtain

$$dH = vdP , \quad \text{grad } P = -(1/v) \text{ grad } (\phi_1 = \phi_2) - \rho \text{ grad } (\phi_1 + \phi_2) .$$

Therefore, the stationary flux merges nicely with the static solution in the subsonic region.

The model outlined above permits us to trace the variations in mass, angular momentum, and energy of a star in the course of evolution during the stage when matter, as well as radiation, is being emitted. The effects of GTR considered above permit us to determine the moment of transition to relativistic collapse.

A necessary condition for performing this program and for a subsequent consideration of the collapse per se is an analysis of the tangential forces which accomplish the transfer of angular momentum from the central part of the star to the escaping matter It should be added that the entire interpretation suggested here has been advanced only recently, and at the present time only preliminary results are available.

We will cite here the results of numerical computations. For a star with mass $M = 10^9 M_\odot$ and angular momentum $K = 10^{68}$ g cm^2 sec^{-1}, the evolution time prior to the beginning of mass shedding amounts to only 50 years. Thereafter, the evolution is determined by the joint action of the emission of radiation and the shedding of mass. In the calculations it was assumed that the magnetic coupling between the escaping matter and the star (the magnetic bond produces the constancy of angular velocity) is broken at a distance[25] of $r_0 = (GM\omega^{-2} \sin \theta)^{1/3}$. The quasistatic evolution ends, and relativistic collapse begins when $T_{center} = 3 \times 10^{6\circ}$ K, a temperature which is still too low for nuclear reactions. At this instant, the ratios of the mass to the initial mass and of the angular momentum to the initial angular momentum are, respectively, $M/M_0 = 0.85$ and $K/K_0 = 0.05$ The period of slow evolution lasts for a time $\tau = 3 \times 10^4$ years. During this period, 10^{59} ergs of energy are released.

25. Note that, if r_0 is small, then the escaping gas does not escape to infinity, but settles into distant, circular orbits around the star.

Appendix to § 11.11D

The process of neutronization described in § 11.8 is a direct consequence of the experimentally investigated interactions which transform protons into neutrons, and conversely. Modern theory has predicted, with considerable confidence, that neutrino-antineutrino pairs can be emitted during any change in an electron's momentum, during a jump of an electron from one orbit to another, and finally, during the annihilation of an electron and a positron (even though none of these processes have been observed in the laboratory):

$$e^- \rightarrow e^{-*} + \nu + \bar{\nu}, \quad e^- + e^+ \rightarrow \nu + \bar{\nu}.$$

The astrophysical consequences of these processes were first pointed out by Pontecorvo (1959). The first computations were performed by Gandel'man and Pinaev (1959) for radiation by an electron which passes near a nucleus, $e^- + Z = e^{-*} + Z + \nu + \bar{\nu}$. However, the neutrino process becomes essential for astrophysics only at temperatures of the order of 5×10^8 ° K and higher, where the annihilation of e^+e^- pairs is important (Chiu 1961*b*, 1964; Masevich, Kotok, Dluzhnevskaya, and Mazani 1965), as are URCA processes with electrons and positrons, such as $e^+ + n \rightarrow p + \nu$, $e^- + p \rightarrow n + \nu$ (Pinaev 1963*a*, *b*). The radiation of ν_μ is negligibly small compared with ν_e when $T < 2 \times 10^{11}$ ° K.

In the theory of the evolution and explosions of stars and quasars, it is important to examine the rate of neutrino losses. For nonrelativistic electrons and positrons, the rate of energy loss is

$$du/dt = 2.8 \times 10^{-40} n_+ n_- \text{ ergs sec}^{-1} \text{ cm}^{-3},$$

where n_- and n_+ are, respectively, the number densities of electrons and positrons per cubic centimeter.

At a nonrelativistic temperature (i.e., when $kT < m_e c^2$), and at a density such that the gas is nondegenerate, the equation of thermodynamic equilibrium yields (see § 14 of chapter 10)

$$n_+ n_- = \frac{4(2\pi m_e kT)^3}{(2\pi\hbar)^6} \exp\left(-\frac{2m_e c^2}{kT}\right), \quad \frac{p_F{}^2}{2m_e} < kT < m_e c^2.$$

The inequality on the right side gives the domain of applicability of the formula for $n_+ n_-$; m_e is the electron mass. Substituting numerical values and continuing to express the temperature in units of 10^9 ° K (T_9), we obtain

$$n_+ n_- = 10.7 \times 10^{58} T \exp(-11.9/T),$$

$$du/dt = 4.8 \times 10^{18} T^3 \exp(-11.9/T) \text{ ergs sec}^{-1} \text{ cm}^{-3},$$

$$3(\rho/\mu_e 10^6)^{2/3} < T < 6 \quad (\text{everywhere } T \equiv T_9).$$

In the high-temperature domain, $kT > m_e c^2$, the number of $e^+ e^-$ pairs grows proportionally to T^3, just as does the number of light quanta; so the number of pairs becomes considerably greater than the number of electrons which neutralize the nuclei. The product $n_+ n_-$ grows as T^6, while the annihilation cross-section grows as T^2. The spectrum of the newly formed neutrinos and antineutrinos has approximately the form

$$E^6 \exp{(-12E/T)}dE , \quad (E \text{ is in MeV}) ,$$

so that the mean energy is of the order of 0.5 T MeV (larger than the energy kT by a factor of 6).[26]

In this situation

$$
\begin{aligned}
n_+ &= n_- = 1.6 \times 10^{28} T^3 \text{ cm}^{-3} , \\
u_{\text{pairs}} &= 1.75 a T^4 = 1.3 \times 10^{22} T^4 \text{ ergs cm}^{-3} , \\
du_\nu/dt &= 4.3 \times 10^{15} T^9 \text{ ergs sec}^{-1} \text{ cm}^{-3} , \\
T &> 6 , \quad T/6 > (\rho/\mu_e 10^6)^{1/3} .
\end{aligned}
\tag{11.11A.1}
$$

Actually, as has been pointed out by Fowler and Hoyle (1964), the last formula is satisfactory starting from $T > 3$ (an exaggeration of 50 percent, rapidly decreasing to 10 percent when $T = 6$).

We recommend the use of the intermediate interpolation formula

$$du_\nu/dt = 10^{14} T^{12.5} \text{ ergs sec}^{-1} \text{ cm}^{-3} , \quad 1 < T < 3 .$$

which satisfactorily fills the interval where the theoretical, asymptotically correct formulae are not applicable.

Finally, in a relativistically degenerate gas, the chemical potential of the electrons[27] is given by the expression $\mu_- = m_e c^2 (\rho/\mu_e \, 10^6)^{1/3}$, and the positron concentration is given by

$$T < 6 , \quad \frac{\mu_-}{m_e c^2} > 1 : \quad n_+ = 1.3 \times 10^{29} T^{3/2} \exp\left[-\frac{1}{T} (6 + \mu_-/m_e c^2) \right] ,$$

$$T > 6 , \quad \frac{\mu_-}{m \, c^2} > T/6 :$$

$$n_+ = 1.5 \times 10^{28} T^3 \exp\left\{ -\frac{1}{T} [6(1 + \mu_-/m_e c^2)] \right\} .$$

26. For comparison, we note that the spectrum of equilibrium photon radiation has a maximum at $\hbar\omega = 4kT$. The maximum for fermions is even higher. For this reason, when we formally write the condition $kT > m_e c^2$, for example, in reality a weaker condition $2kT > m_e c^2$ is sufficient. Thus, for the application of the ultrarelativistic formula, it is sufficient that $T = T_9 > 3$. The spectrum of the escaping neutrinos is not in equilibrium at all as long as the matter remains transparent to the neutrinos. Since the interaction grows with increasing energy, the radiated spectrum is harder than the equilibrium spectrum.

27. Not to be confused with the μ_e in the parentheses, which is the molecular weight per one electron.

For the energy losses, we obtain approximately

$$du_\nu/dt \approx 10^{-40} n_+ n_- (\mu_-/m_e c^2)^2 \,, \qquad\qquad T < 6 \,,$$

$$du_\nu/dt \approx 3 \times 10^{-40} n_+ n_- (\mu_-/m_e c^2)^2 (T/6) \qquad T > 6 \,,$$

so that finally, in the last case,

$$du_\nu/dt = 5 \times 10^{17} T^4 (\rho/\mu_e 10^6)^{5/3} \exp\left(-\frac{1}{T} \{6[(\rho/\mu_e 10^6)^{1/3} + 1]\}\right),$$

$$6 < T < 6\,(\rho/\mu_e 10^6)^{1/3} \,.$$

See Pinaev (1963a, b) for expressions for the energy loss due to the URCA process (in which p,n or nuclei participate); in the case considered here (large stars) it is negligible, but for small masses it may be significant.

Consider now the energy losses in the course of the free-fall contraction of matter with a given initial value of the specific entropy S_0. Let the initial state be such that the radiation energy and the energy of pairs dominates the energy of the particulate plasma. We obtain at once the expressions for $T > 6$, when the $e^+ e^-$ pairs are an important addition to the energy density:

$$u = u_{\text{pairs}} + u_{\text{radiation}} = 2.75\, aT^4 = 2.1 \times 10^{22}\, T^4 \,,$$

$$S = 2.8 \times 10^{22} (T^3/\rho)\,(\text{in units ergs per gram per } 10^{9\,\circ}\,\text{K}) \,.$$

The equation for the rate of change of entropy (by using eq. [11.11A.1]) may be written in the form

$$dS/dt = -(1/\rho T)(du_\nu/dt) = -(1/\rho) 4.3 \times 10^{15}\, T^8 \,.$$

It remains to express the temperature in terms of the entropy and density

$$T = (\rho S/2.8 \times 10^{22})^{1/3}$$

Finally, we obtain

$$dS/dt = -3.6 \times 10^{-45} \rho^{5/3} S^{8/3} \,.$$

If the dependence of density on time, $\rho(t)$, is given—for example, the density variation produced by free fall—then, expressing dt in terms of $d\rho$

$$dt = 450\rho^{-3/2} d\rho \,,$$

we immediately obtain a simple, easily soluble equation

$$dS/d\rho = -1.6 \times 10^{-42} \rho^{1/6} S^{8/3} \,.$$

Astronomers, who are not accustomed to using entropy, should consider the simplification of the derivation of this equation, as compared with the conventional procedure in which the energy and work of the pressure forces must be taken into account. (Notice that they do not enter into the entropy equation!)

Integrating with the initial conditions

$$\rho = 0 \,, \quad S = S_0 \,,$$

we find

$$S = [S_0^{-5/3} + 1.6 \times 10^{-42} \rho^{7/6}]^{-3/5} \,. \qquad (11.11\text{A}.2)$$

If $S(\rho)$ is known, it is easy to write the integral which gives the total loss of energy Δ from one gram of matter in the contraction from $\rho = 0$ to a given density ρ. For this, we express all quantities—S, T, $(du_\nu/dt)dt$— in $\int(1/\rho)(du_\nu/dt)dt$ in terms of ρ and $d\rho$.

If we were to disregard the decrease of the entropy in the course of contraction, then we would obtain

$$\Delta = k_1 \int_0^\rho (T^9/\rho)dt = k_2 S_0^3 \int_0^\rho (\rho^3/\rho)\rho^{-3/2}d\rho \equiv \tfrac{2}{3}k_2 S_0^3 \rho^{3/2} ,$$

$$k_2 = 8.8 \times 10^{-50} , \tag{11.11A.3}$$

i.e., an integral which is divergent at the upper limit when $\rho \to \infty$. When $\rho \to \infty$, the losses would increase infinitely.

However, substituting S from equation (11.11A.2), we obtain a convergent integral

$$\Delta = 8.8 \times 10^{-50} \int_0^\infty [S_0^{-5/3} + 1.6 \times 10^{-42}\rho^{7/6}]^{-9/5}\rho^{1/2}d\rho .$$

Simple calculations yield[28]

$$\Delta = 7 \times 10^4 S_0^{6/7} . \tag{11.11A.4}$$

We write this expression in terms of some effective ρ^* in the same form as equation (11.11A.3):

$$\Delta = \tfrac{2}{3}k_2 S_0^3 \rho^{*3/2} .$$

From here, substituting Δ from equation (11.11A.4), we find ρ^*:

$$\rho^* = 1.64(1.6 \times 10^{-42}S_0^{5/3})^{-6/7} .$$

Obviously, the quantity ρ^* is the density at which a decrease of entropy effectively halts the energy loss.[29]

There is another restriction upon the radiation due to gravitational self-closure. It takes place at a mean density of the order of $\rho_g \approx 2 \times 10^{16}(M/M_\odot)^{-2}$.

Let us retrace now the entire process: A massive star becomes unstable at a critical state, the parameters of which are given in § 14 of chapter 10. In the critical state, the temperature (for $M > 10^4\ M_\odot$) is

$$T = 0.02 \times \rho^{1/3}(M/M_\odot)^{1/6}\ (T \equiv T_9) .$$

28. The dimensionless integral has the following value:

$$\int_0^\infty (1 + z^{7/6})^{-9/5}z^{1/2}dz = 1.4 .$$

29. The loss of energy can increase substantially only when a shock wave is generated by rapid compression (see § 11.4).

The corresponding expression for the entropy (in a state without pairs!) is

$$S = \tfrac{4}{3}a\frac{T^3}{\rho} = 0.97 \times 10^{22}\frac{T^3}{\rho} = 8 \times 10^{16}\left(\frac{M}{M_\odot}\right)^{1/2}\left(\frac{\text{erg}}{10^{9}\text{ }^\circ\text{ K g}}\right).$$

The creation of pairs in itself is the cause of a decrease of γ to a value less than $\tfrac{4}{3}$, i.e., the trigger of the catastrophic collapse. See Table 15 for the critical parameters for $M < 10^4\ M_\odot$.

The contraction takes place adiabatically, i.e., at constant entropy; a situation where the pairs become important constituents of equilibrium is reached before the neutrinos can reduce the entropy noticeably. Thus, in the course of contraction in the range $0.5 < T < 3$, there takes place a transition to the formulae

$$u = 2.75aT^4, \quad S = 2.8 \times 10^{22}T^3/\rho,$$

from which we obtain $T = 0.014\ (M/M_\odot)^{1/6}\rho^{1/3}$.

For adiabatic contraction (without energy loss) to a density that corresponds to gravitational self-closure, $\rho_0 = 1.8 \times 10^{16}(M/M_\odot)^{-2}$, we derive

TABLE 20

MASS RADIATED AS NEUTRINOS DURING COLLAPSE AND THE SPECIFIC PARAMETERS OF THE STAR AT THE MOMENT WHEN THE NEUTRINO RADIATION COMES TO AN EFFECTIVE STOP

PARAMETER	$M/M_\odot$					
	100	10^3	10^4	10^5	10^6	10^8
$\Delta M/M$	0.1	10^{-3}	4×10^{-4}	10^{-5}	3×10^{-7}	10^{-14}
T_9	70	50	36	11	3.6	0.5
ρ^*	2.5×10^{10}	...	...	...	...	...
ρ_0	...	5×10^9	2×10^8	2×10^6	2×10^4	2

an important physical conclusion: The energy of photons and pairs in a unit volume amounts to 0.24 of the rest mass-energy ρc^2. In this state, the temperature is $T_9 = 4000\ (M/M_\odot)^{-1/2}$.

It is impossible to give more energy than is available. As applied to neutrino radiation, this means that under no circumstances can the total radiated energy exceed $0.24\ Mc^2$.

In reality, however, the integration of the equation of energy losses yields considerably smaller values (see Table 20). If the mass of the star is large, $M > 10^4\ M_\odot$, then the neutrino radiation is limited by gravitational self-closure in the regime of practically constant entropy. For $M > 10^6\ M_\odot$ the temperature is such that the creation of pairs becomes exponentially small. The neutrino radiation is always limited by the line of the "last ray" (see § 11.7). Because of the strong dependence of du_ν/dt on ρ, the cutoff at the last ray yields a considerably greater variation in the mass lost than appears at first glance. Zel'dovich and Podurets (1964) have demonstrated that to account roughly for the redshift before the last ray and the cutoff at the last ray, one must multiply the energy loss by a star, as derived from the aforenamed formulae, by a factor of 7×10^{-3}.

When $M < 200 \ M_\odot$, the decreasing entropy due to the neutrino radiation itself restricts the general loss of energy before relativistic self-closure takes place; the transition from one domain to the other is rather gradual.

For the effective density and temperature at which the loss has reached $\frac{1}{2}\Delta$, the following expressions apply:

$$\rho^* \approx 8 \times 10^{11} (M/M_\odot)^{-5/7} , \quad T_9 \approx 100(M/M_\odot)^{-1/14} .$$

The entire Table 20 is an extremely rough approximation. In the domain of small masses $M < 100 \ M_\odot$, it is necessary to take into account the plasma energy and the electron degeneracy. All processes, including gravitational self-closure, must be considered against the background of a truly hydrodynamic solution, which depends on time as well as on the coordinates. Finally, for large densities, other processes of neutrino radiation, including the creation of μ-neutrinos, may turn out to be more important (see § 11.4).

12 STAR CLUSTERS

12.1 GENERAL OVERVIEW AND BASIC EQUATIONS

The theory of star clusters is one of the most rapidly growing branches of astrophysics. Globular star clusters and the various types of galaxies observed by astronomers are nonrelativistic. As a rule, their mean star velocity is not more than ~ 300 km sec^{-1}, a value corresponding to a dimensionless Newtonian potential of $\phi/c^2 \sim v^2/c^2 \sim 10^{-6}$, so that general relativity is not needed.

The theory of the evolution of such a system predicts the evaporation of some of its stars, accompanied by an increase in the density of the remaining cluster; but this process is very slow even on a cosmological time scale. On the other hand, observation reveals objects with great density and tremendous energy output (nuclei of galaxies, Seyfert galaxies, N-type galaxies, QSSs [quasars], and QSGs [radio-quiet quasistellar galaxies]). Some people have suggested that these objects (or perhaps some of them) are very dense star clusters (see, for example, the Hoyle and Fowler [1967] cluster model for QSSs).

Clearly the theoretical problem arises as to what are the properties of dense clusters and what, if any, are the limits to their densities.

One complication at high density is the frequent occurrence of direct star-star encounters. In a normal galaxy (such as our own, for example), the probability of such events is negligible; and, in fact, such encounters are never observed. Such encounters in the nucleus of our Galaxy are obscured from view by the clouds of gas and dust surrounding the nucleus.

On the other side of the problem are GTR effects, including relativistic collapse of the cluster as a whole. The relative importance of encounters and GTR depends upon the number of stars (N) and their size and nature—normal stars, dwarfs, neutron stars, or collapsed stars. There exist conditions under which GTR is the more important; it is these situations which will be discussed in detail here.

The general method of approach to the problem is to discuss first static solutions for a cluster without any account being taken of encounters. Under such an assumption, the cluster is idealized as a set of point masses moving along their trajectories in the collective, smoothed-out, gravitational field of the whole system. The second step is the analysis of the stability of such a static model against collective perturbations. If the model is unstable, it will be destroyed in a time of the order of several traversals of a star across the cluster.

If pairwise encounters of stars are taken into account, there are no static solutions—but if the encounters are rare enough, they will lead to a slow evolution through a sequence of static solutions. There is a far-reaching analogy here to a hot star, which is static and stable as a mechanical system, but undergoes evolution under the action of heat transfer and nuclear reactions. The very important difference from the case of a single star is that a star with $M < 2\,M_\odot$ has a final cold state in which all the evolution ends, whereas star clusters have no such final state, since sooner or later they must inescapably collapse. The problem is whether the time needed is not greater than the cosmological time scale.

Just as was done for the case of the structure of a single star, we begin with a discussion of the nonrelativistic problem.

A stationary (equilibrium) solution for a nonrelativistic star cluster is determined by its gravitational potential $\phi(x, y, z)$ and the distribution function f of its "particles" (i.e., stars) in coordinate and momentum space. The masses of the particles can be different, so that $f = f(x, y, z, p_x, p_y, p_z, m)$. The particle density n and the mass density ρ in coordinate space are

$$n(x, y, z) = \int f\, d^3p\, dm; \; \rho(x, y, z) = \int mf\, d^3p\, dm \, .$$

Henceforth, we shall consider the special case in which the particles all have the same mass m, and thus we shall omit the integration over m. In the nonrelativistic limit this means simply that $\rho = mn$.

The gravitational potential and the distribution function are constrained by two equations: First, the distribution function f, which is independent of time, must be a solution of the kinetic equation describing the motion of a swarm of particles moving in the gravitational potential ϕ. The second equation states that the mass density of the particles generates this potential ϕ.

The kinetic equation is nothing more than the equation of continuity for the flow of particles in the six-dimensional phase space. Their velocity is $v = p/m$ along the space coordinates, and $dp/dt = m\,dv/dt = F = -m\nabla\phi$ along the momentum (p) coordinates. Therefore, in the general case of $f(x, p, t)$

$$\frac{\partial f}{\partial t} = -\left(\frac{p}{m}\cdot\nabla_x\right)f + m[(\nabla_x\phi)\cdot\nabla_p]f = -\frac{p_x}{m}\frac{\partial f}{\partial x} - \ldots + m\frac{\partial\phi}{\partial x}\frac{\partial f}{\partial p_x} + \ldots$$

It can be shown from the kinetic equation that in the absence of friction against surroundings and of encounters between stars, the phase-space density is conserved along the trajectory of each particle.[1] This statement is

---

1. The trajectory in phase space is given by $dx/dt = p_x/m = \partial H/\partial p_x, \ldots, dp_x/dt = -m\partial\phi/\partial x = -\partial H/\partial x$, where $H = p^2/2m + m\phi$ is the Hamiltonian. Mathematically speaking, these equations are the characteristic equations of the partial differential kinetic equation.

called Liouville's theorem; and the density in phase space of which it speaks is related to the entropy.

If the phase-space density (distribution function) is independent of time, it must be constant along every orbit in the potential ϕ. To make this clear, let us imagine an observer located at a point (x, y, z) and counting particles with given p_x, p_y, p_z (to be precise, he counts the number dn in the region between x and $x + dx$, . . . , etc., and divides it by the "volume" dx . . . dp_x . . .). As time passes, groups of particles come into the measuring device from different parts of the trajectory, bringing with them their phase-space density f. By Liouville's theorem

$$f(x, \ldots, p_x, \ldots, t) = f(x_0, \ldots, p_{x_0}, \ldots, t_0) .$$

So, if $f(x, \ldots, p_x, \ldots, t)$ is independent of t at a definite place (x, y, z), we must have $f = $ const. at a definite time t_0 but different $(x_0, \ldots, p_{x_0}, \ldots)$ with the only condition that $(x_0, \ldots, p_{x_0}, \ldots)$ and $(x, \ldots, p_x, \ldots)$ are connected by a free-particle orbit in the potential ϕ.

The possibility of connecting different points by an orbit is limited by the integrals of the particle motion. In the stationary case, with potential ϕ independent of time, the energy of a particle is conserved during its motion. If $E(x_1, p_{x_1}) \equiv E_1 \neq E_2 \equiv E(x_2, p_{x_2})$, then there can be no trajectory connecting the first and second points.

In a spherically symmetric potential the vector angular momentum I of a particle is also conserved; this implies, too, the conservation of its magnitude L. In some special, "degenerate" cases there are other conservation laws. If $E_1 = E_2$ (and $L_1 = L_2$, etc.) so that the conservation laws do not forbid the motion of a particle from the first point to the second point, then as a rule a particle which is initially at the first point will actually reach the the second point after a more or less long travel along its trajectory. This statement is the so-called ergodic theorem. For our purposes we have no need to embellish this statement with ornaments about infinitesimal nearness, sets of measure zero, etc.

The conclusion which is important for the following is that the distribution function f must be a function only of the constants of the motion if the overall picture is to be stationary, i.e., independent of time. This is true whether the motion occurs in an external potential or in the potential due to the particles themselves.

The simplest and most general case is the one in which f depends solely upon the energy E, i.e., only upon $p^2/2m + m\phi$. Suppose that at every space point ϕ is known. For a given $f(E)$ we conclude that: (1) the part of $f(E)$ with $E < m\phi$ gives no contribution (particles with that E do not reach the point under consideration); (2) for $E > m\phi$ the kinetic energy of the particles is $p^2/2m = E - m\phi > 0$, and the phase-space density is independent of the direction of the velocity (at every point the distribution is isotropic in velocity space; all directions are equally populated); (3) the particle density is

$$n(x, y, z) = 4\pi \int f p^2 dp = 4\pi (2m^3)^{1/2} \int\limits_{E=\phi}^{\infty} f(E)(E - m\phi)^{1/2} dE \ .$$

In the important case of spherical symmetry, a possible choice of f is $f(E, L)$. Note that L, the magnitude of a particle's angular momentum, can be written as $L = mv_t r = p_t r$, where v_t and p_t are the tangential components of the velocity and momentum, i.e., the components perpendicular to the radius vector r from the origin (center of symmetry) to the point considered. Thus, $p^2 = p_r^2 + p_t^2$, where p_r is the radial component of the momentum.

In the spherical case $f = f(E, L)$, p_r and p_t are not on the same footing in f, so a distribution is obtained which is anisotropic in the velocities. For example, if $f(E, L) = f_1(E)\delta(L)$, this means that only $p_t = 0$ is allowed; i.e., all the particles are moving radially. A complicated combination of E and L [the precise form of the function depends on the potential $\phi(r)$] can be arranged so that all the particles move in circular orbits. Examples of $f(E, L)$ will be given later.

Here we must emphasize that such an anisotropy of the velocities does not destroy the spherical symmetry (SS) of the overall picture. In a SS picture the radial direction is distinguished (it is the direction, at every point, of the gravitational force); and it is just this direction which is distinguished in the velocity distribution. Imagine the case of a purely radial distribution of velocities: such a system has spherical symmetry after the manner of a hedgehog (rolled up into a ball). In the case of purely circular orbits, the particles move in planes which pass through the center of the system, the normals to these planes being distributed uniformly with all possible orientations, so that the system resembles a spherical ball of yarn.

To finish this discussion of general properties of star clusters (or, mathematically speaking, sets of point masses), let us dwell on the concept of pressure.

The naïve notion of pressure implies the interaction of particles: it focuses attention on the force with which each body acts on its surroundings. The pressure of a gas on a wall is given by the time-averaged force acting during encounters. But the gas pressure can be defined in the middle of a gas by the formula $P = nkT;$ one sees that the mean free path and the probability of encounters do not affect the formula. We may speak about an imaginary wall, arbitrarily oriented at an arbitrary point inside the gas.

The generalization of the pressure is $T_{\alpha\beta}$, the stress tensor. It is defined as the flux of the α-component of momentum through a unit surface area perpendicular to the β-direction. For the simple case of an isotropic velocity distribution of particles, $T_{\alpha\beta} = P\delta_{\alpha\beta}$, where P is the pressure, i.e., the force which would act on an impermeable wall of unit area inserted into the gas.

In this case the force is normal to the wall and independent of its orientation (Pascal's law). Since isotropy of velocities corresponds to $f = f(E)$, one can write

$$P = T_{zz} = \int p_z v_z f d^3p = \tfrac{1}{3}\int pv\,fd^3p = (1/3m)\int p^2 fd^3p$$
$$= \tfrac{2}{3}\int E_{\text{kin}} fd^3p = \tfrac{2}{3}\langle E_{\text{kin}}\rangle n$$
$$= \tfrac{2}{3}4\pi(2m^3)^{1/2}\int f(E)(E - m\phi)^{3/2}dE .$$

But in the anisotropic case $f(E, L)$, with the axes of local Euclidean coordinates taken along the radial (r), meridional (θ), and azimuthal (ϕ) directions, one has

$$T_{rr} = P_r = \int (p_r{}^2/m)f\,d^3p , \qquad T_{\theta\theta} = T_{\phi\phi} = P_t = \tfrac{1}{2}\int (p_t{}^2/m)f\,d^3p ,$$

all other components vanishing identically. For purely radial motion only P_r is nonzero, and for purely circular motion only P_t is nonzero.

Having introduced the pressure (which is complicated and non-Pascalian in the anisotropic case) of a collisionless set of particles, one can consider the overall picture of a star cluster from a new point of view.

It was motion along an orbit that we were tracing above. Now let us take a definite unit volume in space. The total number of particles in the volume element is constant, as granted by the kinetic equation for a stationary cluster integrated over all momenta:

$$\partial n/\partial t = (\partial/\partial t)\int fd^3p = \int \partial f/\partial t\,d^3p = 0 .$$

The total momentum, $Q = \int pfd^3p$, of particles contained in the volume element is also constant; in the isotropic and spherically symmetric case it is, in fact, equal to zero. Apply Newton's second law of motion: the inflow and outflow of momentum must compensate the gravitational force on the particles which are in the element. For a stationary, isotropic solution we obtain from Newton's second law—or, equally well, from manipulations of the kinetic equation:

$$\partial P/\partial r = -\rho\partial\phi/\partial r ; \tag{12.1.1a}$$

and for the stationary, anisotropic case, we obtain

$$\frac{1}{r^2}\frac{\partial(r^2 P_r)}{\partial r} - \frac{2P_t}{r} = -\rho\frac{\partial\phi}{\partial r} . \tag{12.1.1b}$$

Both formulae are written for SS potentials and distributions.

The first formula is already familiar to us from the theory of the equilibrium of a star. To exploit this formula we must know the "equation of state" $P = P(\rho)$. For a given distribution function $f(E)$, the equation of state is readily found. We write it in parametric form:

$$\rho(\phi) = 4\pi(2m^5)^{1/2}\int_{E=m\phi}^{\infty}f(E)(E - m\phi)^{1/2}dE , \tag{12.1.2}$$

$$P(\phi) = \frac{8\pi}{3}(2m^3)^{1/2}\int_{E=m\phi}^{\infty}f(E)(E - m\phi)^{3/2}dE . \tag{12.1.3}$$

The reader will recognize that ϕ is the potential at the radius r where ρ and P are evaluated. However, the result $P = P(\rho)$, obtained after the elimination of ϕ from the expressions for $\rho(\phi)$ and $P(\phi)$, depends only on the form of the function $f(E)$ and is independent of the form of $\phi(r)$. It is easily seen that equations (12.1.2) and (12.1.3) guarantee that the condition of equilibrium, namely,

$$dP/d\phi = -\rho \,,$$

is fulfilled identically. Thus a connection between the theory of the equilibrium of star clusters and the equilibrium of gas spheres is obtained.

Collisionless systems of particles cannot give rise to every conceivable equation of state.

For $f(E) \sim e^{-E/\theta}$, one finds $P = \theta\rho$, corresponding to an adiabatic index $\gamma = 1$ and a polytropic index $n = \infty$ (analogue of an isothermal, ideal gas).

For a step function $[f(E) = \text{const.}, E < E_0; f(E) = 0, E > E_0]$ one obtains $P = \text{const.} \times \rho^{5/3}$ (analogue of a Fermi gas).

For a delta-function $[f = \delta(E - E_0);$ monoenergetic particles], one finds $P = \text{const.} \times \rho^3$ corresponding to $\gamma = 3$, $n = 0.5$.

Greater values of γ ($\gamma > 3$) cannot be achieved, due to the condition that we must have $f(E) \geq 0$ everywhere (no negative ghost stars!).

The equation of equilibrium (12.1.1b) for anisotropic distributions can be understood easily in terms of the forces acting on a small volume element obtained by making radial slices through a thin spherical shell. The transverse pressure acts on the side walls. These walls are not parallel; therefore the pressure P_t (itself, not its derivative) gives a resultant force which tends to lift the element upward. Ancient architects utilized this force to construct arches. In the limiting case of purely circular motion, $P_r = 0$, each shell is an independent arch by itself—it does not lean against the underlying shells and does not feel the pressure of the overlying ones. This picture helps greatly in understanding stability problems for clusters.

12.2 SOLUTIONS FOR NONRELATIVISTIC STAR CLUSTERS

The ideas of the preceding section enable us to find exact self-consistent solutions for collisionless star clusters.

The Emden solutions for a polytropic gas sphere with $P = K\rho^{1+1/n}$ are well known. It is clear from formulae (12.1.2) and (12.1.3) that a power-law distribution function, $f \sim (E_0 - E)^a$ for $E < E_0$ and $f = 0$ for $E > E_0$, gives $\rho \sim (E_0 - m\phi)^{a+3/2}$, $P \sim (E_0 - m\phi)^{a+5/2}$, so that $P \sim \rho^{(2a+5)/(2a+3)}$. Therefore, a given value of a in the distribution function corresponds to $\gamma = (2a + 5)/(2a + 3)$ and $n = a + \frac{3}{2}$ in the polytropic Emden solution. The permitted values of a (in order to keep the integrals for ρ and P finite) are $-1 < a < \infty$, $\frac{1}{2} < n < \infty$. The solutions for $a < 0$, $n < \frac{3}{2}$ are probably unstable (see below).

Solutions for the distribution function f can be obtained in the case of models with $\phi = Ar^{-k}$ and $\rho = Br^{-k-2}$, with either anisotropic or isotropic velocities (Bisnovatyi-Kogan and Zel'dovich 1969a, b). These models need some modification to smooth out the divergence at $r = 0$ and to bring ρ to zero at large r so that the mass will be finite.

In a gas sphere, atoms are distributed according to the Maxwellian distribution $f = g(r)e^{-E/T(r)}$. It is tempting to seek a solution of the same type in the collisionless case. But in this case the heat conductivity and the diffusivity are infinite, so that the temperature T and the factor g must be the same throughout the cluster; i.e., f must be a function of E only. With g and E constant, although the time-independent Liouville equation is automatically satisfied, we cannot obtain any sensible solution for the problem as a whole. The reason is that the Maxwellian distribution contains particles with positive energy which are able to escape the cluster. This difficulty can be phrased in another way. A finite mass will give $\phi = -GM/r \rightarrow 0$ at $r \rightarrow \infty$, which in turn will give a finite density $\rho = (2\pi \, mT)^{3/2}mq$ at $r \rightarrow \infty$, so that $M = 4\pi \int \rho r^2 dr$ diverges at $r \rightarrow \infty$. Therefore, in order to obtain physically realistic solutions, one has to modify the Maxwellian distribution. Numerical solutions have been found for "truncated" Maxwellian distributions (Woolley 1954). For example, one might take

$$f = q \, e^{-E/T}, \quad E < -T/2 \; ;$$
$$f = 0, \quad\quad E > -T/2 \; .$$

For such a function the solution is perfectly well defined—but a truncated Maxwellian distribution is very definitely no longer Maxwellian. We shall return to this point when we discuss the evolution of star clusters. The exact solution with a Maxwellian distribution exists only for infinite mass and radius: $\rho = A/r^2$, $m(r) = 4\pi Ar$. It can also be obtained from equations (12.1.2) and (12.1.3) in the limit $n \rightarrow \infty$ of the polytropic models (Bisnovatyi-Kogan and Zel'dovich 1969b, 1970).

We shall mention only briefly solutions without spherical symmetry. Their astronomical importance is obvious: we observe elliptical galaxies and various forms of spiral galaxies.

One limiting case is that of flat, axisymmetric solutions. A further simplification which is sometimes made considers only circular orbits in the plane of the disk. Even then the problem is not easy, because the connection between the gravitational potential $\phi(r, z)$ and the surface density in the plane, $\mu(r)$ g cm^{-2}, is much more complicated than the connection between ρ and $\phi(r)$ in the spherical case. In particular, an elementary ring between r and $r + dr$, lying in the plane $z = 0$, produces a gravitational field not only outside the ring but also inside it. If a potential $\phi(r, z)$ is given which satisfies

$$\nabla^2\phi = \frac{1}{r}\frac{\partial}{\partial r}\left(r\frac{\partial\phi}{\partial r}\right) + \frac{\partial^2\phi}{\partial z^2} = 0$$

everywhere except at $z = 0$, then the surface density is given by the discontinuity of the normal component of the field across the plane $z = 0$:

$$\left.\frac{\partial \phi}{\partial z}\right|_{z=0+\epsilon} - \left.\frac{\partial \phi}{\partial z}\right|_{z=0-\epsilon} = 4\pi\mu(r) \ .$$

The motion of particles in circular orbits in the plane $z = 0$ is found easily, and so the solution may be completely determined. However, it is already known that such solutions with uniform rotation are unstable.

A special set of solutions utilizes the special property of the gravitational field inside a triaxial ellipsoid filled with matter of constant density ρ_0. The potential is that of a three-dimensional simple harmonic oscillator, with $\phi = \phi_0 + \alpha x^2 + \beta y^2 + \gamma z^2$ so that $\nabla^2\phi = 2(\alpha + \beta + \gamma) = 4\pi G\rho_0$. Here the ratios β/α, γ/α depend upon the ratio of the ellipsoid axes (see Landau and Lifshitz 1962). In this potential not only is the total energy E conserved, but also the "energies along the axes"

$$E_x = m(v_x^2/2 + \alpha x^2) \ , \quad E_y = m(v_y^2/2 + \beta y^2) \ , \quad E_z = m(v_z^2/2 + \gamma z^2)$$

are individually conserved with $E = E_x + E_y + E_z$. The limiting case is a flat ellipse in the plane $z = 0$, with the surface density $\mu = \int \rho_0 dz = $ const. $\times$ $(1 - x^2/a^2 - y^2/b^2)^{1/2}$ (imagine a three-dimensional ellipsoid with initial half-axis a, b, c which is then squeezed down in the z-direction to zero thickness).

A further simplification of this case is the axisymmetric disk of radius r_0 with

$$\mu = \mu_0(1 - r^2/r_0^2) \ , \quad \text{and} \quad \phi(r, 0) = \pi^2 G\mu_0 r_0(1 - r^2/r_0^2) \text{ for } r < r_0 \ .$$

In this case each particle in a circular orbit has the same angular velocity, so that the solution describes rigid rotation of the disk as a whole, even though the disk is held together only by gravity. But there is also another solution with this μ and ϕ: one with perfectly isotropic velocities in the plane, i.e., with equal mean tangential and radial velocities (motion perpendicular to the plane is strictly absent). In this case

$$f = (A - BE)^{-1/2} \ , \quad \mu = \int f dp_x dp_y \ .$$

Intermediate solutions have also been found (Bisnovatyi-Kogan and Zel'dovich 1970).

It is commonly accepted that a spherical cluster has no net angular momentum and, conversely, that a system without net angular momentum is spherical, so that a flattened cluster results from the presence of angular momentum about an axis perpendicular to the plane of the cluster. Probably this is perfectly true for objects we observe in nature. On the other hand, it is easy to construct solutions for collisionless, spherical clusters with angular momentum: one merely has to divide the particles into two groups, those with $L_z > 0$ and those with $L_z < 0$ and then reverse the motion of one of these groups. The density is unaffected and the field remains spherically symmetric. Conversely, one also has completely flat solutions without angular momentum as in the case of the isotropic-disk solution. Therefore, the connection between the shape of the cluster and its angular momentum depends upon the processes of cluster formation and evolution rather than upon the properties of the collisionless solutions by themselves.

A general property of all solutions is the virial theorem, which ensures that all nonrelativistic clusters are gravitationally bound with $\mathfrak{E}_{tot} = -\mathfrak{E}_{kin} = \frac{1}{2}\mathfrak{E}_{grav} < 0$.

12.3 THE STABILITY OF COLLISIONLESS SOLUTIONS

In recent years a large number of very deep and elegant works on the stability problem have appeared (Antonov 1960, 1962; Lynden-Bell 1966; Lynden-Bell and Sanitt 1969; Ipser and Thorne 1968; Ipser 1969a, b; Fackerell 1970). Partly they are induced by progress in electronic-ionic plasma theory; it may be that competition and fashion also play some role in this burst of interest. Here we shall merely quote some crude considerations.

Let us begin with the SS case. The nonrelativistic "gas" whose atoms are stars has an adiabatic index equal to $\gamma = \frac{5}{3}$ ($n = \frac{3}{2}$), quite independently of the details of the distribution function.[2] Therefore, all solutions are stable against a homologous compression, just as a gas sphere with the adiabatic index $\gamma = \frac{5}{3}$ is stable independently of its polytropic index.

In some exact theorems in order to prove stability one has to assume $\partial f/\partial E < 0$; i.e., f must be a decreasing function of energy. There is no proof yet that the violation of this condition (so that $\partial f/\partial E > 0$ in some regions of phase space) always results in an instability; however, it is very probable that this is the case. The situation with $\partial f/\partial E > 0$ is like the population inversion of laser theory: the number of excited particles (with large E) is greater than the number of particles at a lower energy. In plasmas such a situation is definitely unstable against the generation of waves with an appropriate velocity, the energy of the excited-state particles being fed into the traveling wave.

We note that a power-law distribution function of the form $f \propto (E_0 - E)^a$ with $a < 0$ gives $\partial f/\partial E > 0$, so that the corresponding polytropic gas sphere has $n_2 > n_1$ (see § 12.2) and is unstable against convection.

A general theorem connects the stability of a collisionless isotropic star cluster of density $\rho(r)$ and pressure $P_r = P_t = P(r)$ to the stability of a gas sphere with the same density $\rho(r)$ and the same pressure $P(r)$, the adiabatic index of the gas sphere being taken to be $\gamma = (\rho \, dP/dr)/(P \, d\rho/dr)$.

---

2. The adiabatic index γ (not to be confused with the polytropic $\gamma = 1 + 1/n$ of the last section) is defined for a homologous transformation of the distribution function with conserved phase-space density; the volume in momentum space is $V_p \sim p^3 \sim V_x^{-1} \sim \rho$, so that the pressure is $P \sim p^2 V_p \sim \rho^{5/3}$.

The star cluster (assumed to have $\partial f/\partial E < 0$) is stable if the corresponding gas sphere is stable (for details see Lynden-Bell and Sanitt 1969). The theorem proves the stability of polytropic clusters with $\frac{3}{2} < n < 3$, which corresponds to $0 < a < \frac{3}{2}$. A stronger result was obtained by Antonov (1962), who, without using the analogy between clusters and gas spheres, proved that clusters with $f \sim (E_0 - E)^a$ are stable for $0 < a < \frac{7}{2}$. In addition, he was able to show that clusters for which $\partial f/\partial E < 0$ are always stable against nonradial oscillations.

A rough argument that we may use to analyze stability is that a perturbation of the space density of a cluster is always accompanied by a corresponding change in the momentum space occupied by the particles. (By Liouville's theorem, the phase-space density, i.e., the distribution function itself, is always conserved following the motion of a particular group of particles, even in a time-dependent potential.) For example, in the case of a Fermi gas with $f = f_0 = $ const. in a part of phase space with volume $V_p V_x$, and $f = 0$ elsewhere, perturbations keep the space density ρ proportional to the volume in momentum space, $\rho = \int f_0 d^3 p = f_0 V_p \propto V_p$. In the collisionless case, if the initial volume element V_p was spherical, it must be deformed after the perturbation. But in the case of a gas one assumes that collisions always bring V_p back to a spherical form. For a given volume V_p the kinetic energy has a minimum if the volume is a sphere. Therefore, the gas sphere is more easily perturbed, since the perturbation requires less energy than the corresponding perturbation of the equivalent star cluster.[3]

This argument does not work for star clusters with initially anisotropic distribution functions. It is possible that anisotropy of the distribution function can cause instability. However, notwithstanding some investigations of particular cases (Bisnovatyi-Kogan, Zel'dovich, Sagdeev, and Friedman 1969; Marochnik and Suchkov 1969), it is still unclear whether in the gravitational case there exists an analogue of plasma-stream instabilities.

The case of a disk is of considerable importance. Here the general rotation counteracts gravity and stabilizes the disk against total collapse. But if one considers a small part of the disk, the kinetic energy of motion around the center of mass of that part is $\sim m(v - v_{\text{cm}})^2 \sim m r_1^2 = \mu\, r_1^4$, where r_1 is the linear size of the part taken. The self-gravitational energy of this part is $Gm^2/r_1 = G\mu^2 r_1^3$. For the disk as a whole ($r_1 = r_0$), the two quantities are of the same order. But for a small part, $r_1 \ll r_0$, the gravitational energy is always predominant, so every such part is unstable against self-collapse. Of course, these remarks are really an afterthought, since the result has been obtained by exact methods in several papers (Hunter 1963; Toomre 1964).

Let us return to elementary methods of analyzing the stability of a disk in which some chaotic motion is present. The energy of chaotic motion in a

3. In the case of a Fermi gas with a step function $f(E)$, the above reasoning is exact.

given small part is proportional to its mass, $\mu r_1{}^2$; it can stabilize volumes which are small enough; this is a manifestation of the Jeans wavelength, which arises when chaotic motion is introduced. As we pass from purely chaotic motion without angular momentum to purely circular motion, the stability must be lost somewhere. The interaction of stars in the disk with surrounding hot gas can also play an important role in the stability of the disk. There are indications that perhaps the unstable mode that emerges first is of a quadrupole ($l = 2$) type, so that two arms are formed (Bisnovatyi-Kogan, Zel'dovich, Sagdeev, and Friedman 1969). For fundamental work on spiral galaxies see Lin and Shu (1964); also Marochnik and Suchkov (1969). Much of what we have said about a flat disk-like star cluster is applicable also to flat gas disks.

12.4 PHYSICAL CONDITIONS, COLLISIONS, AND EVOLUTION IN STAR CLUSTERS

We now inquire what the physical conditions are in typical clusters, such as are observed in nature, what processes occur in them, and what are the evolutionary trends due to these processes.

A typical large galaxy contains $\sim 10^{11}$ stars, each with $M \sim M_\odot$, in a sphere of radius $R \sim 10$ kpc, which amounts to a volume $V \sim 10^{68}$ cm^3; hence the average number density of stars is $n \sim 10^{-57}$ cm^{-3}, and the gravitational potential at the surface is $\phi_s \sim -5 \times 10^{14}$ cm^2 sec^{-2}. Let us approximate the spatial distribution of all quantities by those of a polytrope with polytropic index n equal to 3. Then

$$\langle \phi \rangle = [6/(5 - n)] \, \phi_s = 3\phi_s = -3 \times 10^{15} \text{ cm}^2 \text{ sec}^{-2} \, ;$$

$$\langle v^2 \rangle = 7.5 \times 10^{14} \text{ cm}^2 \text{ sec}^{-2} \, ;$$

$$\langle v^2 \rangle^{1/2} = 2.7 \times 10^7 \text{ cm sec}^{-1} = 270 \text{ km sec}^{-1} \, .$$

In the center of the cluster,

$$\rho_c = 54 \langle \rho \rangle = 5 \times 10^{-56} \, M_\odot \text{ cm}^{-3} = 10^{-22} \text{ g cm}^{-3} \, ;$$

$$\langle v_c{}^2 \rangle = 1.5 \times 10^{15} \text{ cm}^2 \text{ sec}^{-2} \, ;$$

$$\langle v^2 \rangle^{1/2} = 3.9 \times 10^7 \text{ cm sec}^{-1} = 390 \text{ km sec}^{-1} \, .$$

The interaction between stars can proceed through direct physical encounters, so that the material of both stars suffers the passage of a shock wave. If the average star is assumed to have the dimensions of the Sun, the mean time needed for it to experience such a direct collision with another star is between 10^{27} sec $= 3 \times 10^{19}$ years and 10^{25} sec $= 3 \times 10^{17}$ years, corresponding to the average and central conditions, respectively, in the cluster. Consequently, direct encounters do not constitute an important process in the ordinary regions of ordinary star clusters; the interaction process that we must discuss instead is that of inverse-square "gravitational encounters."

In the Newtonian approximation, the evolution of a system of gravitationally interacting point masses has been analyzed comprehensively by a number of investigators. The rate of evolution was originally computed by Ambartsumyan (1938). Subsequently, the problem was examined by many authors; see, for example, Chandrasekhar (1942, 1943), Gurevich (1954a), Ogorodnikov (1958), Michie and Bodenheimer (1963a, b). For modern developments, see below.

The dominant evolutionary process is gravitational interaction without physical contact during the near passage of two stars. A first approximation to the cross-section for this process may be obtained from the maximum radius of interaction, calculated as the distance r_1 at which $GM^2/r_1 = \frac{1}{2}Mv^2$; this amounts to considering encounters for which the trajectory of a star is deflected through an angle of the order of 90°. For $M = M_\odot$ and $v = 3 \times 10^7$ cm sec^{-1}, we obtain $r_1 = 3 \times 10^{11}$ cm $= 4$ $R_\odot$. Now, it is well known that the joint effect of numerous distant interactions leads to an extra factor in the rate of deflections (i.e., rate of deviations by $\sim$90°) of the order of ln $(R_{cluster}/r_1)$. As a result, the characteristic time for gravitational pairwise encounters is some 300 times smaller than that for direct physical encounters (10^{17}–10^{15} years).

For other N and R but with $M = M_\odot$ for each star, the formula for the mean gravitational deflection time is $t_d = 10^{16}$ years $(N/10^{11})^{1/2}(R/10$ kpc$)^{3/2}$. This is to be compared with the characteristic time of revolution of the star orbits

$$t_r \approx 3 \times 10^7 \text{ years } (N/10^{11})^{-1/2}(R/10 \text{ kpc})^{3/2} .$$

The ratio is $t_d/t_r = N/300$. Hence for ordinary, not-dense galaxies the time of evolution through encounters is so great, compared with the cosmological time scale that all kinds of encounters can safely be neglected. The evolution of individual stars accompanied by the loss of mass in the form of gas (and including explosions), with subsequent motion of the gas, is of primary importance.

For galaxies the following question must be answered: Why are they so similar in form and in velocity distribution to equilibrium configurations, if the elastic gravitational encounters are so slow in establishing equilibrium? There are two possible reasons. One is that perhaps the equilibrium form was taken by the gas cloud previous to star formation. This is probable for globular clusters (Peebles 1969). This reason has also been advanced for objects with greater masses, up to clusters of galaxies (Doroshkevich, Zel'dovich, and Novikov 1967c); but, as pointed out by Sunyaev, this would require gas temperatures so high as to contradict X-ray measurements. The other explanation results from calculations of the collisionless dynamics of star systems which are not in equilibrium. The Vlasov-type (i.e., collisionless) kinetic equation does not produce an entropy increase and so cannot give rise to the establishment of a true equilibrium. However, after several hydrodynamic times, t_r, streams of particles intersect one another, so that multistream velocity distributions result which are difficult to distinguish from smooth distributions (Marochnik 1970; Prendergast and Miller 1969).

12.5 Relativistic Star Clusters

We shall postpone the discussion of the astronomical implications of the theory of relativistic star clusters to the section devoted to quasars.

The theory of static, collisionless, relativistic star clusters is constructed along the same lines as the corresponding Newtonian theory: a distribution function is constructed from the first integrals of the motion of a "particle," so that this distribution function automatically satisfies the kinetic equation. Poisson's equation in the Newtonian case is replaced by the Einstein equations relating the curvature to the stress-energy tensor of the distribution of particles.

In the case of spherical symmetry five types of solutions have been found:

1. Clusters with purely circular orbits (Einstein 1939).

2. Models with step distribution functions in energy; the density and pressure of these solutions coincide with the calculations made by Oppenheimer and Volkoff for a Fermi gas (Zel'dovich and Podurets 1965).

3. Models with truncated Maxwellian distribution functions (Zel'dovich and Podurets 1965; Fackerell 1968).

4. Relativistic polytropic star clusters (Ipser 1969*b;* Fackerell 1970).

5. A set of self-similar solutions with power-law dependence of the redshift upon radius (Bisnovatyi-Kogan and Zel'dovich 1969*b;* Bisnovatyi-Kogan and Thorne 1970).[4]

By virtue of these last models we now believe that, in principle, arbitrarily large central redshifts can be achieved in stable, collisionless models. However, it seems quite unlikely that such models actually occur in nature. Considerable astronomical evidence has accumulated against the idea that they could occur in quasars, for example.

In clusters which are dense enough to display relativistic effects, encounters of various types occur quite frequently. In addition, new features, including gravitational radiation and inelastic coalescence of collapsed stars, are introduced. The problem of the stability of clusters is now more important, since the virial theorem, which ensures the stability of Newtonian clusters against homologous perturbations, is no longer valid.

One can predict that evolutionary processes rapidly drive a dense cluster to the onset of a GTR instability and subsequent collapse. However, there is an important difference from the GTR collapse of a star, since some of the

---

4. We mention in passing general-relativistic disk-type solutions both with angular momentum (Bardeen and Wagoner 1969) and without (Morgan and Morgan 1969).

stars in the cluster can remain in stable orbits even after the core of the cluster collapses. These stars can then produce violent events through encounters even after a frozen, self-closed core has been formed.

In the SS case the metric and the field equations may be taken directly from equations (3.1.1)–(3.1.5), with $g_{\theta\theta} = r^2$ and with all time derivatives set equal to zero. In order to allow for anisotropic velocity distributions we do not assume that $T_{rr} = T_{\theta\theta} = T_{\phi\phi}$, but only that $T_{\theta\theta} = T_{\phi\phi}$ due to spherical symmetry.

The distribution function and the stress-energy tensor will be written in a locally Euclidean coordinate system, oriented in the r-direction, and with x a cylindrical radial coordinate in the plane tangent to the sphere. It is important now to work with momenta rather than with velocities. We have

$$p^2 = p_r^2 + p_x^2 \; ; \quad E^2 = c^2 p_0^2 = (mc^2)^2 + c^2 p^2 \; .$$

When all the particles have the same rest mass, we find that

$$\epsilon = T_{00}(r) = \int p_0 \, f(p_r, p_x, r) \, dp_r \, 2\pi \, p_x \, dp_x \, ,$$

$$P_r = T_{rr}(r) = \int (p_r^2/p_0) \, f(p_r, p_x, r) \, dp_r \, 2\pi \, p_x \, dp_x \, ,$$

$$P_t = T_{xx}(r) = \tfrac{1}{2} \int (p_x^2/p_0) \, f(p_r, p_x, r) \, dp_r \, 2\pi \, p_x \, dp_x \, ,$$

(see Podurets 1970; Fackerell 1968; or Ipser and Thorne 1968 for more general expressions which include a spectrum of rest masses for the particles).

As shown in § 1, integrals of the motion for orbits in static SS metrics are the energy $E_1 = E \, e^{\nu/2}$, as observed by a distant observer, and the angular momentum $L = r p_x$. Therefore, the general solution of the kinetic equation is

$$f = f(E \, e^{\nu/2}, r p_x) \; ;$$

and the solution for an isotropic velocity distribution is

$$f = f(E \, e^{\nu/2}) \; .$$

The previously mentioned case of circular orbits does not need such a sophisticated theory. By the general property of SS fields (both Newtonian and GTR), the field at a given radius r depends only upon the mass inside this radius, $m(r)$. By equation (3.3.2), $e^{-\lambda} = 1 - 2Gm(r)/c^2 r$. For circular motion $T'_{rr} = T_{rr} = 0$ so that $\nu' = (e^\lambda - 1)/r$. (Here a prime denotes a radial derivative.) Thus with given $m(r)$ the functions $\lambda(r)$ and $\nu(r)$ as well as T_{00} and $T_{\theta\theta} = T_{\phi\phi} = T_{xx}$ may be readily determined.

The velocities of particles on circular orbits are readily found either directly or by comparison of T_{00} and T_{xx}:

$$v/c = (2T_{xx}/T_{00})^{1/2} \; .$$

A simple case is that in which the linear velocity is constant throughout the cluster. Inside such a cluster $\lambda = \text{const.}$, $\nu = a \ln r$, $(1 + z) = \text{const.} \times r^{-a/2}$. From this case we learn that, in principle, it is possible to have $z \to \infty$ at the center of a cluster, even though v/c is arbitrarily small.

The second case mentioned previously, that of a Fermi gas, is determined by

$$f = \text{const. for } E \, e^{\nu/2} < E_0$$

and

$$f = 0 \text{ for } E \, e^{\nu/2} > E_0 \,.$$

Obviously, at the surface of the cluster $E = mc^2$, $\nu = \nu_s$, so that $E_0 = mc^2 \exp(\nu_s/2)$. From the work of Oppenheimer and Volkoff we know that the minimum potentials and maximum redshifts at the surface and center of such a cluster are

$$e^{\nu_s/2} = 0.7 \,, \quad e^{\nu_c/2} = 0.4 \,; \quad z_s = 0.4 \,, \quad z_c = 1.5 \,.$$

It is still unknown if the stability limit in the collisionless case is not at greater density (and greater z_c) than the limit of stability for the gas sphere. As pointed out by Zel'dovich (1964d), at still greater z_c there exist collisionless isotropic solutions with total energy exceeding the rest mass (negative binding energy). Their stability is very doubtful; they are largely a mathematical curiosity, illustrating a gross violation of the virial theorem in GTR.

The solutions for a truncated Maxwellian distribution have been given in numerical form by Zel'dovich and Podurets (1965) and by Fackerell (1968). The temperature as measured by local observers for these solutions is proportional to $e^{-\nu/2}$. The temperature divided by the rest mass of a star is a measure of how relativistic the cluster is. The maximum binding energy (which coincides to high accuracy with the stability limit; see Ipser 1969b) is equal to 0.03 Mc^2 and occurs at a central redshift of $z_c = 0.51$.

The relativistic polytropic star clusters (Ipser 1969b; Fackerell 1970), like the Maxwellian clusters, have been constructed largely by numerical integrations. Like their Newtonian counterparts (see § 12.2), they have the same density and pressure distributions as polytropic gas spheres; but their particles move about without collisions.

Finally, the self-similar solutions are found by taking $\lambda = \text{const.}$, $\nu = a \ln r + \beta$, $\rho = \epsilon = \text{const.}/r^2$, $m = \gamma r$. Therefore, the case of circular orbits with constant velocities belongs to this set of solutions. A more general distribution function which produces this same λ, ν, ρ, and m is

$$f(E \, e^{\nu/2}, L) = L^{-2} \psi(E \, e^{\nu/2} L^{-a/2}) \,,$$

where ψ is an arbitrary function. The power of L is chosen to cancel the r-dependence in the argument of ψ. One can choose ψ so as to obtain isotropic solutions. For details see Bisnovatyi-Kogan and Zel'dovich (1969b).

Making order-of-magnitude calculations, one finds that the onset of GTR effects (Newtonian $\phi/c^2 \approx 0.1$ in the center) occurs at huge densities, corresponding to a cluster mass M and radius R such that

$$M/M_\odot \approx 5 \times 10^{11} \, R_{pc} \,.$$

12.5 Relativistic Star Clusters

Let us take as a numerical example the case where $M/M_\odot = 10^{10}$; $R = 0.02$ pc:

$$\langle \rho \rangle = 2 \times 10^{-8} \text{ g cm}^{-3} \; ; \quad \rho_c = 10^{-6} \text{ g cm}^{-3} \; .$$

A star with dimensions of the Sun would experience a collision ten times per year; hence such stars cannot exist at all in such a dense cloud. Collisions of stars in less extreme situations have been considered in connection with QSSs. Neutron stars and collapsed stars have collision times of 10^8–10^9 years in such a cluster, thus giving 100–10 events in the cluster per year; these collisions would produce a large amount of gravitational radiation. It has been assumed that a similar situation in the nucleus of our own galaxy could explain Weber's results (Weber 1969; Braginsky, Zel'dovich, and Rudenko 1969).

Physical encounters between neutron stars should yield considerable energy in the form of both electromagnetic radiation and hot plasma.

In encounters between collapsed stars and in the interaction of neutron stars, the most important effect on their cluster is a rearrangement of the energy distribution. The distribution tends to a fully Maxwellian one without truncation of the high-energy stars. The particles that acquire positive energy are then evaporated from the cluster.[5]

Other processes that affect the evolution are coalescence of stars and gravitational radiation.

Although the total mass of the cluster diminishes during the evolution, the rest mass decreases still more slowly. The density of the cluster grows, and the cluster gradually advances toward the onset of relativistic collapse. For the numerical example given above, the lifetime up to collapse is of the order of 10^8–10^9 years. Therefore, if dense clusters are indeed created by some process, they will not last for very long, and their collapse should be observed! (For further discussion see Zel'dovich and Podurets 1965; Fackerell, Ipser, and Thorne 1969.)

The physical picture of the collapse in this case is an increase of the core mass, accompanied by a transformation of formerly stable orbits into inward spirals, which lead the core and accompanying stars into their collective gravitational radius (cf. chapter 3). The noncircular orbits are important in this process because they connect the different layers of the cluster. A numerical investigation of such collapse is badly needed, but no one up to now has had the fortitude to attempt it.

---

5. It should be pointed out that the particles with enough energy to escape to infinity in a fully Maxwellian distribution constitute a small fraction, of the order of 0.01, of all the particles. The two processes of evaporation and overall relaxation to a Maxwellian distribution depend in the same manner upon N, R, M, etc., but since evaporation occurs about 100 times more slowly than relaxation, a truncated form of Maxwellian distribution actually exists.

13 PHYSICAL PROCESSES IN THE VICINITIES OF RELATIVISTIC OBJECTS AND A COMPARISON WITH OBSERVATIONS

13.1 ACCRETION OF GAS BY NEUTRON STARS AND COLLAPSED STARS

It is very difficult for us to discover what the processes are that produce the observed features of relativistic objects. This difficulty is due in part to our insufficient knowledge of these objects and of the processes which occur on their surfaces—processes that may have little connection with the factors determining the objects' general structure. Another source of difficulty is that relativistic objects do not reside in a vacuum; rather, they are surrounded by an interstellar medium which complicates all observable phenomena enormously.

The stars in galaxies are always surrounded by interstellar gas and dust. At certain stages in their evolution, stars emit matter, either in a continuous flux or by catastrophic explosions. Moreover, the composition of the galaxies may include matter which has never resided in the interior of any star. This matter has reached its current gaseous state as a result of the cosmological expansion of the primordial matter, which was nearly uniform and—according to Friedmann's cosmological solution—had a large density long ago.

Such are the sources of interstellar matter in the galaxies. Observations reveal that the mean density of interstellar gas in the spiral arms of our Galaxy is of the order of 10^{-24} g cm^{-3}; in the galactic center, it may be considerably higher.

Neutron stars and frozen (collapsed) stars can absorb matter from the interstellar medium; i.e., they can draw surrounding matter onto themselves. This process is called *accretion*. We begin our discussion of observable properties of relativistic objects with a treatment of the problem of accretion.

A principal incentive in the study of accretion is the energy release that it produces. Particles falling onto a neutron star hit its surface with kinetic energies equal to 20 or 30 percent of their rest mass-energies; thus the energy released per gram is many times greater than that produced by nuclear reactions. Particles falling onto a frozen star (i.e., "collapsed star" or "black hole") are accelerated to relative velocities approaching the velocity of light. A crucial question is: What fraction of their kinetic energy can be released by radiation to the outside? Possibilities of observing neutron stars and frozen stars are obviously related to these accretion phenomena. As was recently demonstrated by Cameron and Mock (1967), even for accretion onto as noncompact a star as a white dwarf, there is a possibility of producing a sizable flux of X-rays.

In performing qualitative estimates, one must keep in mind that neutron stars and frozen stars may be surrounded by gas of considerably higher density than the average interstellar density: the processes which form these stars may involve catastrophic phenomena, in the course of which part of the mass may get stripped from the surface and deposited in a gas cloud about the star. The energy released by accretion in turn influences the rate of accretion itself; the outpouring radiation interacts with the incident matter and slows it down (light pressure!). For a sufficiently great density of circumstellar matter, this phenomenon results in a self-regulation of the accretion.

Accretion can be analyzed systematically in two limiting approximations: as the motion of individual particles (atoms, molecules, dust particles), and as the motion of a continuous medium. Depending on the prevailing conditions, one approximation or the other may be valid. Magnetic fields can also be very important. The effect of the large-scale magnetic field of the star upon accretion will be discussed in subsequent sections, which deal with electromagnetic phenomena.

A curious, purely relativistic aspect of accretion has been computed by Doroshkevich (1965b): as we have discussed (§ 4.3), the gravitational field of a rotating body in the framework of GTR differs from the field of a nonrotating body of the same mass. As a result of that difference, a rotating body captures primarily those particles whose angular momenta are opposite in direction to its own. Consequently, the accretion of particles distributed isotropically in space will decrease the angular momentum of a rotating body.

The fundamental features of accretion were first analyzed in the nonrelativistic approximation at the end of the 1930's. This analysis remains adequate not only for white dwarfs but also for neutron stars: In a gravitational potential of $(0.2$–$0.3)$ c^2, when the stellar radius is three to four times greater than the Schwarzschild radius r_g, corrections for GTR may be as great as a factor of 2. In the presence of other indeterminacies (primarily with respect to the density of the incident gas), such corrections are not essential. However, the use of GTR is absolutely essential to obtain qualitatively correct conclusions for frozen stars. For better coherence of the exposition, we will begin by recalling some general, well-known facts.

13.2 THE INFALL OF NONINTERACTING PARTICLES

Imagine a large number of particles each with mass m and speed v_∞ far from a star; let their number density be n_∞. Let us consider first the Newtonian approximation. Let the speed v_∞ be small compared with the parabolic velocity at the star's surface, $v_\infty \ll v_p = (2GM/R)^{1/2}$. Then the maximum angular momentum permitting capture is $I_{max} \approx mv_pR$. We are interested in the impact parameter l_{max}, corresponding to infall tangentially onto the star's surface. Expressing it by $I_{max} = mv_\infty l$, we find $l_{max} = R(v_p/v_\infty)$. The flux of particles with $l < l_{max}$ is obviously $j = nv_\infty\pi(l_{max})^2$. Therefore, the rate of fall of mass onto the star is (with $\rho_\infty \equiv mn_\infty$)

$$dM/dt = m\,v_\infty\,n\pi R^2\,v_\infty^{-2}(2GM/R) = \pi r_g{}^2c\,\rho_\infty(c/v_\infty)(R/r_g) \ . \ (13.2.1)$$

The critical GTR value for the angular momentum in accretion onto a frozen star[1] is $2mcr_g$ (see § 3.9), which corresponds to a mass-capture rate of

$$dM/dt = 4\pi\,r_g{}^2c\,\rho_\infty(c/v_\infty) \ . \tag{13.2.2}$$

Thus, one cannot use the Newtonian formula if $R/r_g < 4$. Expression (13.2.2) gives a lower limit on dM/dt and describes accretion not only onto frozen stars but also onto very dense neutron stars—those with $4 > R/r_g > 1.7$ (1.7 is the minimum possible value according to Tsuruta and Cameron 1966). Rewriting expression (13.2.2) in a more convenient form, we have

$$\frac{d(M/M_\odot)}{d(t/10^{10} \text{ years})} = 10^{-13}(\rho_\infty/10^{-24} \text{ g cm}^{-3})(M/M_\odot)^2 \\ \times (v_\infty/10 \text{ km sec}^{-1})^{-1} \ . \tag{13.2.3}$$

There is no reasonable (stationary) answer for particles which are strictly at rest at infinity; if, in fact, a massive object were suddenly put into a region occupied by particles at rest, then the particles would begin to fall radially inward; and one can easily verify that the flux onto the surface would begin to grow according to

$$dM/dt = (64/3\pi)\,\rho_\infty\,GMt \ . \tag{13.2.4}$$

But let us treat the stationary case. How will the density of noninteracting particles vary with distance from the star? Consider the case $r \gg R$; i.e., neglect the removal of particles by collision with the star's surface. Those particles which have an isotropic distribution of velocities at infinity and have energies between E_∞ and $E_\infty + \Delta E$ make up a microcanonical ensemble. According to Liouville's theorem, the volume per particle in phase space is constant: $\Delta\Gamma = 4\pi n^{-1}p^2\Delta p = $ const. Taking $E = p^2(r)/2m + m\phi(r) = $ const.; $\Delta E = p\Delta p = $ const., we find

$$n = n_\infty(p/p_\infty) = n_\infty(1 + GMm/r\,E_\infty)^{1/2} \ . \tag{13.2.5}$$

Let r_c be the "critical radius" at which the kinetic energy of a particle becomes of the order of its gravitational potential energy. Then, according to the above formula, the density is virtually constant for $r > r_c$; but for $r < r_c$ it begins to increase as $n \propto r^{-1/2}$.

---

1. The influence of gravitational radiation on the capture efficiency is proportional to m/M (§ 3.11) and is therefore negligible.

It is important to notice that formula (13.2.5) does not give the full answer to our question of the particle density around the star. In particular, that formula applies only to particles which come in from infinity, i.e., for $E > 0$. However, there can be an unlimited number of particles with $E < 0$ around the star if collisions are ignored. For these particles, moving along elliptic-type orbits, the density is an initial condition which can be specified arbitrarily and independently of equation (13.2.5).

The problem can be defined more concretely in the following manner. Introduce the cross-section for interaction between particles, σ. If σ is small enough, then the distribution of particles with $E > 0$ will be completely determined by the particle distribution at infinity. Moreover, in a steady state with $\sigma \to 0$, the only particles which can hit the star's surface are those with $E > 0$; particles with $E < 0$ and with impact parameters sufficiently small to hit the star will have all been wiped out by the time a steady state is achieved. Consequently, in the limiting case $\sigma \to 0$ the particle flux is given correctly by expression (13.2.1) or (13.2.2).

However, if $\sigma \neq 0$, the flux dM/dt will be different, because collisions will bring the tangential velocities of particles to zero and thereby enable them to fall onto the star. We shall not seek the solution to the problem for arbitrary σ; instead, we shall consider only the opposite limiting case of very large cross-sections, for which the gas particles can be described as a hydrodynamical fluid. In this case the hydrodynamical equations can be used to describe not only accretion of matter onto a star but also the outflow of matter from a star.[2]

13.3   Four Regimes of Hydrodynamical Flow in the Case of Spherical Symmetry

In the hydrodynamical approximation (σ large) we shall describe the infalling gas by the following characteristics: adiabatic index γ, density ρ, pressure P, and sound velocity $a = (\gamma P/\rho)^{1/2}$. Consider a narrow cone with solid angle $d\Omega$ and with apex at the star's center; and write the law of matter conservation for spherically symmetric flow in the form (dS is the flux in $d\Omega$):

$$dS = \rho u r^2 d\Omega = \text{const.} ; \quad u = A/\rho r^2 ;$$
$$A = dS/d\Omega = (1/4\pi)(dM/dt) . \tag{13.3.1}$$

The Bernoulli law, which expresses energy conservation, reads:

$$-GM/r + u^2/2 + [\gamma/(\gamma - 1)](P/\rho) = \text{const.} = B . \tag{13.3.2}$$

Equation (13.3.2) assumes the absence of heat exchange between fluid elements; however, heat exchange can be taken into consideration by suitably modifying γ. The constant B is determined by the boundary conditions; in

2. The outflow of noninteracting particles is of no particular interest.

the case of accretion it is most convenient to specify them at infinity where the gas is at rest:

$$B = [\gamma/(\gamma - 1)](P_\infty/\rho_\infty) . \qquad (13.3.3)$$

However, for outflow it is better to specify them near the star's surface:

$$B = -GM/R + u_R^2/2 + [\gamma/(\gamma - 1)](P_R/\rho_R) . \qquad (13.3.4)$$

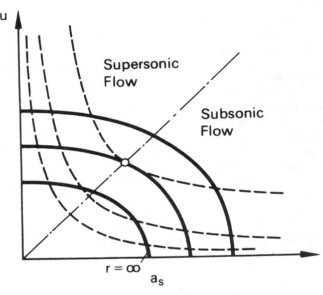

Fɪɢ. 69.—Hydrodynamic accretion. Horizontal coordinate a_s is the velocity of sound in the matter; and u is the radial velocity of the matter. Solid lines are a family of ellipses on which the Bernoulli equation is satisfied; each ellipse corresponds to a particular radius r. Dashed lines are a family of ·hyperbolalike curves, on which the equation of continuity is satisfied for a given r.

In the (a, u)-plane the Bernoulli equation is characterized by the set of ellipses

$$u^2/2 + [1/(\gamma - 1)]a^2 = B + GM/r . \qquad (13.3.5)$$

For given boundary conditions B and a given radius r, u and a must lie somewhere on a particular ellipse. Similarly, the continuity equation is characterized by the set of hyperbolae of fractional power

$$u = (A/\rho_i r^2)(a_i/a)^{2/(\gamma-1)} . \qquad (13.3.6)$$

(This equation follows from combining equation (13.3.1) with the adiabatic law relating sound velocity to density, $a \propto \rho^{(\gamma-1)/2}$.) These ellipses and hyperbolae are shown in Figure 69.

The hyperbolae depend on the parameter A, which is determined for outflow by equations (13.3.1) and (13.3.4), that is, by the conditions near the star's surface. The situation for accretion will be discussed in greater detail below.

It is easy to see[3] that if, for particular boundary conditions and a particular radius, there are two intersection points of the corresponding ellipse with the hyperbola (cf. Fig. 69), then the lower intersection corresponds to subsonic flow and the upper to supersonic flow. If the two curves meet just once, then their intersection point corresponds to $u = a$. If, finally, the curves do not cross at all for any r with given boundary conditions (i.e., a given choice of A), then the mass inflow A has been chosen too large and such flow cannot occur.[4] Excluding this possibility, we have two possibilities depending upon A: (1) the curves cross each other twice (above the 45° line of Figure 69, and beneath it) for every value of r except one, $r = r_s$, where the curves meet tangentially; (2) the curves cross each other twice for every value of r, with no exception.

There are two types of flow corresponding to each picture. Let us enumerate them taking into account that the flow at infinity is necessarily subsonic in the case of accretion (because $u \to 0$) and is necessarily supersonic in the case of outflow (because there $a \to 0$).

Outflow

The two curves meet tangentially if the radius is $r = r_s$; if $r < r_s$, then $u < a$; if $r > r_s$, then $u > a$. The inner layers of the star are at rest; the outer layers move with subsonic velocities; the velocity becomes supersonic at the radius $r = r_s$; and at infinity supersonic flow occurs. This type of outflow can occur if the star has sufficiently large entropy (Bisnovatyi-Kogan and Zel'dovich 1966) and if the star has a corona of the solar type (see Parker 1965b; Noble and Scarf 1965); it can also result from rotational shedding of matter (see § 11.18).

Accretion

The curves intersect tangentially for the radius $r = r_s$; if $r > r_s$, then $u < a$; if $r < r_s$, then $u > a$. The gas is at rest at infinity, and far from the star there is subsonic flow. The point r_s is a transition point between subsonic and supersonic flow; the flow is supersonic near the star. This type of flow will be considered in detail below.

Ejection (Supersonic Outflow Everywhere)

For no value of r do the hyperbola and ellipse of Figure 69 intersect tangentially; for all values of r, $u > a$. Such a type of outflow can occur only as a result of non-hydrodynamical processes of particle acceleration, which

3. The following account is based on the assumption that $1 < \gamma < \frac{5}{3}$; for a discussion of this assumption, see below.

4. This situation can occur only for the case of accretion, i.e., when condition (13.3.3) is used

are not contained in our equations. Such acceleration might be connected, for example, with the star's vibration (Cameron 1965*b*) or with special pulsar processes (Gold 1968), which act near or on the star's surface. This type of flow can be complicated by a breakdown, at all radii, in the conditions for the applicability of the hydrodynamical approximation, and by important effects of magnetic fields which are frozen into the outflowing matter, and which thereby link fluid elements to each other.

Settling (Subsonic Accretion Everywhere)

In this case, again, for no value of r do the curves meet tangentially; but, by contrast with the last case, $u < a$ everywhere. Such flow is possible only if the pressure is sufficiently high near the star's surface.

Let us return to supersonic accretion. To be sure that inflow leads to accretion ($u > a$ at $r < r_s$), we must demand not only the existence of an r_s at which ellipse and hyperbola touch; we must also insist that processes near the star not be able to react back on events near $r = r_s$. Let H be the scale-height of the star's atmosphere, as defined by

$$H = N_0 k T R^2 / G M \mu \,. \qquad (13.3.7)$$

Here N_0 is Avogadro's number, and μ is the mean molecular weight. If

$$H + R < r_s \,, \qquad (13.3.8)$$

then processes near the star do not affect events near r_s, and a transition to supersonic flow does occur at r_s.

13.4 THE CASE OF ACCRETION AND THE VALIDITY OF THE HYDRODYNAMICAL APPROXIMATION

Let us consider the case of accretion. If $r_s \gg R$, then supersonic flow near the star requires a temperature there which satisfies (cf. eq. [13.3.8])

$$T \ll 10^7 (M/R)(R_\odot/M_\odot) \,. \qquad (13.4.1)$$

It will be shown below that the inequality $r_s \gg R$ is readily satisfied by a star with $R \lesssim R_\odot$ and with realistic conditions at infinity. This means that accretion onto a sufficiently dense star must always lead to supersonic flow near the star and a maximal influx of matter.[5] Let us treat this situation in greater detail.

It is easy to show by using equations (13.3.3), (13.3.5), and (13.3.6) that the gas density remains approximately constant from $r = \infty$ down to $r = r_c$, the critical radius where the gravitational potential becomes of the order of a_∞^2:

$$r_c = \delta(\gamma) \, GM/a_\infty^2 \,,$$

$$\delta(\gamma) = \tfrac{1}{2}[2/(5 - 3\gamma)]^{(5-3\gamma)/(3\gamma-3)} \,, \quad \delta(\tfrac{5}{3}) = \tfrac{1}{2} \,, \quad \delta(\tfrac{4}{3}) \approx 1 \,. \qquad (13.4.2)$$

5. For a discussion of the temperature which is developed during accretion onto a neutron star, see § 13.5.

13.4 Validity of the Hydrodynamical Approximation

Recall that a similar situation occurred for noninteracting particles (eq. [13.2.5]). However, at $r < r_c$ the density changes as $\rho \sim r^{-3/2}$ in this case, whereas it changed as $\sim r^{-1/2}$ in the case of noninteracting particles. The $r^{-3/2}$ law results from combining the continuity equation $\rho u r^2 = \text{const.}$ with the radial variation of velocity, $u \sim r^{-1/2}$. (The velocity is practically parabolic, $u \sim r^{-1/2}$ at $r < r_c$, because below $r = r_c$ there is no means for supporting the gas; the pressure is "turned off.")

The radius r_s at which transition to supersonic flow begins is given by

$$r_s = \frac{5 - 3\gamma}{4} GM/a_\infty^2 . \qquad (13.4.3)$$

At this transition radius

$$u_s = a_s = a_\infty [2/(5 - 3\gamma)]^{1/2} . \qquad (13.4.4)$$

To evaluate the radii r_c and r_s numerically, we must know the conditions "at infinity," i.e., at $r \sim 2r_c$. The energy released by the accretion of gas onto a fairly dense star (neutron star or white dwarf) produces ionization out to a distance from the star of roughly 1 parsec (this will be discussed below). Since this distance turns out to be far outside r_c, the matter "at infinity" is ionized, and we must take as the temperature at infinity the super-ionization value $T_\infty = 10^4$–$10^5 \,^\circ$ K.[6] From the above discussion and formulae we then find that

$$r_c \approx 10^{13}\text{--}10^{14} \text{ cm} , \quad r_s \approx 10^{12}\text{--}10^{13} \text{ cm} . \qquad (13.4.5)$$

Here we have assumed $M = M_\odot$ and $\gamma = 1.5$.

Recall that one of the necessary conditions for accretion is $r_s \gg R$. This will be satisfied if $\frac{5}{3} - \gamma \gg 10^{-5}$ in the case of a white dwarf and if $\frac{5}{3} - \gamma \gg 10^{-8}$ for a neutron star. Of course, such inequalities always hold.

By using equations (13.4.3) and (13.4.4), one can readily calculate the flux of matter onto the star

$$\frac{dM}{dt} = 4\pi r_s^2 u_s \rho_s = a(\gamma) r_g^2 c \rho (m_p c^2/kT_k)^{3/2}$$

$$a = \frac{\pi}{4\gamma^{3/2}} \left(\frac{2}{5 - 3\gamma} \right)^{(5-3\gamma)/[2(\gamma-1)]} \qquad (13.4.6)$$

In this formula the quantities ρ, T, and γ should be evaluated near the critical radius r_c. Notice that dM/dt has a rather weak dependence upon γ: $a(\gamma \to 1) \approx 1.5$; $a(\gamma \to \frac{5}{3}) \approx 0.3$; $a(\frac{4}{3}) \approx 1.4$. The singularities in the formula (13.4.6) at $\gamma = 1$ are not physical; the inequality $\frac{5}{3} - \gamma > 0$ always holds. The expression (13.4.6) for the rate of accretion differs significantly from that for the case of noninteracting particles (see § 13.2). For example, the present formula does not contain the radius of the star; this is quite natural because the capacity of the star to accrete matter is governed by the

---

6. Some authors assume $T_\infty = 100^\circ$ K; however, this is invalid because the role of ionization is always essential.

surface area at $r = r_c$, where the "bottom drops out from under the gas" and the gas begins to achieve a parabolic velocity of inflow.

The rate of accretion in the present case is greater than that in the case of noninteracting particles (eq. [13.2.3]) by a factor of $(c/v_\infty)^2 \sim 10^9$! The physical cause of this enormous difference is obvious: in the gas there are frequent collisions between atoms. These collisions limit the increase of the tangential velocities of infalling atoms, but have no significant effect on their velocities of radial infall.

This difference of up to nine orders of magnitude is so large that we should devote several pages to an investigation of the validity of the hydrodynamical approximation. The analysis which follows is due largely to our colleague in Moscow, V. Schwartzman.

It is well known that the mean free path of a particle undergoing Coulomb collisions in a plasma is

$$l = \frac{(kT)^2}{ne^4 L_{\text{Coul}}} \approx 10^{12}(T/10000°)^2(\text{cm}^{-3}/n) . \tag{13.4.7}$$

Here e is the electronic charge and L_{Coul} is Coulomb's logarithm. From this formula it is clear that at $r > r_c$ the mean free path of a particle is always less than or of the order of the region of interest, and, consequently, the hydrodynamical approximation is valid.

One can show easily that at $r < r_c$, $l/r \sim r^{-3(\gamma-7/6)}$. Thus, the validity of the hydrodynamical approximation at $r < r_c$ depends critically on whether $\gamma < 7/6$ or $\gamma > 7/6$. The value of γ is linked directly to the change in temperature during free fall. To find the nature of this link we rewrite the first law of thermodynamics, $dE = -PdV + dQ$ in the form

$$\tfrac{3}{2}R^* \frac{dT}{dt} = R^* \frac{T}{\rho} \frac{d\rho}{dt} - 5 \times 10^{20}T^{1/2}\rho\beta + \frac{dQ_i}{dt} . \tag{13.4.8}$$

Here R^* is the universal gas constant, $R^* = 8.3 \times 10^7$ ergs per mole per degree. The second term on the right-hand side describes the energy loss of 1 gram of plasma due to radiation; $\beta = 1$ for free-free radiation of a completely ionized plasma; the third term describes the energy change due to the interaction between the falling matter and photons coming up from below, and also due to the ionization of heavy elements by collisions with electrons, and to other nonadiabatic processes. In studying equation (13.4.8) it is convenient to reexpress ρ in terms of the total luminosity produced by the accretion and the efficiency ϕ with which the accretion energy is turned into that luminosity:

$$\frac{dT}{dr} = -\frac{T}{r} + 1.5 \times 10^3 \left(\frac{L}{L_0}\right)\left(\frac{0.1}{\phi}\right)\left(\frac{M_\odot}{M}\right) \sqrt{T} \frac{1}{r} \beta - \frac{dQ_i/dt}{u_{\frac{2}{3}}^2 R^*} . \tag{13.4.9}$$

Here L_0 is the maximum luminosity due to accretion that can be achieved by a star with $M = M_\odot$ ($L_0 = 1.3 \times 10^{38}$ ergs sec^{-1}, cf. § 11 of chapter 11);

ϕ is measured in units of c^2 ($0.1\ c^2$ is the typical magnitude of the gravitational potential near the surface of a neutron star, cf. the next section); and $u \equiv -dr/dt \approx 1.2 \times 10^{13}\ r^{-1/2}\ (M/M_\odot)^{1/2}$.

In our search for the link between γ and the temperature change during free fall, let us consider first the case in which one can neglect the role of the last term of equation (13.4.9). In this case the falling gas will cool due to radiation reaction if the temperature "at infinity" is small enough (i.e., if the second term on right-hand side of eq. [13.4.9] dominates). In this case, the cooling law will be $T \approx (A \ln (r/r_0) + T_0^{1/2})^2$ initially; however, since $T_0 \sim 10000°$ K, recombination will quickly come into play, and the temperature will level off at a constant value, $T = \text{const.}$, corresponding to $\gamma = 1$. On the other hand, if the temperature "at infinity" is large, then the role of radiation reaction will diminish continuously during the fall, the first term of equation (13.4.9) will dominate, and the temperature change will approach $T \sim 1/r$, corresponding to $\gamma \to \frac{5}{3}$. The critical temperature separating these two extreme cases is

$$T_k = 2 \times 10^6 (L/L_0)^2 (0.1/\phi)^2 (M_\odot/M)^2 \beta^2\ °\text{K} . \qquad (13.4.10)$$

The existence of two limiting cases, together with the great difference between the values of γ for those cases, shows the importance of taking into account the last term in equation (13.4.9). The general situation, unfortunately, has not yet been investigated. However, by limiting ourselves to a consideration of Compton scattering between the plasma's electrons and the outward-flying photons, we can write the last term of equation (13.4.9) in the form (Zel'dovich and Shakura 1969)

$$-\frac{dQ_i/dt}{\frac{3}{2}R^*u} = 2 \times 10^6 \left(\frac{L}{L_0}\right) (T - T_p) \left(\frac{1}{r^{3/2}}\right), \qquad (13.4.11)$$

where T_p is the photon temperature. By inserting this term into equation (13.4.9) one readily sees that the temperature is practically constant for $r < 3 \times 10^{12}\ (L/L_\odot)^2$, and is equal to T_p. Other processes will obviously have only weak influence on this conclusion, but they will be important in determining what happens at larger radii.

We should note, however, that there is probably a justification, independent of the above considerations, for idealizing the accreting matter as a fluid. This justification is the existence of a magnetic field, frozen into the plasma. More particularly, the Larmor radius for a proton moving with thermal velocity turns out to be smaller than the characteristic length scale of the infalling matter whenever $H > 1.3\ T^{1/2}\ r^{-1}$ gauss. Taking account of the fact that $T \leq T_\infty(r_c/r)$, and assuming $T_\infty = 10000°$ K, $r_c = 10^{14}$ cm, we can rewrite the condition for linkage of the protons to the field lines as

$$H > 10^{-12}(r_c/r)^{3/2}\ \text{gauss} . \qquad (13.4.12)$$

If the field lines were strictly frozen into the plasma, the flow would be smooth and the component of the field perpendicular to the motion would decrease as $H_\phi \sim r^{1/2}$ (if the fall is assumed to be free). This cannot be so, however, because then the radial component of the field would increase as $\sim r^{-2}$, and the field energy $\epsilon = H^2/8\pi$ would quickly become larger, with decreasing radius, than the potential energy of the gas. It is more natural to assume that the energy of the field would be of the order of the thermal energy of the plasma because of the existence of turbulence (Batchelor's theorem); in this case $H \sim r^{-3\gamma/4}$; in any case, the field energy cannot rise faster than the potential energy of the particles (corresponding to $H \sim r^{-5/4}$).[7] However, $H \geq 10^6$ gauss and, therefore, inequality (13.4.12) is valid for all $r;$ the protons are everywhere linked to the field lines; and the hydrodynamical approximation is probably justified.

In the case of fully developed turbulence and of $H^2/8\pi \sim nkT$ the effect of the field on the rate of fall is negligible.

Are there any situations in which frozen-in fields change the picture of accretion? Can one assert for certain that frozen-in fields are never of much importance? These questions are open.

13.5 THE LUMINOSITY DUE TO SYMMETRIC ACCRETION ONTO NEUTRON STARS AND WHITE DWARFS

Let ϕ be the gravitational potential near the surface of the star. By using equation (13.4.6), we can write the luminosity of the star due to accretion in the form

$$L = \phi \frac{dM}{dt} \approx 2 \times 10^{31}(\phi/0.1c^2)(M/M_\odot)^2$$
$$\times (10000°/T_c)^{3/2}n_c \text{ ergs sec}^{-1} . \tag{13.5.1}$$

Here n_c is the number density of interstellar gas atoms in units per cubic centimeter, T_c is the gas temperature (both quantities are measured in the neighborhood of $r = r_c$), and γ is taken equal to $\frac{4}{3}$.

According to equation (13.5.1) the intensity of the radiation produced by accretion onto a star depends on the temperature of the interstellar plasma near the radius r_c at which the infall of matter begins. However, the plasma temperature in turn depends upon the stellar luminosity—i.e., upon the rate of accretion. As a consequence, a peculiar "negative feedback" arises, and diminishes the dependence of the accretion upon the boundary conditions (Schwartzman 1970a). To clarify this point let us introduce the parameter $a \equiv T_c/T_b$ (where T_b is the star's bolometric temperature, $T_b = (L/4\pi\sigma R^2)^{1/4}$), and let us rewrite equation (13.5.1) in the form

$$L \approx 4 \times 10^{29}a^{-12/11}n_\infty{}^{8/11} \text{ ergs sec}^{-1} . \tag{13.5.2}$$

---

7. A definite role can be played by the violation of the frozen-in condition as well as by the annihilation of field lines at knot points due to the mixing up of lines of force and also to gas compression.

Here we have assumed $M = M_\odot$ and $\phi = 0.1\ c^2$ (corresponding to a neutron star). The problem of determining the parameter a is nontrivial.

As has been shown by the work of Zel'dovich and Shakura (1969), the accretion causes the stellar atmosphere to split up into a thin, hot layer where the braking of the infalling matter occurs, plus a comparatively cold inner region with $T = T_b$. Zel'dovich and Shakura calculated the temperature T_e of the thin layer by means of energy balance and found it to be of the order of $10^8\,°$–$10^9\,°$ K. Before their analysis one could only allege that $T_b < T_e < T_{\text{shock}}$, where T_{shock} is the value found from the Hugoniot adiabat. These results show that the radiation spectrum should consist of a superposition of two sources: an approximately equilibrium (i.e., blackbody) one with $T = T_b$, and a bremsstrahlung source with $T \approx T_e$.[8] The ratio of the luminosities of these two sources depends upon the parameters mentioned above; according to the calculations of Zel'dovich and Shakura, for accretion onto a neutron star with total luminosity $L \approx 10^{36}$–10^{37} ergs sec^{-1}, with a path length for the infalling particles of $y_0 = 2$–20 g cm^{-2}, and with $T_e \sim 10^8\,°$–$10^9\,°$ K, the ratio of luminosities L_e/L_b is of the order of $1/100$.

The motion of the star relative to the interstellar gas will also influence the plasma temperature near the critical radius. Asymmetry does not lead, however, to any significant change in the flux of mass onto the star (see § 13.7). The temperature is affected because the motion prevents there from being sufficient time for ionization equilibrium to be reached deep down in the ionization zone.

On the other hand, the fact that the typical velocities of stars are close to the sound velocity in the gas ($10000°$ K corresponds to $a \approx 10$ km sec^{-1}) prevents a significant density gradient—in contrast to the temperature gradient—from arising in the ionization zone. Put differently, the motion causes $n_e \sim n_\infty$. It has been shown by Schwartzman (1970a) that $a \sim 0.1$–10 depending on the density of the interstellar gas ($n \approx 0.1$–3 cm^{-3}), the role of the hot thin layer in producing radiation ($\eta = 1$–10^{-5}), its temperature ($T_e = 10^7\,°$–$10^9\,°$ K), the star's velocity ($v = 10$–100 km sec^{-1}), and so on. The corresponding value of L is 10^{28}–10^{31} ergs sec^{-1}.

Thus, a single neutron star, without a magnetic field, in a state of accretion radiates most of its energy in the ultraviolet range $\lambda_{\max} \sim 150$–900 Å. For peculiarities in the accretion onto neutron stars with magnetic fields see § 13.10.

Consider now the infall of matter onto a white dwarf ($M = M_\odot, R = 10^9$ cm, $\phi = 10^{-4}\ c^2$). In this case the total luminosity L_{tot} due to accretion comes from the thin, hot layer ($T_e \sim 10^7\,°$–$10^8\,°$ K), $\eta \equiv (L_e/L_{\text{tot}}) = 1$–$10^{-2}$ corresponds to $L_{\text{tot}} \sim (10^{26}$–$10^{27})n_\infty$ ergs sec^{-1}; and $\eta \lesssim 10^{-3}$ corresponds to $L_{\text{tot}} \sim 10^{28}\ n_\infty$ ergs sec^{-1}. The corresponding bolometric temperatures are

8. The existence of these two sources was first discussed by Cameron and Mock (1967). A definite role can also be played by Compton scattering of photons which emerge from the lower layers (Zel'dovich and Shakura 1969).

very low, $T_b \sim 500°$–$2000°$ K, so the radiation is in the infrared region. However, to distinguish such a component of luminosity from the normal thermal luminosity of a warm white dwarf would seem to be impossible; one could record, in principle, only the X-radiation from the thin, hot layer.

13.6 THE PROBLEM OF DISCOVERING COLLAPSED STARS

The story is entirely different when gas is accreted by a frozen star. The gas motion is similar in its properties to particle motion in the Schwarzschild gravitational field. Recall that falling particles reach r_g in finite proper time; from the viewpoint of the particles the moment of crossing r_g is by no means special. If a particular particle is followed by another particle and if the distance between them was finite somewhere far from r_g, then it will also remain finite in the neighborhood of r_g. From this, we draw the following conclusion for a stream of particles, i.e., for a gas: in a reference frame that comoves with the gas, the density remains finite. In order of magnitude, the density at the gravitational radius is $\rho_m = \rho_\infty (r_c{}^2 a_\infty / r_g{}^2 c)$—or, equivalently,

$$\rho_m \cong \rho_\infty (c/a_\infty)^3 .$$

When $a_\infty = 10$ km sec^{-1} and $\rho_\infty = 10^{-24}$ g cm^{-3}, we find that $\rho_m \sim 3 \times 10^{-11}$ g cm^{-3}.

This is the density as measured in a frame comoving with the gas. In a stationary frame (one with $r = $ const., in which the gas moves with velocity u) the Lorentz contraction makes the density larger:

$$\rho_{\text{stationary}} = \frac{\rho_m}{(1 - u^2/c^2)^{1/2}} = \frac{\rho_m}{(1 - r_g/r)^{1/2}} .$$

This density diverges as $r \rightarrow r_g$ because the gas velocity approaches the speed of light there (infinite Lorentz contraction).

As viewed by distant observers, the particles only approach asymptotically to r_g after infinite time. No matter how long a distant observer watches the stationary flow onto a frozen star, he will reason that none of the particles which passed him long ago have reached r_g. They all appear to accumulate in a domain adjacent to r_g. In a precise stationary solution for the gas flow, the total number of particles contained between two spheres $r = R_1$ and $r = R_2$ diverges as the lower limit approaches the gravitational radius, $R_1 \rightarrow r_g$:

$$V = 4\pi \int_{R_1}^{R_2} r^2 \sqrt{g_{11}}\, dr , \qquad N = \int_{R_1}^{R_2} \frac{\rho_m}{(1 - r_g/r)^{1/2}}\, r^2 \frac{dr}{(1 - r_g/r)^{1/2}} .$$

The denominator in the integrand diverges as $r - r_g$, so the integral diverges logarithmically. From the viewpoint of the distant observer, the flux can only asymptotically approach a stationary state.[9] The capacity of the layer adjacent to r_g in a stationary solution is infinite; to fill it, an infinite time is required.

Can the radiation emitted by gas falling onto a frozen star at rest reveal its location? This question has been considered by Schwartzman. By virtue of the results of § 13.4, one can neglect the role of radiation reaction in braking the infalling matter if the initial temperature is high enough; in this case the temperature rises during infall with the law $T \sim r^{-1}$. If the initial temperature is small, then energy release during infall will hold $T \approx$ const. $\approx (5-10) \times 10^3$ ° K. This temperature is the value needed for partial ionization of the gas by collisions. From equation (13.4.9) it is easy to verify that heating of the gas during infall occurs in accretion onto a body with mass

$$M < M_{\mathrm{cr}} \approx 10^5 M_\odot n_\infty^{-1} (T'/10000°)^2 . \tag{13.6.1}$$

In the case of matter falling onto an object more massive than this, the condition $T \approx$ const. will be valid. In formula (13.6.1) T' is the gas temperature at the distance of interest; because of the slow increase of T' as one moves in toward the star, when $M \approx M_{\mathrm{cr}}$ the heating of the gas will begin at some particular radius. In the case $M > M_{\mathrm{cr}}$ the luminosity of the object is

$$L = \frac{dM}{dt} \, \mathfrak{R} \int_{r_1}^{r_c} T \, \frac{dn}{n} \approx 10^{34} \left(\frac{M}{10^5 \, M_\odot}\right)^2 \left(\frac{10000}{T_c}\right)^3 n_\infty \text{ ergs sec}^{-1} . \tag{13.6.2}$$

Here r_1 is the radius of the "last ray" (see § 7 of chapter 11), $r_1 \approx r_g$. If $M < M_{\mathrm{cr}}$, then the greater the depth the higher the temperature; and the luminosity of the spherical layer between $\frac{1}{2}r$ and r is given by

$$\begin{aligned} L &\approx r^3 \times 10^{-27} n^2 T^{1/2} \\ &\approx 10^{24} (M/M_\odot)^3 (10000°/T_c)^3 (T/10^{12})^{1/2} n_\infty^2 \text{ ergs sec}^{-1} . \end{aligned} \tag{13.6.3}$$

In this formula we have assumed that the most probable radius for partially ionizing the gas is $r = r_c$ for parameters of interest. The magnitude of T is limited: $T < T_1 = T(r_1)$; T_1 is certainly less than 10^{12} ° K; more probably it is $T_1 \sim 10^{10}$–10^{11} ° K. In the limiting case $\gamma = \frac{5}{3}$ the temperature which develops during the adiabatic compression corresponds in order of magnitude to the kinetic energy of infall, and is thus approximately equal to the temperature of the shock wave which would arise if the gas were to collide

9. The time required to approach the stationary state with given accuracy at a given point of space is greater, the closer this point is to r_g.

with an object at rest. Consequently, the maximum temperature which one calculates without taking radiation into account for the case of gas falling onto a frozen star is the same as that for the case of gas falling onto a neutron star, although in the neutron-star case the gas is brought to rest by a shock wave whereas in the frozen-star case it falls without stopping.

The difference in radiation output in the two cases depends not upon the difference in temperature (the temperatures are about the same) but upon the difference in optical depth: Near a neutron star the hot gas is gathered into a thick layer, but near a frozen star the gas moves into a region of large redshift from which only a small amount of radiation can escape.

It is curious that if $M > M_{\mathrm{cr}}$, then practically all the radiation is in the optical region of the spectrum; but if $M < M_{\mathrm{cr}}$, then only $\sim 10^{-3}$–10^{-4} of the total power output is in the optical region; the main portion of the output is in X-rays and gamma rays.

By combining the expression $L = 10^{38}(M/M_\odot)$ ergs sec^{-1} for the maximum luminosity from accretion (see § 11.11), with formula (13.6.2) for the actual luminosity, one can easily eliminate the mass M and thereby obtain an expression for the maximum luminosity at which radiation pressure on the infalling matter begins to impede the accretion:

$$L_m = 3 \times 10^{47}(T_c/10000°)^{3/2}(10000°/T_1)^{1/4}n_\infty^{-1} . \qquad (13.6.4)$$

The corresponding mass is $M \sim 10^9\, M_\odot$ and the accretion rate is $dM/dt \sim 10^4\, M_\odot$ year^{-1}. For frozen stars more massive than this, the luminosity can increase only very slowly with increasing density of matter at infinity, because the accretion is regulated by light pressure; while for frozen stars less massive (but $\gtrsim 10^4\, M_\odot$) the luminosity decreases catastrophically, tending to the value (13.6.2) (see § 13.9). It is intriguing that the luminosity (13.6.2) is close to the observed luminosities of quasars; the typical estimate of the lifetime is also close to the quasar value of $T \sim 10^5$ years. Notice that the dependence of L_m on the change in temperature during infall is weak.

Can one have confidence in the idealized picture of spherically symmetric accretion described above? Might there be other conditions under which the infall of matter onto a frozen star with $M \sim M_\odot$ will give a greater energy release than is predicted by formula (13.6.3)? The next section is devoted to these questions.

13.7 THE CASE OF ASYMMETRIC GAS FLOW

One obtains a completely different picture even in the case of the infall of noninteracting particles in a Newtonian field, when the mean velocity of the particles relative to the star is nonzero at infinity. The particles approach a spherical surface of radius R at a variety of different angles; the closer a particle's impact parameter is to the maximum one, $l_{\max} = Rv_p/v_\infty$ (see

§ 13.2), the greater is the angle between its trajectory and the normal to the surface.[10]

Expression (13.2.1) for the flux in the noninteracting case is valid here. However, the introduction of a nonzero cross-section, $\sigma \neq 0$, changes the situation. Collisions between particles moving along hyperbolic orbits lead to tangential-velocity losses, to the emission of energy, and to infall onto the star (Fig. 70). The ratio of the energy that escapes from the star as a result of collisions divided by the energy lost into the star increases with increasing σ and is small for small σ.

Visualize a frozen star immersed in gas, which is considered a continuous medium (σ large). Contrary to the postulate of the preceding section, assume that at infinity the gas moves with respect to the star with a velocity

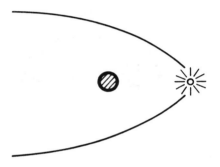

Fig. 70.—Collision of particles with opposite momenta in the field of an attracting center.

u_∞. Consider first the limiting case $u_\infty \gg a$: in the presence of this ordered motion (with velocity u_∞), a consideration of the individual motions of the gas molecules is no longer necessary. Since we can neglect the velocity of sound by comparison with the velocity u_∞, we can also neglect the pressure. However, in the absence of pressure, the motion of a continuous medium does not differ essentially from the motions of solid particles; the equations of hydrodynamics merely represent the equations of particle mechanics in a different notation. In the stationary problem, the flow lines are identical with the trajectories of freely falling particles. Salpeter (1964) has suggested the following general representation of the motion (Fig. 71): gas flows from left to right in the gravitational field, the flow lines become curved, and the gas velocity increases according to the law of energy conservation (Bernoulli equation).

At the tail of the flow, adjoining the star's surface (or the Schwarzschild

10. It is curious that the number of particles crossing a unit surface in a field $\phi = $ const./r is, nonetheless, independent of the crossing angle (for $v_\infty \ll v_p$). The proof of this statement is trivial, it holds in Newtonian theory.

sphere), is the elongated shock front of a shock wave. The shock wave replaces the collisions of particle pairs described above. Upon crossing the shock front, the gas loses the velocity component which is perpendicular to the front. The velocity component that is parallel to the front, i.e., directed radially, remains unchanged.

Applying Kepler's laws, we can find the critical trajectory (*dashed line*) and the critical impact parameter l_k which determine the fate of a gas element. If $l > l_k$, then after a gas element has passed through the shock

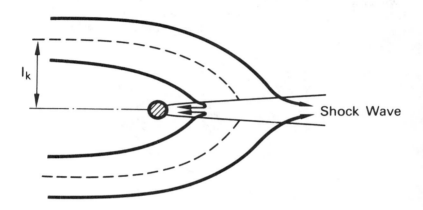

FIG. 71.—Hydrodynamic diagram of accretion. In the shock wave the particles lose the velocity component which is perpendicular to the shock front. For impact parameters less than l_k, the component of radial velocity at the shock front is less than parabolic, and the particle falls into the attracting center.

wave its velocity remains greater than parabolic, and it flows out to infinity; if $l < l_k$, it falls onto the star after passing through the shock wave. It is evident that l_k and, accordingly, the rate of accretion depend on the gas velocity u_∞, just as these quantities depended on the sound velocity in the spherical problem (see above). The important difference from the spherical problem is that in the shock wave there occurs an effective reprocessing of the kinetic energy into heat and light. Salpeter draws the following conclusions: A body, moving at supersonic speed relative to the gas, decelerates in a time so short that accretion cannot appreciably change its mass. Therefore, we should consider motion of the body relative to the gas with a velocity of the order of the sound velocity, i.e., of the order of the chaotic velocity of the molecules and gas clouds. In this case

$$\frac{dM}{dt} \sim a \frac{M^2 n}{u^3 \times 3 \times 10^{11}},$$ (13.7.1)

where u is the velocity in kilometers per second, M is the mass in units of $M_\odot$, n is the density of interstellar gas (H atoms per cm³), t is in years, and a is a pure number in the range $0.1 < a < 1$.

This formula for dM/dt is the same as expression (13.4.7) for the spherically symmetric case, except for the coefficient a. During accretion onto a normal star the main part of the energy emitted is released by collision with the stellar surface. For this reason, taking account of asymmetry for a fixed dM/dt does not change the star's luminosity to any considerable extent (see also § 13.5). However, the case of accretion onto a frozen star is different. There the luminosity of an object with $M \sim M_\odot$, at rest in the surrounding medium, is insignificant. If the collapsed body is in motion, however, then there arises a shock wave in which a sizable part of the rest mass-energy of infalling particles can be emitted; this sharply changes the situation. Unfortunately, reliable computations of this process have not been performed yet.

There is one more case where asymmetry is very important: accretion in binary systems. The incentive for its investigation is obvious: a high gas density at "infinity" can be ensured in this case. In fact, variable stars of the β Lyrae type lose up to 10^{-5} $M_\odot$ year⁻¹; almost half of the mass lost can fall onto the second component under favorable conditions. Gas accretion in close binaries as a probable source of cosmic X-rays was first taken note of by Hayakawa and Matsuoka (1964).

Works by Zel'dovich and Novikov have pointed out that X-rays could be given off by a binary system, one component of which is a neutron or frozen star.

Shklovsky (1967c) argues directly from observational data on Sco XR-1 that the source of the radiation must be a neutron star in a close binary pair.

Such models for X-ray sources have been discussed in detail by Prendergast and Burbidge (1968). In particular, they argue that the gas flow forms a disk around the dense star, with approximately a Keplerian velocity distribution. To fall onto the star, each particle must lose practically all of its angular momentum. The mechanism of angular-momentum loss is bound up with viscosity, which any real gas always has. Viscosity tends to make the angular velocity of the disk uniform; the inner regions of the disk slow down and the outer ones accelerate. Part of the gas will leave the binary system; the rest of it will be accumulated by the dense star. During slow accumulation each particle must radiate half of its potential energy. Therefore, it becomes clear at once how accretion onto a frozen star in a close binary pair can be distinguished from accretion onto a "normal" dense star (e.g., a neutron star or a white dwarf). In the frozen-star case *all* the radiation is emitted from a thin, hot layer; in the "normal" star half of the radiation comes from the disk and half from the stellar surface. In the normal case a significant part of the radiation must have a low-temperature, equilibrium form. Schwartzman and Shakura have noticed that the luminosity of the disk can be very important as a method for detecting a frozen star in a binary pair.

Concrete numerical values will depend on the particular mechanism of viscosity, the inner and outer boundary conditions, and so on. To date, calculations have been made only for the case of accretion onto white dwarfs with the following parameters (Prendergast and Burbidge 1968): $M = 10^{33}$ g, $R = 1.5 \times 10^{10}$ cm, $dM/dt = 2 \times 10^{19}$ g sec^{-1}, $\rho_\infty = 10^{-14}$ g cm^{-3}, $T_\infty = 3 \times 10^4$ ° K. It was assumed that the viscosity is due to turbulence, with the mixing length equal to half the thickness of the disk. The results in terms of dimensionless radius $x = r/R$ are

$$T = 1.4 \times 10^6 x^{-1.05}, \quad \sigma = 10^3 x^{-0.9} \text{ g cm}^{-2},$$

$$L(x_1, x_2) = 3.5 \times 10^{36}(x_1^{-1.45} - x_2^{-1.45}) \text{ ergs sec}^{-1}. \tag{13.7.2}$$

Here σ is the surface density of the disk. Thus, during accretion, the temperature becomes very high; the inner part of the disk gives off X-rays; and the outer part gives radiation in the optical range.

If the conclusions of this work could be extrapolated to other cases, then accretion in a close binary system onto a neutron star or a frozen star would have to give powerful X-radiation. But there is an additional circumstance which one must keep in mind. The creation of neutron stars, and probably also of frozen stars, is connected with a catastrophe of the supernova type; a considerable fraction of the star's primary mass is thrown off. Mass loss from a binary system during such an explosion takes a time much less than the orbital period; consequently, the condition for preservation of the binary system has the form (Huang 1963)

$$\Delta M/M < r/2a. \tag{13.7.3}$$

Here r is the distance between the stars when the mass is ejected; a is the semimajor axis of the orbit; and M is the total mass of both stars. If condition (13.7.3) is violated, the stars fly apart ("slingshot effect").

There are no binaries among the known neutron stars (pulsars) as of September 1969; this testifies in favor of the assumption that, as a rule, $\Delta M/M > \frac{1}{2}$. Are there any exceptions to the rule? Has it any connection to the creation of frozen stars? Instead of answering these questions directly, we can only cite the conclusions of a recent paper by Trimble and Thorne (1969), which studied spectroscopic binary stars with invisible secondary components (this paper was an extension of ideas and analyses first introduced by Zel'dovich and Guseynov 1965c). The conclusions are inauspicious: "Only few, if any, of the systems contain collapsed or neutron-star secondaries."

13.8   ACCRETION AS A FACTOR IN STELLAR EVOLUTION

It has long been known that accretion onto a star which is near the limits of stable equilibrium can bring on an instability and transform the star into another evolutionary stage: a white dwarf into a neutron star; a neutron star into a frozen one.

For a single object with $M \sim M_\odot$, the rate of growth of mass due to accretion from the interstellar medium is $dM/dt \sim 10^{-14}-10^{-16} M_\odot$ year^{-1}, so the effect is unimportant. For binary systems the mass transfer can be large. Notice that in order to ensure a luminosity of $L \sim 10^{37}$ ergs sec^{-1} (this appears to be the typical luminosity of an X-ray source) the flow onto a neutron star must be $\sim 10^{-9} M_\odot$ year^{-1}; and onto a white dwarf, $\sim 10^{-6} M_\odot$ year^{-1}. The corresponding lifetimes are $\sim 10^9$ years and $\sim 10^6$ years. It is possible, however, that bursts of nuclear burning in the hydrogen envelope which is being deposited on the star will eject mass and prolong these lifetimes (Cameron and Mock 1967; Saslaw 1968). Moreover, such burst burning would explain quite naturally the fact that most (or even all) novae are constituents of close binaries.

The mass of a frozen star is unlimited, in principle. Solution of equation (13.4.7) gives the formula $M(t) = M_0/(1 - AM_0 t)$ for the mass increase with time (A is a constant). The formula diverges for some t. The associated "catastrophic accretion" would come into play in a time less than 10^{10} years if

$$M_0 > 10^4 M_\odot (T_e/10000°)^{3/2} n_\infty^{-1} . \tag{13.8.1}$$

In reality, of course, no such catastrophe takes place. When a mass of $M \sim 10^9 M_\odot$ is approached, the energy release comes to be so significant that light pressure limits the rate of subsequent mass accumulation[11]; see Salpeter (1964).

According to equation (13.8.1), frozen stars with $M < 100 M_\odot$ are always far from "catastrophic accretion." Frozen objects with larger masses will be discussed in the chapter on quasars.

13.9   THE ELECTROSTATIC FIELD, ACCELERATION OF POSITRONS DURING ACCRETION, AND GAMMA-RAY EMISSION

A more detailed consideration of accretion when the energy outflow L is of the same order as the critical, self-regulating value L_{cr} reveals interesting details.

The force due to L acts primarily on electrons, whereas gravity acts on protons and nuclei. Should this not cause the electrons to be blown out by the

11. One can say that in this sense frozen stars are "unstable" in the range $10^4 M_\odot < M < 10^9 M_\odot$.

radiation, while the protons fall onto the star unimpeded? Obviously, such a splitting of negative and positive charges would give the star an enormous electric charge. In a stationary state there must arise an electrostatic field which prevents the splitting of charges. This field is effectively a tie between the electrons and protons; the radiation drag acting on the electrons is transferred to the protons by the field. The electrons are practically in equilibrium: the radiation drag and the electrostatic force compensate each other, while inertia and gravitation are negligible for the electrons. The electric potential of a star in equilibrium is

$$\phi_{el} = \frac{m_p c^2}{e} \left(\frac{\phi_{grav}}{c^2}\right) \left(\frac{L}{L_{cr}}\right) = 9.4 \times 10^8 \left(\frac{\phi_{grav}}{c^2}\right) \left(\frac{L}{L_{cr}}\right) \text{ volts .}$$

Schwartzman (1970*b*) has pointed out that this situation is very favorable to the acceleration of positrons, which are created near the surface of the star (see below).

For slowly moving positrons, the electrostatic force and the radiation drag both act in the same direction!

A positron could obtain energy up to $2e\phi_{el}$. Actually, after being accelerated to a relativistic energy of $\sim$5–10 MeV, the positron is braked by stray radiation; its energy does not rise above this limit. Still, the possibility of positron acceleration is an interesting feature of spherically symmetric accretion.

The X-ray emission by the hot electrons (see § 13.5) must be accompanied by gamma radiation due to direct nuclear encounters. Such encounters produce nuclear excitations, neutron ejection with subsequent n-γ capture, and π^0 and π^+ mesons which decay by $\pi^0 \rightarrow 2\gamma$; $\pi^+ \rightarrow \mu^+ \rightarrow e^+$, $e^+ + e^- \rightarrow 2\gamma$.[12]

The calculations of Schwartzman (1970*c*) give an output of gamma luminosity of $(10^{-5}$–$10^{-3})\ L_{tot}$, depending on the gravitational potential near the star's surface.

There is hope that these gamma rays can be detected and even their spectrum measured. Perhaps the redshift of the gamma rays (for example, from $n + p \rightarrow D + \gamma$) could give directly the gravitational potential at the surface of a neutron star.

The gamma-ray spectrum should be useful for distinguishing electromagnetic acceleration (of the pulsar type) from gravitational acceleration during accretion.

13.10 PULSARS

The discovery of pulsating radio sources by Hewish, Bell, Pilkington, Scott, and Collins (1968) is widely known to astronomers and to the public.

---

12. Most of the positrons succeed in avoiding annihilation and are accelerated electrostatically as described above.

The period of repetition of the radio bursts varies from $P = 3.8$ sec in the slowest pulsar to $P = 0.033$ sec in the fastest. Although the periods are given here crudely, they are known with precisions of the order of 10^{-8}–10^{-9}.

In the early months after the discovery, two types of theories were discussed for the clock mechanism which governs the period: the radial oscillations of white dwarfs and the rotation of neutron stars. In those days only pulsars with periods $P \sim 1$ sec were known. A huge amount of work was done on the calculation of oscillation periods of white dwarfs. The idealized theory led to $P_{\text{osc}} \gtrsim 2$ sec; to explain shorter periods the influence of rotation was considered, higher modes of oscillation (up to the fifth) were assumed, etc. The discovery of the pulsar with $P = 0.033$ sec in the Crab Nebula was a *coup de grâce* which put an end to this line of thought. The only surviving theory is now that of a rotating neutron star. A neutron star is the only type of celestial body known to theorists for which the equatorial velocity of rotation at $P = 0.033$ sec is small enough (2000 km sec^{-1} for $R = 10$ km, against 120000 km sec^{-1} for a dwarf with $R = 6000$ km) and the gravitational acceleration is great enough that centrifugal forces do not destroy the body.

One has to assume that each pulsar is radiating a beam of electromagnetic radiation with strong angular dependence. The direction of the beam is rigidly tied to the body of the star and rotates with it. Every time the beam passes through the Earth, the observer sees a burst of radiation.

At the moment when this section was written, it had not been decided whether the beam is like a pencil or like a knife (see Ginzburg, Zheleznyakov, and Zaitsev 1969). It is commonly accepted that the strong angular dependence is linked to the magnetic field of the pulsar. Assume that the field is that of a dipole, whose magnetic axis does not coincide with the axis of rotation. Radiation emitted from the poles along the magnetic lines of force would give a pencil-type beam (perhaps two pencils, of which only one strikes the Earth in most cases). The emission of radiation by particles captured (in a plane) near the magnetic equator gives an example of a knife-type beam.

The total solid angle of the knife beam is 10–20 times greater than the angle subtended by the pencil. Therefore, the two assumptions give different probabilities of observing a pulsar and different estimates of the number of pulsars in our Galaxy.

That the pulsars are in our Galaxy has been inferred from their concentration in the galactic plane, and from measured values of $\int n_e dl$, where n_e is the number density of electrons, and l is the path length between the pulsar and Earth. The $\int n_e dl$ is measured by its effects on the propagation of the pulsar radio waves. The discovery of the pulsar in the Crab Nebula and of the Vela pulsar clinches the galactic nature of all pulsars and suggests that pulsars are the results of supernova explosions.

All of the roughly forty-five pulsars which have been observed are at distances of less than several thousand parsecs, and their ages are estimated (see below) to be typically 10^7 years. From these figures one estimates that our Galaxy contains 10^4 pulsars with the same intrinsic characteristics as those we are able to observe. If pulsars have pencil-type radiation patterns, their number is $\sim 10^5$. Assuming that pulsars have been created at the same rate during all cosmological time ($\sim 10^{10}$ years), but are active only during the first 10^7 years after birth, we estimate further that the total number of pulsars, active and inactive, is of the order of 10^7–10^8 in our Galaxy. The corresponding number of supernova explosions during 10^{10} years in our Galaxy is estimated to be 10^8. Of course, such reasoning is too rough to give preference for one or the other type of radiation pattern.

Very impressive is the enormous "temperature" of the radio radiation. The radiating surface must be small in order to give sharp bursts;[13] the energy output is great, so the effective blackbody temperature, given by $F_\nu = (2\pi kT/\lambda^2)$ ergs (cm² sec sterad)$^{-1}$, is very high: $T \sim 10^{23}\,°$–$10^{27}\,°$, $kT \sim 10^7$–10^{11} ergs.

Such thermal temperatures, and even such energies for individual particles, are out of the question. Therefore, it is necessary that some collective, coherent mechanism be involved in the radio emission. Perhaps this coherent radiation is more easily collimated than normal radiation. But one must remember that the Crab pulsar also emits optical radiation and X-rays; these are also pulsed (and therefore directed) in phase with the radio bursts. For the X-rays a coherent mechanism is not needed (the effective temperature is lower than some 100 keV, and such electrons are surely present); nor is it easy to invent. For literature about the emission mechanism see Ginsburg, Zheleznyakov, and Zaitsev (1969). The source of energy which feeds the radiation (and also feeds the surrounding nebula in the case of the Crab) is usually thought to be the rotation of the neutron star.

Measurements of the deceleration of pulsars (the increase of P with time) are in accord with this idea.

The qualitative idea that rotation of a central body could feed the needed energy into the Crab Nebula was put forward by Wheeler (1966b), and in another context by Kardashev (1964). A crude Newtonian estimate gives, for $M = 1.2\ M_\odot$ and $R = 1.2 \times 10^6$ km, a moment of inertia $I = 1.4 \times 10^{45}$ g cm². The period $P = 0.033$ sec corresponds to an angular velocity of $\omega \sim 200$ sec^{-1}, so that the stored rotational energy in the pulsar today is equal to 3×10^{49} ergs. The observed deceleration is $(1/\omega)(d\omega/dt) = -1/\tau_0 = -1/2340$ years, so that $dE/dt = 2E/\tau_0 = 10^{39}$ ergs sec^{-1}.

Before the discovery of pulsars it was pointed out that the Crab Nebula needs $\sim 10^{38}$ ergs sec^{-1} of energy injection in the form of ultrarelativistic electrons (Shklovsky 1966, 1969a; Haymes, Ellis, Fishman, Rurfoss, and

---

13. The bursts, repeated with a period P, have durations of $\sim 0.03\ P$, but they consist of much shorter subpulses, down to $\sim 10^{-4}$ sec or less. It is not clear whether instrumental and propagational factors are important here.

Tucker 1968). Thus, the problem of the energy balance of the Crab Nebula has found a solution.

The next question is how the transfer of energy and angular momentum occurs. The most plausible idea relies on magnetic dipole radiation (Pacini 1968; Gunn and Ostriker 1969). As shown by Gunn and Ostriker (1969) a magnetic field of the order of 3×10^{12} gauss at the surface gives the needed braking torque. For magnetic dipole radiation

$$dE/dt = I\omega \, (d\omega/dt) = -\text{const.} \, (d^2\mathfrak{M}/dt^2)^2 = -\text{const.} \, \mathfrak{M}^2\omega^4 .$$

With a constant magnetic moment $\mathfrak{M}$ this gives

$$d\omega/dt = -a\omega^3 , \quad (2/\omega^2) - (2/\omega_0^2) = at , \quad \omega = (2at + 1/\omega_0^2)^{-1/2} ;$$

and with $\omega_0^2 \gg \omega^2$,

$$\omega = 1/(2at)^{1/2} , \quad d\omega/dt = -\omega/2t .$$

This law is fairly well confirmed by observations; the initial rotational velocity, which could be different for different pulsars, cancels out. The constant a varies by not more than a factor of about 10 over the life of a pulsar, $dP/dt = 3 \times 10^{-15}/P$, which means $d\omega/dt = 10^{-13}\omega^3$.

For the Crab pulsar, the exact time of birth is known so that the deceleration law can be checked. One finds $d\omega/dt = -\omega/2370$ years $= -\omega/2t$. In actuality, the explosion was in the year 1053 A.D., $t = 916$ years ago. The agreement is not too bad.

There is another source of deceleration—gravitational radiation emitted by the rotating star. Of course, if the star is symmetric around its axis of rotation, it will not radiate. Assume that the star possesses a quadrupole moment which is constant in the corotating frame. Then

$$dE_r/dt = I\omega\dot{\omega} = -\text{const.} \, (d^3Q/dt^3)^2 = -\text{const.} \, Q^2\omega^6 .$$

The corresponding law of deceleration is

$$\omega = \left(at + \frac{1}{\omega_0^4} \right)^{-1/4} \to bt^{-1/4} ,$$

i.e.,

$$d\omega/dt = -\omega/4t ,$$

which is in worse accord with the observations, but with a discrepancy of the opposite sign.

A combination of gravitational and magnetic dipole radiation can be found which matches exactly the age and current deceleration rate of the Crab Nebula's pulsar. By assuming an initial angular velocity of $\Omega_0 = 10^4$, the authors obtain $(L_0)_{\text{grav}} = 3 \times 10^{48}$ ergs sec^{-1}, $(L_0)_{\text{mag}} = 5 \times 10^{45}$ ergs sec^{-1}; and for the current epoch, $L_{\text{grav}} = 2 \times 10^{38}$ ergs sec^{-1}, $L_{\text{mag}} = 10^{39}$ ergs sec^{-1}.

What could cause the necessary quadrupole moment? The magnetic field, inclined at an angle relative to the axis of rotation, will produce some deformation. However, Ostriker and Gunn have shown that the dipole magnetic field needed for the magnetic deceleration is too weak to produce the necessary deformation. One has to assume an internal field of the order of 10^{15} gauss with magnetic field lines hidden inside the star in order to sustain the necessary asymmetry against the force of gravity.[14]

Although the numbers given are plausible, it is obvious that caution is needed; the results are too new to be definitive.

The gravitational radiation escapes completely from the pulsar and its surroundings. In connection with Weber's experiment the important point is the existence of a rigorous upper limit on the action of a pulsar on a resonant detector of gravitational waves; this limit does not depend on any assumption about the star's asymmetry. In the absence of other mechanisms of deceleration, the total energy emitted in gravitational waves per unit angular frequency is obviously

$$\Delta E_\omega/\Delta\omega = \Delta\left(\tfrac{1}{2} I \omega_r^2\right)/\Delta\omega = I \omega_r\Delta\omega_r/\Delta\omega = \tfrac{1}{4}I\omega .$$

Here it is taken into account that the frequency of the emitted waves is twice the angular velocity of rotation ω_r. For the maximum possible $I = 3.5 \times 10^{45}$ g cm^2 of a neutron star, and for the $\omega = 10^4$ of Weber's apparatus, we find

$$\Delta E_\omega/\Delta\omega = \tfrac{1}{4} \times 3.5 \times 10^{45} \times 10^4 = 10^{49} \text{ ergs} .$$

The corresponding averaged flux at a distance of 10 kpc (galactic nucleus) is

$$F_\omega = \frac{\Delta E_\omega}{\Delta\omega 4\pi r^2} = 10^5 \text{ ergs cm}^{-2} \text{ Hz}^{-1} .$$

This is definitely below the limit of Weber's sensitivity.

Slowly rotating pulsars, such as those we observe now including the Crab pulsar, lose their energy largely by electromagnetic processes. We shall not attempt to give a detailed picture of these processes, nor shall we present a detailed theory for the directed radiation. As mentioned above, Ostriker and Gunn calculate the energy emitted by a rotating magnetic dipole. Other authors are considering charged particles and plasma which rotate with the magnetic field out to the radius where the linear velocity of rotation attains the speed of light (Gold 1968, 1969; Goldreich 1969). These two different points of view give the same order of magnitude for the rates of loss of energy and angular momentum. How can one understand this coincidence?

Consider the electromagnetic field in vacuum. It is a static dipole field $|H| \sim |M|/r^3$ in the near zone, i.e., at distances less than the reduced wavelength of the radiation emitted, $r < \lambdabar = c/\omega$, where $\lambdabar = \lambda/(2\pi)$. At greater distances it is an outgoing-wave field, with an energy flux of the

14. The idea of a strong internal magnetic field, much greater than the external dipole field, is familiar for ordinary stars.

order of $q = H^2/4\pi c$. The formula for the radiation is obtained by joining the fields at $r = \lambda$: we find $q = (M/\lambda^3)^2(c/4\pi)$:

$$Q = 4\pi r^2 q = 4\pi\lambda^2(M/\lambda^3)^2(c/4\pi) \sim M^2\omega^4/c^3 \sim c^{-3}(d^2M/dt^2)^2 .$$

(Note that for a rotating dipole $r = \lambda = c/\omega$ coincides with the condition $v = \omega r = c$; we have calculated the Poynting vector at the radius where $v = c$.) If the field carries charged particles, their density is of the order of the energy density of the field itself. Therefore, the energy flux carried by a plasma injected with the help of the field should be equal to the Poynting flux. This explains the coincidence.

If a rotating body loses energy, it always loses angular momentum as well at a corresponding rate: d(angular momentum)$/dt = $ (angular velocity)^{-1}d(energy)$/dt$. A rigorous Newtonian proof is given by Ostriker and Gunn (1969); and the analogous proof in general relativity is given by Hartle (1970).

By ignoring the superconductivity of the neutron star, one estimates that the time required for the magnetic field to decay is $\sim 10^7$ years. This is just the time of the active life of typical pulsars. Schwartzman (1970d) has pointed out that as the magnetic field decreases, particle ejection can be changed into accretion of surrounding matter onto the neutron star. Gas accretion onto a pulsar with a magnetic field is accompanied by Langmuir plasma vibrations, and by a transformation of the energy of such vibrations into electromagnetic radiation. Thus the pulsar radiation, which decays at first, can grow again because of accretion. However, the radiation must finally terminate when the pulsar is $\sim 10^7$ years old.

13.11 THE SUPERFLUIDITY AND SUPERCONDUCTIVITY OF HIGHLY COMPRESSED MATTER, AND THEIR INFLUENCE ON THE BEHAVIOR OF NEUTRON STARS

The interaction of neutrons in the highly compressed matter inside neutron stars is likely to lead to superfluidity. Experimental studies of atomic nuclei reveal that nuclear forces between like particles are attractive: two neutrons in vacuum cannot be bound; but nevertheless, when they are in the fields of other particles, for example in a nucleus, they are paired. The same effect probably occurs in a homogeneous, degenerate nuclear fluid.

Pairs of neutrons are Bose particles and therefore their behavior must be analogous to that of ^{4}He atoms in liquid helium. But as Kapitsa found experimentally, liquid ^{4}He at temperatures below 2° K is a superfluid. In neutron matter, superfluidity may exist at temperatures below ~ 1 MeV $= 10^{10}$ ° K. The same reasoning leads to the prediction that the protons in nuclear matter (whose number is of the order of a few percent of all baryons) will be paired, and will give rise to the phenomenon of superconductivity. A discussion of these properties and of the order of magnitude of the critical temperatures, as well as detailed lists of references, is given in a review article by Ginzburg (1969*a*, *b*). The most recent developments are described by Baym, Pethick, Pines, and Ruderman (1969). Some caution is needed on the region of superfluidity and superconductivity: at subnuclear densities the matter is no longer homogeneous; it consists of individual nuclei. It can be crystalline rather than liquid. It is possible that at the highest densities (somewhat above nuclear), repulsive forces outweigh the attraction. Further investigations are needed to elucidate this question.

What are the anticipated influences of "super"-phenomena on the astrophysical behavior of neutron stars? They can be classified as (1) thermal, (2) magnetic, and (3) hydrodynamical.

1. To begin with, the heat capacity of a superfluid or superconductor is smaller than the heat capacity of matter in a normal Fermi-gas or Fermi-fluid state. The reason is obvious: to excite baryons, one must first break the pair; there is an energy gap Δ, and the heat content at low temperatures will be proportional to $e^{-\Delta k /T}$. However, other sources of heat capacity which are not plagued by the gap (the electrons and acoustical waves in the superfluid) still operate; they reduce the influence of superfluidity on the heat capacity. At the same time, the rate of heat losses due to neutrino emission by the URCA process (i.e., by the interplay between $p + e^- \rightarrow n + \nu$ and $n \rightarrow p + e^- + \bar{\nu}$) is decelerated by the pairing of neutrons and protons. As a whole, however, it is presumed that the cooling of neutron stars is more rapid due to superphenomena.

2. In the laboratory the transition to a superconducting state is accompanied by the magnetic field being pushed outside the specimen (Meissner phenomenon). In a star a corresponding change in the magnetic field over the macroscopic stellar scale would produce a tremendous electric field; such a change is quite impossible. The attainment of superconductivity will change only the microscopic field distribution; instead of a uniform magnetic field, we anticipate the creation of a heterogeneous situation: there will be domains where the superconductivity is destroyed by the magnetic field, and there will also be superconducting domains which carry the electric current. Such a situation is familiar in laboratory: the so-called hard superconductors of the second type (Nb_3Sn for example) support high magnetic fields in just this manner.

The time of decay of the stellar magnetic field is infinite in the superconducting case, but even without superconductivity it was great. Ostriker and Gunn (1969), in connection with pulsars, use $\tau \sim 10^6$ years; but there are much higher values given in the literature—up to 10^{24} years.

3. The influence of superfluidity on hydrodynamic phenomena in equilibrium is presumed to be small, due to the immense dimensions of the star (compared with the nuclear scale). The superfluid is like an ideal liquid

(with $\mathbf{\nabla} \times \boldsymbol{v} = 0$; i.e., without vorticity) only on the microscopic scale. It is known that discrete, quantized vortex lines distributed in the liquid make the macroscopic behavior similar to that of a normal, viscous liquid. The same can be said of convection and convective heat transfer in the star. Still, there is a difference: the transfer of velocity from the superfluid neutron liquid to the proton component (which is tied to the magnetic field) is slow. Consequently, the relaxation of the stars fluid motions to an equilibrium, rigid-body rotation after a change in the rotation velocity can take a long time.

This discussion of hydrodynamic effects may sound pessimistic; but in the case of pulsars, even minute changes in period can be measured. There are very interesting hypotheses which explain the detailed variations in the periods of some pulsars by the superfluidity of the nuclear matter and the rigidity of the outer layers of the neutron star. Even if the effects are small, they are of great intrinsic interest, because they provide a means for studying the properties of matter under extreme conditions which are far outside the realm of possible laboratory physics.

Perhaps the details of the period variations (with jumps, etc., as observed in the Vela and Crab pulsars) will give, with the help of a complete theory, additional information on the mass, density, etc., of the pulsar.

13.12 Magnetic and Magnetohydrodynamic Phenomena in Collapsing Bodies

Ginzburg (1964) was the first to point out that the collapse of a star must be accompanied by a great increase of its magnetic field. Since then several contributions have concentrated on this problem (see below). In analyzing the collapse of stars and of very large gas masses, one must take account of the qualitatively different topologies of the external magnetic fields of these objects (Kardashev 1964). According to observational data and contemporary theories of stellar origin, the magnetic field of a normal star (e.g., the Sun) has a quasi-dipole nature. The topology of such a field is shown in Figure 72. Most of the magnetic field lines close not too far from the star. The topology of the external field of a galaxy or of a metagalactic object is different. According to contemporary ideas (Pikel'ner 1965), the field lines are not closed; they extend practically to infinity, thus connecting the body with the surrounding medium and with other objects.

Magnetic processes in collapse, including phenomena in the surrounding envelope, are rather complex; they have probably been investigated even less than the effects of rotation. Thus, we will limit ourselves to a very general analysis of such processes. Let us begin with the collapse of normal stars. It is known from observations that the surface magnetic fields of normal stars can be as large as $\sim 10^4$ gauss. (For comparison, the Sun's magnetic field is ∼1 gauss.) In all cases, the magnetic energy is considerably smaller than the gravitational energy.

The conductivity of stellar matter is rather high; for stars similar to the Sun, $\sigma_\odot = 10^{16}$ sec^{-1}. Hence, the field attenuation time is (Pikel'ner 1961)

$$t_\odot \approx 4n\sigma_\odot R^2_\odot/c^2 \approx 7 \times 10^{17} \text{ sec} \approx 2 \times 10^{10} \text{ years} ,$$

which is considerably greater than the age of the Sun (∼5 × 10⁹ years). It is clear that the field attenuation time is considerably greater than the characteristic contraction time for any star (Ginzburg 1964). When a normal star is transformed to a neutron star, its conductivity increases enormously.

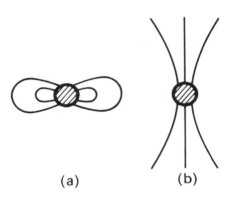

(a) (b)

Fig. 72.—Topology of the magnetic field of (*a*) a star, and (*b*) a metagalactic object

Thus, we can always regard the field as "frozen into" the stellar matter. Under such conditions, $H \sim R^{-2}$ during contraction, and the magnetic energy increases as $E_{\text{magnetic}} \sim H^2R^3 \sim R^{-1}$; i.e., it increases by the same factor as the gravitational energy. In normal stars, as mentioned above, $E_{\text{magnetic}} \ll E_{\text{grav}}$; thus the magnetic field has no influence whatsoever upon the dynamics of the stellar collapse, for the star as a whole.

How will the magnetic field change during relativistic collapse, when a normal star is transformed to a frozen star? This question has been investigated by Ginzburg (1964) and by Ginzburg and Ozernoy (1964).

As we have emphasized, the field is "frozen" into the stellar matter; hence, in contraction to dimensions ∼r_g it must reach the colossal magnitude of ∼10¹⁰–10¹⁴ gauss as seen by an observer comoving with the matter. To a stationary observer, however, the field will appear to behave very differently as $R \to r_g$. To find this behavior, Ginzburg and Ozernoy first considered the static problem.

Visualize a sequence of static, gravitating, magnetic spheres with successively smaller dimensions, and with fixed magnetic flux. Examine their external (dipole) magnetic fields. In the classical theory, the dipole moment d of each sphere is proportional to R: $d = d_0(R/R_0)$, where d_0 is the dipole moment for radius R_0; d approaches zero as $R \to 0$. Ginzburg (1964) has demonstrated that in the relativistic theory, in the limit as $R \to r_g$, the dipole moment varies according to the modified law

$$d = \frac{d_0 r_g}{R_0 \times 3 \ln [r_g/(R - r_g)]} \, .$$

Thus, $d \to 0$ when $R \to r_g$.

Turn now from a sequence of static stars to the variation with time of the magnetic moment of a collapsing star. At the surface, the magnetic field is computed exactly as in the stationary case, so the dipole field in the near zone decreases exactly as before. Knowing the dependence of the star's radius R on time (cf. § 11.6), Ginzburg and Ozernoy find finally that, as $t \to \infty$,

$$d = d_0 r_g^2/(3R_0 ct) \, .$$

[*Note added by K. S. Thorne:* Actually, this result is not correct because in the late stages of collapse, as R nears r_g, every radius r enters the wave zone within a time $\Delta t \sim r_g/c$ after the star's surface passes $R = r$. Thus, a correct analysis cannot avoid dealing with wave-zone phenomena. An analysis taking them into account (Price 1970, 1971a, b) reveals the same decay law for the magnetic field at late times,

$$\text{(field strength at fixed } r) \propto t^{-(2l+3)} \ln t = t^{-5} \ln t \, ,$$

as for nonspherical perturbations in the gravitational field (§ 5.4).]

What is the physical explanation for the attenuation of the magnetic field? The magnetic field is produced by circular electric currents in the star. Since all processes attenuate as seen by the external observer, the rotation speed of the charges, v, approaches 0; since $v \to 0$, the current attenuates, $I \to 0$. This attenuation of the current results in an attenuation of the external magnetic field of a collapsing star, as measured by an observer at fixed Schwarzschild radius r.

The change with time of the magnetic field during contraction produces an electrical eddy field. In the near (non–wave) zone, this eddy field may generate a curent-carrying layer in the plasma that surrounds the star, and/or may generate magnetohydrodynamic waves (Ginzburg and Ozernoy 1964). These processes have never been investigated, so we will limit ourselves to some comments about the distant (wave) zone (Novikov 1964a). Let r_g/c and r_g be the characteristic time and scale of the near-zone phenomena, respectively. Then the wave zone begins at a radius $l > r_g$. Let the external magnetic field of the star include a dipole term with a magnetic moment $d = \Phi R$, where $\Phi = $ const. Estimate the total energy radiated by the external magnetic field during its attenuation. It is clear a priori that the energy radiated will be of the order of the total energy of the magnetic field at $R \sim r_g$, because when $R \sim r_g$, the contraction velocity is of the order c, and there exists no dimensionless parameter in which the radiated energy could be expanded. We will make a more precise calculation under the assumption that, in the process of collapse, the matter falls almost freely. Then

$$R_{,tt} = -GM/R^2 , \quad R_{,t} = (2GM/R)^{1/2} . \tag{13.12.1}$$

The total power radiated is

$$I = (2/3c^3)(d_{,tt})^2 . \tag{13.12.2}$$

Substituting equation (13.12.1) into equation (13.12.2), we get for the instantaneous power when the star is at radius R,

$$I = (\Phi^2 c/6r_g^2)(r_g/R)^4 , \tag{13.12.3}$$

and, for the total energy radiated during the contraction to radius R,

$$E = (\Phi^2/15r_g)(r_g/R)^{2.5} . \tag{13.12.4}$$

If $\Phi \approx 3 \times 10^{21}$ (erg cm)$^{1/2}$, as it is for the Sun, and $r_g \approx 3 \times 10^5$ cm, then

$$I \approx 5 \times 10^{41}(r_g/R)^4 \text{ ergs sec}^{-1} , \tag{13.12.5}$$

$$E \approx 3 \times 10^{36}(r_g/R)^{2.5} \text{ ergs} . \tag{13.12.6}$$

Most of the radiation emerges in a single burst which lasts for a time $\sim r_g/c$.[15] This calculation was performed in the nonrelativistic theory; and the formulae used lose their validity near the Schwarzschild sphere. However, they should yield a correct estimate of the order of magnitude of the power and energy radiated. The total energy radiated in this case is not large. We will see below that the energy radiated is much greater when the contracted body is $\sim 10^5$–$10^8 M_\odot$ rather than $\sim 1\ M_\odot$ as assumed above.

Up to now we have tacitly assumed that the plasma which surrounds the star does not interfere with the outflow of the radiation. The radiation frequencies are low, $\omega \sim c/r_g$, especially for large masses. Even when the plasma has very low density, the natural frequency of plasma vibrations,

$$\omega_0 = (4\pi e^2 n_e/m_e)^{1/2} \approx 2 \times 10^4\ n_e^{1/2} \text{ sec}^{-1} ,$$

is considerably larger than ω. This suggests that radiation might not occur at all. However, if the occurrence of radiation is to be prevented, then, in addition to the condition $\omega > \omega_0$, it is necessary that the maximum possible current which the plasma can support, $I_{max} = n_e e c$ (where n_e is the electron number density and e is the electron charge), be capable of compensating for the variation of the magnetic field in the near zone. Let us calculate the critical value of the density of the surrounding plasma for which radiation no longer occurs (Novikov 1964a). From Maxwell's equations, we obtain

$$|\boldsymbol{\nabla} \times \boldsymbol{H}| = 4\pi\ I_{max}/c = 4\pi n_e e . \tag{13.12.7}$$

---

15. The decay of the pulse is caused by the relativistic freezing of all processes when $R \to r_g$; see above.

13.12 Magnetohydrodynamic Phenomena

We restrict ourselves to order-of-magnitude estimates. Consider the characteristic stage of contraction when $(R - r_g) \sim r_g$. Since $|\nabla \times H| \sim H/r_g$ and $H \sim \Phi/r_g$, we obtain from equation (13.12.7)

$$\Phi_{\text{crit}} \approx 4\pi \, n_e e r_g{}^3 .$$

Thus, if the inequality

$$n_e < \Phi/(4\pi r_g{}^3) , \tag{13.12.8}$$

is satisfied, electromagnetic waves are generated in the wave zone (beginning at distances $L > r_g$), even when $c/r_g \ll \omega_0$. Substituting $\Phi \approx 3 \times 10^{21}$ and $(r_g)_\odot \approx 3 \times 10^5$ cm into equation (13.12.8), we find

$$n_e < \sim 10^{15} \text{ cm}^{-3} . \tag{13.12.9}$$

For comparison, the electron concentration in the solar corona near the surface of the Sun is $\sim 10^8$ cm^{-3}.

Thus, when condition (13.12.9) is satisfied, radiation is emitted by the star in the form of a single pulse, though it may be absorbed by the plasma in the wave zone far from the star. Let us recall again that the total energy radiated by normal stars is relatively small.

Let us now turn our attention from normal stars to the contraction of very massive gas clouds. The gravitational contraction of huge clouds with magnetic fields was presumably responsible for the formation of various types of galaxies and of radio galaxies. This has been particularly emphasized recently in contributions by Piddington (1964) and Pikel'ner (1965). This raises the question whether, under certain conditions, a contracting gas cloud with a magnetic field can become a supermassive star (quasar).

Here we deal with a different magnetic-field topology than in the case of a normal star (see Fig. 72). The lines of force link the contracting cloud to the surrounding medium. At the onset of contraction, the magnetic energy of the cloud is probably already of the order of its rotational energy, and the contraction is accompanied by a continual loss of angular momentum. Contraction causes "constrictions" in the lines of the magnetic field.[16] The processes which occur during this contraction and constriction are only poorly understood, so we will confine ourselves to very rough estimates. In interstellar clouds there exist magnetic fields with an intensity of $\sim 10^{-5}$ gauss. Consider an initial cloud with $M = 10^8 \, M_\odot$. The external regular field will have a strength $H_0 \approx H_1 n^{-1/2} \approx 10^{-7}$ gauss. After equilibrium evolution and collapse down to r_g, the density of the cloud will have changed by approximately twenty-four orders of magnitude (from the initial density of $\sim 10^{-24}$ g cm^{-3} to a final density of ~ 1 g cm^{-3}), and the field will have grown by sixteen orders of magnitude to an intensity $H \approx 10^9$ gauss. The increase of the field energy is accompanied by a growth of the energy of the particles which are frozen onto the field.

Will the magnetic field emit radiation during its contraction? Criterion

16. On this subject, see contributions by Mestel (1959).

(13.12.8), for the emission of radiation in the above case when the initial field at the surface of the cloud is $H_0 \approx 10^{-7}$ gauss, reads

$$n_e < 10^4 (10^8 M_\odot / M)^{7/3} \text{ cm}^{-3} . \qquad (13.12.10)$$

Since it is likely that $H_0 \sim M^{1/2}$, the exponent on the expression in the parentheses might be 3 rather than $\frac{7}{3}$. Thus, the plasma density required to prevent the emission of radiation ranges from $\sim 10^4$ cm^{-3} for $M \approx 10^8 M_\odot$ to 10^{11}–10^{13} cm^{-3} for $M \approx 10^5 M_\odot$.

If the plasma density satisfies criterion (13.12.8), then radiation is emitted. Let us estimate the total power and energy radiated for $M \approx 10^8 M_\odot$, using formulae (13.12.5) and (13.12.6):

$$I = 10^{54} (r_g/R)^4 \text{ ergs sec}^{-1} , \quad E = 5 \times 10^{56} (r_g/R)^{2.5} \text{ ergs} .$$

If H_0 scales with mass as $H_0 \sim M^{-1/2}$, then I and E scale as $I \sim M^{-5/3}$, $E \sim M^{-2/3}$. If H_0 is independent of mass, then $I \sim M^{-2/3}$, $E \sim M^{1/3}$. Thus, the electromagnetic radiation from the external magnetic field is quite impressive for the collapse of superlarge masses, in contrast to the collapse of normal stars; possibly such radiation plays a significant role in the energy balance of quasars during their collapse stage.

13.13 THE STATISTICS OF STARS AT THE ENDPOINT OF STELLAR EVOLUTION

Do astronomical observations yield any indications about the final fates of massive stars, and about the conditions for the creation of neutron stars and frozen stars?

If the collapse of a star produces an explosion which destroys the entire star or almost the entire star, then the explosion, of course, must be visible from a tremendous distance. It is natural to identify such catastrophes with supernova outbursts, which are observed by astronomers. In such an outburst, an energy of the order of 10^{52} ergs or even more is released.

Unfortunately, it is entirely unclear at present just what portion of the mass of the star is emitted in a supernova outburst of Type I or Type II. For details see the review by the Burbidges (1958), and the monograph by Shklovsky (1966).

Assume that every star with a mass that is greater than $\sim 1.5 M_\odot$ terminates its evolution by a supernova outburst; then the number of supernovae in a time interval Δt must equal the number of massive stars which terminate their evolution during that time interval. Let us estimate this number (Hoyle and Fowler 1963; Zwicky 1958; Schwarzschild 1958; Novikov and Ozernoy 1964; Hoyle, Fowler, Burbidge, and Burbidge 1964) and compare it with the observational data.

Assume that each massive star in the process of its evolution does not lose a substantial amount of mass, or at least that the mass losses are not so great as to bring the star below the critical mass. A star spends most of its lifetime on the "main sequence," burning hydrogen in its center (see § 11.1). We recall that the time of evolution in the main-sequence stage (almost the total time of equilibrium evolution) is

$$t \approx 10^{10}(L_\odot/L)(M/M_\odot) \text{ years} .$$

For bright stars on the main sequence, the approximate relationship $L \sim M^3$ is satisfied; therefore, the preceding formula can be written in the form

$$t \approx 10^{10}(M_\odot/M)^2 \text{ years} . \tag{13.13.1}$$

From astronomical observations we determined the number dN of stars on the main sequence per unit volume of space with masses in the interval M to $M + dM$. If the life span, t, of a star in the range M to $M + dM$ is less than the age of the Galaxy ($\sim 10^{10}$ years)[17] then, dividing dN by t, we obtain the frequency of star births averaged over the past t years. This must also be the frequency of star deaths in the given mass range, if the stellar population is stationary. A computation of this sort performed by Salpeter (1955) has yielded

$$d/N = 2 \times 10^{-12}(M/M_\odot)^{-2.4}d(M/M_\odot) \text{ stars pc}^{-3} \text{ year}^{-1} . \tag{13.13.2}$$

All stars with masses that are greater than critical, $M > 1.5\ M_\odot$, have evolution times smaller than the age of the Galaxy. Multiplying equation (13.13.2) by the volume of the Galaxy,[18] $\sim 10^{13}$ pc^3, and integrating over the mass range $M > 1.5\ M_\odot$, we obtain the number of stars in our Galaxy with a mass that is greater than critical, which each year complete their equilibrium evolution:

$$\frac{dN}{dt} = \int_{1.5}^{\infty} 20 \left(\frac{M}{M_\odot}\right)^{-2.4} d\left(\frac{M}{M_\odot}\right) \simeq 7 \text{ stars year}^{-1} . \tag{13.13.3}$$

It follows that, if each massive star were to flare up as a supernova at the end of its evolution, then each year there would be several supernova outbursts in the Galaxy. This is greater by three orders of magnitude than the observed number of outbursts cited by Zwicky (1958) (one supernova per 300 years) on the basis of observations of supernovae in other galaxies. Kukarkin (1965) has pointed out that in spiral galaxies of type Sc the frequency

17. The conclusions are not affected appreciably if we assume that $t_{\text{Galaxy}} = 5 \times 10^9$ years.

18. *Editors' note.*—The rather large effective galactic volume used here (10^{13} pc^3) can be obtained by assuming a spherical galaxy with a radius of 15 kpc. One should use instead a volume corresponding to the mass of the galaxy divided by its mean density. Even for Population II objects the mean distance from the galactic plane, $\bar{z}$, is only 2 kpc (Blaauw 1965); for stars of $M > 1.5\ M_\odot$ Population I is more representative, so that $\bar{z} \lesssim 0.4$ kpc. Then dN/dt is reduced by about a factor of 10 or more from the value, ~ 7 star deaths per year, given in the text. This relaxes the authors' quantitative conditions but does not change their qualitative conclusions. We note that the rate of stellar deaths for all masses $M \gtrsim 5\ M_\odot$ would be about $1/(30 \text{ years})^{-1}$.

of supernova outbursts is greater by a factor of 10–20 than has generally been assumed. The contradiction, nonetheless, remains. If all stars with a mass that is greater than some limit become supernovae at the end of their evolution, then in order to obtain a supernova frequency of $\simeq 0.01$ per year, the limiting mass $M_\odot$ must be of the order of 20 $M_\odot$ (see Stothers 1963). It should be emphasized that estimates of the number of pulsars are in rough agreement with the observed number of supernova outbursts (see § 13.10).

Thus, observations seem to militate against the assumption that, at the end of the evolution of a massive star, a nuclear explosion or some other catastrophe which destroys the star prevents it from being transformed into a frozen star. Another argument for this conclusion is derived from the presence of white dwarfs in binary star systems (see below). Aren't there other processes which might prevent a massive star from being transformed into a neutron star or a frozen star?

Perhaps stars always eliminate their excess mass, not usually by means of catastrophic explosions, but more often by stationary emission of matter from their surfaces or by small-scale discrete ejections of matter in the course of their equilibrium evolution (the Fesenkov [1949] hypothesis for stellar evolution along the main sequence; see also Masevich 1956). The observational data in this area are rather scanty and do not yield a conclusive answer. (A survey has been published by the Burbidges 1958.)

It is significant that there are stars in which an intensive emission of mass from the surface is observed. These stars include the so-called W-R (Wolf-Rayet) stars. However, as a rule, these are very massive stars ($M \sim 10\ M_\odot$); the observations suggest that once their mass has decreased, their rate of mass loss will decrease (Burbidge and Burbidge 1958; Rublev 1964). Therefore, it is not clear whether their masses in the course of time decrease below the critical limit.

Another type of star which emits great amounts of mass is the spectral type Be, with bright lines in its spectrum. Stars of this type rotate rapidly. Their rate of mass loss is estimated at 10^{-6}–$10^{-10}M_\odot$ per year, but these estimates are rather uncertain.[19]

The greatest outflow of matter is observed in stars of the P Cygni type. The supergiant[20] P Cyg itself loses $10^{-5}\ M_\odot$ per year (data by Paifiel cited in the Burbidge [1958] review).

Note that the types of stars discussed above are typically members of stellar associations and unquestionably are young (see the monograph edited by Mikhaylov 1962). It is not known what fraction of their mass they will lose as a result of mass shedding in the course of their future evolution.

19. Let us note that the mass lost by light emission has no relation to the present problem whatsoever. In essence, we are interested in the loss of baryons, rather than of mass.

20. Let us recall that the term "giant" and "supergiant" are used in astronomy to characterize the large luminosity of the star.

13.13   Statistics of Stars

Moreover, astronomers observe some relatively old stars with masses greater than critical which, in the course of their future evolution, presumably will not pass through Be, P Cygni, or W-R stages, and thus cannot lose mass in the manners described above. Therefore, the data on young stars do not have a determining significance for the question of whether frozen stars are "to be or not to be."

As pointed out by Shayn (1943), conditions for the emission of matter from stars are most favorable in the red-giant stage of evolution, when the

TABLE 21

BINARY SYSTEMS IN WHICH ONE COMPONENT IS A WHITE DWARF

Star	Absolute Visual Magnitude	Spectral Type	Mass $(M/M_\odot)$	Period (years)
Sirius A	1.4	Al V	2.28 ⎞	
Sirius B	11.4	A5	0.98 ⎠	49.9
Procyon B	2.6	F5 IV–V	1.76 ⎞	
Procyon C	13.1	. . .	0.65 ⎠	40.6
O² Eridinae B	10.9	B9	0.45 ⎞	
O² Eridinae C	12.5	M4	0.21 ⎠	247.9
L745–46	18.7	M	0.02	10^3
W485	13.9	M6	0.07	7×10^5
−37°6571	2.5	G6V	1.0	3×10^3
LDS 678	12.5	dM5	0.10	3×10^3
LDS 235	6.8	dK3	0.60	10^5
L1405−40	12.4	dM2e	0.10	6×10^4
LDS 455	10.2	K–M	0.20	2×10^4
W672	11.9	sdM6	0.15	5×10^3
LDS 682	8.8	sdM1	0.35	10^5
LDS 749	7.2	sdK4	0.50	2×10^5
L1512−34	11.2	dM5	0.20	10^5

SOURCE.—From Auer and Woolf (1965).
With the exception of the first three systems, only data for the primary (non–white dwarf) component are given. The orbital periods of the first three systems have been measured directly; for the remaining systems, the periods were computed by assuming that the total mass of the system is 1 $M_\odot$ and that the observed separation of the components (angular separation on sky times distance from Earth) is equal to the semimajor axis of the orbit. The masses of the stars in the first three systems were calculated directly from the observed orbital motion; the masses of the primary components of the remaining systems were estimated from their luminosities and the empirical mass-luminosity relation.

dimensions of the outer envelope are great and the acceleration of gravity on the surface is small. However, observations show that the rate of mass loss in this stage is 1.5 orders of magnitude too small to produce a substantial decrease in the star's mass (Rublev 1964). Finally, for the so-called flare stars of type UV Ceti, which emit matter discretely (a flare takes place on an average of once every 1.5 days) observations yield a totally insignificant rate of mass loss (Gershberg 1964):

$$dM/dt \approx 2 \times 10^{-12} \, M_\odot \; \text{year}^{-1} \, .$$

However, from extensive investigations of three white dwarfs (the first three stars of Table 21) which are members of binary systems, it can be demonstrated that under certain conditions a star *can* shed its excess mass and turn into a white dwarf. As was noted in the literature a long time ago, in two of these three systems the mass of the primary (i.e., non–white dwarf) component is greater than that of the secondary (white dwarf); moreover, the primary is a main-sequence star; i.e., it has not evolved very far. However, the greater the mass of a star, the faster it evolves (see formula [13.12.1]).

Since the two stars of each system were formed simultaneously[21] (the probability of capture is negligibly small) and since the less massive one has already been transformed into a white dwarf, the primary component should have completed its evolution long before now. However, this is not the case. From this, we conclude that the secondary component originally had the greater mass, evolved faster, and subsequently lost much of its mass.

Possibly the binary nature of these systems has played a crucial role here, even though the separation of the components is now very great. Another explanation (other than mass loss) may be the nonsimultaneous formation of the components in the cluster where they were created.

When mass loss in a binary system occurs in a time which is considerably smaller than the orbital period, and when the inequality (Huang 1963)

$$\Delta M/M > r/2a$$

is satisfied (where r is the separation of the components when the mass loss takes place, a is the semimajor axis of the orbit, and M is the total mass of the system), then the binary system is destroyed because the velocities of the components exceed escape velocity.[22] For a circular orbit, this ratio is $\frac{1}{2}$. In order to become transformed into a white dwarf, a massive star must lose the major portion of its mass. Auer and Woolf (1965) point out that, if we assume that (a) the initial mass of the secondary star which becomes a white dwarf is greater than 1.5 $M_\odot$, (b) the final mass of the white dwarf is 0.5 $M_\odot$, and (c) the primary component (cf. Table 21) has a mass smaller than 0.5 $M_\odot$, then the mass lost by the system will be greater than half of the initial mass. Consequently, the mass loss must take place in a time larger than an orbital period (i.e., $\tau > 10^4$ years for systems of Table 21) if the system is to hold together. Obviously, this type of reasoning is not as convincing for the last eleven systems of Table 21 as it is for the first three, because the masses of the primary components are smaller than the Chandrasekhar limit and so we have no firm proof that mass loss has occurred at all.

Auer and Woolf (1965) have shown that observations of white dwarfs in

21. There is, however, reason to believe that the process of star formation in clusters extends over time intervals which are of the order of the evolution time of a massive star (see, for example, Masevich and Kotok [1963]).

22. This is the mechanism which Blaauw (1961) has used to explain the origin of certain high-velocity stars.

the Hyades Cluster indicate that the mass loss, which transforms a massive $(M > 2.5\ M_\odot)$ main-sequence star into a white dwarf is not due primarily to rotational mass shedding from the equator during the contraction of the star.

Bisnovatyi-Kogan and Zel'dovich (1966) have suggested that the chief mechanism of mass loss may be a slow, hydrodynamic escape of mass from a star in the late stages of its evolution. Bisnovatyi-Kogan (1967b) has demonstrated that such a mechanism may produce a substantial loss of mass in the course of evolution for stars with $M \gtrsim 15\ M_\odot$.

Does such mass loss always[23] occur and transform a massive star into a white dwarf? The answer to this question can be obtained only by detailed computations and, of course, observations. Auer and Woolf point out arguments in favor of such a possibility.

If a substantial rate of mass loss exists throughout the stages of helium and carbon burning, then the outflow will terminate only after the mass has decreased to $M < 0.5\ M_\odot$, so that the helium and carbon nuclear reactions can no longer take place (see § 10.8). In such a case, the mass loss will automatically reduce the mass of the star below the critical limit, and a white dwarf will be the endpoint of the evolution.

Additional evidence supporting the idea that massive stars are transformed into white dwarfs is the presence of white dwarfs in stellar clusters. In particular, twelve white dwarfs have been discovered in the Hyades (see Eggen and Greenstein 1965), the age of which cluster is such that only stars with $M > 2\ M_\odot$ can have completed their evolution by now.

However, as has been pointed out by Luyten (1952), if all the stars in the Hyades with $M > 2\ M_\odot$ had been transformed into white dwarfs by now, then the total number of white dwarfs in the Hyades would be approximately twenty-three. Even though the list of white dwarfs discovered in the Hyades is definitely not complete, the total number in the cluster in all likelihood is less than twenty-three. The whole situation is quite unclear.

According to extragalactic astronomical observations, the ratio of the mass of a galaxy to its luminosity, M/L, differs from one galactic type to another (see, for example, the review by de Vaucouleur 1962). This ratio varies from 100 (in solar units, $M_\odot/L_\odot = 0.5$ g sec erg^{-1}) for elliptic galaxies, through 10 for spiral galaxies such as our own, to unity for irregular galaxies. The great size of the ratio M/L for elliptic galaxies, as well as the features of their spectra, indicates that they contain relatively few young, bright stars, and relatively great amounts of either nonluminous or only weakly luminous matter.

Generally, in these galaxies there is very little diffuse interstellar matter. If, moreover, there are relatively few white dwarfs in these galaxies, as

23. Or, more precisely, does it always occur for stars with $M < 20\ M_\odot$, where termination of the evolution by a supernova outburst is not observationally acceptable (see the beginning of this section)?

in some clusters of our own Galaxy (see above), then all these things taken together suggest that there may be many stars which are hard to observe (neutron stars and frozen stars) in elliptic galaxies that have progressed far in their evolution.

Galactic X-ray sources, when first discovered, were thought to perhaps be neutron stars. However, Friedman (1964, 1965) measured the diameter of the X-ray source in the Crab Nebula by using a lunar occultation, and found it to be approximately 10^{18} cm. From this astronomers were forced to conclude that the Crab's X-ray source is not a star at all. It appeared for a while that the Crab Nebula had put to rest the hopes of astronomers that neutron stars had finally been discovered. However, in Shklovsky's opinion, the X-radiation from other sources comes from the surfaces or neighborhoods of neutron stars as was previously thought. According to this hypothesis, the mechanism of the source in the Crab Nebula is an "annoying exception." In 1968 the pulsar in the Crab Nebula was discovered; and it is now known to be a source of part of the Crab's X-radiation.

Another peculiarity of X-ray sources, emphasized by Shklovsky, is their obvious concentration in the galactic plane. It follows from this that those we see lie at considerable distances from the Earth—distances of the order of galactic dimensions. In other words, the known X-ray sources constitute a considerable fraction of all sources that exist at present in the Galaxy. Astronomers have discovered approximately twenty sources by now (mid-1967); therefore, the total number is not greater than approximately 100 (possibly 10^3).

The problem of the nature of X-ray sources and of the connection between X-ray sources and neutron stars is now open.

Neutron stars, after the decay of their pulsar emission, and frozen stars are the two types of star which are the most difficult to observe—i.e., which are "invisible." How can they be detected?

Far away from a frozen or a neutron star, at $r \gg r_g$, the gravitational field is the same as it was before the collapse, i.e., during the normal evolution of the star. Consequently, in the dynamics of stellar systems, invisible stars have the same gravitational effects as normal stars. This means that invisible stars in principle can be detected as follows. The mass of a system, for example a globular cluster, can be computed in two ways: by applying the virial theorem to the motions of its visible stars, and by adding up the masses of all the visible stars, gas, and dust. The difference between the masses obtained by these two methods is the mass of the invisible components of the cluster. Note that this includes not only the masses of the invisible star but also the masses of all other forms of matter in the Universe that are difficult to observe—e.g., neutrinos and gravitational waves. These types of matter (neutrinos and gravitational waves) are not concentrated exclusively in galaxies, but fill the Metagalaxy uniformly. Of course, for relatively small systems (star clusters and galaxies), the total mass of neutrinos and gravitons, even if their density is the maximum possible (see Zel'dovich and Smorodinsky 1961) is negligibly small compared with the probable mass of invisible stars; consequently, this correction is not essential.

Let us estimate the relation between the total mass of visible stars in the Galaxy and the total mass of invisible stars, under the most favorable conditions for invisible-star formation (no catastrophes which disrupt a star before it collapses, no mass loss, etc.). To make the estimate, obviously we must divide the total mass of all stars with $M > 1.2\ M_\odot$, which have been created during the entire existence of the Galaxy, by the total mass of stars with $M < 1.2\ M_\odot$. In this estimate, as in our derivation of equation (13.13.3), we must assume that the rate of stellar formation has remained unchanged during the entire lifetime of the Galaxy. Moreover, we must take into account the fact that the minimum mass which, during the life of the Galaxy, is capable of forming into a star from the diffuse medium as a consequence of gravitational contraction is about $0.1\ M_\odot$. Incidentally, our final result would be decreased by only a factor of 3 if we took the minimum mass to be $0.01\ M_\odot$. The ratio we seek can be obtained with the help of relation (13.13.2); it is

$$\frac{M_{\text{invisible}}}{M_{\text{visible}}} = \int\limits_{1.2}^{\infty} M^{-1.4}dM \Big/ \int\limits_{0.1}^{1.2} M^{-1.4}dM = 0.6 \ .$$

Thus, the invisible mass may be equal to a substantial fraction of the visible mass. Unfortunately, the precision with which the masses of stellar systems are known at present is inadequate to permit one to use those masses to discover whether invisible stars actually exist. The rotation of stars and other sources of mass loss could reduce $M_{\text{invisible}}/M_{\text{visible}}$ substantially.

Guseynov and Zel'dovich (1966) have suggested the possibility of discovering invisible stars when they are components in binary systems of which the other star (the "primary" component) is visible. (For more recent work along these lines see Trimble and Thorne 1969.) For a long time astronomers have known of visual binaries and spectroscopic binaries in which the secondary component cannot be seen. Generally, the invisibility of the secondary is explained by an anomalously great luminosity of the primary, or by a weakness of the spectral lines of the secondary (in the case of spectroscopic binaries) or by the assumption that the secondary is itself multiple and is thus underluminous for its mass. From the motion of the visible component, from its spectrum, and from its luminosity one can determine approximately the mass of each component. In some cases the mass of the invisible secondary component turns out to be greater than the mass of the visible primary. In these cases, if the secondary were a single, normal, main-sequence star, then its greater mass would imply for it a luminosity greater than that of the primary. Consequently, the secondary cannot be a normal star, and one is tempted to explain its invisibility by assuming that it is a "frozen" star. In spite of the complexity of this argument, it deserves attention because the discovery of at least one frozen star would have an immense theoretical significance. A single observational discovery, of course, would be better proof of the existence of these objects than any theoretical, statistical, or other indirect considerations can be.

[*Note added by editors:* Recently Lynden-Bell (1969) has argued that explosive activity in quasars and the nuclei of galaxies may leave behind "frozen stars" of masses 10^7–10^8 M; and he has constructed detailed models for the accretion of gas onto such supermassive frozen stars. The spectra of the radiation emitted by the accreting gas and surrounding gas clouds show remarkable resemblances to the spectra of the nucleus of our own Galaxy, and of the nearby galaxies M31, M32, M81, M82, M87, and NGC 4151.

[More recently Wolfe and Burbidge (1970) have given a detailed, optimistic analysis of the possibility that much of the mass of giant elliptical galaxies may be contained in frozen stars—both large ones in the nuclei and small ones in the outer regions of the galaxies.

[A recent contribution by Bardeen (1970) has pointed out that the frozen stars in the real Universe should typically have large amounts of angular momentum ($K \lesssim GM^2/c$) and should thus be described by the Kerr metric rather than the Schwarzschild metric. This has important implications for the efficiency with which accreting matter radiates: the energy released per unit infalling mass can be as large as $0.432c^2$, compared to Salpeter's limit of $0.057c^2$ for the Schwarzschild case.]

14 QUASISTELLAR OBJECTS

14.1 OBSERVED PROPERTIES OF QUASISTELLAR OBJECTS

The discovery of quasistellar radio sources ("quasars" or "QSSs") by Maarten Schmidt in 1963 is the principal factor that generated the current tremendous interest in relativistic astrophysics. The picture was expanded two years later by Sandage's (1965) discovery of radio-quiet quasars, called "quasistellar galaxies" or "QSGs." QSSs and QSGs together are called quasistellar objects (QSOs). A number of original papers and review articles have dealt with the history of quasars, with observational investigations of them, and with theories of their structure. At this point, we will not give a detailed review of the literature on theories of QSOs. The reader is referred to the review by Greenstein (1963), to supplementary comments by Ozernoy (1964), to a collection of articles (in Russian) edited by Pskovsky (1965), to the Transactions of Symposium 29 (Byurakan Observatory), to the proceedings of the Texas Symposia on Relativistic Astrophysics, to the very exhaustive reviews by the Burbidges (E. M. Burbidge 1967, G. R. Burbidge and E. M. Burbidge 1967b), and to Schmidt's (1969) review, which also include extensive bibliographies.

At the present time, there exists no theory of QSOs that is either convincing or generally accepted. The six years since QSOs were discovered have been characterized, on one hand, by the accumulation of more and more data which reveal the complexity and many-sidedness of QSOs; and on the other hand, by a growing realization that long and hard theoretical efforts will be required to account fully for the observational data. In this chapter we will review very briefly the observational data on QSOs, and will then dwell on those attempts at explaining their nature which, in our opinion, are the most promising.

In the optical range, quasars are observed as dim point objects (the brightest and also the closest is 3C 273 whose apparent magnitude is ~13); some of the closest ones have a diffuse jet or a halo. Radio sources with small angular dimensions (sometimes smaller than 10^{-3} seconds; see the review of Cohen 1969) have been identified with these objects. Incidentally, the number of quasars with identified spectra which had been discovered by autumn 1969 was about 120.

The "quasistellar galaxies" discovered by Sandage (1965) are similar to quasars except that they do not have strong radio emission. The number of quasistellar galaxies is dozens of times greater than the number of quasars. The redshifts of about 40 QSGs have been measured.

The total number of QSOs out to the Hubble radius is estimated to be about 10^6 (Schmidt 1969). Below we shall speak mainly about QSSs (quasars).

Along with normal lines, the emission spectra of quasars also contain forbidden lines of highly ionized elements. The widths of these lines are 10–50 Å or more. Some quasars have small resonance bands in absorption.

The spectral lines show large redshifts: 3C 273 has $z = \Delta\lambda/\lambda = 0.158$. In other quasars, z is greater; in autumn 1969 there were twenty known sources with $z > 2$.

There are many arguments against explaining QSO redshifts by strong gravitational fields in the sources (see the next section). Consequently, the redshift must be caused by the recession of the objects at tremendous relative velocities. Objects inside our Galaxy would not be expected to be always moving away from us, especially when one considers the statistical relationship between the apparent magnitude and the redshift. Therefore, one can assume that the redshift is caused by the cosmological expansion of the Metagalaxy, and that quasars are at tremendous distances from us, which can be determined from the redshift. (For a discussion of the alternative possibility, that quasars are "nearby," see § 14.2.) Knowing the distance and the observed flux on Earth, one can determine the total flux of energy from a quasar. Including the infrared radiation (Low and Johnson 1965), this flux is of the order of 10^{47} ergs $\sec^{-1}$, which is greater than the luminosity of the brightest galaxies by two orders of magnitude.

If we assume an approximate steady state of the object's plasma under the forces of radiation and gravity in the region where the continuous spectrum is formed (the emission lines are formed further away from the center, in an external shell) and also consider radiative equilibrium (§ 11.10), we obtain a rough lower limit on the mass of a quasar. For the luminosity $L \approx 3 \times 10^{47}$ ergs $\sec^{-1}$ of 3C 273, this yields $M = 3 \times 10^9\ M_\odot$ (Zel'dovich and Novikov 1964*b*). Analyses of the physical conditions in the shell of the quasar, where the emission lines originate, have been given by Shklovsky (1964); Greenstein and Schmidt (1964, 1965); and Burbidge, Burbidge, Hoyle, and Lynds (1966). For a review of more recent work see the book of Burbidge and Burbidge (1967*b*).

One astonishing property of quasars is the variability of their optical brightness, discovered simultaneously by Soviet astronomers (Sharov and Efremov 1963) and American astronomers (Smith and Hoffleit 1963). Upon investigation, many quasars have turned out to be variable. According to Penston and Cannon (quoted by Schmidt 1969) about 20 percent of all quasars show great variability in brightness. The brightness of a quasar sometimes changes more or less regularly with a period of several years; and large variations in the brightness (approximately by 60 percent) in times of the order of a week are occasionally observed (see Schmidt 1966 and papers cited in Schmidt's 1969 review). This means that the linear dimensions of the radiating surface are not greater than a light week! If one assumes, in turn, that the variations in brightness occur throughout the object as a whole, then it follows that the quasar mass must be less than $10^{11}\, M_\odot$; otherwise it would be inside its gravitational radius!

An analysis of the continuous spectrum, as well as the presence of radio emission, polarization, and infrared radiation, favors a nonthermal explanation of the radiation (see bibliography in the references above). A second and no less remarkable property of quasars is their radio variability (Scholomitsky 1965). Even though other observers have not confirmed Sholomitsky's conclusions as to the variability of CTA 102, they have discovered variability in other objects. The variations in the radio brightness have a number of fascinating properties. At different wavelengths, the intensity varies in different manners, sometimes with derivatives of opposite signs; see the review of Kellermann and Pauliny-Toth (1968).

The variation of the radio brightness indicates that the radio radiation, like the visible, is emitted from a very small region. This in turn imposes restrictions on the possibilities of explaining the radiofrequency emission by the synchrotron mechanism because one would observe the phenomenon of reabsorption of radiation in a comparatively dense plasma (see Slysh 1963). A probable radiation mechanism in this case may be the radiation of a coherent plasma (Ginsburg and Ozernoy 1966).

A third remarkable property of quasars is that they are not typically found in clusters of galaxies, whereas most galaxies are members of clusters. This is even more remarkable because Seyfert galaxies, whose centers are similar to quasars on a reduced scale, are found in clusters just as are normal galaxies.

Recently Bahcall, Schmidt, and Gunn (1969) have carefully examined the sky around the six QSOs of smallest known redshift. They discovered that of these six QSOs, five are located in small clusters of galaxies. For one of these clusters the redshifts of four galaxies could be measured, and were found to agree with the redshift of the associated QSO (B264, with $z = 0.095$). However, as those authors emphasized, this quasar is not a typical QSO, and it may in fact be a N-type galaxy. On the other hand, there are definitely no clusters around the QSSs 3C 273 or 3C 48 (Sandage and Miller 1966).

Very recently, variability has been found in the light output of the nuclei of several Seyfert and N-type galaxies. The fact that this variability is of the same kind as one finds in quasars further strengthens the resemblance between active galactic nuclei and quasars, and it is a strong argument in favor of the cosmological interpretation of quasars.

Another recent observational discovery is the presence of systems of absorption lines with different redshifts in the same quasar. For example, in

PKS 0237−23, which has $z_{em} \approx 2.223$, there are systems of absorption lines with $z = 1.364, 1.513, 1.656, 1.671, 2.202$, and probably also 1.595 and 1.955. Whenever z_{abs} is not the same as z_{em}, then almost always $z_{abs} < z_{em}$. In those few cases which violate this rule, the difference $z_{abs} - z_{em}$ is small. It is natural to suppose that absorption lines with $z_{abs} < z_{em}$ arise not in the vicinity of the quasar, but rather in the passage of its light through objects which lie between us and it.

A very peculiar observation is this: of twenty quasars with $z > 1.90$, four have emission lines with z in the range between 1.955 and 1.956; moreover, of eighteen different absorption-line redshifts found in quasars thus far, six of them lie between 1.94 and 1.97. This concentration near 1.95 for the absorption-line redshifts was first noticed by Burbidge and Burbidge (1967b). Although the statistical significance of these absorption-line coincidences is controversial, they probably testify, together with the emission-line redshifts, in favor of the existence of a Λ-term in the Einstein equations (see chapter 1), and in favor of an essential influence of the Λ-term upon the expansion of the Universe (Petrosian, Salpeter, and Szekeres 1967; Shklovsky 1967b; Kardashev 1967). For a further discussion of this point see Vol. II on "Cosmology."

Of considerable importance for understanding quasars may be the observation that in the envelopes where the emission lines are formed the chemical composition is the same as that which we are accustomed to finding in stars and in galactic nebulae (Shklovsky 1964).

Also of great importance is observational data about the structure of quasars. Some light on this has been provided by studies with long-baseline interferometers (see Kellermann *et al.* 1968).

14.2 THEORIES OF QUASISTELLAR OBJECTS

In the literature about quasars many factors have been mentioned which may play a role in the energy balance, in the stability of quasars, and in the observable external picture.

Let us enumerate here the major contributing elements and factors[1] that have been mentioned at various times:

1. Hot plasma, including plasma that is in local equilibrium with radiation.

2. Magnetic fields including small-scale fields, and fields on the scale of the entire quasar.

3. Relativistic particles.

4. Gas clouds in turbulent motion.

5. Rotation of the quasar as a whole.

6. Stars aggregated together into systems with surrounding gas. These

---

1. Antiquasars from antigravitating antimatter are beyond the scope admissible in this book.

can be classified as follows: (*a*) stars that explode like supernovae; (*b*) stars that collide and eject gas; (*c*) neutron stars that create a gravitational field and gravitational redshift; (*d*) neutron stars as an energy source in accretion; and (*e*) frozen stars behaving like (*c*) and (*d*).

Sometimes the possible role of an unusually powerful gravitational field is also mentioned.

The observational data definitely indicate the presence of plasma, magnetic fields, and relativistic particles; this is particularly demonstrated by the polarization of the radiation and by the radiofrequency radiation (radio emission) from quasars. It is not impossible that quasistellar galaxies, i.e., objects that are radio-quiet (that do not emit radiofrequency radiation) but are optically similar to quasars, contain fewer relativistic particles and have weaker fields than quasars. A systematic comparison of quasars and quasistellar galaxies has apparently not yet been made.

Observations of the breadth of the emission lines indicate the existence of gas velocities of the order of 3000 km sec^{-1}.

An understanding of the nature of quasars requires an explanation of the source of their tremendous energies. The typical ages of quasars are probably about 10^6 years, since some of them have ejected clouds of radiating particles which are now $\sim 10^6$ light-years away from the center. (Such particles could not travel outward faster than light.) If a quasar emits 10^{47} ergs sec^{-1} over 10^6 years, the total energy radiated is $\sim 10^{60}$ ergs. Astronomers hope that the same type of source which powers quasars, or an analogous one, provides the total energy supply for powerful radio galaxies such as Cygnus A (10^{60} ergs) and also causes the explosions which occur in the centers of some galaxies (Ambartsumyan 1964).

In chapters 10 and 11, it was demonstrated that a quasar cannot be a supermassive star with a nuclear source of energy and without macroscopic motions of matter, as was initially assumed by Hoyle and Fowler (1963*a*). (The supermassive-star theory of Hoyle and Fowler gave birth in the Soviet literature to the term "superstar" (*sverkhzvezda*), which is just one of the numerous names for quasars that did not gain general acceptance.) Other attempts to use nuclear energetics were also unjustified (see the comments by Ozernoy on Greenstein's review). It was noted above that the maximum possible yield of nuclear energy is $\sim 8 \times 10^{-3}$ mc^2, whereas the yield of gravitational energy in principle may be as large as mc^2. In 1961 Ginzburg suggested gravitational collapse of gas as an energy source in radio galaxies (see also Shklovsky 1962). It is only necessary to find an appropriate mechanism for the translation of the kinetic energy of the collapsing mass into other forms of energy. In preceding sections, we considered some possible mechanisms. In the case of spherical or near-spherical collapse, the yield of energy cannot be substantial because of the effect of self-closure. The mechanism of gravitational radiation (Fowler 1964*a;* Shklovsky and Kardashev 1964; Zel'dovich and Novikov 1964*b*) is not adequate either, because gravitational waves interact with matter hardly at all. Therefore, gravitational waves can be considered as only a channel by which energy flows away from the system. Accretion of matter might release a sufficient amount of energy. Also, a substantial role in the quasar phenomenon might be played by rotation and turbulence, which can stabilize objects against collapse and by magnetic and magnetohydrodynamic processes. We will analyze these ideas further below.

Before the cosmic microwave radiation was discovered, experimentally verifying the "big bang" cosmological model (vol. II), Novikov (1964*b*) and Ne'eman (1965) suggested that quasars and explosions in galactic centers might be small regions of the Universe whose cosmological expansion had been retarded until recently. However, as has been demonstrated by Zel'dovich and Novikov (1966*a, b*), in a hot model of the Universe such an interpretation encounters a number of difficulties.

Finally, let us call attention to an attempt by Hoyle and Burbidge (1966) to construct a model which circumvents a number of difficulties in the explanations of quasars. Their suggestion is that the quasars are in the neighborhood of our Galaxy, that they are very small and radiate relatively weakly, and that they were ejected with velocities near that of light by an explosion in the center of an active galaxy (or several explosions in several such galaxies). This is a "local hypothesis" to be contrasted with the "cosmological hypothesis." It was pointedly remarked at the Byurakan Symposium in 1966 that from this viewpoint powerful radio galaxies and Seyfert galaxies should also have a local nature because there we encounter the same energy problems as in quasars. A critique of the local theory has been given by Setti and Woltjer (1966).

Up to this point we have discussed primarily the problem of the tremendous energy release. But the quasars which we observe are, perhaps, stationary objects; and for the explanation of their nature we must construct a stationary model which explains all the observational data recounted in the preceding section.

The theories of quasars can be divided into two groups. In the first group a quasar is considered to be a single body of gas (a massive superstar). In the second, a quasar is regarded as a system of stars flying about in a gaseous nebula.

The emission of radio waves by quasars is the most convincing proof that they contain magnetic fields with high-energy electrons moving in their fields. The unique shapes of the glowing jets which emerge from the centers of some quasars also indicate the important role of magnetic fields.

The large number of investigations of the equilibrium and evolution of supermassive stars without a consideration of magnetic fields and macroscopic motions resulted essentially in negative results. Let us summarize those results (see chapters 9 and 10). A spherically symmetric stellar mass of $M > 10^5 \, M_{\odot}$, without rotation or turbulence, becomes unstable at a rather low density and moderate temperature (at a moment when the gravitational potential is still small and nuclear energy cannot yet be released). Loss of stability is followed by collapse, in which gravitational energy is converted into thermal and kinetic energy of matter; but none of these forms of energy escape from the star; all are buried in the gravitational field of the star after its self-closure. A consideration of rotation changes these conclusions and permits us to explain the release of the energy observed in quasars.

Analyses of configurations with asymmetric motions reveal that such motions can lead to the release of sufficient amounts of energy in the form of thermal radiation and kinetic energy of jets. However, to establish the true picture of quasars, one must consider magnetohydrodynamic effects because one cannot ignore the direct observational data on magnetic fields. Therefore, we will begin with an analysis of the possible role of the magnetic field.

Ginzburg and Ozernoy (1964) have calculated the behavior of the magnetic field of a collapsing star. The field is amplified in the course of contraction because the field lines are frozen in. The relationship between the magnetic and gravitational energy does not change until the relativistic stage:

$$H \sim (1/R^2) , \qquad W_m \sim H^2 R^3 \sim (1/R) .$$

Ginzburg and Ozernoy emphasize that the magnetic energy accounts for only a small fraction of the total energy of the star, and has no effect upon the dynamics of the main gas mass. Inclusion of the effects of GTR shows (cf. § 13.6) that the gravitational self-closure is accompanied by a swallowing of that part of the magnetic field of the star which is adjacent to the Schwarzschild surface. In this process the part of the external magnetic field which was anchored in the star rapidly vanishes. In the plasma surrounding the star, complex magnetohydrodynamic phenomena may arise.

Kardashev (1964) has considered a mechanism for amplifying the magnetic field of a rotating plasma cloud during its gravitational contraction. The relative motion of various parts of the cloud is accompanied by a shearing of the magnetic field lines and by an amplification of the field. Kardashev assumes that the energy of the field is comparable in order of magnitude to the gravitational energy of the cloud, and he analyzes under these circumstances the formation of magnetohydrodynamic waves in the plasma during the cloud's rapid contraction. Problems of magnetohydrodynamic phenomena, and specifically an explanation of the periodic variation of quasar brightness in the framework of synchrotron-radiation theory, have been analyzed by Ozernoy (1965a, 1966b).

A logical conclusion of this line of thought is an idea which was formulated most precisely by Layzer (1965) and has been analyzed by Ozernoy (1966a). Here a quasar is construed as a body in which gravitation is balanced primarily by chaotic and turbulent magnetic fields.

It is well known that the stress tensor of a magnetic field corresponds to repulsion in directions perpendicular to the field lines and to pressure along the field. Therefore, a chaotic field in which all directions are uniformly represented creates on the average a repulsion corresponding to a mean pressure which equals one-third the energy density. The characteristic relation $\langle p \rangle = \frac{1}{3}\epsilon$ is always correct on the average for a stationary electromagnetic field; it is correct for an aggregate of electromagnetic waves in vacuum (i.e., for photon gas); and it is correct for a chaotic magnetic field in a plasma, supported by currents flowing through the plasma.

Before we proceed to establish the details of a magnetoturbulent quasar model, let us explore the general relationships between the mass of a magnetoturbulent, supermassive star and the current creating the magnetic field H, which is required for equilibrium. In order of magnitude $GM^2/R = H^2R^3$, $H = MG^{1/2}/R$. The current is found from the equation $\nabla \times H = 4\pi j = 4\pi n e(v/c)$, where e is the electron charge in electrostatic units, n is the electron number density, and v is the mean velocity of the electrons which produce the current. If we substitute $|\nabla \times H| \sim H/r$, where r is the characteristic length scale, and if we write $M = Nm_p = nR^3m_p$, where N is the total number of nucleons in the star and m_p is the proton mass, then we obtain

$$v = c[(R/r)(Gm_p{}^2/e^2)]^{1/2} .$$

Thus, the expression for v involves the ratio of the gravitational interaction of two protons to their electrostatic interaction:

$$(Gm_p{}^2/e^2) = 10^{-36} ; \quad v = 10^{-18}(R/r)^{1/2}c .$$

It is precisely because the gravitational interaction is so small, compared with the electrostatic interaction, that a very small ordered motion of the particles of one sign with respect to the particles of the other sign can produce a repulsion which balances gravity (cf. the computation at the end of chapter 5).

This model of the object is incomplete in two respects. First, it is difficult for a magnetic field to balance gravity not only on the average but also locally at each point. The magnetic force acting upon a unit volume of the plasma [which equals $(H \times j) \sim H \times (\nabla \times H)$] is not a potential force. The gravitational force, $\rho\nabla\phi$, is a potential force when $\rho = $ const., or $\rho = \rho(\phi)$. It is not clear whether one can attain a precise equality of the two forces at each point for an arbitrary distribution of density. Second, if the only forces involved are gravitational and magnetic, then the body on the whole is in a state of neutral stability since the magnetic field has an adiabatic index of $\gamma = \frac{4}{3}$ (with respect to a homologous contraction of the body).

These difficulties are resolved if, following Layzer and Ozernoy, we assume that along with the magnetic field there exist turbulent macroscopic motions of the matter, and that the kinetic energy of these motions is of the order of the magnetic energy.

The turbulent motion occurs with a nonrelativistic velocity; its adiabatic index is $\frac{5}{3}$. Therefore, the body on the whole has an adiabatic index which is a mean of $\frac{4}{3}$ and $\frac{5}{3}$, and is in stable equilibrium. Thus, here we assume the mechanism for stellar stabilization as in § 11.17).

As energy is dissipated, the body contracts slowly. If it were not for the dissipation, then during the contraction the kinetic energy would increase faster than the magnetic energy. The perturbed equilibrium is restored by the transformation of kinetic energy into magnetic energy as a result of further tangling and stretching of the magnetic field lines. The current distributions which create the magnetic field may be unstable with respect to the pinch effect, i.e., in certain regions the plasma may contract due to the attraction of parallel currents. Then electric fields will arise and will accelerate individual groups of charged particles; such phenomena have also been observed during experiments involving discharges in rarefied plasmas. The body considered turns out to be a powerful source of cosmic rays, in which the energy put into each particle is proportional to the particle's charge (the particles move with equal velocities in the given electric field). Collapse finally sets in when GTR effects become important in the contraction; according to Layzer, the collapse is accompanied by an ejection of part of the mass. During the ejection, the magnetic field lines are stretched and the ejected matter takes the form of individual jets, in which the plasma is strongly tied to the frozen-in magnetic field and moves outward along the jet.

Layzer's work is basically descriptive in nature; it includes very few quantitative estimates, and those which are included are not reliable. This does not diminish its value: its main stipulations—the important role of the magnetic field in the total energy balance,[2] the slow evolution of the quasi-equilibrium state, and the generation of relativistic particles—agree well with our observational knowledge of the phenomena that take place in quasars. Detailed computations have been made by Ozernoy (1966*a*).

A problem that still remains unclear is the rate of evolution and the space dimensions of the magnetoturbulence. We recall that Batchelor's fundamental theorem on the equality of magnetic and kinetic energies in a turbulent, highly conductive fluid has not yet been proved. It is not clear just how fast the turbulence will attenuate. Presumably the energy dissipation is rather great. If so, and if the energy dissipation takes place on a time scale shorter than 10^3 or 10^4 times the estimates of § 11.17 for dissipation of turbulence without a magnetic field, then the dissipation is too fast and the suggested model cannot be valid.

It is possible that the greatest energy is associated with the largest-scale turbulence. If so, then another approach is possible, namely, an idealization which assumes stationary, ordered magnetic fields and ordered motions in the gravitational field. One version corresponds to convection with ascent at the equator and descent at the poles. Observational data—especially the

---

2. This point, as we have mentioned, was also contained in Kardashev's work (1964).

organized ejection of one or two jets and the regularity of the variations in brightness—seem to support this type of idealization. Another possible version involves an axisymmetrical configuration with a toroidal magnetic field and rotation of matter about the axis of symmetry; the equilibrium state corresponds to the minimum energy for a given distribution in the matter of the angular momentum per unit mass and of the magnetic flux. In the configuration of minimum energy the angular velocities of the various jets might be different. However, the fact that the energy is minimal indicates a suppression of turbulence by the magnetic field, which in turn will produce an increase in the lifetime of such a state.

The authors still feel that the most realistic hypothesis of the physical nature of a quasar, viewed as a unified entity, is that in which the quasar is construed as a hot, supermassive star, stabilized mainly by rotation;[3] i.e., the picture described in §§ 11.13–11.16 and 11.18. Nonrelativistic motions are necessary to stabilize the star. However, any turbulence present probably decays rapidly into thermal energy and heats up the plasma. If the turbulence is not being continually generated, it soon disappears and cannot stabilize the star. Mechanisms for generating large-scale turbulence, except for differential rotation, are not known. But if turbulence is being generated by rotation, most of the energy will be contained in the rotational motion. Therefore a supermassive-star model of a quasar must be stabilized chiefly by rotation. Naturally, turbulence and the magnetic field may also play a substantial role. They are especially important for a correct understanding of the external manifestations of a quasar—i.e., its observed properties—as we have emphasized previously.

Let us next consider models of quasars as stellar systems. For example, quasars might be compact galaxies in the process of formation or in the process of dying (i.e., collapse).

The assumption of the existence of stars within quasars is not based on direct observations. Rather, this assumption is supported by the fact that the chemical composition of the luminescent plasma in a quasar is that of matter which has been exposed to nuclear reactions in stars. In a condensation of primordial prestellar matter which has not been exposed to the stellar forge, the gas would have consisted of only hydrogen and helium.

Considerations, based primarily on the virial theorem, indicate that in a quasar there must take place a nonrelativistic motion of matter. This assumption is the general foundation of the theories considered in the first part of this section, which involve turbulent or rotational motion of plasma. However, on the basis of the same considerations one can assume also the motion of stars contained in the nucleus of the quasar.

On the assumption that stars are present in the composition of the quasar,

3. Morrison (1969) has stressed the possible role of rotation in the visual appearance of a quasar.

it is necessary to consider the balance between the creation of gas in explosions and collisions of stars, and the consumption of gas by the formation of new stars, by gas accretion onto stars, and by gas ejection from the quasar to the outside.

Spitzer and Saslaw (1966) assumed that the gas is consumed primarily by the formation of new stars; however, under other temperature and density conditions this conclusion may be invalid.

Along with accretion, during a rapid motion of a star relative to the gas, ablation—i.e., a blowing away of material from the star's surface—is also possible. In a dense accumulation of stars and gas, normal stars are frequently destroyed, whereas neutron stars, on account of their small diameter, can survive much longer. The process of natural selection may guarantee that many neutron stars will be included in the composition of a quasar.

As we have noted, the accretion of gas onto the surface of a neutron star is a powerful source of energy.

The theory of compact galaxies was described in the preceding chapter.

Hoyle and Fowler (1967) have proposed a quasar model which attempts to explain the large z by a gravitational redshift. According to this model, a quasar is a compact galaxy consisting perhaps of neutron stars or collapsed objects with $M \approx 10^{12}$–10^{13} $M_\odot$, and $R \sim 10^{18}$–10^{19} cm. The galaxy is transparent to light. A gas cloud in the center emits the line radiation; the gravitational potential at the center is considerably greater than that at the surface. In the opinion of Hoyle and Fowler, the gravitational potential is responsible for the redshift, and may give $z > 2$. However, the equilibrium structure of the galaxy cannot be isothermal or polytropic, since all isothermal and polytropic star clusters with $z_{central} > 0.5$ are unstable against gravitational collapse (Ipser 1969b; see § 11.12). For other remarks critical of this model see Schmidt (1969).

It is not at all necessary to link the idea of a compact galaxy to the idea of a gravitational redshift for the quasar's spectral lines, as did Hoyle and Fowler (1967). Instead, we can build a quasar model based on a compact galaxy whose gravitational potential is small everywhere (see, e.g., Colgate 1967). The observation of forbidden lines supports the existence of a rarefied, extended quasar shell, which can only be located in a domain of small gravitational potential. The mass of a quasar may be due primarily to stars immersed in the gas. The hot gas would be stable against gravitational collapse in the galaxy's gravitational field. The redshift of the spectral lines in this case would be due to the Hubble expansion (cosmological redshift).

Let us conclude with some comments on the evolution of quasars. There exist many observational and theoretical arguments supporting the idea that quasars have a relatively short life span (the average estimates are grouped around a time period of 10^5–10^6 years). In this case it is necessary to understand the form which the quasar matter takes in its latent state, prior to its flare-up in the form of a quasar.

A hypothesis that there exists a rather dense accumulation of stars prior to the flare-up is conceivable. Its evolution (on account of evaporation and collisions) is slow at first, but eventually it acquires a catastrophic nature. It is entirely conceivable that we are observing quasars not at the moment of a catastrophe, but upon the completion of processes that are related to the destruction of normal stars. It has been pointed out, for example, that 3C 273 is in a practically steady state in which its luminosity varies quasi-regularly: these ordered oscillations do not agree with the picture of random explosions of supernovae, or of collisions. Finally, from the assumed short active life of a quasar, it follows that the density of extinct quasars must exceed the density of observable quasars by many times ($\sim 10^5$!). The predictable distance between us and the closest expired quasar (or quasistellar object) must be 100–300 times smaller than the distance to 3C 273, i.e., several megaparsecs.

At the present time, there are neither observational nor theoretical indications as to possible methods of discovering and identifying extinct quasars. Unquestionably, this problem is extremely important.

REFERENCES

Abell, G. O. 1965, *Ann. Rev. Astr. and Ap.*, **3**, 1.

Abrikosov, A. A. 1954, *Astr. Zh.*, **31**, 112.

Alfvén, H. 1964, preprint (Stockholm).

———. 1965, *Astr. Zh.*, **42**, 873.

Alfvén, H., and Fälthammar, G.-G. 1963, *Cosmical Electrodynamics* (London: Oxford University Press).

Alfvén, H., and Klein, O. 1962, *Ark. f. Fys.*, **23**, 187.

Aller, L. H. 1961, *Abundance of the Elements* (New York: Wiley-Interscience).

Aller, L. H., and McLaughlin, D. B. (ed.). 1965, *Stellar Structure* (Chicago: University of Chicago Press).

Alpher, R., Bethe, H. A., and Gamow, G. 1948, *Phys. Rev.*, **73**, 803.

Alpher, R., and Herman, R. 1948, *Phys. Rev.*, **74**, 1737.

———. 1950, *Rev. Mod. Phys.*, **22**, 153.

———. 1953, *Ann. Rev. Nucl. Sci.*, **2**, 1.

Al'tschuler, L. V. 1965, *Uspekhi Fiz. Nauk*, **85**(2), 197.

Ambartsumyan, V. A. 1938, *Uchenye zapiski Leningrad Gosud. Univ.*, **22**, 19.

———. 1958, *Rapport 11 Conseil de Physique Solvay* (Brussels: Stoops).

———. 1960, *Nauchnyye trudy* [Collected works], Vol. **2** (Yerevan: Akademiya Nauk Armyanskoy SSR).

———. 1964, *Rapport 13 Conseil de Physique Solvay* (Brussels: Stoops).

Ambartsumyan, V. A., and Saakyan, G. S. 1960, *Astr. Zh.*, **37**, 193.

———. 1963, *Voprosy kosmogonii* [Problems of Cosmogony], **9**, 91.

Ames, W. L., and Thorne, K. S. 1967, *Ap. J.*, **151**, 659.

Anand, S. P. S. 1965, *Nature*, **207**, 1345.

Anders, E. 1959, *Ap. J.*, **129**, 327.

Anderson, B. A., Donaldson, W., Palmer, H., and Rowson, B. 1965, *Nature*, **205**, 375.

Anderson, J. L., and Gautreau, R. 1966, *Phys. Letters*, **20**, 24.

Antonov, V. A. 1960, *Astr. Zh.*, **37**, 918 (English transl. in *Soviet Astr.—AJ*, **4**, 859, 1961).

———. 1962, *Vestnik Leningrad Gosud. Univ.*, **7**, 135.

Argan, P. E., Mantovani, G., Marazzini, P., Piazzoli, A., and Scannicchio, D. 1965, *Nuovo Cimento Suppl.*, **3**, 245.

Arnett, W. D. 1966, *Canadian J. Phys.*, **44**, 2553.

———. 1967, *ibid.*, **45**, 1621.

———. 1968, *Ap. J.*, **53**, 341.

References

Arnett, W. D. 1969, *Ap. Spa. Sci.*, **5**, 180.

Auer, L. H., and Woolf, N. J. 1965, *Ap. J.*, **142**, 182.

Baade, W. 1963, *Evolution of Stars and Galaxies* (Cambridge: Cambridge University Press).

Baade, W., Christy, R., Burbidge, G., Fowler, W. A., and Hoyle, F. 1956, *Pub. A.S.P.*, **68**, 296.

Baade, W., and Zwicky, F. 1934, *Proc. Nat. Acad. Sci.*, **20**, 255.

Baglin, A. 1964, *C.R.*, **258**, 5801.

――――. 1965, *ibid.*, **260**, 2424.

――――. 1966, *Ann. d'Ap.*, **29**, 103.

Bahcall, J. N., and Salpeter, E. E. 1965, *Ap. J.*, **142**, 1677.

――――. 1966, *Ap. J.*, **144**, 847.

Bahcall, J. N., Schmidt, M., and Gunn, J. E. 1969, *Ap. J.* (*Letters*), **157**, L77.

Bahcall, J. N., and Wolf, R. A. 1965*a*, *Phys. Rev.*, **140**, B1445, B1452.

――――. 1965*b*, *Ap. J.*, **142**, 1254.

Barber, D., Donaldson, W., Miley, G. K., and Smith, M. 1966, *Nature*, **209**, 753.

Bardeen, J. M., 1968, *Bull. Am. Phys. Soc.*, **13**, 41.

――――. 1968*b*, unfinished manuscript (see Hartle 1970).

――――. 1970*a*, *Nature*, in press.

――――. 1970*b*, *Ap. J.*, **162**, 71.

Bardeen, J. M., and Anand, S. P. S. 1966, *Ap. J.*, **144**, 953.

Bardeen, J. M., and Wagoner, R. V. 1969, *Ap. J.* (*Letters*), **158**, L65.

Barenblatt, G. I., and Zel'dovich, Ya. B. 1958, *Doklady Akad. Nauk*, **118**, 671.

Baym, G., Pethick, C., Pines, D., and Ruderman, M. 1969, *Nature*, **224**, 872.

Baz', A. I., Gol'dansky, V. I., and Zel'dovich, Ya. B. 1965, *Uspekhi Fiz. Nauk*, **85**(3), 445.

Bechi, C., Gallinaro, G., and Morpurgo, G. 1965, *Nuovo Cimento*, **39**, 409.

Belinfante, F. G. 1966, *Physics Letters*, **20**, 25.

Belinsky, V. A., and Khalatnikov, I. M. 1969, *Zh. Eksp. Teoret. Fiz.*, **56**, 1700.

Bellamy, E. H., Hofstadter, R., and Lakin, W. 1968, *Phys. Rev.*, **166**, 1391.

Berestetsky, V. B., Lifshitz, E. M., and Pitaevsky, L. P. 1969, *Quantum Mechanics, Part 2, the Relativistic Theory* (Moscow: Nauka 1968)

Bertotti, B. 1966, *Proc. Roy. Soc. London*, *A*, **294**, 195.

Bird, G. F. 1964, *Rev. Mod. Phys.*, **36**, 717.

Birkhoff, G. 1923, *Relativity and Modern Physics* (Cambridge, Mass.: Harvard University Press).

Bisnovatyi-Kogan, G. S. 1966, *Astr. Zh.*, **44** (1), 89.

――――. 1967, *Doklady na XIII kongresse MAS* (Report at the XIIIth International Congress of the IAU).

――――. 1968*a*, *Astrofizika*, Vol. **4**, 89.

References

———. 1968*b*, *Astron Zh.*, **45**, 74.

———. 1969*a*, *Astrofizcia*, **5**, 425.

———. 1969*b*, *ibid.*, **5**, 223.

———. 1970, *ibid.*, (in press).

Bisnovatyi-Kogan, G. S., and Kazhdan, Ya. M. 1966, *Astr. Zh.*, **43**(4), 761.

Bisnovatyi-Kogan, G. S., and Seidov, Z. F. 1969, *Astrofizika*, **5**, 243.

———. 1970, *Astr. Zh.*, **47**, 139.

Bisnovatyi-Kogan, G. S., and Thorne, K. S. 1970, *Ap. J.*, **160**, 885.

Bisnovatyi-Kogan, G. S., and Zel'dovich, Ya. B. 1966, *Astr. Zh.*, **43**(6), 1200.

———. 1968, *ibid.*, **45**, 241.

———. 1969*a*, *Astrofizika*, **5**, 425.

———. 1969*b*, *ibid.*, **5**, 223.

———. 1970, *ibid.* **6**, 387.

Bisnovatyi-Kogan, G. S., Zel'dovich, Ya. B., and Novikov, I. D. 1967, *Astr. Zh.*, **44**, 525.

Bisnovatyi-Kogan, G. S., Zel'dovich, Ya. B., Sagdeev, R. Z., and Friedman, A. M. 1969, *Zh. Prikladnoi y Tekhnicheskii Fizika*, **3**, 3.

Blaauw, A. 1961, *Bull. Astr. Inst. Netherlands*, **15**, 265.

———. 1965, in *Stars and Stellar Systems*, Vol. **5**, *Galactic Structure*, ed. A. Blaauw and M. Schmidt (Chicago: University of Chicago Press).

Bondatev, B. V., Voyevodski, A. V., Dadykin, V. L., Maslow, V. N., Milcheev, S. P., Sborshchikov, V. G., and Stepanov, V. J. 1968, *Izvest. Akad. Nauk, Fiz. Ser.*, **32**, 531.

Bondi, H. 1961, *Cosmology* (2d ed.; London and New York: Cambridge University Press).

———. 1964, *Proc. Roy. Soc. London*, *A*, **282**, 303.

Bondi, H., and Gold, T. 1948, *M.N.R.A.S.*, **108**, 252.

Bondi, H., and Lyttleton, R. A. 1959, *Proc. Roy. Soc. London*, *A*, **252**, 313.

———. 1961, three lectures, BBC (London).

Bonnor, W. B. 1957, *M.N.R.A.S.*, **117**, 104.

Bogorodsky, A. F. 1962, *Uravneniya polya Einsteina i ikh primeneniye v astronomii* (Einstein's equations of the field and their application in astronomy) (Kiev: Izdatel'stvo Kievskogo universiteta).

Borst, L. 1950, *Phys. Rev.*, **78**, 807.

Bowyer, S., Byram, E. T., Chubb, T. A., and Friedman, H. 1964, *Nature*, **201**, 1307.

Boyer, R. H., and Lindquist, R. W. 1967, *J. Math. Phys.*, **8**, 265.

Boyer, R. H., and Price, T. G. 1965, *Proc. Cambridge Phil. Soc.*, **61**(2), 531.

Brachmachary, R. L. 1965, *Nuovo Cimento*, **2**, 850.

Braginsky, S. I. 1964, *Zh. Eksp. Teoret. Fiz.*, **47**, 2178.

Braginsky, V. B. 1965, *Uspekhi Fiz. Nauk*, **86**, 433.

———. 1966, *Pis'ma* [Letters], *Zh. Eksp. Teoret. Fiz.*, **3**, 69.

Braginsky, V. B., Zel'dovich, Ya. B., Martynov, V. K., and Migulin, V. V. 1967, *Zh. Eksp. Teoret. Fiz.*, **52**, 29.

References

Braginsky, V. B., Zel'dovich, Ya. B., Martynov, V. K., and Migulin, V. V. 1968a, *ibid.*, **54**, 91.

———. 1968b. [Ed. note: reference unknown.]

Braginsky, V. B., Zel'dovich, Ya. B., and Rudenko, V. N. 1969, *Pisma* [Letters]; *Zh. Eksp. Teoret. Fiz.*, **10**, 437.

Brill, D. R., and Cohen, J. M. 1966, *Phys. Rev.*, **143**, 1011.

Brill, D. R., and Hartle, J. B. 1964, *Phys. Rev.*, **135**, B271.

Bronstein, M. P. 1934, *Phys. Zs. Sowjetunion*, **2**, 100.

———. 1936, *ibid.*, **9**, 140.

Brueckner, K. A., Gammel, J. L., and Kubis, J. T. 1960, *Phys. Rev.*, **118**, 1095.

Burbidge, E. M. 1967, *Ann. Rev. Astr. and Ap.*, **5**, 399.

Burbidge, E. M., and Lynds, C. R. 1967, *Ap. J.*, **147**, 388.

Burbidge, G. R., and Burbidge, E. M. 1958, *Hdb. d. Phys.*, **51**, 131.

———. 1966, *Ap. J.*, **143**, 271.

———. 1967a, *Ap. J.* (*Letters*), **148**, L107.

———. 1967b, *Quasistellar Objects* (San Francisco: Freeman).

Burbidge, G. R., Burbidge, E. M., Hoyle, F., and Lynds, C. R. 1966, *Nature*, **210**, 774.

Burbidge, G. R., and Hoyle, F. 1956, *Nuovo Cimento*, **4**, 558.

Burke, W. L. 1969, Ph.D. thesis, California Institute of Technology (available from University Microfilms, Ann Arbor, Michigan).

———. 1970, *Phys. Rev.* (in press).

Burke, W. L., and Thorne, K. S. 1969, in *Relativity*, M. Carmeli, S. I. Fickler, L. Witten eds. (New York: Plenum Press), p. 209.

Cameron, A. G. W. 1959a, *Ap. J.*, **130**, 884.

———. 1959b, *ibid.*, p. 895.

———. 1959c, *ibid.*, p. 916.

———. 1965a, *Nature*, **205**, 787.

———. 1965b, *ibid.*, **206**, 1342.

———. 1969, *Comments Ap. and Space Sci.*, **1**, 172.

Cameron, A. G. W., and Mock, M. 1967, *Nature*, **215**, 464.

Carter, B. 1966a, *Phys. Rev.*, **141**, 1242

———. 1966b, *Phys. Letters*, **21**, 423.

———. 1968, *Phys. Rev.*, **174**, 1559.

Chandrasekhar, S. 1933, *M.N.R.A.S.*, **93**, 390.

Chandrasekhar, S. 1939, *Stellar Structure* (Chicago: University of Chicago Press).

———. 1942, *Principles of Stellar Dynamics* (Chicago: University of Chicago Press).

———. 1943, *Rev. Mod. Phys.*, **15**, 1.

———. 1964a, *Phys. Rev. Letters*, **12**, 114, 437.

———. 1964b, *Ap. J.*, **140**, 1517.

———. 1965, *Phys. Rev. Letters*, **14**, 241.

References

Chandrasekhar, S., and Esposito, F. P. 1970, *Ap. J.*, **160**, 153.

Chandrasekhar, S., and Tooper, R. F. 1964, *Ap. J.*, **139**, 1396.

Chau, W.-Y. 1967, *Ap. J.*, **147**, 664.

Chechetkin, V. M. 1969, *Astr. Zh.*, **46**, 202.

Chiu, H. Y. 1961a, *Ann. Phys.*, **15**, 1.

———. 1961b, *ibid.*, **16**, 321.

———. 1961c, *Phys. Rev.*, **123**, 1040.

———. 1964, *Ann. Phys.*, **26**, 354.

———. 1966a, *Phys. Rev. Letters*, **17**, 712.

———. 1966b, *Stellar Evolution* (New York: Plenum Press).

Chiu, H. Y., and Salpeter, E. E. 1964, *Phys. Rev. Letters*, **12**, 413.

Chupka, W. A., Schiffer, J. P., and Stevens, C. M. 1966, *Phys. Rev. Letters*, **17**, 60.

Clayton, D. D. 1968, *Principles of Stellar Evolution and Nucleosynthesis* (New York: McGraw-Hill Book Co.).

Clifford, F. E., and Tayler, P. J. 1965, *Mem. R.A.S.*, **69**, 21.

Cohen, J. M., and Brill, D. R. 1968, *Nuovo Cimento*, **56B**, 209.

Cohen, M. H. 1969, *Ann. Rev. Astr. and Ap.*, **7**, 619.

Colgate, S. A. 1967, *Ap. J.*, **150**, 163.

Colgate, S. A., and McKee, C. 1969, *Ap. J.*, **157**, 623.

Colgate, S. A., and White, R. H. 1966, *Ap. J.*, **143**, 626.

Collins, R. A. 1968, *Nature*, **217**, 709.

Conklin, E. K., and Bracewell, R. V. 1967, *Phys. Rev. Letters*, **18**, 614.

Cowling, T. G. 1951, *Ap. J.*, **114**, 272.

Cox, A. N., Stewart, J. N., and Eilers, D. D. 1965, *Ap. J. Suppl.*, No. 94, **11**, 1.

Danby, G., Gaillard, J. M., Goulianois, K., Lederman, L. M., Misty, M., Schwarzt, M., and Steiberg, J. 1962, *Phys. Rev. Letters*, **9**, 36.

Dashevsky, F. M., and Slysh, V. I. 1965, *Astr. Zh.*, **42**, 863.

Dashevsky, V. M., and Zel'dovich, Ya. B. 1964, *Astr. Zh.*, **41**, 1071.

Davies, R. D., and Jennison, R. C. 1964, *M.N.R.A.S.*, **128**, 123.

Davis, R., Jr., Harmer, D. S., and Hoffman, K. C. 1968, *Phys. Rev. Letters*, **20**, 1205.

de la Cruz, V., Chase, J. E., and Israel, W. 1970, *Phys. Rev. Letters*, **24**, 423.

DeGrasse, R. W., Hogg, D. C., Ohm, E. A., and Scovil, H. E. D. 1959, *J. Appl. Phys.*, **30**, 2013.

DeWitt, B. 1967, *Phys. Rev.*, **160**, 1113.

DeWitt, C., and DeWitt, B. (eds.) 1964, *Relativity, Groups, and Topology* (New York: Gordon & Breach).

Dicke, R. H. 1961a, *Sci. Am.*, **205**, 84.

———. 1961b, *Phys. Rev. Letters*, **7**, 352.

———. 1962, *Phys. Rev.*, **126**, 1580.

———. 1964, in *Gravitation and Relativity*, ed. H. Y. Chiu and W. F. Hoffman (New York: W. A. Benjamin).

References

Dicke, R. H. 1969, *Gravitation and the Universe*, Jayne Lectures for 1969 (Philadelphia: American Phil. Soc.).

Dicke, R. H., Peebles, P. J. E., Roll, P. G., and Wilkinson, D. T. 1965, *Ap. J.*, **142**, 414.

Dirac, P. A. M. 1937, *Nature*, **139**, 323.

———. 1938, *Proc. Roy. Soc. London, A*, **165**, 199.

Djatshenko, V. F., Zel'dovich, Ya. B., Imshennik, V. S., and Paleichuk, V. V. 1968, *Astrofizika*, **4**, 159.

Dmitriev, N. A. 1962, *Zh. Eksp. Teoret. Fiz.*, **42**, 772.

Dmitriev, N. A., and Kholin, S. A. 1963, *Voprosy kosmogonii* [Problems of cosmogony], **9**, 254.

Dmitriev, N. A., and Zel'dovich, Ya. B. 1963, *Zh. Eksp. Teoret. Fiz.*, **45**, 1150.

———. 1964, *ibid.*, **18**, 793.

Domogazky, G. B. 1969, Sci. inform. astr. council USSR Acad. Sci., No. 13.

Domogazky, G. B., and Zazepin, G. T. 1969a, Proc. Int. Seminar on Neutrino Physics and Neutrino Astrophysics, Moscow, p. 198.

———. 1969b, Proc. 2d Int. Conf. on Cosmic Rays, Budapest.

Doroshkevich, A. G. 1965a, *Astr. Zh.*, **43**, 105.

———. 1965b, *Astrofizika*, **1**, 255.

———. 1966, *ibid.*, **2**, 37.

———. 1967, *ibid.*, **3**, 175.

Doroshkevich, A. G., and Novikov, I. D. 1964, *Doklady Akad. Nauk*, **154**, 809.

Doroshkevich, A. G., and Syunyaev, R. A. 1967, *Astr. Zh.*, Vol. **44**, No. 6.

Doroshkevich, A. G., and Zel'dovich, Ya. B. 1963, *Astr. Zh.*, **40**, 807.

Doroshkevich, A. G., Zel'dovich, Ya. B., and Novikov, I. D. 1965, *Zh. Eksp. Teoret. Fiz.*, **49**, 170.

———. 1967a, *Pis'ma* [Letters], *Zh. Eksp. Teoret. Fiz.*, **5**, 119.

———. 1967b, *Zh. Eksp. Teoret. Fiz.*, **53**, 644.

———. 1967c, *Astr. Zh.*, **44**, 295.

———. 1967d, *Astr. Circ.*, No. 442.

———. 1967e, preprint, Institut Prikladnoy Matematiki.

Drever, R. P. W. 1962, *Phil. Mag.*, **6**, 683.

Durney, B., and Roxburgh, I. W. 1965, *Nature*, **208**, 1304.

———. 1967, *Proc. Roy. Soc. London, A*, **296**, 189.

Dyson, F. G. 1963, *Interstellar Communication* (New York: Benjamin).

Eddington, A. S. 1925, *Relativitäts Theorie in mathematischer Behandlung* (Berlin).

———. 1926, *The Internal Constitution of the Stars* (Cambridge: Cambridge University Press).

Eggen, O. J., and Greenstein, J. L. 1965, *Ap. J.*, **141**, 83.

Ehlers, J. 1965, *Acad. Wiss. Mainz Abh. Math. Nat. Kl.*, No. 11.

References

————. 1967 (ed.), *Relativity Theory and Astrophysics*, Lectures in Applied Mathematics, Vols. **8, 9, 10** (Providence, R.I.: Am. Math. Soc.).

Einstein, A. 1917, *Sitzgsber. Preuss. Akad. Wiss.* **1**, 142.

————. 1918, *ibid.*, **1**, 154.

————. 1939, *Ann. Math.*, **40**, 922.

————. 1950, *The Meaning of Relativity* (Princeton, N.J.: Princeton University Press).

————. 1965, *The Collected Works* (in Russian) (Moscow: Nauka).

Einstein, A., and Rosen, N. J. 1937, *Frankl. Inst.*, **223**, 43.

Einstein, A., and Strauss, E. 1945, *Rev. Mod. Phys.*, **17**, 120.

Emden, R. 1907, *Gaskugeln* (Leipzig: Teubner).

Emin-Zade, T. A. 1959, *Doklady Akad. Nauk Azerb. SSR.*, **15**, 1005.

————. 1964, *Trudy Shemakhinskoy Obs.* (Baku, USSR), **3**, 12.

Eötvös, R. V. 1890, *Math. Naturwiss. Ver. Ungarn*, **8**, 65.

————. 1891, *Beibl. Ann. Phys.*, **15**, 688.

————. 1896, *Ann. d. Phys.*, **59**, 354.

————. 1922, *ibid.*, **68**, 11.

Erez, G., and Rosen, N. 1959, *Bull. Res. Council Israel*, **F8**, 47.

Everitt, C. W. F., Fairbank, W. M., and Hamilton, W. O. 1970, in *Relativity*, ed. M. Carmeli, S. I. Fickler, and L. Witten (New York: Plenum Press), p. 145.

Fackerell, E. D. 1968, *Ap. J.*, **153**, 643.

————. 1970, *Ap. J.*, **160**, 859.

Fackerell, E. D., Ipser, J. R., and Thorne, K. S. 1969, *Comments Ap. and Space Sci.*, **1**, 134.

Fadeev, 1967, Reference unknown.

————. 1968, in *Proc. Fifth Int. Conf. Gravitation and the Theory of Relativity* (Tbilisi).

Faulkner, J., Hoyle, F., and Narlikar, J. V. 1964, *Ap. J.*, **140**, 1100.

Fesenkov, V. G. 1949, *Astr. Zh.*, **26**, 67.

Feynman, R. P. 1948, *Phys. Rev.*, **74**, 939.

————. 1963, *Acta Phys. Polon.*, **24**, 697.

————. 1967, *Uspekhi Fiz. Nauk*, **91**, 29.

Field, G. 1959, *Ap. J.*, **129**, 525.

————. 1964, *Nature*, No. 4934, **202**, 786.

Field, G. B., and Henry, R. C. 1964, *Ap. J.*, **140**, 1002.

Field, G. B., Solomon, P. M., and Wampler, E. J. 1966, *Ap. J.*, **145**, 351.

Fletcher, J. D., Clemens, R., Matzner, R., Thorne, K. S., and Zimmerman, B. A. 1967, *Ap. J. (Letters)*, **148**, L91.

Finkelstein, D. 1958, *Phys. Rev.*, **110**, 965.

Finlay-Freundlich, E. 1945, *M.N.R.A.S.*, **105**, 237.

————. 1947, *ibid.*, **107**, 268.

Finzi, A. 1965, *Phys. Rev. Letters*, No. 15, **15**, 599.

Finzi, A., and Wolf, R. A. 1968, *Ap. J.*, **153**, 835.

Flamm, L. 1916, *Zs. f. Phys.*, **17**, 448.

Fok, V. A. 1961, *Teoriya prostranstva, vremeni i tyagoteniya* (Theory of Space, Time, and Gravitation) (2d ed.; Moscow: Fizmatgiz).

Foss, J., Garelick, D., Komme, S., Lobar, W., Osborne, L. S., and Uglum, J. 1967, *Phys. Letters*, **25B**, 166.

Fowler, R. H. 1926, *M.N.R.A.S.*, **87**, 114.

Fowler, W. A. 1964a, *Rev. Mod. Phys.*, **36**, 545, 1104.

———. 1964b, chapter in *High Energy Astrophysics*, L. Gratton ed. (New York: Academic Press).

———. 1965, *Quasi-stellar Sources and Gravitational Collapse*, ed. J. Robinson, A. Schild, and E. E. Schucking (The Dallas Conference, Dec. 1963) (Chicago: University of Chicago Press).

———. 1966, *Ap. J.*, **144**, 180.

Fowler, W. A., Burbidge, E. M., Burbidge, G. R., and Hoyle, F. 1965, *Ap. J.*, **142**, 423.

Fowler, W. A., and Hoyle, F. 1964, *Ap. J.*, *Suppl.*, No. 91, **9**, 201.

Fraley, G. C. 1968, *Ap. and Space Sci.*, **2**, 96.

Frank-Kamenetsky, D. A. 1959, *Fizicheckiye protsessy vnutri zvëzd* (Physical processes in the stellar interior) (Moscow: Fizmatgiz).

———. 1963, *Astr. Zh.*, **40**, 455.

Frenkel, J. 1928, *Zeit. f. Phys.*, **50**, 234.

Friedman, H. 1965, *Science*, Vol. **147**, No. 3656.

Friedman, H., 1964, preprint.

Friedmann, A. A. 1922, *Z. f. Phys.*, **10**, 377.

———. 1924, *ibid.*, **21**, 326.

———. 1966, *Izbrannye trudy* [selected papers] (Moscow).

Friedmann, G. 1964, *Scientific American*, **210**(6), 36.

Galkin, L. S. 1961, *Tezisy pervoy Sovetskoy gravitatsionnoy konferentsii* [Thesis of the first Soviet conference on gravitation] (Moscow).

Gamow, G. 1946, *Phys. Rev.*, **70**, 572.

———. 1948, *Nature*, **162**, 680.

———. 1949, *Rev. Mod. Phys.*, **21**, 367.

Gandel'man, G. M. 1962, *Zh. Eksp. Teoret. Fiz.*, **43**, 131.

Gandel'man, G. M., and Frank-Kamenetsky, D. A. 1956, *Doklady Akad. Nauk*, **107**, 811.

Gandel'man, G. M., and Pinaev, V. S. 1959, *Zh. Eksp. Teoret. Fiz.*, **37**, 1072.

Garmire, G., Leong, C., and Sreekanton, B. V. 1968, *Phys. Rev.*, **166**, 1280.

Gelfand, J. M., and Fomin, S. V. 1967, *Variazionnoe i Schislenie*, Moscow.

Geroch, R. P. 1966, *Phys. Rev. Letters*, **17**, 445.

Gertsenshteyn, M. Ye. 1966a, *Zh. Eksp. Teoret. Fiz.*, **51**, 129.

———. 1966b, *ibid.*, **51**, 1127.

Gershberg, R. Ye. 1964, *Tezisy symposiuma Peremennyyee zvezdy i zvezdnaya evolyutsiya* ("Variable stars and stellar evolution") (Moscow, November 1964).

References

Gershteyn, S. S., and Zel'dovich, Ya. B. 1966, *Pis'ma* [Letters], *Zh. Eksp. Teoret. Fiz.*, **4**, 174.

Ginzburg, V. L. 1961*a*, *Astr. Zh.*, **38**, 380.

———. 1961*b*, *Propagation of Electromagnetic Waves in Plasma* (New York: Gordon & Breach).

———. 1964, *Doklady Akad. Nauk*, **156**, 43.

———. 1968, *Uspekhi Fiz. Nauk*, **95**, 91.

———. 1969*a*, *Usp. Fiz. Nauk*, **97**, 601.

———. 1969*b*, *J. Statist. Phys.*, Vol. 1.

Ginzburg, V. L., and Krizhnits, D. A. 1964, *Zh. Eksp. Teoret. Fiz.*, **47**, 2006L.

Ginzburg, V. L., and Ozernoy, L. M. 1964, *Zh. Eksp. Teoret. Fiz.*, **47**, 1030.

———. 1965, *Astr. Zh.*, **42**, 943.

———. 1966, *Izvest. vysshikh uchebnykh zavedeniy (Radiofizika)*, **9**, 221.

Ginzburg, V. L., and Syrovatsky, S. I. 1964*a*, *Uspekhi Fiz. Nauk*, **84**(2), 201.

———. 1964*b*, *Doklady Akad. Nauk*, **158**, 808.

———. 1964*c*, *The Origin of Cosmic Rays* (London: Pergamon Press).

———. 1965, *Space Sci. Rev.*, **4**, 267.

Ginzburg, V. L., Zheleznyakov, V. V., and Zaitsev, V. V. 1969, *Ap. and Space Sci.*, **4**, 464.

Gödel, K. 1949, *Rev. Mod. Phys.*, **21**, 447.

Gold, T. 1962, *Recent Developments in General Relativity*, (New York: Pergamon Press), p. 225.

———. 1968, *Nature*, **218**, 731.

———. 1969, *ibid.*, **221**, 25.

Goldreich, P. 1969, *Proc. Australian Acad. Sci.*, **1**, 227.

Gordon, J. M. 1967, *A.J.*, **44**, 1146.

Gould, R. J., and Ramsay, W. 1966, *Ap. J.*, **144**, 587.

Gradshteyn, I. S., and Ryzhik, I. M. 1965, *Table of Integrals, Series, and Products* (New York: Academic Press).

Grasberg, E. K., and Nadezhin, D. K. 1969*a*, *Astr. Zh.*, **46**, 745.

———. 1964*b*, *Sci. Inf. Astr. Council USSR Acad. Sci.*, No. 13.

Greenstein, J. L. 1963, *Sci. Am.*, **209** (6), 54.

———. 1969, *Comments Ap. and Space Sci.*, **1**, 82.

Greenstein, J. L., and Schmidt, M. 1964, *Ap. J.*, **140**, 1.

———. 1965, in *Quasi-Stellar Radio Sources and Gravitation Collapse* (ed. J. Robinson, A. Schild, and E. L. Schucking) (Chicago: University of Chicago Press), p. 175.

Greenstein, G. S., Truran, J. W., and Cameron, A. G. W. 1967, *Nature*, **213**, 871.

Gombas, P. 1950. *Theorie und Losungsmethoden des Mehrteilchenproblems der Wellenmechanik* (Basel).

Grishchuk, L. P. 1967, *Astr. Zh.*, **44**, 1097.

References

Gross, and Yakif, 1966, reference unknown.

Gunn, J. E., and Ostriker, J. P., 1969, *Nature*, **221**, 454.

Gunn, J. E., and Peterson, B. A. 1965, *Ap. J.*, **142**, 1633.

Gupta, S. N. 1950, *Proc. Phys. Soc.*, **A63**, 681.

———. 1952, *ibid.*, **A65**, 608.

Gurevich, L. E. 1954*a*, *Voprosy kosmogonii* [Problems of cosmogony], **2**, 151.

———. 1954*b*, *ibid.*, **3**, 94.

Gurevich, L. E., and Lebedinsky, A. I. 1955, *Trudy chetvertogo soveshchaniya po voprosam kosmogonii*, 147 (Moscow).

Gurovich, V. Ts. 1965, *Astr. Zhur.*, **42**, 974.

Guseynov, O. Kh. 1965, *Tezisy II vsesoyuznoy konferentsii po gravitatsii* (Tbilisi).

———. 1968, *Astr. Zh.*, **45**, 985.

Guseynov, O. Kh., and Zel'dovich, Ya. B. 1966, *Astr. Zh.*, **43**, 313.

Hagedorn, R. 1965, *Nuovo Cimento Suppl.*, **3** (2), 147.

Hagihara, Y. 1931, *Japanese J. Astr. and Geoph.*, **8**, 67.

Hamada, T., and Salpeter, E. E. 1961, *Ap. J.*, **134**, 683.

Hansen, C. J., and Tsuruta, S. 1967, *Canadian J. Phys.*, **45**, 2823.

Hansen, C. J., and Wheeler, J. C. 1969, *Ap. and Space Sci.*, **3**, 464.

Härm, R., and Schwarzschild, M. 1965, *Ap. J.*, **142**, 855.

Harrison, B. K., Thorne, K. S., Wakano, M., and Wheeler, J. A. 1965, *Gravitation Theory and Gravitational Collapse* (Chicago: University of Chicago Press).

Harrison, B. K., Wakano, M., and Wheeler, J. A. 1958, in Onzième Conseil de Physique Solvay, *La Structure et l'evolution de l'univers* (Brussels: Stoops).

Harrison, E. R. 1967, *Phys. Rev. Letters*, **18**, 1011.

Hartle, J. B. 1967, *Ap. J.*, **150**, 1005.

———. 1970, *ibid.* (in press).

Hartle, J. B., and Sharp, D. H. 1967, *Ap. J.*, **147**, 317.

Hartle, J. B., and Thorne, K. S. 1968, *Ap. J.*, **153**, 807.

Hawking, S. W. 1966*a*, *Phys. Rev. Letters*, **17**, 144.

———. 1966*b*, *Ap. J.*, **145**, 544.

———. 1966*c*, *Proc. Roy. Soc. London*, A, **295**, 490.

———. 1966*d*, *Singularities and the geometry of spacetime*, an essay submitted for the Adams prize, Cambridge.

———. 1967, *Proc. Roy. Soc. London*, A, **300**, 187.

Hawking, S. W., and Ellis, G. F. R. 1965, *Phys. Rev. Letters*, **17**, 246.

———. 1968, *Ap. J.*, **152**, 25.

Hawking, S. W., and Penrose, R. 1970, *Proc. Roy. Soc. London*, A, **314**, 529.

Hawking, S. W., and Tayler, R. T. 1966, *Nature*, **209**, 1278.

Hayakawa, S., and Matsuoka, M. 1964, *Prog. Theoret. Phys. Suppl.*, No. 30, p. 204.

Hayashi, C. 1950, *Prog. Theoret. Phys.*, **5**, 224.

―――. 1966, in *Stellar Evolution*, ed. R. Stein, and A. G. W. Cameron (New York: Plenum Press), p. 193.

Haymes, R. C., Ellis, V., Fishman, G. J., Kurfess, J. D., and Tucker, W. H. 1968, *Ap. J. (Letters)*, **151**, L9.

Heckmann, O. 1942, *Theorien der Kosmologie* (Berlin).

Heckmann, O., and Schucking, E. 1955, *Zs. f. Ap.*, **38**, 95.

―――. 1956, *Z. Ap.*, **40**, 81.

―――. 1959, in *Hdb. d. Phys.*, Vol. **53**, (Berlin: Springer-Verlag).

―――. 1962, chapter 11 of *Gravitation: an introduction to current research*, ed. L. Witten (New York: John Wiley and Sons).

Hernandez, W. 1967, *Phys. Rev.*, **153**, 1359.

Hewish, A., Bell, S. J., Pilkington, J. D., Scott, P. F., and Collins, R. A. 1968, *Nature*, **217**, 709.

Hilbert, D. 1917, *Gött. Nachr., Math. Ann.*, Vol. **92**.

Hoffmeister, E., Kippenhahn, R., and Weigert, A. 1964a, *Zs. f. Ap.*, **59**, 242.

―――. 1964b, *ibid.*, **60**, 57.

Hoyle, F. 1948, *M.N.R.A.S.*, **108**, 372.

―――. 1949, *ibid.*, **109**, 365.

―――. 1958a, in Collected Works, *Magnitnaya gidrodinamika*, Vol. **37** (Moscow: Gosatomizdat).

―――. 1958b, In *La structure et l'evolution de l'univers*, 11 Conseil de Physique Solvay (Brussels: Stoops), p. 66.

―――. 1960a, *M.N.R.A.S.*, **120**, 256.

―――. 1960b, *Voprosy kosmogonii* [Problems of cosmogony], **7**, 15.

Hoyle, F., and Burbidge, G. R. 1966, *Ap. J.*, **144**, 534.

Hoyle, F., Burbidge, G., and Burbidge, E. M. 1964, Preprint.

Hoyle, F., and Fowler, W. A. 1960, *Ap. J.*, **132**, 565.

―――. 1963a, *Nature*, **197**, 533.

―――. 1963b, *M.N.R.A.S.*, **125**, 169.

―――. 1965, preprint. (Reference unknown).

―――. 1967, *Nature*, **213**, 373.

Hoyle, F., Fowler, W. A., Burbidge, G., and Burbidge, E. M. 1964, *Ap. J.*, **139**, 909.

Hoyle, F., and Ireland, J. G. 1961, *M.N.R.A.S.*, **122**, 35.

Hoyle, F., Narlikar, J. V., and Wheeler, J. A. 1964, *Nature*, **203**, 914.

Hoyle, F., and Tayler, R. J. 1964, *Nature*, **203**, 1108.

Huang, S. S. 1963, *Ap. J.*, **138**, 471.

Hubble, E. 1929, *Proc. Nat. Acad. Sci.*, **15**, 168.

Hughes, V. W., Robinson, H. G., and Vertran-Lopez, V. 1960, *Phys. Rev. Letters*, **4**, 342.

Humason, M., Mayall, N., and Sandage, A. 1956, *Ap. J.*, **61**, 97.

Hund, V. 1936, *Ergebn. exakt. Naturwiss.*, **15**, 189.

Hunter, C. 1962, *Ap. J.*, **136**, 594.

―――. 1963, *M.N.R.A.S.*, **126**, 299.

References

Hurley, M., and Roberts, P. H. 1965, *Ap. J. Suppl.*, No. 96, 11, 95.

Iben, I. 1964, *Ap. J.*, **140**, 1631.

———. 1965, *Quasi-Stellar Sources and Gravitational Collapse* (ed. J. Robinson, A. Schild, and E. L. Schucking) (Chicago: University of Chicago Press), p. 67.

———. 1967, *Ann. Rev. Astr. and Ap.*, **5**, 571.

Imshennik, V. S., and Chechitkin, V. M. 1969, (See *Astr. Zh.* **47**, 729).

Imshennik, V. S., and Nadezhin, D. K. 1964, *Astr. Zh.*, **41**, 829.

———. 1965, *ibid.*, **42**, No. 6, 1154.

———. 1967, *ibid.*, **44**, 377.

Imshennik, V. S., Nadezhin, D. K., and Pinaev, V. S. 1966, *Astr. Zh.*, **43**, 1215.

———. 1967, *ibid.*, **44**, 768.

Infel'd, L., and Plebansky, J. 1960, *Motion and Relativity* (New York: Pergamon Press).

Inman, C. L. 1965, *Ap. J.*, **141**, 187.

Ipser, J. R. 1969a, *Ap. J.*, **156**, 509.

———. 1969b, *ibid.*, **158**, 17.

Ipser, J. R., and Thorne, K. S. 1968, *Ap. J.*, **154**, 251.

Irvine, W. M. 1961, doctoral thesis, Harvard University.

Isaacson, R. A. 1968a, *Phys. Rev.*, **166**, 1263.

———. 1968b, *ibid.*, **166**, 1272.

Israel, W. 1966, Preprint Series A, No. 10, Vol. 2 (Canada).

———. 1967, *Phys. Rev.*, **164**, 1776.

———. 1968, *Comm. Math. Phys.*, **8**, 245.

Ivanenko, D. D., and Sokolov, A. A., 1947, *Vest. Moscow Gos. Univ.*, **8**, 103.

Ivanova, L. N., Imshennik, V. S., and Nadezhin, D. K. 1967, preprint; subsequently published in *Sci. Inf. Astr. Council USSR Acad. Sci.*, **13** (1969).

Ivanova, L. M., Imshennik, V. S., Nadezhin, D. K., and Chechetkin, V. M. 1969, *Proc. Int. Seminar on Neutrino Physics and Neutrino Astrophysics* (Moscow), p. 180.

Jackson, J. D. 1962, *Classical Electrodynamics* (New York: John Wiley & Sons).

Jacobs, K. C. 1968, *Ap. J.*, **153**, 661.

———. 1969, *ibid.*, **155**, 379.

James, R. A. 1964, *Ap. J.*, **140**, 522.

Jeans, J. H. 1929a, *Astronomy and Cosmology* (London and New York: Cambridge University Press), p. 345.

———. 1929b, *Phil. Trans. Roy. Soc.*, **A199**, 1.

Jordan, P. 1937, *Naturwiss.*, **25**, 513.

———. 1938, *ibid.*, **26**, 417.

———. 1955, *Schwerkraft und Weltall* (Braunschweig: Viehweg).

Jugaku, J., Sargent, W. L. W., and Greenstein, J. L. 1961, *Ap. J.*, **134**, 783.

Kalitkin, N. N. 1960, *Zh. Eksp. Teoret. Fiz.*, **38** (5), 1534.

References

Kaplan, S. A. 1949a, *Zh. Eksp. Teoret. Fiz.*, **19**, 951.

―――. 1949b, *Uchenyye zapiski L'vovskogo Universiteta*, **15**, No. 4, 101.

―――. 1963, *Fizika zvezd* (Moscow: Fizmatgiz).

Kaplan, S. A., and Lupanov, G. A. 1965, *Astr. Zh.*, **42**, 299.

Kaplan, S. A., and Pikel'ner, S. B. 1963, *Mezhzvezdnaya sreda* (Interstellar medium) (Moscow: Fizmatgiz).

Karachentsev, I. D. 1965, *Astrofizika*, **1**, 303.

Kardashev, N. S. 1964, *Astr. Zh.*, Vol. **41**, 807.

―――. 1967, *Astr. Circ.*, No. 430.

Kardashev, N. S., and Sholomitsky, G. B. 1965, *Astr. Circ.*, No. 336.

Kasha, V., and Stefanski, R. 1968, *Phys. Rev.*, **172**, 1297.

Kaufman, M. 1965, *Nature*, **207**, 4998, 736.

Kellermann, K. I., Clark, B. G., Bare, C. C., Rydbeck, O., Elldér, J., Hansson, B., Kollberg, E., Hoglund, B., Cohen, M. H., and Jauncey, D. L. 1968, *Ap. J. (Letters)*, **153**, L209.

Kellermann, K. I., and Pauliny-Toth, L. 1968, *Ann. Rev. Astr. and Ap.*, **6**, 417.

Kenderdine, S., Ryle, M., and Pooley, G. G. 1966, *M.N.R.A.S.*, **134**, 189.

Kerr, R. P. 1963, *Phys. Rev. Letters*, **11**, 237.

―――. 1965, in *Quasi-stellar Sources and Gravitational Collapse*, ed. J. Robinson, A. Schild, and E. L. Schucking (Chicago: University of Chicago Press), p. 99.

Khalatnikov, I. M. 1965, *Zh. Eksp. Teoret. Fiz.*, **48**, 261.

―――. 1969 (in press).

Kholopov, P. N. 1965, *Astr. Zh.*, **42**, 369.

Kibble, T. W. B. 1965, in *High Energy Physics and Elementary Particles* (Trieste Summer Seminar).

―――. 1968, *Uspekhi Fiz. Nauk*, **96**, 497.

Kippenhahn, R., and Weigert, A. 1967, *Zs. f. Ap.*, **65**, 253.

Kippenhahn, R., Thomas, H. C., and Weigert, A. 1965, *Zs. f. Ap.*, **61**, 242.

Kirzhnits, D. A., and Polyachenko, V. L. 1964, *Zh. Eksp. Teoret. Fiz.*, **46**, 755.

―――. 1964, *ibid.*, **46**, 255.

―――. 1964, *ibid.*, **19**, 514.

Klimov, Ya. 1963, *Astr. Zh.*, **40**, 874.

Kobzarev, I. Yu., and Okum, L. B. 1962, *Zh. Eksp. Teoret. Fiz.*, **43**, 1904.

Kohler, A. 1966, *Ap. J.*, **146**, 488.

Kompaneets, A. S. 1956, *Zh. Eksp. Teoret. Fiz.*, **31**, 876.

Kompaneets, A. S., and Chernov, A. S. 1964, *Zh. Eksp. Teoret. Fiz.*, **47**, 1939.

Kopvilem, U. Kh., and Nagibarov, V. N. 1965, *Pis'ma* [Letters], *Zh. Eksp. Teoret. Fiz.*, **2**, 529.

Kopysov, Yu. S., and Kus'min, V. A. 1968, *Izvest. Acad. Nauk (Phys. Ser.)*, **32**, 1790.

Krat, V. A. 1950, *Figury ravnovesiya nebesnykh tel* (Moscow: Gostekhizdat).

References

Kraushaar, W., and Clark, G. 1962, *Phys. Rev. Letters*, **8**, 106.

Kraveov, V. A. 1965, *The Masses of Atoms and Binding Energies of Nuclei* (Moscow).

Kruskal, M. 1960, *Phys. Rev.*, **119**, 1743.

Kukarkin, B. V. 1965, *Astrofizika*, **1**, No. 4, 465.

Kundt, W. 1968, *Springer Tracts in Modern Phys.*, **47**, 111.

Kurt, V. G., and Syunyaev, R. A. 1967*a*, *Kosmicheskie issledovaniya*, **5**, 573.

———. 1967*b*, *Pis'ma* [Letters], *Zh. Eksp. Teoret. Fiz.*, **5**, 299.

———. 1967*c*, *Astr. Zh.*, **44**, 6.

Landau, L. D. 1932, *Phys. Zs. Sowjetunion*, **1**, 285.

———. 1938, *Nature*, **141**, 333.

Landau, L. D., and Lifshitz, E. M. 1954, *Mekhanika sploshnysch sred* (Moscow: Gostekhizdat).

———. 1962, *The Classical Theory of Fields* (2d ed.; Reading, Mass.: Addison-Wesley Publishing Company) .

———. 1964, *Statisticheskaya fizika* (Moscow: Fizmatgiz).

Landau, L. D., and Pomeranchuk, I. Ya. 1955, *Doklady Akad. Nauk*, **102**, 489.

Landau, L. D., and Stanyukovich, K. P. 1945, *Doklady Akad. Nauk*, **46**, 399.

Langer, W. D., and Cameron, A. G. W. 1969, *Ap. and Spa. Sci.* (in press).

Layzer, D. 1954, *Astr. J.*, **59**, 170.

———. 1963, *Ap. J.*, **137**, 351.

———. 1964, *Ann. Rev. Astr. and Ap.*, **2**, 341.

———. 1965, *Ap. J.*, **141**, 837.

Leacock, R. A., Beavers, W. J., and Daub, C. T. 1968, *Ap. J.*, **151**, 1179.

Lebedinsky, A. I. 1954, *Voprosy kosmogonii* [Problems of cosmogony], **2**, 5.

Ledoux, P. 1941, *Ap. J.*, **94**, 537.

Lee, T. D., and Yang, C. N. 1955, *Phys. Rev.*, **98**, 1501.

Lemaître, G. 1933, *Ann. Soc. Sci. Bruxelles*, **A53**, 51.

Lifshitz, E. M. 1946, *Zh. Eksp. Teoret. Fiz.*, **16**, 587.

Lifshitz, E. M., and Khalatnikov, I. M. 1960*a*, *Zh. Eksp. Teoret. Fiz.*, **39**, 149.

———. 1960*b*, *ibid.*, **39**, 800.

———. 1963, *Uspekhi Fiz. Nauk*, **80**(3), 391.

Lifshitz, E. M., Sudakov, V. V., and Khalatnikov, I. M. 1961, *Zh. Eksp. Teoret. Fiz.*, **40**, 1847.

Lin, C. C., and Shu, F. 1964, *Ap. J.*, **140**, 646.

Longair, M. S. 1966, *M.N.R.A.S.*, **133**, 421.

Lorentz, H. A. 1915, *Theory of Electrons*, 2d ed. (reprinted by Dover Publications, New York, 1952).

Low, F. J., and Johnson, H. L. 1965, *Ap. J.*, **141**, 336.

Lozinskaya, T. A., and Kardashev, N. S. 1963, *Astr. Zh.*, **40**, 209.

Luyten, W. J. 1952, *Ap. J.*, **116**, 283.

Lynden-Bell, D. 1966, in *The Theory of Orbits in the Solar System and in*

References

Stellar Systems: IAU Symposium No. 25, ed. G. Contopoulos (New York: Academic Press), chap. 14.

———. 1969, *Nature*, **223**, 690.

Lynden-Bell, D., and Sannit, N. 1969, *M.N.R.A.S.*, **143**, 167.

Markov, M. A. 1966, *Zh. Eksp. Teoret. Fiz.*, **51**, 878.

Marochnik, L. S. 1970, *Astr. Zh.*, **47**, 46.

Marochnik, L. S., and Suchkov, A. A. 1969, *Astr. Zh.*, **46**, 319.

Martynov, D. Ya. 1965, *Kurs obshchey astrofiziki* (Moscow: Fizmatgiz).

Masevich, A. G. 1956, *Astr. Zh.*, **33**, 216.

Masevich, A. G., and Kotok, E. V. 1963, *Astr. Zh.*, **40**, 659.

Masevich, A. G., Kotok, E. V., Dluzhnevskaya, O. B., and Mazani, A. 1965, *Astr. Zh.*, **42**, 334.

Mathews, J., and Sandage, A. 1962, *Pub. A.S.P.*, **74**, 406.

Mayer, C. U. 1968, N.R.L. Space Research Seminar (Washington), p. 73.

May, M. M., and White, R. H. 1964, private communication.

———. 1966, *Phys. Rev.*, **141**, 1232.

McCrea, W., and Milne, E. 1934, *Quart. J. Math.*, **5**, 73.

McKellar, A. 1941, *Pub. Dominion Ap. Obs.*, Vol. **7**, No. 15.

McVittie, G. C. 1956, *General Relativity and Cosmology* (New York: John Wiley & Sons).

———. 1959a, *Hdb. d. Phys.*, Vol. **53** (Berlin: Springer-Verlag).

———. 1959b, *Report on Rouaymont's Conference* (Paris).

———. 1962a, *General Relativity and Cosmology* (Urbana: University of Illinois Press).

———. 1962b, *Fact and Theory in Cosmology* (New York: Macmillan).

———. 1962c, *Phys. Rev.*, **128**, 2871.

Meltzer, D. W., and Thorne, K. S. 1966, *Ap. J.*, **145**, 514.

Mestel, L. 1959, *M.N.R.A.S.*, **119**, 223.

———. 1965, In *Stellar Structure* (Chicago: University of Chicago Press), p. 297.

Mestel, L., and Ruderman, M. 1967, *M.N.R.A.S.*, **136**, 27.

Metzner, A. W. 1963, *J. Math. Phys.*, **4**, 1194.

Michel, F. C. 1964, *Phys. Rev.*, **B133**, 329.

———. 1965, in *Quasi-Stellar Sources and Gravitational Collapse* (ed. J. Robinson, A. Schild, and E. L. Schucking) (Chicago: University of Chicago Press), p. 75.

Michie, R. W., and Bodenheimer, P. H. 1963a, *M.N.R.A.S.*, **125**, 269.

———. 1963b, *ibid.*, **126**, 278.

Mikhaylov, A. A. (ed.); 1962, *Kurs astrofiziki i zvezdnoy astronomii*, Vol. **2** (Moscow: Fizmatgiz).

Miller, R. H., Prendergast, K. H., and Quirk, W. J. 1970, *Ap. J.*, **161**, 903.

Milne, E. 1935, *Relativity, Gravitation, and World-Structure* (London and New York: Oxford University Press).

Milne, E. 1948, *Kinematic Relativity* (London and New York: Oxford University Press).

Minkowski, R. 1964, *Ann. Rev. Astr. and Ap.*, **2**, 247.

Mironovsky, V. N. 1965a, *Zh. Eksp. Teoret. Fiz.*, **48**, 358.

———. 1965b, *Astr. Zh.*, **42**, 977.

Misner, C. W. 1963, *J. Math. Phys.*, **4**, 924.

———. 1967a, *Nature*, **214**, 40.

———. 1967b, *Ap. J.*, **151**, 451.

———. 1967c, *Phys. Rev. Letters*, **19**, 533.

———. 1970, *Phys. Rev.*, in press.

Misner, C. W., and Wheeler, J. A. 1957, *Ann. Phys.*, **2**, 525.

Monaghan, J., and Roxburgh, I. W. 1965, *M.N.R.A.S.*, **131**, 13.

Morgan, T., and Morgan, L. 1969, *Phys. Rev.*, **183**, 1097.

Morrison, R. 1967 (see "Preludes in Theoretical Physics," Amsterdam 1966, p. 347–57).

———. 1969, *Ap. J. (Letters)*, **157**, L73.

Muhleman, D. O., Ekers, R. D., and Fomalont, E. B. 1970, *Phys. Rev. Letters*, **24**, 1377.

Nadezhin, D. K. 1968, *Astr. Zh.*, **45**, 1166.

Nadezhin, D. K., and Chechetkin, V. M. 1969, *Astr. Zh.*, **46**, 270.

Nadezhin, D. K., and Frank-Kamenetsky, D. A. 1962, *Astr. Zh.*, **39**, 1003.

———. 1964a, *ibid.*, **41**, 842.

———. 1964b, *Voprosy kosmogonii* [Problems of cosmogony], **10**, 154.

———. 1965, *Astr. Zh.*, **42**, 290.

Narlikar, J. V. 1963, *M.N.R.A.S.*, **126**, 203.

Ne'eman, Y., 1965, *Ap. J.*, **141**, 1303.

Nemeth, J., and Sprung, D. W. L. 1968, *Phys. Rev.*, **176**, 1496.

Neumann, S., and Scott, E. 1959, *Hdb. d. Phys.*, Vol. **53** (Berlin: Springer-Verlag).

Newman, E., Tamburino, L., and Unti, T. 1963, *Math. Phys.*, **4**, 915.

Neymann, J., Page, E., and Scott, E. 1961, *A.J.*, **66**, 533.

Noble, L., and Scarf, F. 1965, *Ap. J.*, **141**, 1479.

Nordtvedt, K. 1968a, *Phys. Rev.*, **169**, 1014.

———. 1968b, *ibid.*, p. 1017.

———. 1970, *Icarus*, **12**, 1.

Novikov, I. D. 1961, *Astr. Zh.*, **38**, 564.

———. 1962a, *Vestnik*, Moscow State University, ser. 3, No. 5, p. 90.

———. 1962b, *ibid.*, No. 6, p. 61.

———. 1962c, *Soobshcheniya GAISH, Gosud. Astron. Inst. im. Shternberga* (Sternberg State Astronomical Inst.), No. 120, 42.

———. 1963, *Astr. Zh.*, **40**, 772.

———. 1964a, *Astr. Circ.*, No. 290.

———. 1964b, *Astr. Zh.*, **41**, 1075.

———. 1964c, *Zh. Eksp. Teoret. Fiz.*, **46**, 686.

References

————. 1964d, *Soobshcheniya GAISH* (Sternberg State Astronomical Inst.), No. 132, esp. pp. 3 and 43.

————. 1966a, *Pis'ma* [Letters], *Zh. Eksp. Teoret. Fiz.*, **3**, No. 5, 223.

————. 1966b, *Astr. Zh.*, **43**, 911.

————. 1967, preprint.

————. 1969, *Zh. Eksp. Teoret. Fiz.*, **57**, 949.

Novikov, I. D., and Ozernoy, L. M. 1963, *Doklady Akad. Nauk SSSR*, **150**, 1019.

Novikov, I. D., and Ozernoy, L. M. 1964, preprint FIAN, A-17.

Novikov, I. D., and Syunyaev, R. A. 1967, *Astr. Zh.*, **44**, 320.

Novikov, I. D., and Zel'dovich, Ya. B. 1965, *International Conference on Relativistic Theories of Gravitation*, **1**, (London).

————. 1966, *Supplemento al Nuovo Cimento, Serie, I*, **4**, 810.

————. 1967, *Ann. Rev. Astr. and Ap.*, **5**, 627.

Ogorodnikov, K. F. 1958, *Dinamika zvezdnykh sistem* (Moscow: Fizmatgiz).

Ohm, E. 1961, *Bell Syst. Tech. J.*, **40**, 4.

Okun', L. B. 1966, *Uspekhi Fiz. Nauk*, **89**, 603.

Oort, J. 1958, in *La structure et l'evolution de l'univers*, 11 Conseil de Physique Solvay (Brussels: Stoops).

Öpik, E. I. 1957, in *Nuclear Processes in Stars* (Moscow: Inostr. Lit.), p. 105.

Oppenheimer, J. R., and Snyder, H. 1939, *Phys. Rev.*, **56**, 455.

Oppenheimer, J. R., and Volkoff, G. M. 1938, *Phys. Rev.*, **55**, 374.

Ostriker, J. P., and Axel, L. 1969, in *Low-Luminosity Stars* (New York: Gordon and Breach), p. 357.

Ostriker, J. P., and Bodenheimer, P. 1968, *Ap. J.*, **151**, 1089.

Ostriker, J. P., and Gunn, J. E. 1969, *Ap. J.*, **157**, 1395.

Ozernoy, L. M. 1964, *Uspekhi Fiz. Nauk*, **83**, 565.

————. 1965a, *Doklady Akad. Nauk*, **163**, 50.

————. 1965b, *Doklad na 11 Vsesoyuznoy gravitatsionney konferentsii* (Tbilisi).

————. 1966a, *Astr. Zh.*, **43**, 300.

————. 1966b, dissertation (Moscow: GAISH).

————. 1967a, in *Trudy symposiuma Peremennyye zvezdy i zvezdnaya evolyutsiya* (Moscow: Znaniye).

————. 1967b, in *Trudy symposiuma Peremennyye zvezdy i zvezdnaya evolyutsiya* (Moscow: Znaniye).

Ozernoy, L. M., and Chertoprud, V. E. 1966, *Astr. Zh.*, **43**, 20.

————. 1967, *ibid.*, **44**, 537.

Pacholczyk, A. G. 1965, *Ap. J.*, **142**, 805.

Pacini, F. 1968, *Nature*, **219**, 145.

Paczinski, B. 1966, *Acta Astr.*, **16**, 231.

Pariysky, Yu. N. 1967, *Astr. Zh.*, **44**, No. 5.

Parker, E. N. 1965a, *Phys. Rev. Letters*, **14**, 55.

————. 1965b, *Ap. J.*, **141**, 1463 (see also *Space Sci. Rev.*, **4**, 666 [1965]).

References

Parker, P. D., Bahcall, J. N., and Fowler, W. A. 1964, *Ap. J.*, **139**, 602.

Partridge, R. B., and Peebles, P. J. E. 1967a, *Ap. J.*, **147**, 868.

———. 1967b, *ibid.*, **148**, 377.

Pauli, W. 1958, *Theory of Relativity* (New York: Pergamon).

Pauli, W., and Fierz, M. 1939, *Proc. Roy. Soc.*, **173**, 212.

Peebles, P. J. E. 1965, *Ap. J.*, **142**, 1317.

———. 1966, *Phys. Rev. Letters*, **16**, 410.

———. 1967, *Ap. J.*, **147**, 859.

———. 1969, *Ap. J.*, **157**, 1075.

Peierls, S. E., Singwi, K. S., and Wroe, D. 1952, *Phys. Rev.*, **87**, 46.

Penrose, R. 1965, *Phys. Rev. Letters*, **14**, 57.

———. 1967, *An Analysis of the Structure of Spacetime*, 1966 Adams Prize Essay (unpublished; preprinted by Princeton University Department of Physics).

Penzias, A. A., and Wilson, R. W. 1965, *Ap. J.*, **142**, 419.

Peres, A. 1960, *Progr. Theoret. Phys., Japan*, **24**, 149.

Peters, P. C., and Mathews, J. 1963, *Phys. Rev.*, **131**, 435.

Petrosian, V., Salpeter, E., and Szekeres, P. 1967, *Ap. J.*, **147**, 1222.

Petrov, A. Z. 1961, *Prostranstva Einsteina* (Moscow: Fizmatgiz).

———. (ed.) 1965, *Gravitatsiya i otnositel'nost'*.

———. 1966, *Novyye metody v obshchey teorii otnositel'nosti* (Moscow: Nauka).

Piddington, J. H. 1964, *M.N.R.A.S.*, **128**, 345.

Pikel'ner, S. B. 1961, *Osnovy kosmicheskoy elektrodinamiki* (Moscow: Fizmatgiz); English transl.: *Fundamentals of Cosmic Electrodynamics* (Washington: NASA, 1964) (2d Russian ed.; Moscow: Nauka, 1966).

———. 1963, *Astr. Zh.*, **40**, 601.

———. 1965, *ibid.*, **42**, 3.

Pikel'ner, S. B., and Vaynshteyn, L. A. 1966, *Pis'ma* [Letters], *Zh. Eksp. Teoret. Fiz.*, **4**, 307.

Pinaev, V. S. 1963a, *Voprosy kosmogonii* [Problems of cosmogony], **9**, 176.

———. 1963b, *Zh. Eksp. Teoret. Fiz.*, **45**, 548.

Pinaeva, G. V. 1964, *Astr. Zh.*, **41**, 25.

Podurets, M. A. 1964a, *Doklady Akad. Nauk*, **154**, 300.

———. 1964b, *Astr. Zh.*, Vol. **41**, No. 6.

———. 1970, *ibid.*, Vol. **47** (in press).

Pontecorvo, B. 1959, *Zh. Eksp. Teoret. Fiz.*, **36**, 1615.

———. 1963, *Voprosy kosmogonii* [Problems of cosmogony], **9**, 132.

Pontecorvo, B., and Smorodinsky, Ya. A. 1961, *Zh. Eksp. Teoret. Fiz.*, **41**, 239.

Poveda, A., and Woltjer, L. 1968, *A.J.*, **73**, 65.

Prendergast, K. H., and Burbidge, G. R. 1968, *Ap. J.* (*Letters*), **151**, L83.

Price, R. 1970, unpublished PhD thesis, California Institute of Technology.

———. 1971a, *Phys. Rev.* (in press).

————. 1971*b*, *Phys. Rev.* (in press).

Price, R., and Thorne, K. S. 1969, *Ap. J.*, **155**, 163.

Pskovsky, Yu. P. 1960, *Astr. Zh.*, **37**, 1056.

———— (ed.) 1965, *Nablydatel'nye osnovy kosmologii* (Moscow: Izd-vo Inostrannoy literatury).

Pustovoyt, A. P., and Bautin, A. V. 1964, *Zh. Eksp. Teoret. Fiz.*, **46**, 1386.

Pustovoyt, A. P., and Gertsenshteyn, M. E. 1962, *Zh. Eksp. Teoret. Fiz.*, **42**, 163.

Rainich, G. Y. 1924, *Proc. Nat. Acad. Sci.*, **10**, 124.

————. 1925, *Trans. Amer. Math. Soc.*, **27**, 106.

Ramsey, W. H. 1950, *M.N.R.A.S.*, **110**, 325.

Rashevsky, P. K. 1964, *Rimanova geometriya i tensornyy analiz* (Moscow: Nauka), 2d ed.

Raychaudhury, A. 1955, *Phys. Rev.*, **98**, 1123.

Rees, M. J., and Sciama, D. W. 1966, *Ap. J.*, **145**, 6.

————. 1967, *ibid.*, **147**, 353.

Reeves, H. 1965, preprint.

Regge, T., and Wheeler, J. 1957, *Phys. Rev.*, **108**, 1063.

Reines, R., and Wood, R. M. 1965, *Phys. Rev. Letters*, **14**, 20.

Robertson, H. 1933, *Rev. Mod. Phys.*, **5**, 62.

Robertson, H. 1965, in *Science in Progress*, ed. G. A. Baitsell (2d Ser.; New Haven, Connecticut: Yale University Press).

Robertson, H., and Noonan, T. 1968, *Relativity and Cosmology* (Philadelphia: W. B. Saunders Co.).

Roll, P. G., Krotkoy, R., and Dicke, R. H. 1964, *Ann. Phys.*, **26**, 442.

Rosenfeld, A., Barbaro-Galtneri, A., Barkas, W., Bastien, P., Kirz, J., and Roos, M. 1965, (see *Rev. Mod. Phys.*, **36**, 977, 1964).

Rosenfeld, L. 1930, *Zs. f. Phys.*, **64**, 589.

————. 1937, *Ann. Inst. Henri Poincare.*

Roxburgh, I. W. 1965, *Nature*, **207**, 363.

Rozen, G. 1964, *Phys. Rev.*, **136**, 2798.

Rublev, S. V. 1964, Trudy simposiuma *Peremennyye zvezdy i zezdnaya evolyutsiya* (Moscow), esp. p. 8.

Rylov, Yu. A. 1961, *Zh. Eksp. Teoret. Fiz.*, **40**, 1755.

Saakyan, G. S. 1961, *Izv. Akad. Nauk Armyanskoy SSR*, **14**, 117.

————. 1962, *Astr. Zh.*, **39**, 1014.

————. 1965, *Tezisy II Vsesoyuznoy konferentsii po gravitatsii* (Tbilisi).

————. 1968, reference unknown.

Saakyan, G. S., and Vartanyan, Yu. L. 1963, *Soobshcheniya Byuarakanskoy Obs.*, **33**, 55.

————. 1964, *Astr. Zh.*, **41**, 193.

Sakharov, A. D. 1965, *Zh. Eksp. Teoret. Fiz.*, **49**, 345.

————. 1966, *Pis'ma* [Letters], *Zh. Eksp. Teoret. Fiz.*, **3**, 439.

————. 1967*a*, *ibid.*, **5**, 32.

Sakharov, A. D. 1967*b*, *Doklady Akad. Nauk*, **177**, 70.

Salpeter, E. E. 1955, *Ap. J.*, **121**, 161.

———. 1961, *ibid.*, **134**, 669.

———. 1964, *ibid.*, **140**, 796.

Sandage, A. R. 1962, in *Problems of Extragalactic Research*, ed. G. C. McVittie (New York: Macmillan), p. 359.

———. 1965, *Ap. J.*, **141**, 1560.

Sandage, A. R., and Miller, W. C. 1966, *Ap. J.*, **144**, 1240.

Sandage, A. R., and Wyndham, J. D. 1965, *Ap. J.*, **141**, 328.

Sargent, W. L. W., and Jugaku, J. 1961, *Ap. J.*, **134**, 777.

Sargent, W. L. W., and Searle, L. 1966, preprint.

Saslaw, W. C. 1968, *M.N.R.A.S.*, **138**, 337.

Sato, H. 1966, *Prog. Theoret. Phys.*, **35**, 241.

Savedoff, M. P. 1963, *Ap. J.*, **138**, 291.

Schatzman, E. 1954, *Voprosy kosmogonii* [Problems of cosmogony], **3**, 227.

———. 1958, *White Dwarfs* (New York: Interscience).

———. 1959, *IAU Symp. No. 10*, p. 129.

———. 1962, *Ann. d'ap.*, **25**, 1.

———. 1963, in *Star Evolution* (New York: Academic Press), p. 389.

Schmidt, M. 1963, *Ap. J.*, **136**, 164.

———. 1966, *Symposium of the IAU*, No. 29, Byurakan.

———. 1969. *Ann. Rev. Astr. and Ap.*, **7**, 527.

Schucking, E., and Heckmann, O. 1958, *11 Conseil de physique Solvay* (Brussels: Stoops).

Schwartzman, V. F. 1970*a*, *Astron. Zh.*, **47**, 824.

———. 1970*b*, *Astrofizika*, **6**, 309.

———. 1970*c*, *ibid.*, Vol. 6, 123.

———. 1970*d*, *Radiofizika*, **13**, 12.

Schwarzschild, K. 1916, *Berl. Ber.*, Vol. **189**.

Schwarzschild, M. 1958, *Structure and Evolution of the Stars* (Princeton, N.J.: Princeton University Press).

Schwarzschild, M., and Härm, R. 1959, *Ap. J.*, **129**, 637.

———. 1965, *Ap. J.*, **142**, 855.

Sciama, P. W. 1962, *M.N.R.A.S.*, **123**, 317.

Seidov, Z. F. 1967, *Astrofizika*, **3**, 189.

Seielstad, G. A., Sramek, R. A., and Weiler, K. W. 1970, *Phys. Rev. Letters*, **24**, 1373.

Setti, G., and Woltjer, L. 1966, *Ap. J.*, **143**, 838.

Shapiro, I. 1965, *Phys. Rev. Letters*, **13**, 789.

———. 1968, *ibid.*, **20**, 1265.

Sharov, A. S., and Efremov, U. N. 1963, *Inf. Bull. Variable Stars*, No. 23, Comm. 27 of IAU.

Shayn, G. A. 1943, *Bull. Abastumanskoy Ap. Obs.*, **7**, 83.

Shepley, L. C. 1964, *Proc. Nat. Acad. Sci.*, **52**, 1403.

Shirokov, M. F., and Fisher, I. Z. 1962, *Astr. Zh.*, **39**, 899.

Shklovsky, I. S. 1956, *Cosmic Radio Waves* (Cambridge, Mass.: Harvard University Press).

——. 1962, *Astr. Zh.*, **39**, 591.

——. 1963*a*, *Astr. Circ.*, No. 250.

——. 1963*b*, *ibid.*, No. 256.

——. 1964, *ibid.*, p. 801.

——. 1965*a*, *ibid.*, **42**, 287.

——. 1965*b*, *Doklady Akad. Nauk*, **160**(1), 54.

——. 1966, *Sverkhnovye zvezdy* (Moscow: Nauka).

——. 1967*a*, *Astr. Circ.*, No. 401.

——. 1967*b*, *ibid.*, No. 429.

——. 1967*c*, *Astr. Zh.*, **44**, 930.

——. 1967*d*, *Ap. J. (Letters)*, **148**, L1.

——. 1969*a*, *Astr. Circ.*, No. 495.

——. 1969*b*, *Astr. Zh.*, **46**, 446.

Shklovsky, I. S., and Kardashev, N. S. 1964, *Doklady Akad. Nauk*, **155**, 1039.

Sholomitsky, G. B. 1965, *Inform. Bull.*, *Variable Stars*, No. 83, Commission 27 of IAU.

Silk, J. 1968, *Ap. J. (Letters)*, **151**, L59.

Slysh, V. I. 1963, *Nature*, **199**, 682.

Smirnov, Yu. N. 1964, *Astr. Zh.*, **41**, 1084.

Smith, S. F., and Havas, P. 1965, *Phys. Rev.*, **138**, No. 2B, 495.

Smith, H. J., and Hoffleit, D. 1963, *Nature*, **198**, 650.

Spitzer, L. J., and Saslaw, W. C. 1966, *Ap. J.*, **143**, 400.

Stanyukovich, K. P. 1955, *Neustanovivshiesya dvizheniya sploshnoy sredy* (Moscow: Gostekhizdat).

——. 1965. *Gravitatsionnoye pole i elementarnyye chastitsy* (Moscow).

Sterne, T. E. 1933, *M.N.R.A.S.*, **93**, 736, 770.

Stothers, R. 1963, *Ap. J.*, **138**, 1085.

Stothers, R. and Simon, N. 1968, *Ap. J.*, **152**, 233.

Stover, R. W., Moran, J. J., and Trishka, J. W., 1967, *Phys. Rev.*, **164**, 1599.

Struve, O. 1950, *Stellar Evolution* (Princeton: Princeton University Press).

Synge, J. L. 1960, *Relativity—The General Theory* (New York: John Wiley & Sons).

Syunyaev, R. A. 1966, *Astr. Zh.*, **43**, 1237.

——. 1968, *Soviet Phys. Doklady*, **179**, 588.

Takarada, K., Sato, H., and Hayashi, C. 1966, *Progr. Theoret. Phys.*, **36**, 504.

Taub, A. H. 1951, *Ann. Math.*, **53**, 472.

Thirring, H. 1918, *Zs. f. Phys.*, **19**, 33.

——. 1921, *ibid.*, **22**, 29.

Thirring, W. 1961, *Ann. Phys.*, **16**, 96.

References

Thorne, K. S. 1967, in *High Energy Astrophysics*, Vol. **3**, ed. C. DeWitt, E. Schatzman, and P. Véron (New York: Gordon & Breach).

———. 1968, *Ap. J.*, **148**, 51.

———. 1969a, *ibid.*, **158**, 1.

———. 1969b, *ibid.*, p. 997.

———. 1970, in *Proceedings of Course 47 of International School of Physics "Enrico Fermi"* (New York: Academic Press).

Thorne, K. S., and Campolattaro, A. 1967, *Ap. J.*, **149**, 591.

———. 1968, *ibid.*, **152**, 673.

Tolman, R. C. 1930, *Phys. Rev.*, **35**, 875.

———. 1934a, *Proc. Nat. Acad. Sci.*, **20**, 169.

———. 1934b, *Relativity, Thermodynamics, and Cosmology* (Oxford: Clarendon Press).

———. 1949, *Rev. Mod. Phys.*, **21**, 374.

Toomre, A. 1964, *Ap. J.*, **139**, 1217.

Trautman, A. 1966, in *Perspectives in Geometry and Relativity*, ed. B. Hoffman (Bloomington: Indiana University Press).

Trautman, A., Pirani, F. A. E., and Bondi, H. 1965, *Lectures on General Relativity* (Englewood Cliffs, N.J.: Prentice-Hall).

Treder, H. J. 1968, *Int. J. Theoret. Phys.*, **1**, 167.

Trimble, V. L., and Thorne, K. S. 1969, *Ap. J.*, **156**, 1013.

Tsuruta, S., and Cameron, A. G. W. 1965, *Canadian J. Phys.*, **43**, 2056.

———. 1966, *ibid.*, **44**, 1895.

Tsuruta, S., Wright, J., and Cameron, A. G. W. 1965, *Nature*, **206**, 1137.

Tucker, W. H. 1968, *Ap. J. (Letters)*, **151**, L9.

Utiyama, R. 1956, *Phys. Rev.*, **101**, 1597.

Van Horn, H. M. 1968, *Ap. J.*, **151**, 227.

Vishveshwara, C. V. 1968, *J. Math. Phys.*, **9**, 1319.

Vladimirov, Ya. S. 1963, *Zh. Eksp. Teoret. Fiz.*, **45**, 251.

de Vaucouleur, G. 1962, in *Structure of Stellar Systems* (Moscow: Inostr. Lit.), p. 376.

Wagoner, R. V. 1965, *Phys. Rev.*, **138**, B1583.

———. 1969, *Ann. Rev. Astr. and Ap.*, **7**, 553.

Wagoner, R. V., Fowler, W. A., and Hoyle, F. 1967, *Ap. J.*, **148**, 3.

Wallerstein, G. 1962, *Phys. Rev. Letters*, **9**, 143.

Westervel't, P. 1966, *Pis'ma* [Letters], *Zh. Eksp. Teoret. Fiz.*, **4**, 333.

Weber, J. 1961, *General Relativity and Gravitational Waves* (New York: Interscience).

———. 1967, *Phys. Rev. Letters*, **18**, 498.

———. 1968, *ibid.*, **20**, 1307.

———. 1969, *ibid.*, **22**, 1320.

———. 1970, *ibid.* **25**, 180.

Weinberg, S. 1962, *Phys. Rev.*, **128**, 1457.

References

Welch, W. J., Keachie, S., Thornton, D. D., and Wrixon, G. 1967, *Phys. Rev. Letters*, **18**, 1068.

Weyl, H. 1917, *Ann. Phys.*, **54**, 117

———. 1919, *ibid.*, **59**, 185.

Weymann, R. 1965, *Phys. Fluids*, **8**, 2, 12.

———. 1966, *Ap. J.*, **145**, 566.

———. 1967, *ibid.*, **147**, 887.

Wheeler, J. A. 1955, *Phys. Rev.*, **97**, 511.

———. 1958, *La structure et l'evolution de l'univers*, 11 Conseil de Physique Solvay (Brussels: Stoops).

———. 1960, *Neutrinos, Gravitation, and Geometry* (Bologna).

———. 1966a, in *Gravitation and Relativity*, ed. H. Y. Chiu and W. F. Hoffman (New York: W. A. Benjamin), p. 195.

———. 1966b, *Ann. Rev. Astr. and Ap.*, Vol. 4, 393.

———. 1968, in *Proceedings of 1967 Battelle Math-Physics Rencontres* (New York: W. A. Benjamin).

Wheeler, J. A., and Feyman, R. P. 1945, *Rev. Mod. Phys.*, **17**, 157.

Whitrow, G. J., and Yallop, J. B. D. 1964, *M.N.R.A.S.*, **137**, 301.

Wildhack, W. A. 1940, *Phys. Rev.*, **57**, 81.

Wilkinson, D. T., and Partridge, R. B. 1967, *Nature*, **215**, 719.

Will, C. M. 1971, *Ap. J.*, in press.

Wolfe, A. M., and Burbidge, G. R. 1970, *Ap. J.*, **161**, 419.

Woltjer, X. L. 1966, *Blueshifts in Quasi-Stellar Objects*, preprint.

Woolley, R.v.d.R., 1954, *M.N.R.A.S.*, **114**, 191.

Zaitsev, V. V. 1969, *Ap. and Space Sci.*, **4**, 464.

———. 1964, *M.N.R.A.S.*, **128**, 221.

———. 1965, *Adv. Astr. and Ap.*, **3**, 241.

Zel'dovich, Ya. B. 1957, *Zh. Eksp. Teoret. Fiz.*, **33**, 4, 991.

———. 1959, *ibid.*, **37**, 569.

———. 1960, *ibid.*, **38**, 4, 1123.

———. 1961, *ibid.*, **41**, 1609.

———. 1962a, *ibid.*, **42**, 1667.

———. 1962b, *ibid.*, **42**, 641.

———. 1962c, *ibid.*, **43**, 1037.

———. 1962d, *ibid.*, **43**, 1561.

———. 1963a, *Voprosy kosmogonii* [Problems of cosmogony], **9**, 157.

———. 1963b, *Uspekhi Fiz. Nauk*, **80**, 357.

———. 1963c, *Voprosy kosmogonii*, **9**, 240.

———. 1963d, *Atomnaya energiya*, **14**, 92.

———. 1963e, *Voprosy kosmogonii*, **9**, 36.

———. 1963f, *ibid.*, **9**, 232.

———. 1963g, *Astr. Circ.*, No. 250.

———. 1964, *M.N.R.A.S.*, **128**, 221.

Zel'dovich, Ya. B. 1964*a*, *Doklady Akad. Nauk*, **155**, 67.

———. 1964*b*, *Astr. Zh.*, **41**, 19.

———. 1964*c*, *ibid.*, **41**, 873.

———. 1964*d*, preprint.

———. 1965, *Adv. Astr. and Ap.*, **3**, 241.

———. 1965*a*, *Pis'ma* [Letters], *Zh. Eksp. Teoret. Fiz.*, **1** (3), 40.

———. 1965*b*, *Zh. Eksp. Teoret. Fiz.*, **48**, 986.

———. 1965*c*, *Astr. Zh.*, **42**, No. 2, 283.

———. 1965*d*, *Uspekhi Fiz. Nauk*, **86** (2), 303.

———. 1966, *ibid.*, **89** (4), 647.

———. 1967*a*, *Pis'ma* [Letters], *Zh. Eksp. Teoret. Fiz.*, **6**, 922.

———. 1967*b*, *ibid.*, **6**, 1050.

———. 1968, *Uspekhi* (English trans.), 95, 209.

———. 1969, *Zh. Eksp. Teoret. Fiz. (SSSR)*, **56**, L6.

Zel'dovich, Ya. B., and Guseynov, O. Kh. 1965*a*, *Doklady Akad. Nauk*, **162**, 791.

———. 1965*b*, *Pis'ma* [Letters], *Zh. Eksp. Teoret. Fiz.*, **1** (4), 11.

———. 1965*c*, *Ap. J.*, **144**, 841.

Zel'dovich, Ya. B., Kompaneets, A. S., and Rayzer, Yu. P. 1958, *Pis'ma* [Letters], *Zh. Eksp. Teoret. Fiz.*, **34**, 1278, 1447.

Zel'dovich, Ya. B., Kurt, V. G., and Syunyaev, R. A. 1967 (see *Zh. Eksp. Teoret. Fiz.*, **55**, 278, 1968).

Zel'dovich, Ya. B., and Novikov, I. D. 1964*a*, *Uspekhi Fiz. Nauk*, **84** (3), 377.

———. 1964*b*, *Doklady Akad. Nauk*, **155**, 1033.

———. 1965, *Uspekhi Fiz. Nauk*, **86** (3), 477.

———. 1966*a*. *Einsteinovskiy sbornik*, Vol. 1.

———. 1966*b*, *Astr. Zh.*, **43**, 758.

———. 1966*c*, *IAU Symp.*, *No. 29*, Byurakan (USSR).

———. 1967*a*, *Astr. Zh.*, Vol. **44**, No. 3.

———. 1967*b*, *Pis'ma* [Letters], *Zh. Eksp. Teoret. Fiz.*, **6**, 722.

Zel'dovich, Ya. B., Novikov, I. D., and Syunyaev, R. A. 1966, *Astr. Circ.*, No. 371.

Zel'dovich, Ya. B., Okun', L. B., and Pikel'ner, S. B. 1965, *Uspekhi Fiz. Nauk*, **87**, 113.

Zel'dovich, Ya. B., and Podurets, M. A. 1964, *Doklady Akad. Nauk*, **156**, 57.

———. 1965, *Astr. Zh.*, **42**, No. 5, 963.

Zel'dovich, Ya. B., and Rayzer, Yu. P. 1966, *Fizika udarnykh voln* (Moscow: Nauka).

Zel'dovich, Ya. B., and Shakura, N. J. 1969, *Astr. Zh.*, **46**, 225.

———, and Smorodinsky, Ya. A. 1961, *Zh. Eksp. Teoret. Fiz.*, **41**, 907.

———. 1966, reference unknown.

Zel'manov, A. L. 1944, master's thesis, Moscow State University.

———. 1948, *Doklady Akad. Nauk*, **61**, 993.

————. 1956, *ibid.*, **107,** 815.

————. 1959a, *ibid.*, **124,** 1030.

————. 1959b, *Trudy shestogo soveshchaniya po voprosam kosmogonii* (Moscow).

Zorn, J. C., Chamberlain, G. E., and Hughes, V. W. 1963, *Phys. Rev.*, **129,** 2566.

Zwicky, F. 1958, *Hdb. d. Phys.*, **51,** 766.

————. 1959, *ibid.*, Vol. **53.**

Zwicky, F., and Rudnicki, I. 1963, *Ap. J.*, **131,** 707.

AUTHOR INDEX

Index

515

SUBJECT INDEX

Accretion of matter onto white dwarfs, neutron stars, and frozen stars, 433–52; X-rays produced by, 433, 443, 446, 449–52; spherical infall of noninteracting particles, 434–35; hydrodynamical equations for spherical flow, 435–40; validity of the hydrodynamical approximation, 440–42; luminosity produced by spherical accretion onto white dwarfs and neutron stars, 442–44, 446; luminosity produced by spherical accretion onto frozen stars, 444–46; nonspherical, hydrodynamical accretion, 446–50; self-regulation of spherical accretion, 446, 451; accretion in binary system and emission of X-rays, 449–51; acceleration of positrons during accretion, 452; gamma rays produced by, 452

Adiabatic index: for completely ionized gas, 212; for neutral, monatomic gas, 212; for partially ionized gas, 212–14; for radiation in equilibrium with hot matter, 216–17; for electron-positron pairs in equilibrium with hot matter, 219–21, 341–42; for neutrinos in equilibrium with hot matter, 221; for matter undergoing nuclear dissociation, 223–24, 341–42; summarized for wide range of density and temperature, 224; for cold, ideal neutron gas, 283

Anticollapse, 114–15, 119, 478

Antiquasars made of antigravitating antimatter, 476

Axially symmetric, static systems in general relativity, 130–34

Baryonic charge, 156

Binary-star systems: gravitational waves emitted by, 45–51, 106–10; effect of mass transfer on evolution, 340, 451; mass flow between stars and generation of X-rays, 449–50; search for frozen stars and neutron stars in binary systems, 450, 471–72; white dwarfs in, 467–69; disruption of by mass loss, 468

Binding energy of stars. *See* Mass defect

Black holes. *See* Frozen stars

Brans-Dicke theory of gravity, 78–79

Capture of a particle falling in Schwarzschild field, 103–10

Cepheid variable stars, 340

Chandrasekhar limit, 234, 272

Chemical potentials, for nuclear reactions, 207

Chronometric invariants, 19–22

Cold matter. *See* Equation of state for cold matter

Collapsed stars. *See* Frozen stars

Conservation laws, mathematical formulation of, 25–28

Contravariant components of a tensor, 17

Convection in stars: instability against, 253–54, 300; in general relativity, 270; in supermassive stars, 300; in surface layers of a white dwarf, 333

Coordinates, curvilinear, 16–19. *See also* Reference frame

Cosmic microwave radiation, xv

Cosmic rays, produced by supernovae, 363

Cosmological constant, 28–33, 72

Covariant components of a tensor, 17

Critical states for onset of stellar collapse, 309–16. *See also* Gravitational collapse, relativistic; Stellar stability against adiabatic, radial perturbations

Curvature of spacetime: equivalence principle motivates, 9; inertial frames cannot exist globally because of, 12; mathematical formulation of, 22–23; motivated by self-inconsistency of flat-space gravity theories, 68–70. *See also* General theory of relativity

Curvature tensors: Riemann, 23; Ricci, 23; Einstein, 52. *See also* Curvature of spacetime; Gaussian curvature; General theory of relativity

Curvature of 3-space, 16

Deflection of light and radio waves by sun's gravity, 35, 79, 92, 105

Degenerate electron gas: equation of state for, in ideal, noninteracting case, 163–67; equation of state for, corrections at $\rho \gg 10^6$ g cm^{-3}, 167–69, 176; equation of state for, corrections at $\rho < 10^6$ g cm^{-3}, 169–74, 176